The Mathematical Theory of Selection, Recombination, and Mutation

WILEY SERIES IN MATHEMATICAL AND COMPUTATIONAL BIOLOGY

BÜRGER—The Mathematical Theory of Selection, Recombination, and Mutation

CHAPLAIN/SINGH/MCLACHLAN—On Growth and Form: Spatio-temporal Pattern Formation in Biology

CHRISTIANSEN—Population Genetics of Multiple Loci

CLOTE/BACKOFEN—Computational Molecular Biology: An Introduction

DIEKMANN/HEESTERBEEK—Mathematical Epidemiology of Infectious Diseases: Model Building, Analysis and Interpretation

Reflecting the rapidly growing interest and research in the field of mathematical biology, this outstanding new book series examines the integration of mathematical and computational methods into biological work. It also encourages the advancement of theoretical and quantitative approaches to biology, and the development of biological organization and function.

The scope of the series is broad, ranging from molecular structure and processes to the dynamics of ecosystems and the biosphere, but unified through evolutionary and physical principles, and the interplay of processes across scales of biological organization.

Topics to be covered in the series include:

- Cell and molecular biology
- Functional morphology and physiology
- Neurobiology and higher function
- Genetics
- Immunology
- Epidemiology
- Ecological and evolutionary dynamics of interacting populations

A fundamental research tool, the *Wiley Series in Mathematical and Computational Biology* provides essential and invaluable reading for biomathematicians and development biologists, as well as graduate students and researchers in mathematical biology and epidemiology.

The Mathematical Theory of Selection, Recombination, and Mutation

R. Bürger

University of Vienna, Austria

JOHN WILEY & SONS, LTD
Chichester • New York • Weinheim • Brisbane • Singapore • Toronto

Baffins Lane, Chichester,
West Sussex, PO19 1UD, England

National 01243 779777
International (+44) 1243 779777

e-mail (for orders and customer service enquiries): cs-books@wiley.co.uk

Visit our Home Page on http://www.wiley.co.uk or http://www.wiley.com

Other Wiley Editorial Offices

John Wiley & Sons, Inc., 605 Third Avenue,
New York, NY 10158-0012, USA

Wiley-VCH Verlag GmbH
Pappelallee 3, D-69469 Weinheim, Germany

Jacaranda Wiley Ltd, 33 Park Road, Milton,
Queensland 4064, Australia

John Wiley & Sons (Asia) Pte Ltd, 2 Clementi Loop #02-01,
Jin Xing Distripark, Singapore 129809

John Wiley & Sons (Canada) Ltd, 22 Worcester Road,
Rexdale, Ontario, M9W 1L1, Canada

Library of Congress Cataloging-in-Publication Data

Bürger, Reinhard
The mathematical theory of selection, recombination, and mutation / Reinhard Bürger.
p. cm – (Wiley series in mathematical and computational biology)
Includes bibliographical references and index.
ISBN 0-471-98653-4 (alk. paper)
1. Population genetics—Mathematical models. I. Title. II. Series.

QH455 .B87 2000
576.5'8'015118—dc21

00-043345

British Library Cataloguing in Publication Data

A catalogue record for this book is available from the British Library

ISBN 0-471-98653-4

Produced from PostScript files supplied by the author.
Printed and bound in Great Britain by Antony Rowe, Chippenham.
This book is printed on acid-free paper responsibly manufactured from sustainable forestry in which at least two trees are planted for each one used for paper production.

Contents

Series Preface

Theoretical biology is an old subject, tracing back centuries. At times, theoretical developments have represented little more than mathematical exercises, making scant contact with reality. At the other extreme have been those works, such as the writings of Charles Darwin, or the models of Watson and Crick, in which theory and fact are intertwined, mutually nourishing one another in inseparable symbiosis. Indeed, one of the most exciting developments in biology within the last quarter-century has been the integration of mathematical and theoretical reasoning into all branches of biology, from the molecule to the ecosystem. It is such a unified theoretical biology, blending theory and empiricism seamlessly, that has inspired the development of this series.

This series seeks to encourage the advancement of theoretical and quantitative approaches to biology, and to the development of unifying principles of biological organization and function, through the publication of significant monographs, textbooks and synthetic compendia in mathematical and computational biology. The scope of the series is broad, ranging from molecular structure and processes to the dynamics of ecosystems and the biosphere, but it is unified through evolutionary and physical principles, and the interplay of processes across scales of biological organization.

The principal criteria for publication, beyond the intrinsic quality of the work, are substantive biological content and import, and innovative development or application of mathematical or computational methods. Topics will include, but not be limited to, cell and molecular biology, functional morphology and physiology, neurobiology and higher function, immunology and epidemiology, and the ecological and evolutionary dynamics of interacting populations. The most successful contributions, however, will not be so easily categorized, crossing boundaries and providing integrative perspectives that unify diverse approaches; the study of infectious diseases, for example, ranges from the molecule to the ecosystem, involving mechanistic investigations at the level of the cell and the immune system, evolutionary perspectives as viewed through sequence analysis and population genetics, and demographic and epidemiological aspects at the level of the ecological community.

The objective of the series is the integration of mathematical and computational methods into biological work; hence the volumes published should be of interest both to fundamental biologists and to computational and mathematical scientists, as well as to the broad spectrum of interdisciplinary researchers that comprise the continuum connecting these diverse disciplines.

Simon Levin

Preface

During the past 25 years, population geneticists and evolutionary biologists have made great efforts to explore the influence of evolutionary and genetic factors on the distribution and dynamics of quantitative characters under selection. Because such characters are typically determined by several or many gene loci, a large number of theoretical investigations have been performed which aim at a better understanding of the equilibrium and dynamical properties of various population-genetic models involving multiple loci. At the same time, and in part motivated by the theoretical developments, numerous empirical and experimental approaches have been undertaken that have substantially increased our knowledge about the biology and genetics of quantitative traits.

The main purpose of this book is to provide a fairly unified and self-contained treatment of the mathematical theory of genetic multilocus systems subject to the forces of selection, mutation, and recombination. The emphasis is on models that are of relevance in quantitative genetics and evolutionary biology. This is also reflected by the long final chapter, which is dedicated to a presentation and discussion of several topics that have been attracting the interest of many evolutionary geneticists during the last few decades, and for which the mathematical models in this book have an important bearing.

In contrast to other fields in biology, mathematical thinking and methodology has become entrenched in population genetics since its very beginning, as is witnessed by the classical work of Weinberg, Fisher, Haldane, Wright, Kimura, and others. Indeed, important developments in mathematics, and particularly in statistics, have their roots in this field. Therefore, it will not be surprising that the mathematical analysis of the models studied in this book is often not simple, although to a large extent only calculus, linear algebra, and some basic probability theory are needed. The book is aimed at the level of graduate students or researchers. In general, I had a reader in mind, who is interested in the theoretical aspects of evolutionary biology and has a fair knowledge of basic mathematics and a certain fluency in algebraic manipulations. A general treatment of some topics, however, requires knowledge from the qualitative theory of differential- and difference equations, from the theory of nonnegative matrices, from operator theory, or about generating functions and probability distributions. The relevant concepts from these fields, but also some results about random genetic drift in finite populations, are summarized in the Appendix. This may be most useful for readers having a background in mathematics or physics. Other readers may simply believe the results. Indeed, some sections are written primarily for mathematically oriented readers. These sections, as well as those requiring substantial mathematical skills, are indicated by an asterisk. Also, the beginning and end of more technical derivations that can be skipped at a first reading are signified by the signs $\triangleright$ and $\triangleleft$, respectively. I did not provide proofs for all of the results, but rather, of course, for those that are new; proofs are also given if they are simpler than published proofs, or if they are instructive. Therefore, the level of mathematical sophistication fluctuates, as does the amount of detail given in mathematical arguments or derivations. To make the mathematical parts of the book more easily accessible for readers with only basic mathematical abilities, I have in many instances summarized and highlighted the main results of a section in bulleted lists, so that parts of the material can be skipped by readers interested only in the results. Some of the chapters end with a summary or conclusions which may provide a guide to the most important results.

Equations, figures, tables, results, theorems, etc., are numbered within each chapter, and are not preceeded by the chapter number. Also, cross references within a chapter do not indicate the chapter number.

The book opens with a chapter that introduces the basic concepts of mathematical population genetics. Its first section summarizes the necessary basic genetics and contains references to introductory or complementary texts. In the following sections, the most important evolutionary and genetic forces, such as random mating, selection, recombination, mutation, or random genetic drift, and their influence on the genetic composition of a population are studied in isolation. This chapter may serve as an introduction to the field for students from the mathematical or physical sciences, and for biology or genetics students who are already acquainted with population genetics, but feel a need for a theoretically oriented and rigorous introduction.

Chapter II begins with an introduction to the classical two-locus models and then develops the general theory of multilocus models. Among others, it contains a treatment of the theory of the average effects and the components of genetic variance of a quantitative trait that does not assume Hardy-Weinberg equilibrium. It ends with an analysis of the evolutionary dynamics of multilocus systems with arbitrary recombination frequencies that are governed by very general selection regimes.

The next two chapters are devoted to the theory of mutation-selection models at a single locus. Whereas in Chapter III models with a finite number of alleles are treated, as well as the stepwise-mutation model that allows for an infinite sequence of possible alleles, Chapter IV presents a general account of models allowing for a continuum of possible alleles. This also constitutes the most mathematically advanced part of the book.

In Chapter V, a general theory for analyzing both the dynamics of the distribution of an additive quantitative trait under selection and the distribution of multilocus genotypes that determine that trait is set out. It is formulated in terms of multivariate cumulants and generating functions. In Chapter VI, various models are investigated that have been devised to explore the potential of stabilizing selection to maintain genetic variation in an additive quantitative trait in the absence or presence of recurrent mutation. In all these chapters, populations are assumed to be sufficiently large so that random genetic drift can be ignored.

Whereas the first six chapters are focused on the analysis of mathematical models, the emphasis in Chapter VII is more on the biological problems and the conclusions that can be drawn from the analyses of the models. This chapter has been written with a broader readership in mind, and it is mathematically much less sophisticated than the others. To a certain extent, it is a review of topics that have attracted the attention of researchers in evolutionary genetics during the past few decades, and therefore is less self-contained. However, it also contains new results and, perhaps, some new perspectives. It deals comprehensively with the problem by which mechanisms and to what extent genetic variation in quantitative traits can be maintained. Further, the role of genetic variation for the response to directional selection is studied, as well as some models concerning the evolutionary consequences of recurrent deleterious mutations. In contrast to previous chapters, random genetic drift is not ignored here, and in several contexts the role of small population size is explored.

Chapter I, the first section of Chapter II on two-locus models, and the first two sections of Chapter III on mutation-selection models can be used for an introductory course to theoretical population genetics.

This book could never have been written without the support and motivation that I have received from numerous colleagues and friends. Throughout the years, I have benefitted enormously from interactions with Wilfried Gabriel, Josef Hofbauer, Russell Lande, Michael Lynch, Oliver Mayo, Thomas Nagylaki, Karl Sigmund, and, last but not least, Günter Wagner. Many thanks go to Warren Ewens, Jim Fry, Sergey Gavrilets, Alexander Gimelfarb, Josef Hofbauer, Peter Keightley, Christoph Krall, Oliver Mayo, Vladimir Passekov, and Claus Vogl for useful comments on earlier drafts of some of the chapters, and to Ellen Baake and Russell Lande, who read a large fraction of the manuscript and provided a great number of helpful suggestions. I am especially indebted to Thomas Nagylaki for his most thorough and detailed reading of a major part of the manuscript and the innumerable improvements he suggested. Clearly, I take the full responsibility for all remaining errors, obscurities, and deficiences, in particular, because I did not always follow the advice of my colleagues.

I am grateful to the International Institute of Applied Systems Analysis (IIASA) in Laxenburg, Austria, and Ulf Dieckmann for financial support and the possibility to work for two months in a wonderful atmosphere.

Finally, I would like to express my sincere thanks to Christine Vigne for much support during the arduous and lengthy process of finalizing this project, and my deep gratitude to my parents for their life-long assistance, and for providing the background that paved my way to an academic career.

Reinhard Bürger
Vienna, May 2000

I

Elementary Population Genetics

1. INTRODUCTION

Population genetics is concerned with the study of the genetic composition of populations. This composition may be changed by segregation, selection, mutation, recombination, mating structure, migration, and other genetic, ecological, and evolutionary factors. Therefore, in population genetics these mechanisms and their interactions and evolutionary consequences are investigated. Traditionally, population genetics has been applied to animal and plant breeding, to human genetics, and more recently to ecology and conservation biology. It also has important interfaces with molecular biology, systematics, natural history, mathematics, statistics, and computing. One of the main subjects is the investigation of the mechanisms that generate and maintain genetic variability in populations, and the study of how this genetic variation, shaped by environmental influences, leads to evolutionary change, adaptation, and speciation. Therefore, research in population genetics relies on empirical observations, on experiments, and on theoretical considerations. In particular, population genetics provides the basis for understanding the evolutionary processes that have led to the diversity of life we encounter and admire.

Since so many factors interact and determine the evolutionary fate of a population, a proper understanding of the relevant processes requires a good deal of abstraction in planning experiments and in devising mathematical models. A good mathematical model, as well as a good experiment, takes into account the relevant biological mechanisms for studying a particular phenomenon and disregards the less relevant ones. As in other sciences, good model building must rest on an adequate knowledge of the basic biological reality and requires a clear formulation of the underlying hypotheses. The process of abstraction that is involved entails generality which, sometimes, may appear to be unnecessary. However, general methods or models, devised to study a particular phenomenon, may reveal the essence and the underlying structure more clearly and can often be applied to questions not anticipated before. Examples include the study of gene families and microsatellites, where classical population genetic modeling has been successfully applied.

Mathematical models and methods have a long history in population genetics, tracing back to Gregor Mendel, who used elementary mathematics to calculate the expected frequencies of the genes in his experiments. Francis Galton and the biometricians, notably Karl Pearson, developed new statistical methods to describe the distribution of trait values in populations and to predict their change between generations.

The foundations of modern population genetics were laid by the work of Ronald A. Fisher, J.B.S. Haldane, and Sewall Wright, who reconciled Mendelism with Darwinism during the second and third decades of the twentieth century. They demonstrated that the theory of evolution by natural selection, proposed by Charles Darwin (1859), can be justified on the basis of genetics as governed by Mendel's laws. The work of Fisher, Haldane, and Wright was highly mathematical for the biology of that time and was properly understood by only a small number of people. Nevertheless, their influence was enormous and they set the standards for mathematical modeling and for rigor of theoretical investigations for the subsequent decades.

Prior to 1900, the year when Mendel's work was rediscovered and then rapidly accepted, the hereditary mechanisms were unknown. Darwin believed in blending inheritance, according to which the hereditary material itself blended. However, as already noted by Darwin, blending inheritance produces uniformity and destroys variation that is so ubiquitous. In modern terms, heritable variance would be halved in each generation of random mating with blending inheritance (Fisher 1930). Therefore, one half of the heritable variance maintained in a population would have to arise anew in each generation. There were controversial lines of thought about the nature of this huge amount of new variation and its consequences for evolution. The 'gradualists', to which Darwin and the biometricians adhered, considered the changes across generations as gradual and incremental, whereas the 'saltationists' (e.g., T.H. Huxley and Galton) held that evolutionary changes occurred in 'jumps' of considerable magnitude. Much of the scientific dispute about Darwin's theory of evolution originated from the ignorance of the true hereditary mechanisms.

Despite the early work of Yule (1902), Hardy (1908), and Weinberg (1908), who showed that under the particulate mode of inheritance proposed by Mendel (1866), genetic variability is preserved under random mating, it was not before 1918 that the synthesis between genetics and the theory of evolution through natural selection began to take shape through Fisher's (1918) work (see Provine (1971) for a detailed account of the history of population genetics).

Today, the hereditary mechanisms have been firmly established and our knowledge about the molecular biology of the genes is rapidly increasing. Mutations are known to be the ultimate source of genetic variability, and many different processes at the chromosomal and molecular level have been identified that generate mutations. On the phenotypic level, the role of selection in shaping evolutionary change has been amply documented, whereas on the molecular level, a significant amount of neutral evolution appears to take place, its extent still being disputed. Nevertheless, there remain many open problems, some of which are qualitative in nature and some quantitative. Questions concerning the processes involved in speciation events or in the evolution of sex belong to the first class, whereas questions concerning the prediction of the expected evolutionary change of a population subject to selection belong to the second class. Such predictions are highly nontrivial, unless confined to one or a few generations, because there exist many different forms of selection and the response to selection depends on the pattern and amount of genetic variability in the population. This variation, however, is a function of many genetic details (such as number of genes determining a trait, mutational properties, degree of linkage), of the demography (population size, mating structure), and of the selective forces acting. Therefore, the genetic variability may change from one generation to the next.

One of the main purposes of this book is to provide the mathematical theory for predicting the evolutionary change under the combined action of selection, mutation, and recombination, and to present applications to several topics of evolutionary interest, in particular, related to the evolution of quantitative traits. The first chapter, which is rather elementary, introduces fundamental concepts of population genetics and studies several evolutionary mechanisms in isolation, whereas later chapters are concerned with their interactions. Let us now summarize some of the very basic genetic knowledge that is required. For concise and lucid introductory texts to population and evolutionary genetics, the reader is referred to Crow (1986) or Maynard Smith (1998). As a comprehensive introduction to general genetics, the textbook of Hartl (1994) can be recommended.

Mendel's (1866) prime achievement was the recognition of the particulate nature of the hereditary determinants, now called genes. A gene may have different forms, called *alleles*. From his experiments with peas he concluded that genes are present in pairs, one member of each pair having been inherited from the maternal parent, the other from the paternal. The allelic composition is called the *genotype*, and the set of observable properties derived from the genotype is called the *phenotype*. Thus, supposing that there are two alleles $\mathcal{A}_1$ and $\mathcal{A}_2$, there are three possible genotypes, $\mathcal{A}_1\mathcal{A}_1$, $\mathcal{A}_1\mathcal{A}_2$, and $\mathcal{A}_2\mathcal{A}_2$. In the first and third case, the organism's genotype is *homozygous* (for $\mathcal{A}_1$ or $\mathcal{A}_2$, respectively), in the second case it is *heterozygous*. In general, the genotypes $\mathcal{A}_1\mathcal{A}_2$ and $\mathcal{A}_2\mathcal{A}_1$ cannot be distinguished. When the phenotype of the heterozygote $\mathcal{A}_1\mathcal{A}_2$ is the same as one of the homozygotes, say $\mathcal{A}_1\mathcal{A}_1$, allele $\mathcal{A}_1$ is called *dominant* and $\mathcal{A}_2$ is called *recessive*.

Mendel's first law states that when pure-bred (homozygous) strains are crossed, the hybrid progeny constituting the F_1 generation (the letter F stands for filial) are uniform (their genotype being $\mathcal{A}_1\mathcal{A}_2$ if the parents were $\mathcal{A}_1\mathcal{A}_1$ and $\mathcal{A}_2\mathcal{A}_2$) and usually express one of the two phenotypes (the one controlled by the dominant allele). According to Mendel's second law, recessive characters, which are masked in the heterozygous F_1, reappear in the F_2 in the proportion $1 : 3$ of the dominant character. This leads to the *Principle of Segregation*, stating that each reproductive cell (*gamete*) contains only one of the two alleles and that each gamete is equally likely to contain either one. The separation of the paired alleles from one another and their distribution to different cells, the gametes, is called *segregation* and occurs during *meiosis*. Meiosis is the process of formation of gametes from somatic cells. At mating, two reproductive cells fuse and form a *zygote* (fertilized egg), which contains the full (diploid) genetic information.

Mendel also performed experiments with pure-bred lines that differed in two characters, round versus wrinkled seed shape and yellow versus green color. From previous experiments he knew that 'round' was dominant over 'wrinkled' and 'yellow' dominant over 'green' because their F_2 ratios were $3 : 1$ each. The F_1 seeds, from crosses of lines having round and yellow seeds with lines having wrinkled and green seeds, were all round and yellow. In F_2 progeny from the dihybrid cross, all four phenotypes reappeared, approximately in the proportions 9/16 'round yellow', 3/16 'wrinkled yellow', 3/16 'round green', and 1/16 'wrinkled green'. Thus, the proportion of the four phenotypes is $9 : 3 : 3 : 1$, as expected when two pairs of alleles segregate independently, so that the $3 : 1$ ratios are combined at random. This is called Mendel's third law or the *Principle of Independent Assortment*.

Since the 1940s it has been known that the genetic material is *deoxyribonucleic acid* (*DNA*). It consists of four bases: adenine (A), guanine (G), thymine (T), and cytosine (C). Each base is linked to a sugar and a phosphate group, yielding a *nucleotide*. The nucleotides are arranged along two chains to form a double-stranded helix in which the pairings A–T and G–C between the strands are formed. Therefore, all the genetic information is contained in each of the two strands. Three bases code for one amino acid, which are the building blocks of polypeptide chains and proteins. A gene typically represents a contiguous region of DNA coding for one polypeptide chain. Its position along the DNA is called the *locus*, and a particular sequence there is called an allele. Thus, two genes at the same locus, sampled from a population, may or may not be of the same allelic type. A double-stranded helix of DNA forms the backbones of the *chromosomes*, which are contained in the nucleus of each cell. In *diploid* organisms (higher plants and animals) chromosomes form homologous pairs, each one inherited from one parent. The exceptions are the *sex chromosomes*, which are involved in the genetic determination of sex. Usually, this is one pair of chromosomes which differ from each other, one called the X-chromosome, the other the Y-chromosome. In all mammals, in *Drosophila*, and in many other species and taxa, but not in birds, XX is female and XY is male. The term *autosome* is used for chromosomes that are not sex chromosomes. The number of different chromosomes per nucleus is characteristic of each species.

Any heritable change in the genetic material is called a *mutation*. Mutations are the ultimate source of genetic variability, and form the raw material upon which selection can act. Although the term mutation includes changes in chromosome structure and number, the vast majority of genetic variation is caused by gene mutations. Modern genetics has revealed that at the molecular level (gene) mutations occur in many different ways, for instance as base substitutions, in which one pair of nucleotides is replaced by another, as insertions or deletions of DNA, as inversions of sequences of nucleotides, or as transpositions. The latter are mainly caused by transposable elements changing their position from one site to another. For many population-genetic models, however, the molecular origin of a mutant is not necessarily of relevance. What often counts is only the rate at which mutations occur and a mutant's effect on fitness or, more generally, on the character under consideration. Typically, spontaneous mutation rates per locus per generation are of the order of 10^{-4} to 10^{-6}, and genomic mutation rates summed over all loci may be on the order of one per generation, but can vary substantially between species.

During meiosis, different chromosomes assort independently and *crossing over* between two homologous chromosomes may occur. Consequently, the newly formed gamete contains maternal alleles at one set of loci and paternal alleles at the complementary set. This process is called *recombination*. Since it leads to random association between alleles at different loci, recombination has the potential to combine favorable alleles of different ancestry in one gamete and to break up combinations of deleterious alleles. These properties are thought to confer a substantial evolutionary advantage to sexual species relative to asexuals.

The mating pattern may have a substantial influence on the evolution of gene frequencies. The simplest and most important mode is *random mating*. This means that matings take place without regard to ancestry or the genotype under consideration. It seems to occur frequently in nature. For example, among humans, matings within a population appear to be random with respect to blood groups and allozyme phenotypes, but are nonrandom with respect to height. Random mating conserves allele frequencies and, after one generation, genotypic frequencies.

Selection occurs when individuals of different genotype leave different numbers of progeny because they differ in their probability to survive to reproductive age (*viability*), in their mating success, or in their average number of produced offspring (*fertility*). Darwin (1859) recognized and documented the central importance of selection as the driving force for adaptation and evolution. Since selection affects the entire genome, its consequences for the genetic composition of a population may be complex. Selection is measured in terms of *fitness* of individuals, i.e., by the number of progeny contributed to the next generation. There are different measures of fitness, and it consists of several components because selection may act on each stage of the life cycle.

In the subsequent sections of this introductory chapter, the exposition repeatedly follows, more or less closely and without further notice, one of the texts of Crow and Kimura (1970), Ewens (1979), and Nagylaki (1992). For material complementary to this chapter, the mathematically oriented reader may consult the books of Ewens, which is mainly focused on stochastic models, and Nagylaki, which presents derivations of many important, primarily deterministic, models from basic principles and their analysis. The book of Hartl and Clark (1997) provides a good biological background to the theory of population genetics.

2. THE HARDY–WEINBERG LAW

With the blending theory of inheritance, variation in a population declines rapidly, and this was one of the arguments against Darwin's theory of evolution. With Mendelian inheritance there is no such dilution of variation, as was shown independently by the famous British mathematician Hardy and the German physician Weinberg. In fact, only two years after the rediscovery of Mendelian heredity, Yule (1902) had pointed out that the ratio $1:2:1$ of the frequency of genotypes $\mathcal{A}_1\mathcal{A}_1$, $\mathcal{A}_1\mathcal{A}_2$, $\mathcal{A}_2\mathcal{A}_2$, as obtained in the F_2 generation of a cross of $\mathcal{A}_1\mathcal{A}_1$ and $\mathcal{A}_2\mathcal{A}_2$ individuals, persisted in further random bred generations. Castle (1903) extended this to other gene frequencies. The general principle, on which these observations are based, and which is now called the Hardy–Weinberg Law, was discovered independently by Hardy (1908) and Weinberg (1908). We first derive the simple original version of this law, where at a given gene locus only two alleles, $\mathcal{A}_1$ and $\mathcal{A}_2$, occur. Thereafter, we shall treat some extensions.

2.1 TWO ALLELES

We consider a random-mating population with discrete, nonoverlapping generations (as in annual plants and many insects) that is either *monoecious* (i.e., every individual has both male and female sexual organs, as in most plants and some animals) or *dioecious* (i.e., admits two sexes) with initially identical genotype frequencies in both sexes. Then the relative frequencies of the genotypes $\mathcal{A}_1\mathcal{A}_1$, $\mathcal{A}_1\mathcal{A}_2$, and $\mathcal{A}_2\mathcal{A}_2$ in the population can be described by one set of variables, labeled P, $2Q$, and R, respectively, and $P + 2Q + R = 1$. The heterozygous genotype $\mathcal{A}_1\mathcal{A}_2$ has been assumed to be unordered, so that $2Q$ is the combined frequency of the ordered genotypes $\mathcal{A}_1\mathcal{A}_2$ and $\mathcal{A}_2\mathcal{A}_1$. We assume, furthermore, that the population is so large that gene and genotype frequencies may be treated as deterministic, and relative frequency can be identified with probability.

We want to derive the frequencies of the three genotypes $\mathcal{A}_1\mathcal{A}_1$, $\mathcal{A}_1\mathcal{A}_2$, and $\mathcal{A}_2\mathcal{A}_2$ in the next generation. This can be achieved by calculating the frequencies of all possible matings and their offspring produced. For example, with random mating (with respect to the locus under consideration), the probability of the mating $\mathcal{A}_1\mathcal{A}_1 \times \mathcal{A}_1\mathcal{A}_2$ is $4PQ$, because $\mathcal{A}_1\mathcal{A}_1$ can be male or female (and $\mathcal{A}_1\mathcal{A}_2$, thus, female or male), and the probabilities of the genotypes $\mathcal{A}_1\mathcal{A}_1$ and $\mathcal{A}_1\mathcal{A}_2$ are P and $2Q$, respectively. According to Mendel's laws, half of the progeny of such a mating are $\mathcal{A}_1\mathcal{A}_1$ and half are $\mathcal{A}_1\mathcal{A}_2$. Table 2.1 summarizes all possibilities.

Table 2.1 Mating table

Mating	Mating prob.	Cond. prob. of progeny		
		$\mathcal{A}_1\mathcal{A}_1$	$\mathcal{A}_1\mathcal{A}_2$	$\mathcal{A}_2\mathcal{A}_2$
$\mathcal{A}_1\mathcal{A}_1 \times \mathcal{A}_1\mathcal{A}_1$	P^2	1	0	0
$\mathcal{A}_1\mathcal{A}_1 \times \mathcal{A}_1\mathcal{A}_2$	$4PQ$	$\frac{1}{2}$	$\frac{1}{2}$	0
$\mathcal{A}_1\mathcal{A}_1 \times \mathcal{A}_2\mathcal{A}_2$	$2PR$	0	1	0
$\mathcal{A}_1\mathcal{A}_2 \times \mathcal{A}_1\mathcal{A}_2$	$4Q^2$	$\frac{1}{4}$	$\frac{1}{2}$	$\frac{1}{4}$
$\mathcal{A}_1\mathcal{A}_2 \times \mathcal{A}_2\mathcal{A}_2$	$4QR$	0	$\frac{1}{2}$	$\frac{1}{2}$
$\mathcal{A}_2\mathcal{A}_2 \times \mathcal{A}_2\mathcal{A}_2$	R^2	0	0	1

Therefore, the frequency of $\mathcal{A}_1\mathcal{A}_1$ homozygotes in the next generation is[1]

$$\begin{aligned} P' &= P^2 \cdot 1 + 4PQ \cdot \tfrac{1}{2} + 2PR \cdot 0 + 4Q^2 \cdot \tfrac{1}{4} + 4QR \cdot 0 + R^2 \cdot 0 \\ &= P^2 + 2PQ + Q^2 = (P+Q)^2 \end{aligned} \tag{2.1a}$$

and, similarly,

$$2Q' = 2PQ + 2PR + 2Q^2 + 2QR = 2(P+Q)(Q+R) \tag{2.1b}$$

and

$$R' = Q^2 + 2QR + R^2 = (Q+R)^2 \,. \tag{2.1c}$$

Here we have assumed that no mutation occurs and that no evolutionary forces, such as viability selection, differential fertility, geographical dispersal, or separate sexes, change the genotype frequencies. By substituting P', Q', and R' into the right-hand sides of (2.1a)–(2.1c), and observing the fact that $P + 2Q + R = 1$, we obtain after another generation of random mating

$$P'' = (P' + Q')^2 = (P+Q)^2 = P' \tag{2.2a}$$

[1] Unless stated otherwise, a prime will always signify the next generation.

and, similarly,

$$Q'' = Q' \qquad \text{and} \qquad R'' = R' \ . \tag{2.2b}$$

Thus, the genotype frequencies established after one generation of random mating are maintained under random mating in all subsequent generations.

Now let us consider the gene frequencies p and $q = 1 - p$ of the alleles $\mathcal{A}_1$ and $\mathcal{A}_2$. Since all the genes in $\mathcal{A}_1\mathcal{A}_1$ individuals, and half the genes in $\mathcal{A}_1\mathcal{A}_2$ individuals, are $\mathcal{A}_1$ genes, therefore $p = \frac{1}{2}(2P + 2Q)$ and similarly $q = Q + R$. Hence, we can rewrite the equations (2.1) as

$$P' = p^2, \quad 2Q' = 2pq, \quad R' = q^2 \ . \tag{2.3}$$

These are the famous Hardy–Weinberg proportions, and the Hardy–Weinberg Law states that after one generation of random mating, the genotype frequencies remain constant and can be expressed in terms of the allele frequencies according to (2.3). In particular, the allele (gene) frequencies remain constant and no genetic variability is lost by random mating.

2.2 THE CASE OF k ALLELES

Next, we generalize the Hardy–Weinberg Law to the case of k alleles, as first obtained by Weinberg (1909). We denote the alleles by $\mathcal{A}_i$, $i = 1, \ldots, k$, the frequency of the ordered genotype $\mathcal{A}_i\mathcal{A}_j$ by P_{ij} so that the frequency of the unordered genotype $\mathcal{A}_i\mathcal{A}_j$ is $P_{ij} + P_{ji} = 2P_{ij}$. Then the frequency of allele $\mathcal{A}_i$ in the population is

$$p_i = \sum_{j=1}^{k} P_{ij} \ . \tag{2.4}$$

The unordered genotype $\mathcal{A}_i\mathcal{A}_j$ $(i \neq j)$ can result from the unordered matings $\mathcal{A}_i\mathcal{A}_k \times \mathcal{A}_l\mathcal{A}_j$. It is convenient to classify these matings according to the number (and kind) of heterozygous genotypes involved: $\mathcal{A}_i\mathcal{A}_i \times \mathcal{A}_j\mathcal{A}_j$, $\mathcal{A}_i\mathcal{A}_k \times \mathcal{A}_j\mathcal{A}_j$, $\mathcal{A}_i\mathcal{A}_i \times \mathcal{A}_l\mathcal{A}_j$, $\mathcal{A}_i\mathcal{A}_j \times \mathcal{A}_i\mathcal{A}_j$, and $\mathcal{A}_i\mathcal{A}_k \times \mathcal{A}_l\mathcal{A}_j$, where $k \neq i$, $l \neq j$, and $(k,l) \neq (j,i)$. These matings occur with probabilities $2(P_{ii}P_{jj})$, $2(2P_{ik} \cdot P_{jj})$, $2(P_{ii} \cdot 2P_{lj})$, $2P_{ij} \cdot 2P_{ij}$, $2(2P_{ik} \cdot 2P_{lj})$, respectively. The conditional probabilities that an offspring of such a mating is of genotype $\mathcal{A}_i\mathcal{A}_j$ are 1, $\frac{1}{2}$, $\frac{1}{2}$, $\frac{1}{2}$, and $\frac{1}{4}$, respectively. Therefore, the total probability of an $\mathcal{A}_i\mathcal{A}_j$ genotype in the next generation is

$$2P'_{ij} = 2P_{ii}P_{jj} + 2\sum_{k \neq i} P_{ik}P_{jj} + 2\sum_{l \neq j} P_{ii}P_{lj} + 2P_{ij}^2 + 2\sum_{k \neq i} \sum_{\substack{l \neq j \\ (k,l) \neq (j,i)}} P_{ik}P_{lj} \ .$$

Since this is exactly $2\sum_k \sum_l P_{ik}P_{lj} = 2(\sum_k P_{ik})(\sum_l P_{il})$, it follows that

$$P'_{ij} = p_i p_j \qquad \text{for every } i \text{ and } j \ , \tag{2.5}$$

because the case $i = j$ can be proved in a similar but easier manner.

A population in which the genotype frequencies satisfy these equations is said to be in Hardy–Weinberg equilibrium. A mathematically trivial but biologically important consequence of (2.5) is that gene frequencies remain constant across generations, i.e.,

$$p'_i = p_i \quad \text{for every } i \ . \tag{2.6}$$

If random union of gametes is posited, as may be realistic for some marine organisms, then (2.5) follows from the definition of random union.

2.3 SEPARATE SEXES

Hardy–Weinberg equilibrium is also attained, though delayed by one generation, in a population with separate sexes and different initial genotypic frequencies at an autosomal (not sex-linked) locus. Let P_{ij} and Q_{ij} be the frequencies of the ordered genotype $\mathcal{A}_i\mathcal{A}_j$ of males and females, respectively. The gene frequencies in the two sexes are $p_i = \sum_j P_{ij}$ and $q_i = \sum_j Q_{ij}$. An argument similar to that leading to (2.5) shows that after one generation of random mating the genotypic frequencies in the two sexes are equal and

$$P'_{ij} = Q'_{ij} = \tfrac{1}{2}\left(p_i q_j + p_j q_i\right) \ , \tag{2.7}$$

and the gene frequencies are

$$p'_i = q'_i = \tfrac{1}{2}\left(p_i + q_i\right) \ . \tag{2.8}$$

Thus, as shown above, another generation of random mating yields Hardy–Weinberg ratios

$$P''_{ij} = Q''_{ij} = p'_i q'_j \ . \tag{2.9}$$

2.4 X-LINKAGE

In most higher organisms sex is determined by a pair of non-homologous chromosomes, the sex chromosomes. One sex has chromosomes XX, the other XY. We assume that the males are the heterogametic sex XY and females are XX. Genes carried on the X-chromosome are called X-linked and genes on the Y-chromosome are said to be Y-linked. Formally, the dynamics of Y-linked genes is identical to that in haploid populations. X-linked loci are of considerable importance in human genetics, and we shall now investigate the validity of the Hardy–Weinberg Law for such genes.

Let the relative frequency of the allele $\mathcal{A}_i$ be p_i in males and q_i in females. The frequency of ordered genotypes $\mathcal{A}_i\mathcal{A}_j$ in females is denoted by Q_{ij}. Since a male inherits its gene from the mother, we have

$$p'_i = q_i \ . \tag{2.10}$$

Under the assumption of random mating, the genotype frequencies among females in the next generation are

$$Q'_{ij} = \tfrac{1}{2}\sum_l \left(p_i Q_{jl} + p_j Q_{il}\right) = \tfrac{1}{2}\left(p_i q_j + p_j q_i\right) \ , \tag{2.11}$$

because $q_i = \sum_j Q_{ij}$. It follows that

$$q_i' = \tfrac{1}{2}(p_i + q_i) \,. \tag{2.12}$$

Therefore, the frequency of $\mathcal{A}_i$ in the male gene pool satisfies the recursion relation

$$p_i'' = \tfrac{1}{2}(p_i' + p_i) \,. \tag{2.13}$$

Let $x_i = \frac{1}{3}(p_i + 2q_i)$ denote the average frequency of $\mathcal{A}_i$ in the entire population, and let $y_i = p_i - q_i$ be the difference between male and female gene frequencies. Then (2.10) and (2.12) imply $x_i' = x_i$ and $y_i' = -\frac{1}{2}y_i$. Therefore, $x_i(t) = x_i(0)$ and $y_i(t) = y_i(0)(-\frac{1}{2})^t$. It follows that

$$p_i(t) = x_i(0) + \tfrac{2}{3}(-\tfrac{1}{2})^t y_i(0) \,, \tag{2.14a}$$

$$q_i(t) = x_i(0) - \tfrac{1}{3}(-\tfrac{1}{2})^t y_i(0) \,. \tag{2.14b}$$

The same result is, of course, obtained by directly solving (2.13). Equations (2.14) show that the allele frequencies converge to Hardy–Weinberg proportions,

$$p_i = q_i = x_i(0) \quad \text{and} \quad Q_{ij} = x_i(0)x_j(0) \,, \tag{2.15}$$

in an oscillatory manner. Convergence is rapid, but no longer occurs in one or two generations.

For the rest of this book we shall exclusively be concerned with autosomal loci and refer the reader to Nagylaki (1992) for an elaborate treatment of mutation and selection acting on X-linked loci.

In subsequent sections, we shall frequently encounter the Hardy–Weinberg Law. Apart from its fundamental consequences concerning the maintenance of genetic variability, it is of considerable technical importance, because if Hardy–Weinberg proportions obtain, the dynamical and equilibrium behavior of a population can be investigated in terms of the k variables p_i instead of the $k(k+1)/2$ variables P_{ij}, $i \le j$ (given that gene frequencies in both sexes are identical).

3. CORRELATION BETWEEN RELATIVES AND COMPONENTS OF VARIANCE

The resemblance between offspring and their parents or, more generally, between relatives, is not only a popular subject of conversation, but was used by humans long before the era of genetics for successful animal breeding. Historically, the calculation of the correlation between relatives and its comparison with data played a key role in the reconciliation of biometry and Mendelism. In this section we will develop a quantitative approach to measure the resemblance between relatives.

A particularly simple way of quantifying the resemblance between relatives is to calculate how many genes on average they have in common. More precisely, we want to calculate what fraction of their genes is identical by descent. We consider a randomly mating, diploid population. According to Mendelian inheritance, parents pass 50% of their genes to their offspring. Therefore, a parent and its offspring have $\frac{1}{2}$ of their genes in common. Now consider two siblings A and B. Every maternal gene that has been transmitted to A has a 50% chance of also being inherited to B. Hence, A and B share $\frac{1}{2}$ of their maternal genes, on average, the same being valid for paternal genes. Therefore, the average fraction of genes that are identical by descent is $\frac{1}{2}$ for two siblings. This is just the fraction of genes shared by a parent and its offspring. Similar considerations show that, on average, half siblings have $\frac{1}{4}$ and first cousins have $\frac{1}{8}$ of their genes in common.

Now let us investigate a more elaborate model, namely that of a quantitative character. Our aim is to develop some of the basic notions of quantitative genetics.

3.1 AVERAGE EFFECTS AND GENETIC VARIANCE

Quantitative characters are traits that exhibit continuous or almost continuous variation, and can be measured on a metric scale. Typical quantitative characters are weight, height, various morphological measurements, yield, or fitness. Usually such traits are influenced by a large number of loci, often with small effects. It is generally not useful to describe the population genetics of such characters by gene frequencies, because they are difficult or impossible to measure. Instead, a phenotypical approach and statistical methods are needed, and the description will be in terms of the distribution of the character. The most important quantities to describe a probability distribution are its mean value, measuring the location, and its variance, measuring the dispersion. Often, these quantities can be accurately estimated from data. In this section, we consider the simplest case, in which a character is determined by a single locus and environmental influences are disregarded. More realistic models will be investigated later in this book.

We consider a diploid population that is not necessarily in Hardy–Weinberg equilibrium, and denote by $P_{ij} (= P_{ji})$ the frequency of the ordered genotype $\mathcal{A}_i\mathcal{A}_j$ and by p_i the frequency of allele $\mathcal{A}_i$ (2.4). Let G_{ij} be the *genotypic value* (the measurement) of $\mathcal{A}_i\mathcal{A}_j$ individuals. Then the *mean genotypic value* (the population mean) is

$$\overline{G} = \sum_{i,j} G_{ij} P_{ij} \ . \tag{3.1}$$

The deviation of G_{ij} from the mean is denoted by

$$g_{ij} = G_{ij} - \overline{G} \ , \tag{3.2}$$

and is called the *average excess* of the genotype $\mathcal{A}_i\mathcal{A}_j$. The *genotypic variance*, or (*total*) *genetic variance*, is defined as

$$\sigma_G^2 = \sum_{i,j} g_{ij}^2 P_{ij} \ . \tag{3.3}$$

Although, genotypic values and the total genetic variance are fundamental and measurable quantities, they do not necessarily reflect the 'evolutionary potential' of a population. The reason is that parents pass on their genes to the next generation, but not their genotypic values.

Example 1. As an illustration, consider a population in Hardy–Weinberg proportions with two alleles $\mathcal{A}_1$ and $\mathcal{A}_2$, at frequencies p and $1-p$, respectively. Then the three possible genotypes have frequencies p^2, $2p(1-p)$, and $(1-p)^2$. Let us assume complete dominance, i.e., the genotypic values of the genotypes are $G_{11} \neq G_{12} = G_{22}$, and consider two individuals, one with genotype $\mathcal{A}_1\mathcal{A}_2$, the other $\mathcal{A}_2\mathcal{A}_2$. Hence, they have identical genotypic values, $G_{12} = G_{22}$. Suppose that each of these individuals is mated randomly with a number of other individuals. What will be the expected genotypic value of each's offspring? With the help of Table 2.1, and remembering that the probabilities that a given individual mates randomly with an $\mathcal{A}_1\mathcal{A}_1$, $\mathcal{A}_1\mathcal{A}_2$, or $\mathcal{A}_2\mathcal{A}_2$ individual are p^2, $2p(1-p)$, or $(1-p)^2$, respectively, we calculate the expected genotypic value, $\overline{G}'_{12}$, among the offspring of $\mathcal{A}_1\mathcal{A}_2$ as follows:

$$\begin{aligned}\overline{G}'_{12} &= p^2(\tfrac{1}{2}G_{11} + \tfrac{1}{2}G_{12}) + 2p(1-p)(\tfrac{1}{4}G_{11} + \tfrac{1}{2}G_{12} + \tfrac{1}{4}G_{22}) \\ &\quad + (1-p)^2(\tfrac{1}{2}G_{12} + \tfrac{1}{2}G_{22}) \\ &= \tfrac{1}{2}pG_{11} + (1 - \tfrac{1}{2}p)G_{22} \; . \end{aligned} \tag{3.4a}$$

The expected genotypic value among offspring of the $\mathcal{A}_2\mathcal{A}_2$ individual, however, is

$$\overline{G}'_{22} = G_{22} \; , \tag{3.4b}$$

which differs from $\overline{G}'_{12}$ unless $p = 0$. Thus, the expected genetic measurement of an offspring does not depend in a simple way on the genotypic value of its parent. ◇

Therefore, we wish to assign to each allele $\mathcal{A}_i$ a value γ_i that measures the average contribution of $\mathcal{A}_i$ to the character and represents the 'part' of the genotypic value that determines the "genetic potentiality of the individual" (Fisher 1958, p. 33). We adopt a procedure based on a least-squares approximation devised by Fisher (1918, 1930, 1941). Although this seems arbitrary at the moment, it leads to a natural formulation for the correlation between relatives, and paves the way for understanding the response of quantitative traits to selection. Let us write

$$g_{ij} = \gamma_i + \gamma_j + \vartheta_{ij} \; , \tag{3.5}$$

where γ_i is called the *average effect* of $\mathcal{A}_i$ on the character and ϑ_{ij} is the *dominance deviation*. The alleles $\mathcal{A}_i$ and $\mathcal{A}_j$ are said to act *additively* (on the given scale) if $\vartheta_{ij} = 0$. Otherwise, one speaks of dominance of effects. The least-squares procedure suggested by Fisher requires the minimization of the *dominance variance*

$$\sigma_{\mathrm{D}}^2 = \sum_{i,j} \vartheta_{ij}^2 P_{ij} = \sum_{i,j} (g_{ij} - \gamma_i - \gamma_j)^2 P_{ij} \tag{3.6}$$

with respect to the γ_i. The idea behind this is, of course, to approximate the genotypic value G_{ij} as closely as possible by a linear expression, $\overline{G} + \gamma_i + \gamma_j$, in the sense that the expected value of the squared deviations ϑ_{ij}^2 is minimized. Differentiation of σ_{D}^2 with respect to γ_i leads to the conditions

$$\sum_j \vartheta_{ij} P_{ij} = \sum_j (g_{ij} - \gamma_i - \gamma_j) P_{ij} = 0 \quad \text{for every } i \; . \tag{3.7}$$

From the definition (3.2) of g_{ij}, (3.5), and (2.4), we obtain

$$0 = \sum_{i,j} g_{ij} P_{ij} = 2 \sum_{i} \gamma_i p_i + \sum_{i,j} \vartheta_{ij} P_{ij} \ .$$

Hence, (3.7) implies that the mean of the average effects is

$$\sum_{i} \gamma_i p_i = 0 \ . \tag{3.8}$$

The average excess g_i of $\mathcal{A}_i$ is defined by $g_i p_i = \sum_j g_{ij} P_{ij}$ and, because of (3.5) and (3.7), satisfies the relation

$$g_i p_i = \gamma_i p_i + \sum_{j} \gamma_j P_{ij} \ . \tag{3.9}$$

Therefore, the average effects, γ_i, can be found as the solution of the system of linear equations given by (3.9). If all allele frequencies satisfy $p_i > 0$, then the average effects are uniquely determined subject to the condition (3.8); see Chapter II.3.6. It is important to note that average effect and average excess depend on the gene frequencies and thus on the current genetic composition of the population. Therefore, they may change from one generation to the next.

If the population is in Hardy–Weinberg proportions, then (3.9) together with (3.8) inform us that average effect and average excess coincide, i.e.,

$$\gamma_i = g_i = \sum_{j} g_{ij} p_j \ . \tag{3.10}$$

The part of the total genetic variance that can be accounted for by the average effects of the alleles is called the *additive genetic*, or *genic, variance*. In view of (3.8), it is defined as

$$\sigma_{\mathrm{A}}^2 = \sum_{i,j} (\gamma_i + \gamma_j)^2 P_{ij} \ . \tag{3.11}$$

With Hardy–Weinberg proportions, simple algebra verifies that this is equivalent to the representation

$$\sigma_{\mathrm{A}}^2 = 2 \sum_{i} \gamma_i^2 p_i \ . \tag{3.12}$$

Finally, we obtain the following fundamental decomposition of the (total) genetic variance

$$\begin{aligned} \sigma_{\mathrm{G}}^2 &= \sum_{i,j} (\gamma_i + \gamma_j + \vartheta_{ij})^2 P_{ij} \\ &= \sum_{i,j} (\gamma_i + \gamma_j)^2 P_{ij} + \sum_{i,j} \vartheta_{ij}^2 P_{ij} + 2 \sum_{i,j} (\gamma_i + \gamma_j) \vartheta_{ij} P_{ij} \\ &= \sigma_{\mathrm{A}}^2 + \sigma_{\mathrm{D}}^2 \ , \end{aligned} \tag{3.13}$$

where the last sum in the middle line is zero because of (3.7). The dominance variance σ_{D}^2 is typically much less than the additive genetic variance σ_{A}^2 and vanishes in the absence of dominance.

A trivial, but important, consequence of the Hardy–Weinberg Law is that under random mating the genetic variance and its components remain constant in the absence of forces such as selection, mutation, or random genetic drift.

Example 2. Let us illustrate the notions just introduced for the case of two alleles, assuming Hardy–Weinberg proportions. Denote the frequency of $\mathcal{A}_1$ by p and the genotypic contributions of $\mathcal{A}_1\mathcal{A}_1$, $\mathcal{A}_1\mathcal{A}_2$, and $\mathcal{A}_2\mathcal{A}_2$ by G_{11}, G_{12}, and G_{22}, respectively. Then the mean genotypic value becomes

$$\overline{G} = G_{22} + 2(G_{12} - G_{22})p + 2\vartheta p^2 \ , \tag{3.14}$$

where $\vartheta = \frac{1}{2}(G_{11} + G_{22}) - G_{12}$. From (3.10), the average effects of $\mathcal{A}_1$ and $\mathcal{A}_2$ are calculated to be $\gamma_1 = -(1-p)(G_{22} - G_{12} - 2\vartheta p)$ and $\gamma_2 = p(G_{22} - G_{12} - 2\vartheta p)$, respectively, and, from (3.5), the dominance deviations are $\vartheta_{11} = 2\vartheta(1-p)^2$, $\vartheta_{12} = -2\vartheta p(1-p)$, and $\vartheta_{22} = 2\vartheta p^2$. Thus, the additive genetic variance works out to be

$$\begin{aligned}\sigma_{\mathrm{A}}^2 &= 2p(1-p)[\tfrac{1}{2}(G_{22} - G_{11}) + \vartheta(1-2p)]^2 \\ &= 2p(1-p)(\gamma_1 - \gamma_2)^2 \ ,\end{aligned} \tag{3.15}$$

and the dominance variance reduces to

$$\sigma_{\mathrm{D}}^2 = 4\vartheta^2 p^2 (1-p)^2 \ . \tag{3.16}$$

Under the assumption of random mating, $\gamma_1 - \gamma_2$ is known as the (average) effect of allelic substitution. Returning to the example at the beginning of this subsection, but assuming arbitrary dominance relations, the expected genotypic value among offspring of an $\mathcal{A}_1\mathcal{A}_2$ individual with genotypic value $G_{12} = \overline{G} + \gamma_1 + \gamma_2 + \vartheta_{12}$ is calculated to be

$$\overline{G}'_{12} = \overline{G} + \tfrac{1}{2}(\gamma_1 + \gamma_2) \quad \text{and} \quad \overline{G}'_{22} = \overline{G} + \gamma_2 \ . \tag{3.17a}$$

◇

Thus, the average offspring differs from the population mean by half the additive effect of its parent, irrespective of dominance effects. Therefore, the average effects may be considered to measure the genetic and evolutionary 'essential' properties of a genotype. We shall see later in this book that it is, indeed, the additive genetic variance that is the major determinant for the response of the mean genotypic value to selection. The sum of the average effects of an individual is called its *breeding value*. For instance, $\gamma_1 + \gamma_2$ is the breeding value of $\mathcal{A}_1\mathcal{A}_2$. Since this is twice the expected deviation of its offspring mean genotypic value from the population mean, it is a readily measurable quantity. For an arbitrary number of alleles, (3.17a) can be generalized and becomes

$$\overline{G}'_{ij} = \sum_k (G_{ik} + G_{jk})p_k = \overline{G} + \tfrac{1}{2}(\gamma_i + \gamma_j) \ . \tag{3.17b}$$

If the population is not in Hardy–Weinberg proportions, then the relations (3.17) become more complicated. For example, the second equation in (3.17a) has to be replaced by $\overline{G}'_{22} = \overline{G} + \gamma_2 + \vartheta_{22}(P_{12}^2 - P_{11}P_{22})/P_{12}$, as the reader is invited to prove.

3.2 CORRELATIONS BETWEEN RELATIVES

The appropriate measure for the expected resemblance between individuals with respect to a given quantitative character is the correlation coefficient. Suppose for some trait we have measurements, G_1 and G_2, of pairs of relatives, for instance, fathers and their sons. In our simple model the mean values and the genetic variances are identical in both groups, i.e., they are $\overline{G}$ and σ_G^2. Then the correlation is

$$\mathrm{Corr}(G_1, G_2) = \frac{\mathrm{E}[(G_1 - \overline{G})(G_2 - \overline{G})]}{\sigma_G^2}\,, \tag{3.18}$$

where E denotes the expected value. We shall derive this correlation for parents and offspring.

▷ [2] To this end, we calculate the probability that a parent has genotype $\mathcal{A}_i\mathcal{A}_j$ and its offspring has $\mathcal{A}_k\mathcal{A}_l$. Obviously, this is the frequency p_ip_j of the parent's genotype times the conditional probability that the offspring is of genotype $\mathcal{A}_k\mathcal{A}_l$, given that its parent is $\mathcal{A}_i\mathcal{A}_j$. One obtains the following four cases:

$$\begin{aligned}
\Pr(\mathrm{P} = \mathcal{A}_i\mathcal{A}_i, \mathrm{O} = \mathcal{A}_i\mathcal{A}_j) &= p_i^2 \cdot p_j\,,\\
\Pr(\mathrm{P} = \mathcal{A}_i\mathcal{A}_j, \mathrm{O} = \mathcal{A}_i\mathcal{A}_j) &= 2p_ip_j \cdot (\tfrac{1}{2}p_i + \tfrac{1}{2}p_j) = p_i^2p_j + p_ip_j^2\,, \quad j > i\,,\\
\Pr(\mathrm{P} = \mathcal{A}_i\mathcal{A}_j, \mathrm{O} = \mathcal{A}_i\mathcal{A}_k) &= 2p_ip_j \cdot \tfrac{1}{2}p_k\,, \quad k \neq j\,, j > i\,,\\
\Pr(\mathrm{P} = \mathcal{A}_i\mathcal{A}_j, \mathrm{O} = \mathcal{A}_j\mathcal{A}_k) &= 2p_ip_j \cdot \tfrac{1}{2}p_k\,, \quad k \neq i\,, j > i\,.
\end{aligned}$$

Then the covariance between parent and offspring is

$$\begin{aligned}
\mathrm{E}[(G_\mathrm{P} - \overline{G})(G_\mathrm{O} - \overline{G})] = &\sum_{i,j} g_{ii}g_{ij}p_i^2p_j + \sum_i \sum_{j:j>i} g_{ij}^2(p_i^2p_j + p_ip_j^2)\\
&+ \sum_i \sum_{j:j>i} \sum_{k\neq j} g_{ij}g_{ik}p_ip_jp_k + \sum_i \sum_{j:j>i} \sum_{k\neq i} g_{ij}g_{jk}p_ip_jp_k\,,
\end{aligned}$$

and this is the same as

$$\sum_{i,j} g_{ij}^2p_j^2p_i + 2\sum_{i,j}\sum_{k\neq j} g_{ij}g_{ik}p_jp_kp_i = \sum_i \Big(\sum_j g_{ij}p_j\Big)^2 p_i = \tfrac{1}{2}\sigma_\mathrm{A}^2\,.$$

◁

Therefore, the parent-offspring correlation is

$$\mathrm{Corr}(\text{parent, offspring}) = \frac{1}{2}\frac{\sigma_\mathrm{A}^2}{\sigma_G^2}\,. \tag{3.19}$$

In our treatment we have, among others, neglected environmental influences, so that the genotypic variance σ_G^2 equals the phenotypic variance. Therefore, we may identify the ratio $\sigma_\mathrm{A}^2/\sigma_G^2$ with the so-called *heritability in the narrow sense*, or simply *heritability*, h^2, and (3.19) can be expressed as

$$\mathrm{Corr}(\text{parent, offspring}) = \tfrac{1}{2}h^2\,. \tag{3.20}$$

[2] The beginning and end of more technical derivations or proofs that can be skipped at a first reading are signified by ▷ and ◁.

This is a much more general formula for the parent-offspring correlation than (3.19), because the notion of heritability is readily extended to more general contexts including multiple loci and environmental effects.

In a similar, though more complicated manner, the following (and many more) correlations between relatives can be calculated:

$$\mathrm{Corr(full\ sibs)} = \tfrac{1}{2}h^2 + \tfrac{1}{4}\delta^2 \,, \tag{3.21a}$$

where $\delta^2 = \sigma_{\mathrm{D}}^2/\sigma_G^2$, as well as

$$\mathrm{Corr(uncle,\ niece)} = \mathrm{Corr(half\ sibs)} = \mathrm{Corr(parent,\ grandchild)} = \tfrac{1}{4}h^2 \,, \tag{3.21b}$$

and

$$\mathrm{Corr(first\ cousins)} = \tfrac{1}{8}h^2 \,. \tag{3.21c}$$

Early calculations of the correlation between certain relatives under various restrictive assumptions are due to Pearson (1904), Yule (1906), and Weinberg (1909). Considerable extensions and new methods were developed by Fisher (1918), Wright (1922), Malécot (1948), and Kempthorne (1955). It is this paper of Fisher that marks the beginning of the synthesis of Darwinism, biometry, and Mendelism, though the first steps in this direction are due to Yule and Weinberg. The reader is referred to Crow and Kimura (1970), Nagylaki (1992), and Lynch and Walsh (1998) for a much more complete and general treatment of this important topic, as well as for historical remarks and further references.

4. NONRANDOM MATING

The derivation of the Hardy–Weinberg Law rests on the assumption of a randomly mating population. Random mating is probably the most important mating pattern in nature, and most of the theory in this book is based on its supposition. However, nonrandom mating is also frequent and may have a profound influence on the genetic composition of a population. There are various forms of nonrandom mating. *Inbreeding* occurs if matings between related individuals are more frequent than between randomly chosen ones. The most extreme form of inbreeding is self-fertilization, as observed in many plants and some animals. One speaks of *assortative mating*, if individuals of similar phenotype mate more frequently than randomly chosen ones, whereas *disassortative mating* occurs when like phenotypes mate less frequently than expected with random mating. The latter phenomenon is much rarer than the first. Good examples, however, are provided by plants (e.g., tobacco) that have evolved self-incompatibility mechanisms.

4.1 INBREEDING

Let us first consider self-fertilization to demonstrate that, in the absence of selection, inbreeding reduces the proportion of heterozygous individuals without changing the allele frequencies. Under self-fertilization, all offspring of the homozygotes are again of the same genotype, and $\frac{1}{4}$ of $\mathcal{A}_1\mathcal{A}_2 \times \mathcal{A}_1\mathcal{A}_2$ matings are $\mathcal{A}_1\mathcal{A}_1$, another $\frac{1}{4}$ are $\mathcal{A}_2\mathcal{A}_2$, and $\frac{1}{2}$ are $\mathcal{A}_1\mathcal{A}_2$. Thus, denoting the frequency of $\mathcal{A}_i\mathcal{A}_j$ by P_{ij}, in the next generation we have

$$P'_{11} = P_{11} + \tfrac{1}{4}P_{12}\,, \quad P'_{12} = \tfrac{1}{2}P_{12}\,, \quad P'_{22} = P_{22} + \tfrac{1}{4}P_{12}\,. \tag{4.1}$$

Hence, the proportion of heterozygotes, the so-called *heterozygosity*, is reduced by 50% each generation. For other forms of inbreeding, such as sib mating, the heterozygosity also declines geometrically in the limit of long time, though at a slower rate.

For a more general treatment we introduce the *inbreeding coefficient* (Wright 1922), but use an alternative, probabilistic approach (Cotterman 1940, Malécot 1948). It is based on the notion of identity by descent. An individual in a population derived from some ancestral population can be homozygous for a given locus in two ways. The two homologous genes may be *identical in state*, i.e., of the same allelic type, or they may be *identical by descent*, i.e., derived from the same gene in the ancestral population. Since we ignore mutation, identity by descent implies identity in state, but not vice versa. An individual with two homologous genes that are identical by descent is called *autozygous*, otherwise it is called *allozygous*. Hence, an autozygous individual is always homozygous, but an allozygous individual may be homozygous or heterozygous. The inbreeding coefficient, F, is defined as the probability that the individual is autozygous for the locus in question.

A randomly chosen individual with inbreeding coefficient F is autozygous with probability F and allozygous with probability $1 - F$. It will carry allele $\mathcal{A}_i$ with probability p_i. If it is autozygous, it is necessarily of type $\mathcal{A}_i\mathcal{A}_i$; thus the frequency of autozygotes is Fp_i. If it is allozygous, its genes are stochastically independent and it carries allele $\mathcal{A}_j$ at the other chromosome with probability p_j. Therefore, it is of (ordered) genotype $\mathcal{A}_i\mathcal{A}_j$ with probability p_ip_j. Hence, the ordered genotype frequencies in the population are

$$P_{ii} = Fp_i + (1-F)p_i^2\,, \tag{4.2a}$$

$$P_{ij} = (1-F)p_ip_j\,, \quad \text{if } i \neq j\,. \tag{4.2b}$$

If there is no inbreeding ($F = 0$), this yields the Hardy–Weinberg proportions. In a completely inbred population ($F = 1$), there are no heterozygotes and the homozygotes occur with frequency p_i. Equations (4.2) also show that inbreeding alone does not change the allele frequencies because

$$\sum_j P_{ij} = Fp_i + (1-F)p_i = p_i\,. \tag{4.3}$$

The heterozygosity in an inbred population is given by

$$H = \sum_{i,j:\, i\neq j} P_{ij} = (1-F)H_0\,, \tag{4.4}$$

where $H_0 = \sum_{i,j:\, i\neq j} p_ip_j$ is the heterozygosity with random mating. This shows that the inbreeding coefficient measures the fraction by which inbreeding reduces the heterozygosity relative to a random-mating population, i.e.,

$$F = \frac{H_0 - H}{H_0}\,. \tag{4.5}$$

Example. Let us consider the effects of inbreeding on the genetic variance for the two-allele model of Example 2 in Section 3.1. Instead of assuming Hardy–Weinberg proportions, the genotype frequencies are given by (4.2). Then a simple calculation shows that the mean genotypic value, as a function of F, is

$$\overline{G}(F) = (1-F)\overline{G}(0) + F\overline{G}(1)\,, \tag{4.6}$$

where $\overline{G}(0)$ is the mean value with random mating ($F = 0$), as given by (3.14), and $\overline{G}(1) = G_{11}p + G_{22}(1-p)$ is the mean with complete inbreeding ($F = 1$). A further straightforward calculation yields

$$\sigma_G^2(F) = (1-F)\sigma_G^2(0) + F\sigma_G^2(1) + F(1-F)[\overline{G}(0) - \overline{G}(1)]^2 \tag{4.7}$$

for the total genetic variance. Here, $\sigma_G^2(0)$ and $\sigma_G^2(1)$ are the genetic variances under random mating and complete inbreeding, respectively. Without dominance ($\vartheta = 0$), there is no effect of inbreeding on the mean, i.e., $\overline{G}(0) = \overline{G}(1) = G_{11}p + G_{22}(1-p)$, and the genetic variance becomes

$$\sigma_G^2(F) = (1+F)\sigma_G^2(0) \tag{4.8}$$

because, in this case, $\sigma_G^2(1) = 2\sigma_G^2(0) = (G_{22} - G_{11})^2 p(1-p)$. Thus, inbreeding increases the genetic variance. ◇

The inbreeding coefficient can be estimated from pedigrees. For methods of computation, as well as for much more detailed accounts of inbreeding, we refer to Nagylaki (1992). Certain forms of inbreeding in multilocus systems are treated by Christiansen (1999). We note in closing that inbreeding may profoundly affect gene frequencies if there is selection because, through the increased homozygosity in an inbred population, deleterious recessive alleles are more frequently exposed to selection than in a comparable randomly mating population.

4.2 ASSORTATIVE MATING

Since individuals with similar phenotypes will usually have similar genotypes, the general consequences of inbreeding and assortative mating could be expected to be similar. This is true only to a certain extent, because inbreeding affects all segregating genes whereas assortative mating affects only those genes that are associated with the phenotypic trait in question. In nature, assortative mating occurs primarily for traits that have a polygenic basis. This makes the analysis difficult. We shall illustrate the consequences of assortative mating for the simple case where the phenotype is determined by the genotype at one locus. If different genotypes code for different phenotypes, perfect (imperfect) assortative mating is equivalent to (partial) self-fertilization. Therefore, let us consider the more interesting case of complete dominance, i.e., the genotypes $\mathcal{A}_1\mathcal{A}_1$ and $\mathcal{A}_1\mathcal{A}_2$ have the same phenotype, whereas $\mathcal{A}_2\mathcal{A}_2$ has a different one. Thus, $\mathcal{A}_1$ is (completely) dominant.

Let ρ denote the fraction of individuals mating assortatively and assume that this fraction is the same for both phenotypes. Denote the (ordered) genotype frequencies by P_{ij} and the allele frequencies of $\mathcal{A}_1$ and $\mathcal{A}_2$ by p and $q = 1 - p$, respectively. Then the contribution of the random-mating group of individuals to the progeny of type $\mathcal{A}_1\mathcal{A}_1$, $\mathcal{A}_1\mathcal{A}_2$, and $\mathcal{A}_2\mathcal{A}_2$ will be $(1-\rho)p^2$, $(1-\rho)2pq$, and $(1-\rho)q^2$, respectively. The group of assortatively mating individuals with genotype $\mathcal{A}_2\mathcal{A}_2$ will contribute ρP_{22} to the progeny of the same type and make no contribution to other types. It remains to calculate the contribution of the group of assortatively mating individuals with the dominant phenotype, i.e., with genotype $\mathcal{A}_1\mathcal{A}_1$ or $\mathcal{A}_1\mathcal{A}_2$. Their proportion in the total population is $\rho(1-P_{22})$, and the frequencies of $\mathcal{A}_1$ and $\mathcal{A}_2$ in this group are $p/(1-P_{22})$ and $P_{12}/(1-P_{22})$, respectively. Therefore, this group's contribution to $\mathcal{A}_1\mathcal{A}_1$ progeny is $\rho(1-P_{22})[p/(1-P_{22})]^2$, to $\mathcal{A}_1\mathcal{A}_2$ progeny $\rho(1-P_{22})[p/(1-P_{22})][P_{12}/(1-P_{22})]$, and to $\mathcal{A}_2\mathcal{A}_2$ progeny $\rho(1-P_{22})[P_{12}/(1-P_{22})]^2$. Putting all this together gives, after trivial simplifications, the recursion equations for the genotype frequencies:

$$P'_{11} = (1-\rho)p^2 + \rho\,\frac{p^2}{1-P_{22}}\,, \tag{4.9a}$$

$$P'_{12} = (1-\rho)pq + \rho\,\frac{pP_{12}}{1-P_{22}}\,, \tag{4.9b}$$

$$P'_{22} = (1-\rho)q^2 + \rho\,\frac{q^2 + P_{22}(p-q)}{1-P_{22}}\,. \tag{4.9c}$$

An immediate consequence of these equations is that the allele frequencies p and q remain constant.

For complete assortative mating ($\rho = 1$), (4.9b) implies that the heterozygosity $H = 2P_{12}$ evolves according to

$$H' = H\frac{p}{1-P_{22}} = H\frac{p}{p+H/2}\,. \tag{4.10}$$

This approaches 0 as time increases, but very slowly. For instance, if $p(t=0) = \frac{1}{2}$, then $H(t=0) = \frac{1}{2}$, and the heterozygosity in generation n is $1/(n+2)$. If $\rho < 1$, the heterozygosity still decreases, but reaches a positive equilibrium value. This is in contrast to inbreeding, where the heterozygosity decreases geometrically to zero.

For more detailed studies of assortative and disassortative mating the reader is referred to Crow and Kimura (1970), Jacquard (1974), and Nagylaki (1992).

5. RANDOM GENETIC DRIFT

In previous sections, population size was assumed to be sufficiently large to equate the probability of sampling an allele with its relative frequency. Here we shall briefly explore the consequences of finite population size. In a finite population, changes in allele frequencies must be viewed as a stochastic process, because of variation in the number of offspring produced by different individuals, and because of the stochastic nature of segregation. The stochasticity introduced through such random effects is called *random genetic drift*. The most widely used model for studying finite populations is the so-called *Wright–Fisher model*, explicitly written down by Wright (1931) and implicitly employed by Fisher (1922, 1930). Here we shall introduce it in its simplest version, in which it is assumed that there is only one sex and no mutation, selection, geographic dispersal, etc.

Consider a diploid, monoecious population of fixed size N with two selectively neutral alleles, $\mathcal{A}_1$ and $\mathcal{A}_2$, at a certain locus. Then there are $2N$ genes in the population and we denote the number of $\mathcal{A}_1$ genes in generation t by $X(t)$. In the Wright–Fisher model, it is assumed that the $2N$ genes in generation $t+1$ are obtained from the $2N$ parental genes in generation t by sampling with replacement. This will be a good approximation if allelic proportions are preserved under reproduction and the number of gametes produced is sufficiently high that removing $2N$ gametes randomly does not change the relative frequencies in the gamete pool. Then $X(t+1)$ is a binomial random variable with index $2N$ and parameter $X(t)/(2N)$. More precisely, given that $X(t) = i$, the probability π_{ij} that $X(t+1) = j$ is

$$\pi_{ij} = \binom{2N}{j} \left(\frac{i}{2N}\right)^j \left(1 - \frac{i}{2N}\right)^{2N-j} , \tag{5.1}$$

for $i, j = 0, 1, 2, \ldots, 2N$. The π_{ij} are called the transition probabilities and the matrix (p_{ij}) is the transition matrix of the associated Markov chain $X(\cdot)$. Knowledge of the transition matrix allows one to calculate the probability distribution of $X(t)$ for every generation t if the (probability distribution of the) initial state $X(0) = X_0$ is known, because

$$\Pr[X(t+1) = j] = \sum_{i=0}^{2N} \Pr[X(t) = i]\pi_{ij} . \tag{5.2}$$

For this model let us derive some simple facts about the evolution of a finite population. First, we obtain from (5.2) by using $\sum_j j\pi_{ij} = i$,

$$\mathrm{E}[X(t+1)] = \mathrm{E}[X(t)] = \cdots = \mathrm{E}[X_0] , \tag{5.3}$$

where E denotes the expectation. (We write $\mathrm{E}[X_0]$ because we do not need to know the initial state with certainty, only its distribution.) Similarly, using $\sum_j j^2\pi_{ij} = i + (1 - 1/(2N))i^2$, we get

$$\mathrm{E}[(X(t+1))^2] = \mathrm{E}[X(t)] + \left(1 - \frac{1}{2N}\right) \mathrm{E}[(X(t))^2] . \tag{5.4}$$

Equation (5.3) shows that, on average, allele frequencies remain constant. However, because of random fluctuations, any given population will not maintain a constant frequency of $\mathcal{A}_1$. Indeed, (5.4) implies that the expected heterozygosity decreases geometrically to zero, i.e., if $p(t) = X(t)/(2N)$ and $H(t) = 2p(t)[1 - p(t)]$, then

$$\mathrm{E}[H(t+1)] = \left(1 - \frac{1}{2N}\right) \mathrm{E}[H(t)] . \tag{5.5}$$

Thus, random genetic drift eliminates all heterozygotes from the population. Since the population is random mating, this implies that one of the alleles becomes fixed. Once an allelic type is lost, it cannot be reintroduced into the population because this model ignores mutation.

It is easy to calculate the probability of fixation of, say, allele $\mathcal{A}_1$. Obviously, (5.5) implies that $\lim_{t\to\infty} \Pr[X(t) = i] = 0$ for $i = 1, \ldots, 2N-1$. Therefore,

$$\begin{aligned} \mathrm{E}[X_0] &= \lim_{t\to\infty} \mathrm{E}[X(t)] = \lim_{t\to\infty} \sum_{i=0}^{2N} i \Pr[X(t) = i] \\ &= \lim_{t\to\infty} 2N \Pr[X(t) = 2N] \,, \end{aligned}$$

which shows that the probability of fixation of $\mathcal{A}_1$ is

$$\Pr[\mathcal{A}_1 \text{becomes fixed}] = \frac{\mathrm{E}[X_0]}{2N} \,. \tag{5.6}$$

Thus, if p_0 is the initial frequency of an allele, its fixation probability is also p_0. (A rigorous argument would use the Optional Stopping Theorem; see Karlin and Taylor 1975).

There is also another, more intuitive approach to find the fixation probability of $\mathcal{A}_1$ (Ewens 1979). Note that eventually every gene in a population is descended from one unique gene in the initial generation. The probability that such a gene is $\mathcal{A}_1$ is simply the initial fraction of such genes. This must, therefore, be the fixation probability of $\mathcal{A}_1$.

Finally, we note the close relationship between inbreeding and random genetic drift, caused by the fact that random genetic drift leads to fixation of one allele. During this process an increasing number of genes will become identical by descent, and an increasing amount of inbreeding will occur. Denoting the initial heterozygosity by H_0, and applying (4.5) and (5.5), we infer that the inbreeding coefficient in a finite random-mating population is

$$F = F(t) = 1 - \left(1 - \frac{1}{2N}\right)^t \,. \tag{5.7}$$

We refer to Ewens (1979, 1990), Kingman (1980), and Nagylaki (1992) for treatises of many important topics in the theory of finite populations; see also Appendix E.

6. MUTATION

We shall employ a simple concept of mutation, sufficient for most purposes in population genetics theory, by designating any change from one allelic type to another a mutation. In this section, we assume all mutations to be neutral, i.e., all have the same fitness. Let us consider k alleles, $\mathcal{A}_1, \ldots, \mathcal{A}_k$, at a gene locus and label their frequencies by $p_1, \ldots, p_k$. We assume that the population is sufficiently large to ignore random genetic drift. For $i \neq j$ we denote the probability that an $\mathcal{A}_i$ gene has an $\mathcal{A}_j$ offspring by the mutation rate μ_{ij}. We shall use the convention $\mu_{ii} = 0$ for every i. Then the fraction of $\mathcal{A}_i$ genes that do not mutate is $1 - \sum_j \mu_{ij}$, and $\mathcal{A}_j$ genes give rise to a mutant $\mathcal{A}_i$ with probability μ_{ji}. Therefore, the frequency p_i' of $\mathcal{A}_i$ in the next generation is

$$p_i' = p_i\left(1 - \sum_j \mu_{ij}\right) + \sum_j p_j \mu_{ji} \,. \tag{6.1}$$

We call (6.1) the *(pure) mutation equation*. Due to the convention $\mu_{ii} = 0$, the index i may or may not be excluded in the above summations.

Linear algebra shows that there exists a unique equilibrium if all mutation rates are positive, and that convergence to this equilibrium occurs at a geometric rate (see Chapter III.2 for a more general result).

Example. Let us illustrate this for the simple case of two alleles. Denoting the mutation rate from $\mathcal{A}_1$ to $\mathcal{A}_2$ by μ, the reverse mutation rate by ν, and the frequency of $\mathcal{A}_1$ by p, the recursion (6.1) reduces to

$$p' = p(1 - \mu - \nu) + \nu \ . \tag{6.2}$$

If μ or ν is positive, there exists a unique equilibrium frequency (obtained from the condition $p' = p$). It is given by

$$\hat{p} = \frac{\nu}{\mu + \nu} \ .^{3} \tag{6.3}$$

The recursion equation (6.2) can be solved explicitly and, using (6.3), its solution can be expressed as

$$p(t) - \hat{p} = (p_0 - \hat{p})(1 - \mu - \nu)^t \ , \tag{6.4}$$

where $p_0 = p(0)$ is the initial frequency of $\mathcal{A}_1$. This shows that convergence to equilibrium occurs at a geometric rate, but is very slow because $\mu + \nu$ is typically very small. If the locus contributes additively to a quantitative trait,, such that the effects of $\mathcal{A}_1\mathcal{A}_1$, $\mathcal{A}_1\mathcal{A}_2$, and $\mathcal{A}_2\mathcal{A}_2$ are $-\gamma$, 0, and γ, respectively, the equilibrium genetic variance is

$$\sigma_G^2 = 2\gamma^2 \frac{\mu\nu}{(\mu + \nu)^2} \tag{6.5}$$

(cf. Section 3.1). Thus, depending on the ratio μ/ν, the genetic variance may attain any value between 0 and $\frac{1}{2}\gamma^2$. ◇

7. RECOMBINATION

According to the Hardy–Weinberg Law, the genotype frequencies attain an equilibrium value after one generation of random mating if gene loci are considered separately. This is no longer true for genotypes with respect to two or more loci considered jointly. Consider two loci, $\mathcal{A}$ and $\mathcal{B}$, each with two alleles, $\mathcal{A}_1$, $\mathcal{A}_2$, and $\mathcal{B}_1$, $\mathcal{B}_2$. Then there are ten possible genotypes. If, for instance, in the initial generation only the genotypes $\mathcal{A}_1\mathcal{B}_1/\mathcal{A}_1\mathcal{B}_1$ and $\mathcal{A}_2\mathcal{B}_2/\mathcal{A}_2\mathcal{B}_2$ are present, then in the next generation only these double homozygotes, as well as the two double heterozygotes $\mathcal{A}_1\mathcal{B}_1/\mathcal{A}_2\mathcal{B}_2$ and $\mathcal{A}_1\mathcal{B}_2/\mathcal{A}_2\mathcal{B}_1$ will be present. After further generations of random mating, all other genotypes will occur, but not immediately at their equilibrium frequencies. Of course, the formation of gametic types other than $\mathcal{A}_1\mathcal{B}_1$ or $\mathcal{A}_2\mathcal{B}_2$ requires that recombination between the two loci occur. Disequilibrium with respect to two or more loci is called *linkage disequilibrium*, or gametic phase disequilibrium. It is equivalent to statistical dependence of allele frequencies between loci.

---

[3] Throughout this book, a hat, ˆ, signifies an equilibrium.

For a rigorous treatment, we consider more generally two loci, each with an arbitrary number of alleles. Let the frequencies of the alleles $\mathcal{A}_i$ at the $\mathcal{A}$ locus be denoted by p_i and those of the alleles $\mathcal{B}_j$ at the $\mathcal{B}$ locus by q_j. Let the frequency of the gamete $\mathcal{A}_i\mathcal{B}_j$ be P_{ij}, so that $p_i = \sum_j P_{ij}$ and $q_j = \sum_i P_{ij}$. In general, these allele frequencies are no longer sufficient to describe the genetic composition of the population. *Linkage equilibrium* is defined as the state in which

$$P_{ij} = p_i q_j \tag{7.1}$$

holds for every i and j. Otherwise the population is said to be in linkage disequilibrium.

Let the parameter r denote the *recombination frequency*, or *recombination rate*, between the two loci. This is the probability that a recombination event (crossing over) occurs between them. The value of r usually depends on the distance between the two loci along the chromosome. Loci with $r = 0$ are called completely linked (and may be treated as a single locus) and loci with $r = \frac{1}{2}$ are called unlinked. The maximum value of $r = \frac{1}{2}$ typically occurs for loci on different chromosomes, because then all four gametes are produced with equal frequency $\frac{1}{4}$. Thus, the recombination rate satisfies $0 \le r \le \frac{1}{2}$.

Given P_{ij}, we want to find the gametic frequencies P'_{ij} in the next generation after random mating. The derivation of the recursion equation is based on the following basic fact of Mendelian genetics: an individual with genotype $\mathcal{A}_i\mathcal{B}_j/\mathcal{A}_k\mathcal{B}_l$ produces gametes of parental type if no recombination occurs (with probability $1 - r$), and recombinant gametes if recombination between the two loci occurs (with probability r). Therefore, the fraction of gametes $\mathcal{A}_i\mathcal{B}_j$ and $\mathcal{A}_k\mathcal{B}_l$ is $\frac{1}{2}(1 - r)$ each, and that of $\mathcal{A}_i\mathcal{B}_l$ and $\mathcal{A}_k\mathcal{B}_j$ is $\frac{1}{2}r$ each. From these considerations, we see that the frequency of gametes of type $\mathcal{A}_i\mathcal{B}_j$ in generation $t+1$ produced without recombination is $(1-r)P_{ij}$, and that produced with recombination is rp_iq_j because of random mating. Thus,

$$P'_{ij} = (1 - r)P_{ij} + rp_iq_j \ . \tag{7.2}$$

This shows that the gene frequencies are conserved, but the gamete frequencies are not, unless the population is in linkage equilibrium, (7.1). Commonly, linkage disequilibrium between alleles $\mathcal{A}_i$ and B_j is measured by the parameter

$$D_{ij} = P_{ij} - p_iq_j \ . \tag{7.3}$$

The D_{ij} are often called simply linkage disequilibria, although no single D_{ij} is a complete measure of linkage disequilibrium. From (7.2) and (7.3) we infer that

$$D'_{ij} = (1 - r)D_{ij} \tag{7.4}$$

and, hence,

$$D_{ij}(t) = (1 - r)^t D_{ij}(0) \ . \tag{7.5}$$

Therefore, unless $r = 0$, linkage disequilibria decay at the geometric rate $1 - r$ and linkage equilibrium is approached gradually without oscillation. With unlinked loci, $r = \frac{1}{2}$, linkage disequilibrium is halved each generation.

For two alleles at each locus, it is more convenient to label the frequencies of the gametes $\mathcal{A}_1\mathcal{B}_1$, $\mathcal{A}_1\mathcal{B}_2$, $\mathcal{A}_2\mathcal{B}_1$, and $\mathcal{A}_2\mathcal{B}_2$ by x_1, x_2, x_3, and x_4, respectively. A simple

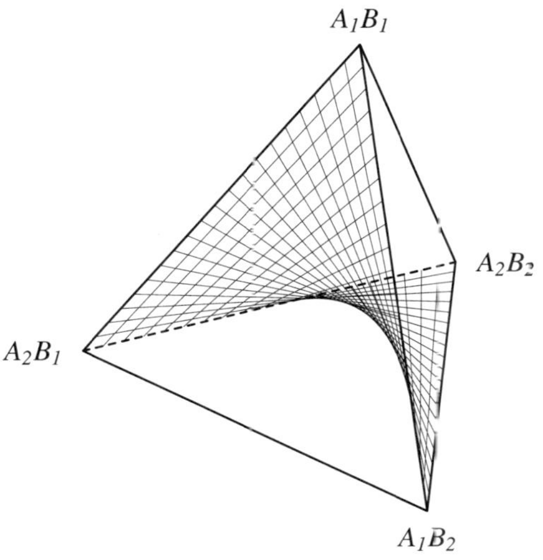

Figure 7.1 The tetrahedron represents the state space of the two-locus two-allele model. The vertices correspond to fixation of the labeled gamete, and frequencies are measured by the (orthogonal) distance from the opposite boundary face. Thus, at the center of the simplex all gametes have frequency $\frac{1}{4}$. The two-dimensional surface is the linkage-equilibrium manifold corresponding to the states in linkage equilibrium, $D = 0$. The states of maximum linkage disequilibrium, $D = \pm\frac{1}{4}$, are the centers of the edges connecting $\mathcal{A}_1\mathcal{B}_2$ to $\mathcal{A}_2\mathcal{B}_1$ and $\mathcal{A}_1\mathcal{B}_1$ to $\mathcal{A}_2\mathcal{B}_2$.

calculation reveals that in this case the difference of the frequency of coupling genotypes, $\mathcal{A}_1\mathcal{B}_1/\mathcal{A}_2\mathcal{B}_2$, and repulsion genotypes, $\mathcal{A}_1\mathcal{B}_2/\mathcal{A}_2\mathcal{B}_1$,

$$D = x_1x_4 - x_2x_3 \ , \tag{7.6}$$

satisfies

$$D = D_{11} = -D_{12} = -D_{21} = D_{22} \ . \tag{7.7}$$

Thus, the recursion equations for the gamete frequencies, (7.2), may be rewritten as

$$\begin{aligned} x_1' &= x_1 - rD \ , \\ x_2' &= x_2 + rD \ , \\ x_3' &= x_3 + rD \ , \\ x_4' &= x_4 - rD \ . \end{aligned} \tag{7.8}$$

The two-locus gametic frequencies may be represented geometrically by the points in a tetrahedron, because $x_1 + x_2 + x_3 + x_4 = 1$. The set of quadruples (x_1, x_2, x_3, x_4), $x_i \geq 0$, satisfying this constraint is called the three-dimensional *simplex*, and denoted by S_4 [cf. (9.24)]. The subset where $D = 0$ forms a two-dimensional manifold and is called the linkage equilibrium, or Wright, manifold. It is displayed in Figure 7.1.

It follows from (7.5) that, if $r > 0$, all solutions of (7.8) converge to the linkage-equilibrium manifold along straight lines, because the allele frequencies, $x_1 + x_2$ and $x_1 + x_3$, remain constant, and sets of the form $x_1 + x_2 =$ const. represent planes in this geometric picture. In the present simple model, the linkage-equilibrium manifold is invariant under the dynamics (7.8). With selection or mutation, this is generally not the case.

If there are more than two loci, linkage disequilibria among any group of at least two loci have to be considered. However, it can again be proved that under random mating all gametic combinations eventually reach equilibrium proportions. The rate of decay can be shown to be $1 - r_{\min}$, where $r_{\min}$ is the smallest of all two-locus recombination fractions. In particular, higher-order disequilibria decay faster than the two-locus disequilibria. The two-locus results were first derived by Robbins (1918), but the simple derivation given above is due to Malécot (1948). The multilocus case was first treated by Geiringer (1944). We shall investigate multilocus-multiallele models in Chapters II and V, where we shall encounter other measures of linkage disequilibrium.

8. POPULATION GROWTH

The size of any population changes in the course of time as a consequence of events such as birth, death, migration, etc. Here, we shall briefly describe simple deterministic models for the growth of a population that have been used in population genetics and ecology, and which will form the basis of many of the models considered in this book. We shall assume, as before, that the population is sufficiently large that random effects can be neglected.

8.1 DISCRETE, NONOVERLAPPING GENERATIONS

Let us consider a population with discrete and nonoverlapping generations. This is the case in annual plants and many insects where the parents die after reproduction, or in experiments where the parents are no longer counted. Time is measured in units of generations. This model is widely used in population genetics because of its simplicity and because it is often a useful first approximation.

Let $N(t)$, $t = 0, 1, 2, 3, \ldots$, be the number of individuals in generation t. If the average number of offspring per individual is W, where W is a compound measure of reproductive success and survival, the population size at generation t can be calculated from that of generation 0 according to

$$N(t) = WN(t-1) = W^2N(t-2) = \cdots = W^tN(0) \; . \tag{8.1}$$

It will often be convenient to write the fitness W as $W = 1 + s$. Then the change $\Delta N(t) = N(t+1) - N(t)$ in population size can be written as

$$\Delta N(t) = sN(t) \; , \tag{8.2}$$

and the population increases if $s > 0$ and decreases if $s < 0$.

Now let us consider a population that is composed of k types of individuals, with fitness values W_i, $i = 1, \ldots, k$. Let $n_i = n_i(t)$ denote the number of individuals of type i in generation t and let $N = N(t) = \sum_i n_i(t)$. We define the *mean fitness* of the population by

$$\overline{W} = \overline{W}(t) = \sum_i W_i \frac{n_i(t)}{N(t)} \; . \tag{8.3}$$

The number of self-reproducing (asexual) individuals of type i in the next generation is $n_i' = W_i n_i$, and the population size is, using (8.3),

$$N' = \sum_i n_i' = \sum_i W_i n_i = \overline{W} N . \tag{8.4}$$

In general, mean fitness $\overline{W}$ will change from generation to generation, since the n_i will change. Iterating (8.4), we obtain

$$N(t) = \overline{W}(t-1)N(t-1) = \cdots = \prod_{\tau=0}^{t-1} \overline{W}(\tau) \cdot N(0) , \tag{8.5}$$

generalizing (8.1).

8.2 A CONTINUOUS-TIME MODEL

Assume that in a population of size $N = N(t)$ an average of $b\,\Delta t\,N(t)$ progeny are born during an infinitesimal time interval Δt and $d\,\Delta t\,N(t)$ individuals die in the same time span. The parameter b is then called the birth rate and d the death rate. The total change in population number during the time interval Δt is

$$\Delta N(t) = N(t + \Delta t) - N(t) = (b - d)\,\Delta t\,N(t) . \tag{8.6}$$

As $\Delta t \to 0$, one obtains[4]

$$\dot{N} = \frac{\mathrm{d}N}{\mathrm{d}t} = mN , \tag{8.7}$$

where $m = b - d$ is called the *Malthusian parameter*, or the *intrinsic growth rate*. The differential equation (8.7) has the solution

$$N(t) = e^{mt} N(0) , \tag{8.8}$$

so that the population grows exponentially. This model applies, for instance, to bacterial growth in an unlimited environment. Time can be measured in arbitrary units as long as m is measured in the reciprocal of that unit.

The relation between the Malthusian parameter m and the fitness W is, by comparison of (8.1) and (8.8),

$$W = e^m \quad \text{or} \quad m = \ln W . \tag{8.9}$$

If W is nearly 1, and thus s is small, we have $\ln W = \ln(1+s) \approx s$, so that $m \approx s$.

Example. Let us consider three populations. Population 1 has discrete, nonoverlapping generations and fitness $W = 1.2$, implying that the population increases by 20%

---

[4] Throughout this book, we denote the time derivative $\frac{\mathrm{d}x}{\mathrm{d}t}$ of any function of time, $x = x(t)$, by $\dot{x}$. As usual, the argument t will be suppressed in such differential equations, i.e., we write $\dot{x} = f(x)$ instead of $\dot{x}(t) = f(x(t))$.

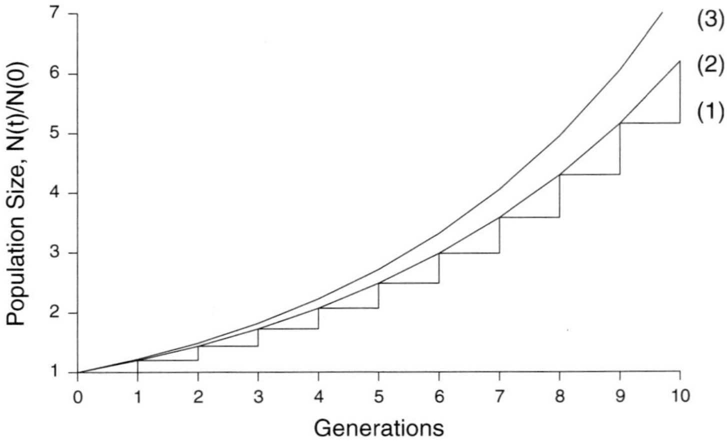

Figure 8.1 Growth of populations in discrete and continuous time (after Crow and Kimura 1970). The lines labeled (1), (2), and (3) refer to the populations described in the text. One unit in continuous time corresponds to one generation.

each generation. Equation (8.1) shows that $N(t) = 1.2^t N(0)$. Population 2 grows continuously with $m = \ln 1.2 \approx 0.182$. According to (8.8), we have $N(t) = e^{0.182t} N(0)$. Population 3 also grows continuously, but with $m = s = 0.2$, so that $N(t) = e^{0.2t} N(0)$. Figure 8.1 displays the growth of these three populations. ◇

Next let us assume that there are k types in the population with different Malthusian fitness parameters m_i and absolute frequencies $n_i = n_i(t)$, $i = 1, \ldots, k$. Then the total population size is $N = \sum_i n_i$ and it changes according to

$$\dot{N} = \sum_i \dot{n}_i = \sum_i m_i n_i = \overline{m} N \ , \tag{8.10}$$

where $\overline{m} = \sum_i m_i n_i / N$ is the mean (Malthusian) fitness of the population. Let $p_i = n_i/N$ denote the (relative) frequency of type i in the population. Then a simple calculation shows that the dynamics of type frequencies is given by

$$\dot{p}_i = p_i(m_i - \overline{m}) \ . \tag{8.11}$$

Clearly, $\overline{m}$ depends on time and, in general, will change in time. Let us calculate this change and remember that the m_i are constants, but n_i and N depend on t. Then we obtain from (8.11)

$$\begin{aligned} \dot{\overline{m}} &= \sum_i m_i \dot{p}_i \\ &= \sum_i (m_i - \overline{m})\, \dot{p}_i \\ &= \sum_i p_i \,(m_i - \overline{m})^2 = \sigma^2 \ , \end{aligned} \tag{8.12}$$

where σ^2 is the variance of the fitness values m_i. If the different types in the population correspond to different alleles (or genes), then σ^2 is the genic variance in fitness. Thus the rate of change in fitness, measured in Malthusian parameters, is equal to the genic variance in fitness. In particular, since σ^2 is a variance and therefore always nonnegative, the mean Malthusian fitness $\overline{m}$ is a nondecreasing function of time. This is a very special case of Fisher's (1930) *Fundamental Theorem of Natural Selection.* In subsequent sections and chapters, we shall encounter much more general versions of this theorem.

The above continuous-time model is very simplistic in several respects. For instance, it would be more realistic to assume that the death rate of an individual or the probability of giving birth to an offspring depend on age. This leads to age-structured population models. Simple population-genetic models incorporating age structure are presented in Crow and Kimura (1970) and Nagylaki (1992). There it is shown that once the population has reached a stable age distribution, the gene-frequency dynamics is again given by (8.11). Readers interested in models of age-structured and physiologically structured populations are referred to Metz and Diekmann (1986) and Charlesworth (1994).

8.3 POPULATION REGULATION

The discrete- and continuous-time models treated above lead to exponential population growth if $\overline{W} > 1$ or $\overline{m} > 0$. Obviously, no population can grow exponentially forever, because the larger a population grows the fewer its resources become. Eventually, this limitation of resources will reduce the growth rate. In our continuous-time model (8.7), the growth rate $\dot{N}/N$ of the population was independent of population size, i.e., $\dot{N}/N = m$. A simple form of density-dependent growth is obtained by assuming that the growth rate is of the form $\rho\,(1 - N/K)$, with positive constants ρ and K. This yields the so-called *logistic equation*

$$\dot{N} = \rho N\left(1 - \frac{N}{K}\right) . \tag{8.13}$$

The quantity ρ is the intrinsic growth rate, the rate at which the population would grow if the resources were unlimited. The term $-\rho N^2/K$ reduces this growth and reflects competition within the population.

The qualitative behavior of the differential equation (8.13) is easy to analyze. First, it is obvious that $N(t) = 0$ for all $t > 0$ if $N(0) = 0$, and $N(t) = K$ if $N(0) = K$. Next, if $0 < N(t) < K$ then $\dot{N} > 0$ and the population size increases and converges to K. If $N(t) > K$ then $\dot{N} < 0$ and the population size decreases and converges to K. The quantity K is called the *carrying capacity.* In this simple model, it is a globally asymptotically stable equilibrium. Figure 8.2 displays the typical dynamics of population size for different initial conditions.

Actually, (8.13) can be solved explicitly (for example by the method of separation of variables), and its solution is

$$N(t) = \frac{KN(0)e^{\rho t}}{K + N(0)\left(e^{\rho t} - 1\right)} . \tag{8.14}$$

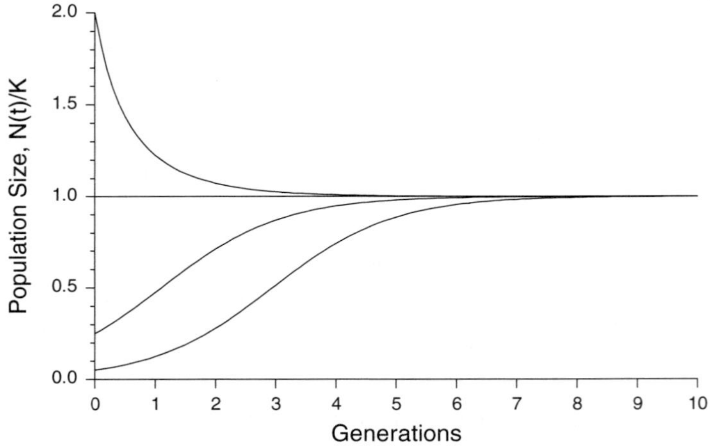

Figure 8.2 Logistic population growth in continuous time according to (8.14).

Now let us consider k types in a population whose growth rates $\dot{n}_i/n_i$ are not constant but decrease as N increases. In analogy with (8.13), we assume that

$$\frac{\dot{n}_i}{n_i} = \rho_i - \bar{\rho}\,\frac{N}{K}\,, \tag{8.15}$$

where $\bar{\rho} = \sum_i \rho_i p_i$ and $p_i = n_i/N$. Then easy calculations show that the total population grows according to (8.13) with ρ replaced by $\bar{\rho}$ and that

$$\dot{p}_i = p_i\,(\rho_i - \bar{\rho})\ . \tag{8.16}$$

Therefore, the dynamics of type frequencies p_i is as simple as without population regulation [cf. (8.11)].

More generally, in exactly the same way it can be shown that for growth rates of the form $\dot{n}_i/n_i = \rho_i + f(n_1, \dots, n_k)$, so that population regulation acts identically on all types, the dynamics of type frequencies is again given by (8.16). Therefore, the simple equation (8.11) is more general than its derivation suggests, and this is one of the reasons for the widespread use of equations of this kind in population genetics, where the main interest centers on the dynamics of the proportion of various genetic types within a population. Other forms of population regulation are discussed in Crow and Kimura (1970).

8.4 THE DISCRETE LOGISTIC EQUATION

There has been a good reason for considering first population regulation in continuous time. Even the simplest density-dependent population-growth models in discrete time may lead to very complicated and counter-intuitive behavior of the population size. Let N and N' denote population size in two subsequent generations and let us assume, as above, that the rate of increase $(N' - N)/N$ is of the form $\rho(1 - N/K)$. However, then $N > K(\rho + 1)/\rho$ implies that $N' < 0$. To obtain a biologically useful model, we define

$$N' = \begin{cases} N\left[1+\rho\left(1-\frac{N}{K}\right)\right], & \text{if } N \le K\frac{\rho+1}{\rho}, \\ 0, & \text{otherwise}. \end{cases} \tag{8.17}$$

This type of equation has been subject to intensive mathematical research because of its complex dynamical properties. Illuminating and more biologically oriented treatments can be found in May (1976) and Hofbauer and Sigmund (1998). Very briefly, the following happens: if $0 < \rho \le 1$, the population size converges monotonically to the carrying capacity K; if $1 < \rho \le 2$, convergence still occurs, but with damped oscillations around K. If $\rho > 2$, the equilibrium $N = K$ becomes unstable and the population size eventually oscillates periodically, first with period two (for $\rho < \sqrt{6} \approx 2.45$); then oscillations of all periods 2^n occur step by step, and eventually periodic oscillations of every period emerge (the last period is three and occurs first for $\rho = \sqrt{8} \approx 2.83$). Finally, there exist values of ρ (e.g., $\rho = 3$) where the dynamics of $N(t)$ is completely irregular, as if stochastic. For instance, if the interval $[0, K(\rho+1)/\rho]$ is divided into two equal parts, a lower and an upper half, and if we are given any finite, or even infinite, sequence of zeros and ones, we can find a point whose trajectory visits the lower and the upper interval in the corresponding sequence. Two points, however, even if extremely close, may have very different trajectories. This kind of dynamic behavior is called chaos. For $\rho > 3$, the population goes extinct within a finite number of generations.

The reason for this 'strange' behavior is that for large ρ the density-dependent population regulation (8.17) is very strong and, in contrast to the continuous-time model (8.13), operates with a delay of one generation. Further discrete-time models of population regulation, some of them perhaps biologically more realistic and exhibiting well-behaved dynamics like the continuous-time logistic equation, may be found in May and Oster (1976).

9. SELECTION AT A SINGLE LOCUS

Selection is the major driving force of evolution. It occurs when genotypes in a population differ in their fitnesses, i.e., in their viability, mating success, or fertility and, therefore, leave different numbers of progeny. The basic mathematical models of selection were developed and investigated in the 1920s and early 1930s by Fisher (1930), Wright (1931), and Haldane (1932). Here, we shall mainly be concerned with the consequences of selection caused by differential viability for the evolution of a population. After a short introductory section about selection in haploid populations, our presentation concentrates on the diploid one-locus case in discrete time, treating Fisher's Fundamental Theorem of Natural Selection as well as more refined results about the dynamics and equilibrium behavior of this model.

9.1 ASEXUAL HAPLOID POPULATIONS

In an asexually reproducing haploid population, we consider one gene locus at which k alleles $\mathcal{A}_1, \ldots, \mathcal{A}_k$ occur. Individuals carrying allele $\mathcal{A}_i$ are assumed to have fitness W_i, where we define fitness as the product of viability (the probability that an offspring survives to reproductive age) and fertility (the average number of offspring). These alleles correspond to the types in Section 8.1. Let p_i denote the (relative) frequency of allele $\mathcal{A}_i$, so that $\sum_i p_i = 1$. The frequency of $\mathcal{A}_i$ in the next generation is $p_i' = n_i'/N'$, with $n_i' = W_i n_i$, and (8.4) implies

$$p_i' = p_i \frac{W_i}{\overline{W}} , \tag{9.1}$$

where $\overline{W} = \sum_j W_j p_j$ is the mean fitness of the population. Given the initial frequencies $p_1(0), \ldots , p_k(0)$, the explicit solution of the recursion (9.1) is

$$p_i(t) = \frac{p_i(0)W_i^t}{\sum_j p_j(0)W_j^t} , \quad i = 1, \ldots , k , \tag{9.2}$$

as is easily seen by solving $n_i' = W_i n_i$ [cf. (8.1)] and subsequent normalization.

If one allele, say $\mathcal{A}_1$, has higher fitness than every other allele, then $(W_j/W_1)^t \to 0$ for $j \geq 2$ as $t \to \infty$, and (9.2) implies that $p_1(t) \to 1$ as $t \to \infty$, provided $p_1(0) > 0$. Therefore, in the long run, the fittest allele will be fixed and all others will be lost.

We do not further investigate (9.1), because below we shall see that (9.1) occurs as a special case of the diploid selection dynamics.

9.2 DIPLOID POPULATIONS

We assume discrete and nonoverlapping generations, random mating, and that genotype frequencies are the same in both sexes (as is the case if individuals are monoecious, or if they are dioecious with the same viabilities in both sexes and the same sex ratio in all matings). Suppose that at an autosomal locus the alleles $\mathcal{A}_1, \ldots , \mathcal{A}_k$ can occur. We count individuals at the zygote stage and denote the (relative) frequency of $\mathcal{A}_i\mathcal{A}_i$ homozygotes by P_{ii} and that of (unordered) $\mathcal{A}_i\mathcal{A}_j$ heterozygotes by $2P_{ij}$ (cf. Section 2.2). Then the frequency of the allele $\mathcal{A}_i$ is

$$p_i = \sum_j P_{ij} . \tag{9.3}$$

Since mating is at random, the genotype frequencies P_{ij} are in Hardy–Weinberg proportions (2.5). Let us suppose that selection acts solely through differential viabilities, and denote the fitness (viability) of $\mathcal{A}_i\mathcal{A}_j$ individuals by W_{ij}. The fitnesses W_{ij} satisfy $W_{ij} \geq 0$ and $W_{ij} = W_{ji}$, because they belong to the same (unordered) genotype $\mathcal{A}_i\mathcal{A}_j$. Then the frequency of $\mathcal{A}_i\mathcal{A}_j$ genotypes among adults that have survived selection is

$$P_{ij}^* = \frac{W_{ij}P_{ij}}{\overline{W}} = \frac{W_{ij}p_ip_j}{\overline{W}} , \tag{9.4}$$

where

$$\overline{W} = \sum_{i,j} W_{ij}P_{ij} = \sum_{i,j} W_{ij}p_ip_j = \sum_i W_ip_i \tag{9.5}$$

is the mean fitness and

$$W_i = \sum_j W_{ij}p_j \tag{9.6}$$

is the *marginal fitness* of allele $\mathcal{A}_i$. In particular, the frequency of $\mathcal{A}_i$ after selection is $p_i^* = \sum_j P_{ij}^* = W_i p_i / \overline{W}$. Because of random mating, the allele frequency p_i' among zygotes of the next generation is also p_i^*, so that allele frequencies evolve according to the *selection equation*

$$p_i' = p_i \frac{W_i}{\overline{W}} \quad \text{for } i = 1, \ldots, k \,. \tag{9.7}$$

This recursion equation preserves the relation

$$\sum_i p_i = 1 \,, \tag{9.8}$$

and describes the evolution of allele frequencies at a single autosomal locus in a diploid population.

The structure of (9.7) is the same as for asexual selection (9.1); however, W_i now is no longer a constant but is given by (9.6). We also note that the right-hand side of (9.7) remains unchanged if all fitness values W_{ij} are multiplied by the same constant. It is often convenient to introduce such a scaling, and to use *relative fitness* instead of *absolute fitness*, which is measured by the expected number of progeny of individuals of a given genotype. In particular, evolution of the gene frequencies in the population is independent of the growth rate of the population, and constant population size can be assumed (cf. also Section 8).

If fitnesses are *multiplicative*, i.e., if constants v_i, $i = 1, \ldots, k$, exist such that $W_{ij} = v_i v_j$ for every i, it is easily shown that (9.7) reduces to (9.1), with v_i and $\overline{v}$ instead of W_i and $\overline{W}$. Formally, the haploid selection dynamics (9.1) can therefore be considered as a special case of the diploid dynamics (9.7). However, it should be noted that the mean fitness $\overline{W}$ in the diploid (multiplicative) model is $\overline{W} = \overline{v}^2$, where $\overline{v}$ corresponds to the haploid mean fitness.

The consideration of differential fertilities leads to a much more complex model, because fertilities have to be assigned to each mating pair $\mathcal{A}_i\mathcal{A}_j \times \mathcal{A}_k\mathcal{A}_l$. Such a general model is derived in Nagylaki (1992), where some of its features are explored. For multiplicative fertilities, this model reduces to the present one, (9.7). Otherwise, its analysis requires following the change of genotype frequencies across generations. A lucid account of the mathematical properties of the diallelic pure fertility equation can be found in Hofbauer and Sigmund (1988, 1998), who also provide further references. For some multilocus results, we refer to Christiansen (1999).

9.3 THE CASE OF TWO ALLELES

It is instructive to consider the diallelic case in some detail, because it exhibits several of the basic properties of the multiallelic case, but can be analyzed by elementary means. We shall write p and $1-p$ instead of p_1 and p_2. We use relative fitnesses and assume $W_{11} = 1$, $W_{12} = 1 - hs$, and $W_{22} = 1 - s$, where s is called the *selection coefficient* and h describes the degree of dominance. The allele $\mathcal{A}_1$ is called *dominant* if $h = 0$, partially dominant if $0 < h < \frac{1}{2}$, *recessive* if $h = 1$, and partially recessive if $\frac{1}{2} < h < 1$. The expressions *additivity*, or *no dominance*, refer to $h = \frac{1}{2}$. From (9.6), the marginal fitnesses of the two alleles are

$$W_1 = 1 - hs + hsp \quad \text{and} \quad W_2 = 1 - s + s(1-h)p \,, \tag{9.9}$$

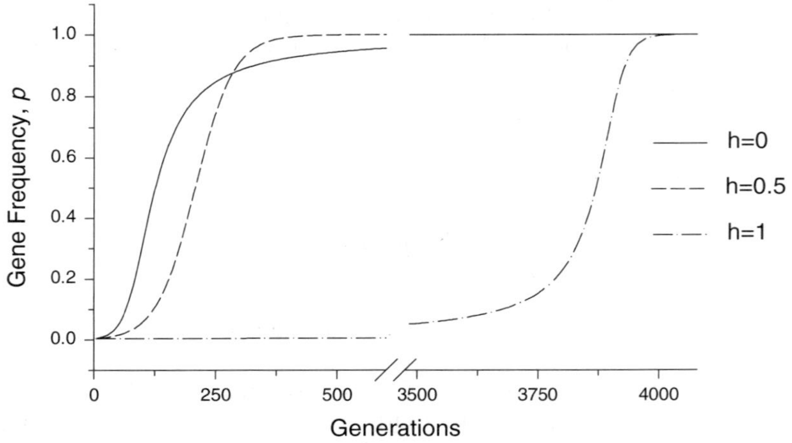

Figure 9.1 Selection of a dominant ($h = 0$, solid line), intermediate ($h = 1/2$, dashed), and recessive ($h = 1$, dash-dotted) allele. The initial frequency is $p_0 = 0.005$ and the selective advantage is $s = 0.05$.

and the mean fitness is

$$\overline{W} = 1 - s + 2s(1-h)p - s(1-2h)p^2 \ . \tag{9.10}$$

It is easily verified that the allele-frequency change from one generation to the next can be written as

$$\Delta p = p' - p = \frac{p(1-p)}{2\overline{W}} \frac{\mathrm{d}\overline{W}}{\mathrm{d}p} \tag{9.11a}$$

$$= \frac{p(1-p)s}{\overline{W}} [1 - h - (1-2h)p] \ . \tag{9.11b}$$

For our analysis we exclude the trivial case of no selection ($s = 0$) and assume $s > 0$. First we observe that $p = 0$ and $p = 1$ are always equilibria of the gene-frequency dynamics (9.11). This is also biologically obvious because we have ignored evolutionary forces such as mutation and migration that could introduce new (or lost) alleles into the population. Any other equilibrium must be a critical point of $\overline{W} = \overline{W}(p)$ and, since $\overline{W}$ is quadratic in p, there can be at most one further equilibrium. There exists a third equilibrium if h is such that $1 - h - (1-2h)p = 0$ for some $0 < p < 1$. This can occur if and only if either $h > 1$ or $h < 0$. In both cases, the equilibrium frequency is

$$\hat{p} = \frac{1-h}{1-2h} \ , \tag{9.12}$$

and $\hat{p}$ is a polymorphic equilibrium, because both alleles are present in the population with nonzero frequency.

The investigation of the selection equation (9.11) requires consideration of three cases.

(*i*) $0 \leq h \leq 1$. Then $\overline{W}$ is an increasing function of p for $p \in [0,1]$ and it follows that $p(t) \to 1$ as $t \to \infty$. Thus, the favored allele $\mathcal{A}_1$ eventually goes to fixation and selection removes all genetic variability. Two particularly simple special cases are that of multiplicative fitnesses, i.e., h such that $1 - s = (1 - hs)^2$, leading to the asexual dynamics and its explicit solution (9.2), and that of *additive* fitnesses (or no dominance), where $h = 1/2$ and $\overline{W}$ is a linear function of p. Although the precise value of h does not influence the eventual results of selection, it has a significant influence on the rate of evolution toward equilibrium. As shown by Figure 9.1, an initially rare, advantageous allele, that is dominant or intermediate, sweeps through the population much faster than a recessive allele. The obvious reason is that a rare allele occurs almost exclusively in heterozygotes, where recessiveness hides it from selection. For analytical results concerning the rate of convergence toward equilibrium in the diallelic case, the reader may consult Nagylaki (1992, Chapter 4.2).

(*ii*) $h < 0$. In this case of *overdominance*, or *heterozygote advantage*, the mean fitness function $\overline{W}(p)$ is concave and the equilibrium $\hat{p}$ (9.12) is the (local) maximum of $\overline{W}$. Using (9.12), we can write (9.11b) as

$$\Delta p = \frac{sp(1-p)}{\overline{W}}(1-2h)(\hat{p}-p) \,. \tag{9.13}$$

As $0 < sp(1-p)(1-2h)/\overline{W} < 1$ for $0 < p < 1$, it follows that convergence to $\hat{p}$ occurs and is monotone, i.e., nonoscillatory. Therefore, the polymorphic equilibrium $\hat{p}$ is *globally asymptotically stable* (see Appendix A for a precise definition).

(*iii*) $h > 1$. In this case of *underdominance*, the polymorphic equilibrium $\hat{p}$ is unstable because fitness is minimized there. Indeed, (9.13) shows that the sign of Δp is the same as that of $p - \hat{p}$, and that $p(t)$ converges monotonically to 0 if $0 < p(0) < \hat{p}$, and to 1 if $\hat{p} < p(0) < 1$. Therefore, $\hat{p}$ is an *unstable* equilibrium and the outcome of evolution depends on the initial state $p(0)$. As in case (i), selection eventually removes any genetic variability and the population becomes *monomorphic*.

9.4 THE FUNDAMENTAL THEOREM OF NATURAL SELECTION

The results above imply that for two alleles mean fitness increases from generation to generation and remains unchanged only at equilibrium states. This is a special case of Fisher's Fundamental Theorem of Natural Selection, which he formulated as follows:

> The rate of increase in fitness of any organism at any time is equal to its genetic variance in fitness at that time. (Fisher 1930)

What Fisher called the genetic variance is now called the genic, or additive genetic, variance (see below and Section 3). Since a variance is always nonnegative, this implies that mean fitness is nondecreasing. Fisher's statement of the theorem was not based on an explicit dynamical model for gene-frequency change and gave rise to ambiguities. The classical interpretation has been that it is a theorem about the (rate of) increase of mean fitness, stating that this increase is (approximately) equal to the additive genetic variance; cf. (9.18) below. However, as we shall see later, mean fitness is increasing only under restrictive assumptions such as random mating, a single locus, etc., whereas Fisher viewed it as a very general theorem. Ewens (1989) argued convincingly that Fisher considered only a certain partial change in mean fitness and, Ewens provided an interpretation for which Fisher's statement of the theorem holds very generally. The biological relevance of this partial change in mean fitness, however, is not obvious. We shall return to this and related interpretations in Chapter II.6.5. Whenever in this text we refer without further specification to the Fundamental Theorem of Natural Selection (often abbreviated as FTNS), we mean its classical interpretation (9.18) as a theorem about the change of mean fitness caused by selection.

The first proofs that mean fitness is nondecreasing under the general multiallele dynamics (9.7) were given by Scheuer and Mandel (1959), Mulholland and Smith (1959), Atkinson *et al.* (1960), and Kingman (1961a). As noted by Nagylaki (1977), a particularly simple proof is obtained by invoking an inequality of Baum and Eagon (1967); see Appendix A.1. Indeed, a straightforward calculation shows that $\partial\overline{W}/\partial p_i = 2W_i$. Then (9.7) can be written as

$$p_i' = p_i \frac{\partial\overline{W}}{\partial p_i} \Big/ \sum_j p_j \frac{\partial\overline{W}}{\partial p_j} \,. \tag{9.14}$$

Therefore, by Theorem A.5, *mean fitness is nondecreasing,*

$$\overline{W}' = \sum_{i,j} W_{ij} p_i' p_j' \geq \overline{W} \,, \tag{9.15}$$

and $\overline{W}' = \overline{W}$ holds only at equilibria of (9.7), i.e., if and only if $p_i' = p_i$ or, equivalently, if $p_i(W_i - \overline{W}) = 0$ for every i.

One of the important consequences of this result is, as discussed in more detail in Section 9.5, that the population will evolve steadily to an equilibrium, and that the equilibrium will be asymptotically stable.

So far, we have proved only that mean fitness is nondecreasing, but Fisher stated that the rate of increase is given by the genic variance. Therefore, let us calculate the change of mean fitness $\Delta\overline{W} = \overline{W}' - \overline{W}$:

$$\begin{aligned}
\Delta\overline{W} &= \overline{W}^{-2} \sum_{i,j} p_i p_j W_{ij}(W_i W_j - \overline{W}^2) \\
&= \overline{W}^{-2} \sum_{i,j} p_i p_j W_i W_j (W_{ij} - \overline{W}) \\
&= \overline{W}^{-1}\sigma_A^2 + \overline{W}^{-2} \sum_{i,j} p_i p_j (W_i - \overline{W})(W_j - \overline{W})(W_{ij} - \overline{W}),
\end{aligned} \tag{9.16}$$

where

$$\sigma_A^2 = 2\sum_i p_i (W_i - \overline{W})^2 \tag{9.17}$$

is the additive genetic, or genic, variance in fitness [cf. (3.12)].

For additive fitnesses, or absence of dominance, there are constants v_i such that $W_{ij} = v_i + v_j$ for every i and j. Then a simple calculation reveals that the last term in (9.16) vanishes, and

$$\Delta\overline{W} = \sigma_A^2/\overline{W} \tag{9.18}$$

is obtained. This can also be shown in the haploid case.

For general fitness values, the second term on the right-hand side of (9.16), below denoted by R, does not vanish. However, upper and lower bounds for R can be derived.

Following Nagylaki (1991), we decompose $W_{ij} - \overline{W} = (W_i - \overline{W}) + (W_j - \overline{W}) + \vartheta_{ij}$. Then R becomes

$$R = \overline{W}^{\,-2} \sum_{i,j} p_i p_j (W_i - \overline{W})(W_j - \overline{W})\vartheta_{ij} \ , \tag{9.19}$$

and the Cauchy–Schwarz inequality gives

$$\overline{W}^{\,2}\,|R| \leq \left[\sum_{i,j} p_i p_j \vartheta_{ij}^2\right]^{1/2} \left[\sum_{i,j} p_i p_j (W_i - \overline{W})^2 (W_j - \overline{W})^2\right]^{1/2} = \tfrac{1}{2}\sigma_{\mathrm{D}}\sigma_{\mathrm{A}}^2 \ ,$$

where σ_{D} is the standard deviation of the dominance effects of fitness; cf. (3.6). Further, we denote the largest and smallest fitness coefficients among the W_{ij} by $W_{\max}$ and $W_{\min}$, exclude lethality by assuming $W_{\min} > 0$, and define the selection coefficient s by $s = (W_{\max} - W_{\min})/W_{\min}$. Then $\sigma_{\mathrm{D}} \leq \sigma_{\mathrm{G}} \leq sW_{\min}$ and $|R| \leq \frac{1}{2}s\sigma_{\mathrm{A}}^2/\overline{W}$. Therefore, (9.16) becomes

$$\Delta\overline{W} = \frac{\sigma_{\mathrm{A}}^2}{\overline{W}}(1 + E) \ , \tag{9.20}$$

where the relative error E obeys the estimate

$$|E| = \left|R\overline{W}/\sigma_{\mathrm{A}}^2\right| \leq \tfrac{1}{2}s \tag{9.21}$$

(Nagylaki 1991). For $s < 2$, (9.20) and (9.21) yield the lower bound

$$\Delta\overline{W} \geq (1 - \tfrac{1}{2}s)\sigma_{\mathrm{A}}^2/\overline{W} \tag{9.22a}$$

for the change in mean fitness. For $s > \frac{1}{2}$, the bound

$$\Delta\overline{W} \geq \tfrac{3}{4}\sigma_{\mathrm{A}}^2/W_{\max} \ , \tag{9.22b}$$

derived by Lyubich *et al.* (1976, 1980) (cf. Lyubich 1992, Chapter 9.2) is better.

For weak selection ($s \ll 1$), (9.20) and (9.21) imply the asymptotic form of the Fundamental Theorem of Natural Selection,

$$\Delta\overline{W} = \sigma_{\mathrm{A}}^2/\overline{W} + O(s^3)\ .^{5} \tag{9.23}$$

If, in addition, $W_{\max}$ is normalized to one, then $\overline{W} \approx 1$ and $\Delta\overline{W} \approx \sigma_{\mathrm{A}}^2$.

9.5 EQUILIBRIA AND DYNAMICS

Although mean fitness is a nondecreasing function, the equilibrium structure of the diploid selection equation (9.7) may be complicated, ranging from the completely degenerate case of no selection, when every state is an equilibrium, to multiple locally

[5] The order (Landau) symbol O is defined as follows: $f(x) = O(g(x))$ as $x \to a$ means $f(x) = g(x)h(x)$, for some function $h(x)$ that is bounded in an interval containing $x = a$. For instance, $f(s) = g(s) + O(s^n)$ if $|f(s) - g(s)|/s^n \leq$ const as $s \to 0$. If $\lim_{x\to a} f(x)/g(x)| = 0$, this is often written as $f(x) = o(g(x))$ as $x \to a$.

stable equilibria, and to selection regimes with a unique globally asymptotically stable equilibrium. Here we shall summarize some of the main results. For definitions and properties of the basic mathematical concepts, the reader is referred to Appendix A.

For the present purpose it is useful to consider the recursion relation (9.7) as a discrete dynamical system on the *simplex*

$$S_k = \{\boldsymbol{p} = (p_1, \dots, p_k) \in \mathbf{R}^k : \sum_i p_i = 1,\ p_i \geq 0,\ i = 1, \dots, k\} , \tag{9.24}$$

where $\mathbf{R}^k$ denotes the k-dimensional Euclidean space. The state of the population in a given generation is represented by the (column) vector of allele frequencies, $\boldsymbol{p} = (p_1, \dots, p_k) \in S_k$, the state in the next generation by $\boldsymbol{p}' = (p_1', \dots, p_k')$, where the p_i' are calculated according to (9.7). Obviously, if $\boldsymbol{p} \in S_k$, then $\boldsymbol{p}' \in S_k$. Thus, the map $\boldsymbol{p} \to \boldsymbol{p}'$ leaves the simplex S_k invariant and defines a discrete dynamical system on S_k. This map can be iterated, and the sequence $\boldsymbol{p}$, $\boldsymbol{p}'$, $\boldsymbol{p}'' = (\boldsymbol{p}')', \dots, \boldsymbol{p}^{(n)}, \dots$ is called the *orbit*, or *trajectory*, of $\boldsymbol{p}$. An orbit characterizes the evolution of the allele frequencies in the population.

The simplex S_k is a $(k-1)$-dimensional convex subset of $\mathbf{R}^k$. Its vertices, $\boldsymbol{e}_i = (0, 0, \dots, 1, 0, \dots, 0)$, the 1 being the ith component, correspond to the monomorphic states of the population with allele $\mathcal{A}_i$ fixed ($p_i = 1$) and other alleles absent. The interior of S_k consists of all $\boldsymbol{p}$ with $p_i > 0$ for every $i = 1, \dots, k$, i.e., of all completely polymorphic states. For every proper subset $J \subset \{1, \dots, k\}$, the set of all $\boldsymbol{p}$, with $p_i = 0$ for $i \in J$, is called a (boundary) face of S_k.

Occasionally, it will be necessary to indicate that marginal and mean fitness depend on $\boldsymbol{p}$, in which case we will write $W_i(\boldsymbol{p})$ and $\overline{W}(\boldsymbol{p})$, respectively. We shall denote the *fitness matrix*, i.e., the $k \times k$ matrix of fitness values W_{ij}, by W. Then $W_i(\boldsymbol{p}) = (\mathsf{W}\boldsymbol{p})_i$ (the ith component of the vector $\mathsf{W}\boldsymbol{p}$) and $\overline{W}(\boldsymbol{p}) = \boldsymbol{p}^\top \mathsf{W} \boldsymbol{p} = \sum_i p_i (\mathsf{W}\boldsymbol{p})_i$.[6]

First, we investigate the equilibria, or fixed points, of (9.7). By definition, these are the points $\hat{\boldsymbol{p}}$ satisfying $\hat{\boldsymbol{p}}' = \hat{\boldsymbol{p}}$. From (9.7), we get the equilibrium conditions

$$p_i(W_i - \overline{W}) = 0 , \qquad 1 \leq i \leq k , \tag{9.25}$$

whence it follows that $\hat{\boldsymbol{p}}$ is an equilibrium if and only if $W_i(\hat{\boldsymbol{p}}) = \overline{W}(\hat{\boldsymbol{p}})$ for every i with $\hat{p}_i > 0$. The monomorphic states $\boldsymbol{e}_i$ are always equilibrium points. A completely polymorphic equilibrium $\hat{\boldsymbol{p}}$ is an equilibrium in the interior of S_k. Therefore, it satisfies $\hat{p}_i > 0$ for every i, and is characterized by the equations

$$W_i(\hat{\boldsymbol{p}}) = \overline{W}(\hat{\boldsymbol{p}}) \qquad \text{for every } i = 1, \dots, k . \tag{9.26}$$

A simple characterization of equilibria in terms of critical points of $\overline{W}$ is the following (Mandel, 1959):

• **9.1** *A state $\hat{\boldsymbol{p}}$ is an equilibrium if and only if it is a critical point for the restriction of mean fitness $\overline{W}(\boldsymbol{p})$ to the minimal face of S_k that contains $\hat{\boldsymbol{p}}$.*

▷ For the proof, it is clearly sufficient to assume that $\hat{\boldsymbol{p}}$ is an interior equilibrium. The critical points of $\overline{W}(\boldsymbol{p})$ subject to the constraint $\sum_i p_i = 1$ are found by the method of

[6] Throughout this book, the superscript $^\top$ denotes transposition of a vector or a matrix.

Lagrange multipliers, i.e., by determining the critical points of the auxiliary function $g(\boldsymbol{p}) = \overline{W}(\boldsymbol{p}) - \lambda \sum_i p_i$. These are given by $\partial g/\partial p_i = 2W_i(\boldsymbol{p}) - \lambda = 0$, $i = 1, \ldots, k$. It follows that $0 = \sum_i p_i(\partial g/\partial p_i) = 2\overline{W} - \lambda$. Therefore, $W_i = \overline{W}$ for every i, and (9.26) yields the assertion. ◁

An immediate, but important, consequence of (9.15) is:

• **9.2** *A state $\hat{\boldsymbol{p}}$ is an equilibrium if and only if $\overline{W}'(\hat{\boldsymbol{p}}) = \overline{W}(\hat{\boldsymbol{p}})$. Otherwise, $\Delta\overline{W} > 0$ and the mean fitness is strictly increasing along orbits.*

Therefore, $\overline{W}$ is a strict Lyapunov function (cf. Appendix A) for the selection dynamics (9.7) and, hence, the central tool for deriving stability results.

Concerning the number of possible equilibria, we have already noted that all monomorphic states are equilibrium points. Thus, *there are at least k equilibria*. In general, continua of equilibria may exist. However, if the fitness matrix W is such that only a finite number equilibria exists, an upper bound is easily derived:

• **9.3** *If the number of equilibria is finite, then it is less or equal than $2^k - 1$.*

▷ Indeed, let $S \subseteq \{1, \ldots, k\}$ be nonempty, and denote the smallest element of S by $i(S)$. We consider the system of k linear equations

$$\begin{aligned} W_i(\boldsymbol{p}) - W_{i(S)}(\boldsymbol{p}) &= 0\,, \quad i \in S,\ i > i(S)\,, \\ p_i &= 0\,, \quad i \notin S\,, \\ \sum_{i \in S} p_i &= 1\,. \end{aligned} \tag{9.27}$$

This linear system has zero, one, or infinitely many solutions. Since there exist $2^k - 1$ nontrivial subsets S, our assertion is proved. ◁

A simple example of a system with exactly $2^k - 1$ equilibria is obtained, if we choose $W_{ii} = 1$ for every i, and $W_{ij} = 0$ for $i \neq j$. Then the equilibria are exactly the centers of each face of S_k. Elementary linear algebra tells us that (9.27) admits exactly one solution if the principal submatrix W_S of W is nonsingular, i.e., if $\det(\mathsf{W}_S) \neq 0$. ($\mathsf{W}_S$ is the matrix formed by the rows and columns of W that correspond to S.) However, this solution does not necessarily satisfy $p_i \geq 0$ for every i. Therefore, the number of possible equilibria is finite and, hence, bounded by $2^k - 1$, if all principal submatrices W_S are nonsingular. Otherwise, lines or, more generally, linear manifolds of equilibria may exist.

Of particular evolutionary interest are stable equilibria. To avoid pathologies, we assume $W_{ii} > 0$ for every i. This implies $\overline{W}(\boldsymbol{p}) > 0$ for $\boldsymbol{p} \in S_k$, and invariance of every boundary face. Therefore, alleles which are present initially do not disappear in finite time (but they may disappear as $t \to \infty$). Two consequences of the fact that mean fitness is strictly increasing along orbits, except at equilibrium points, are the following:

• **9.4** 1. *An interior equilibrium, i.e., a completely polymorphic equilibrium, is asymptotically stable if and only if it is an isolated local maximum of $\overline{W}$.*

2. *Every orbit converges to the set of equilibrium points. If there is only a finite number of equilibria, every orbit converges to exactly one equilibrium point.*

By more sophisticated methods, much stronger results can be derived (cf. Kingman 1961b, Lyubich *et al.* 1980, Losert and Akin 1983, Lyubich 1992). We state some of the most important ones without proof:

- **9.5** 1. *An equilibrium point is stable if and only if it is a local, not necessarily isolated, maximum of* $\overline{W}$.
2. *If an interior equilibrium exists, it is stable if and only if, counting multiplicities, the fitness matrix* W *has exactly one positive eigenvalue.*
3. *If an asymptotically stable interior equilibrium exists, then* every *orbit starting in the interior of* S_k *converges to that equilibrium.*
4. *If the matrix* W *has* i *positive eigenvalues, at least* $(i-1)$ *alleles will be absent at a stable equilibrium.*
5. *Every orbit converges to some, uniquely determined, equilibrium point (even if stable linear manifolds of equilibria exist).*

A fairly detailed treatment of the dynamics and the equilibrium behavior of the discrete-time selection model is contained in Lyubich (1992, Chapter 9). Proofs of the above results as well as of many more results may be found there. Other important references, with additional results, include Karlin (1984) and Nagylaki (1992).

A natural question is about the maximum possible number of coexisting stable equilibria, but this is unsolved except for special cases (Broom *et al.* 1993, Hofbauer and Sigmund 1998, Chapter 19). As we have seen above, there can be at most one asymptotically stable, completely polymorphic equilibrium, which then is globally stable relative to all completely polymorphic initial states. Vickers and Cannings (1988) showed that in a system of k alleles there can be at most two stable equilibria involving $k-1$ alleles. However, they gave an example of a system of four alleles having three stable equilibria, one with three alleles, the others with two alleles. An important result that can be proved is the following (cf. Kingman 1961b, Karlin and Lessard 1984, Vickers and Cannings 1988):

- **9.6** *If* $\hat{\boldsymbol{p}}$ *is a stable equilibrium when there are* $(k-1)$ *alleles and* $\hat{p}_i > 0$ *for every* i, *and if a new allele is introduced which causes the equilibrium to be unstable, then there is a unique stable equilibrium in the enlarged space.*

Therefore, if new alleles are introduced sequentially into a population, such that the time between introductions is sufficient to allow convergence to equilibrium, then the future equilibria are uniquely determined, given the sequence of introduction. This result does not hold if two or more alleles are introduced within a short period. At an externally stable boundary equilibrium (see Appendix A.2), $\hat{\boldsymbol{p}}$, an allele $\mathcal{A}_k$ not present at this equilibrium will die out if introduced at low frequency. This occurs if and only if $W_k(\hat{\boldsymbol{p}}) \leq \overline{W}(\hat{\boldsymbol{p}})$.

10. THE CONTINUOUS-TIME SELECTION MODEL

Most higher animal species have overlapping generations because birth and death occurs continuously in time. This, however, may lead to substantial complications if one wishes to derive a continuous-time model from biological principles. By contrast, discrete-time models can frequently be derived straightforwardly from simple biological assumptions. However, if selection is weak, a continuous-time model can often be derived as an approximation to a corresponding one in discrete time. One of the advantages of models in continuous time is that they lead to differential equations, instead of difference equations, and usually these are easier to analyze because the formalism of calculus is available.

10.1 THE BASIC DIFFERENTIAL EQUATIONS

A rigorous derivation of the differential equations describing gene-frequency change under selection in a diploid population with overlapping generations is a formidable task and requires a complex model. A complete formulation of a continuous selective model without age structure was first given by Nagylaki and Crow (1974). It includes a general mating system, differential fertilities of matings, and differential mortality of genotypes. Here, we shall not repeat this derivation. Instead, we shall motivate the basic equations using assumptions that can be justified only within a more complex model. For a comprehensive and illuminating exposition of selection models in continuous time, the reader is referred to Nagylaki (1992).

Let $N = N(t)$ denote the population size, $n_i = n_i(t)$ the number of $\mathcal{A}_i$ alleles, and $p_i = n_i/N$ their relative frequency. Further, let b_{ij} stand for the birth rate of the genotype $\mathcal{A}_i\mathcal{A}_j$ (i.e., the rate at which $\mathcal{A}_i\mathcal{A}_j$ gives birth, which generally does not equal the rate at which it is born), and d_{ij} for its death rate. The difference $m_{ij} = b_{ij} - d_{ij}$ is the Malthusian fitness parameter of $\mathcal{A}_i\mathcal{A}_j$. Then, denoting by P_{ij} the (ordered) frequency of $\mathcal{A}_i\mathcal{A}_j$, the increase in $\mathcal{A}_i$ genes due to $\mathcal{A}_i\mathcal{A}_i$ parents during the time interval Δt is $NP_{ii}m_{ii}\Delta t$. The increase in $\mathcal{A}_i$ genes due to contributions from $\mathcal{A}_i\mathcal{A}_j$ genotypes is $NP_{ij}m_{ij}\Delta t$. Hence, for $\Delta t \to 0$, we obtain

$$\dot{n}_i = \sum_j NP_{ij}m_{ij} . \tag{10.1}$$

Now we assume that genotypes are in Hardy–Weinberg proportions at any time t, i.e., $P_{ij} = p_ip_j$. We define the marginal fitness of allele $\mathcal{A}_i$ as

$$m_i = \sum_j m_{ij}p_j , \tag{10.2}$$

and the mean fitness of the population as

$$\overline{m} = \sum_i m_ip_i = \sum_{i,j} m_{ij}p_ip_j . \tag{10.3}$$

Then (10.1) can be rewritten as

$$\dot{n}_i = m_in_i , \tag{10.4}$$

and the total population size changes according to

$$\dot{N} = \sum_i m_in_i = \overline{m}N . \tag{10.5}$$

By simple differentiation, we obtain the continuous-time selection dynamics of allele frequencies

$$\dot{p}_i = p_i\,(m_i - \overline{m}) \ , \quad \text{for } i = 1, \dots , k \ . \tag{10.6}$$

This is the analogue of the discrete-time selection dynamics (9.7). Its state space is again the simplex S_k. The equilibria are obtained from the condition $\dot{p}_i = 0$ for every i. The reader may note that the dynamics remains unchanged if the same constant is added to each m_{ij}.

If fitnesses are additive, i.e., if constants ν_i exist such that $m_{ij} = \nu_i + \nu_j$ for every i and j, then (10.6) reduces to the haploid dynamics (8.11).

The above 'derivation' (cf. Kimura 1958) is by no means rigorous. First of all, mating was ignored. Secondly, it was supposed that Hardy–Weinberg proportions obtain at all times. In general this is not true because selection acts continuously and distorts them. (There was no such problem in the model with discrete generations, because under random mating the zygotes are in Hardy–Weinberg equilibrium, and allele frequencies were counted only at this stage of the life cycle.) Thirdly, the definition of birth rates is problematic, and more complicated indeed, because they depend on the genotype frequencies P_{ij} and the fertilities of matings and, hence, are not necessarily constant. In the biologically explicit model of Nagylaki and Crow (1974), allele-frequency change is also given by (10.6), but since the m_i depend on the genotype frequencies P_{ij}, (10.6) provides a complete description of the dynamics only if the P_{ij} are specified as functions of the allele frequencies. The differential equations for the P_{ij}, however, are more complicated than (10.6), because genes are transmitted, not genotypes. For weak selection and random mating, it can be proved that the genotype frequencies evolve toward approximate Hardy–Weinberg equilibrium within a few generations, and that the dynamics of allele frequencies can be approximated to order $O(s)$ (s representing the strength of selection) by (10.6) with Hardy–Weinberg proportions imposed (see Nagylaki 1992, Chapter 4.10). Further results about the fertility-selection model may be found in Akin and Szucs (1994).

10.2 DISCRETE- VERSUS CONTINUOUS-TIME MODELS

It is, to a certain extent, a matter of taste whether to use discrete-time or continuous-time models. Discrete-generation models are particularly useful for comparing theory with experimental results obtained from artificially bred populations, where generations can be kept discrete. However, often a continuous-time formulation is desirable, and usually the resulting differential equations are easier to analyze and their behavior is sometimes more intuitive. An extreme example is provided by the logistic growth equation of Section 8. Fortunately, the dynamical and the equilibrium properties of corresponding discrete- and continuous-time models for gene-frequency change are often very similar. Here we shall justify this statement for the selection dynamics (9.7) and (10.6). Later we shall encounter analogous results in other situations.

First, the difference equation (9.7) and the differential equation (10.6) have the same equilibria, provided there are constants a and $s > 0$ such that

$$W_{ij} = a + s m_{ij} \qquad \text{for every } i, j \ . \tag{10.7}$$

This is obvious by comparing (9.25) with (10.6), upon noting that (10.7) implies

$W_i = a + sm_i$ and $\overline{W} = a + s\overline{m}$. As a consequence, the results in Section 9.5 concerning the number of equilibria apply without change to the dynamics (10.6).

Secondly, for weak selection the discrete model (9.7) can be approximated by the continuous model (10.6); cf. Nagylaki (1992, p. 99). Assume that W_{ij} is given by (10.7) with $a = 1$, rescale time according to $\tau = st$, so that s may be interpreted as generation length, and set $q_i = q_i(\tau) = p_i(t)$, where p_i satisfies the difference equation (9.7). Then we have

$$\frac{d}{d\tau}q_i = \lim_{s\to 0}\frac{1}{s}\left[q_i(\tau+s) - q_i(\tau)\right] = \lim_{s\to 0}\frac{1}{s}\left[p_i(t+1) - p_i(t)\right] .$$

From (9.7) and (10.7), we obtain $p_i(t+1) - p_i(t) = sp_i(t)(m_i - \overline{m})/(1 + s\overline{m})$ and, therefore, $\dot{q}_i = q_i(m_i - \overline{m})$. This proves the assertion because $\Delta p_i \approx s\dot{q}_i = sp_i(m_i - \overline{m})$.

We shall use discrete-time models whenever possible. However, in a number of cases, stronger results can be obtained for continuous-time models. Under such circumstances, we shall employ a continuous-time model and regard it as an approximation to the corresponding discrete-time model.

10.3 GRADIENTS AND THE FUNDAMENTAL THEOREM

The Fundamental Theorem of Natural Selection provides a good example for the assertion that continuous-time models are often easier to analyze. Indeed, from the symmetry property of the fitness coefficients ($m_{ij} = m_{ji}$), we obtain from (10.3) and (10.6), and in analogy to (8.12),

$$\dot{\overline{m}} = 2\sum_{i,j} m_{ij}p_j\dot{p}_i = 2\sum_i m_i\dot{p}_i = \sigma_{\mathrm{A}}^2 , \tag{10.8}$$

where σ_{A}^2 is the additive genetic variance in Malthusian fitness; cf. (9.17). This proof is much simpler than for discrete generations, and shows that Fisher's statement is exactly true for the continuous-time Hardy–Weinberg dynamics (10.6) which, however, is only an approximation. Therefore, the results • **9.1** to **9.6** of Section 9.5 about the number and stability of equilibria carry over to the continuous-time model with the obvious modifications of substituting $\overline{m}$ for $\overline{W}$ and $\dot{\overline{m}}$ for $\Delta\overline{W}$.

Kimura (1958) proposed a maximum principle that refines the FTNS. It states that the gene-frequency changes caused by natural selection, $\dot{p}_i$, are such that they maximize the rate of increase in mean fitness, $\dot{\overline{m}}$, among all possible gene-frequency changes, ξ_i, that obey the constraint $\sum_i \xi^2/p_i = \frac{1}{2}\sigma_{\mathrm{A}}^2$. This statement has little content, unless some biological justification can be given to the constraint. This was provided by Ewens (1992) and will be discussed in a more general context in Chapter II.6.6.

From a purely mathematical point of view, it can be shown that Kimura's Maximum Principle is equivalent to the fact that the selection dynamics (10.6) is a *gradient system* with mean fitness $\overline{m}$ playing the role of the *potential*, provided a new distance metric is introduced on the simplex S_k. The intuitive reason why a non-Euclidean metric may be more adequate to describe gene-frequency change is that a small change in the frequency of a rare gene (i.e., near the boundary of the simplex) is evolutionarily more significant than a change of the same magnitude of an abundant gene.

Roughly speaking, a gradient system is a system of ordinary differential equations of the form

$$\dot{\boldsymbol{x}} = -\nabla V(\boldsymbol{x}) , \tag{10.9}$$

where $\boldsymbol{x} = (x_1, \ldots, x_k)$ is a function of t, V is a sufficiently smooth real-valued function on $\mathbf{R}^k$, called the *potential function*, and

$$\nabla V(\boldsymbol{x}) = \left(\frac{\partial V}{\partial x_1}, \ldots, \frac{\partial V}{\partial x_k} \right)^{\top} \tag{10.10}$$

is the gradient vector of first-order partial derivatives. Gradient systems have important properties: all equilibria are obtained from the condition $\dot{V}(\boldsymbol{x}) = 0$, and there are no cycles or more complex equilibrium states; all trajectories converge to a set of equilibrium points; trajectories are perpendicular to the level surfaces of the potential V (see Appendix A.3 for details).

The selection dynamics (10.6) is a more general gradient system on the simplex S_k. To show this, we define the indicator variables

$$f_i(k,l) = \begin{cases} 1 , & \text{if } k = l = i , \\ \frac{1}{2} , & \text{if } k \neq l, \text{ and } k = i \text{ or } l = i , \\ 0 , & \text{otherwise} , \end{cases} \tag{10.11}$$

that associate to the genotype $\mathcal{A}_k\mathcal{A}_l$ the values 1, $\frac{1}{2}$, or 0, depending on the number of $\mathcal{A}_i$ alleles this genotype contains. Then the expectation of f_i with respect to the frequency distribution of genotypes is

$$\overline{f_i} = \sum_{k,l} f_i(k,l) P_{kl} = p_i , \tag{10.12}$$

and the covariance of f_i and f_j, denoted by $g^{ij} = \mathrm{Cov}[f_i, f_j]$, is calculated to be

$$g^{ij} = p_i(\delta_{ij} - p_j) . \tag{10.13}$$

Here, δ_{ij} is the Kronecker delta defined by $\delta_{ij} = 1$, if $i = j$, and $\delta_{ij} = 0$, otherwise. We shall denote the (symmetric, positive definite) matrix formed by the entries g^{ij} by $\mathsf{G}_{\boldsymbol{p}}$ (see Appendix A.3). With this notation and the simple identity $\frac{\partial \overline{m}}{\partial p_j} = 2m_j$, the selection dynamics (10.6) can be rewritten in the form of a generalized gradient system, i.e.,

$$\dot{p}_i = \tfrac{1}{2} \sum_{j=1}^{k} g^{ij} \frac{\partial \overline{m}}{\partial p_j} , \tag{10.14a}$$

or

$$\dot{\boldsymbol{p}} = \tfrac{1}{2} \mathsf{G}_{\boldsymbol{p}} \nabla \overline{m} . \tag{10.14b}$$

We shall call gradient systems on S_k of this form *Svirezhev–Shahshahani gradients*, because they are gradient systems with respect to the metric introduced by Svirezhev (1972); cf. also Timofeeff-Resovsky and Svirezhev (1970), and Shahshahani (1979).

This distance metric assigns to a vector $\boldsymbol{d}$ of gene-frequency changes from $\boldsymbol{p}$ to $\boldsymbol{p}+\boldsymbol{d}$ (i.e., $\sum_i d_i = 0$) the length

$$||\boldsymbol{d}||_{\boldsymbol{p}} = \sqrt{\boldsymbol{d}^\top \mathrm{G}_{\boldsymbol{p}}^{-1}\boldsymbol{d}} = \sqrt{\sum_i d_i^2/p_i}\ , \tag{10.15}$$

where the matrix $\mathrm{G}_{\boldsymbol{p}}^{-1}$ is given in (A.17). The close relation between the expression on the right-hand side and Kimura's constraint may be noted.

Kimura's Maximum Principle can now be stated as follows: the vector of gene-frequency changes, $\dot{\boldsymbol{p}}$, points into the direction of steepest ascent in the 'mean-fitness landscape' which is given by the generalized gradient vector $\mathrm{G}_{\boldsymbol{p}}\nabla\overline{m}$. This means that with respect to the Svirezhev–Shahshahani metric, $\dot{\boldsymbol{p}}$ it is perpendicular to the level surfaces of $\overline{m}$. Moreover, because the matrix $\mathrm{G}_{\boldsymbol{p}}$ is positive definite, it follows immediately from (10.14b) that mean fitness is nondecreasing:

$$\dot{\overline{m}} = (\nabla\overline{m})^\top\dot{\boldsymbol{p}} = (\nabla\overline{m})^\top \mathrm{G}_{\boldsymbol{p}}\nabla\overline{m} \geq 0\ .$$

A multilocus generalization of Kimura's principle is treated in Chapter II.6.6.

We shall encounter generalized gradients at other occasions as a means of proving convergence and stability results. Some of their properties, as well their relation to measures of genetic distance and to random genetic drift, are summarized in Appendices A.3 and E.3. Further results may be found in Shahshahani (1979), Akin (1979), Svirezhev and Passekov (1990), and Hofbauer and Sigmund (1998).

A simple calculation reveals that (10.14a) can be expressed as

$$\dot{p}_i = \mathrm{Cov}[f_i, m]\ , \tag{10.16}$$

where m is the random variable $m(\mathcal{A}_k\mathcal{A}_l) = m_{kl}$. This is the covariance formula of Li (1967) and Price (1970). It holds under much more general assumptions (see Lessard 1997), and is a special case of Robertson's Secondary Theorem of Natural Selection (see Chapter II.5, II.6).

In the case of two alleles, it is convenient to write $p = p_1$ and $q = p_2 = 1-p$. Then (10.14a) becomes

$$\dot{p} = \frac{pq}{2}\left(\frac{\partial\overline{m}}{\partial p} - \frac{\partial\overline{m}}{\partial q}\right)\ , \tag{10.17a}$$

or after substituting $q = 1-p$ in $\overline{m}$,

$$\dot{p} = \frac{p(1-p)}{2}\frac{\mathrm{d}\overline{m}}{\mathrm{d}p}\ . \tag{10.17b}$$

This representation of gene-frequency change goes back to Wright (1935b), and is equivalent to (9.11a) if $\overline{m}$ is replaced by $\ln\overline{W}$ and $\dot{p}$ by $\overline{W}\Delta p$. It was Wright who proposed and advocated the idea of an *adaptive topography* in which evolution takes place like a hill-climbing process, and mean fitness plays the role of a potential-like function. This, however, is no longer the case if multiple loci contribute to fitness (see Chapter II).

It is of historical interest to point out that Fisher opposed the view that evolution is guided by maximizing $\overline{m}$ (or $\overline{W}$). In 1956, in a letter to Kimura (see Bennett 1983, pp. 229–230), he wrote

> . . . I had, of course, considered the relation between such a situation and that in which a potential function existed, for my mathematical education lay in the field of mathematical physics. As you realize, I preferred to develop the theory without this assumption, which of course in another aspect is a restriction.

One of his main concerns was that "this virtual function" will be changed by environmental and other factors that are not specified in the formulation of the Fundamental Theorem. As we shall discuss in Chapter II.6.5, and as was elucidated by Price (1972), Ewens (1989), and Lessard (1997), Fisher's understanding of the Fundamental Theorem was quite different from that of most other population geneticists.

II

Selection at Two or More Loci

Many, if not most, traits of evolutionary significance are determined by multiple loci. Most early investigations assumed approximate linkage equilibrium, an assumption that allows us to study the loci independently. This view has been questioned, in particular, by Franklin and Lewontin (1970), who proposed that extensive linkage disequilibrium may occur if several loci contribute to fitness. Since then, a substantial body of mathematical literature about multilocus models has accumulated, although to a large extent, it has been concerned with two-locus models.

Even if there are not many known traits that are determined by two loci, the investigation of two-locus systems is important as they constitute the simplest type of model in which the interaction between selection and linkage can be studied analytically. Indeed, two-locus models have provided substantial insight into the phenomena that can occur if several loci interact, and they clearly indicate the limitations of the one-locus theory.

This chapter is devoted to the study of true multilocus systems. It is not the objective to give a comprehensive account of the theory. Instead, this chapter is focussed on those aspects of the theory that are of relevance for quantitative and evolutionary genetics. Much of the classical theory was concerned with existence and properties of equilibria under specific assumptions about the fitnesses of genotypes (mainly about single-locus genotypes and their interaction). Whereas this approach has yielded considerable theoretical insight into evolutionary properties of two- and multilocus systems, in practice gene frequencies and fitness effects of single- or few-locus genotypes are hard or impossible to measure. Ideally, a theory of population genetics would be needed, in which the observables from experiment are the entities of theory (Franklin and Lewontin 1970). Such a formalism has not yet been achieved; but substantial progress has been made towards it. For instance, methods from quantitative genetics have been merged with population genetics theory and, to a certain extent, the dynamics of polygenic traits under selection is now describable in terms of (observable) moments or cumulants of the distribution of genotypic or phenotypic values, instead of gene frequencies. This theory is treated in Chapter V. Considerable progress has also been made in deriving properties of multilocus systems without specific assumptions about the fitnesses of genotypes. For example, assuming weak selection or weak epistasis, approximate formulas for the change in mean fitness and in the mean of an arbitrary character can be derived. It is primarily this kind of theory to which this chapter is devoted.

Throughout this chapter, an infinitely large population size is assumed and random genetic drift is ignored. Most of the theory is derived for randomly mating diploid populations with discrete, nonoverlapping generations. It is supposed that sexes need not be distinguished and that selection is acting only through differential viabilities.

As an introduction, some of the now classical aspects of two-locus theory are treated. This comprises three types of two-locus two-allele models: the additive, the multiplicative, and the symmetric viability model. These have played a central role in the development of multilocus theory and display several typical features of multilocus systems. In Section 2, the general multilocus model is introduced, the recursion relations for the gametic frequencies are derived, and the case of no selection is reviewed. Section 3 deals with quantitative traits determined by an arbitrary number of loci. It is devoted to the theory of average effects and the partitioning of the total genetic variance of a trait into components of evolutionary interest. Also, the decomposition of covariance between a trait and fitness is derived. This theory is developed without the assumption of random mating. In Section 4, the relationship between properties of a multilocus system and those of its subsystems, defined by some subset of loci, is investigated. Evolution in multilocus systems in the absence of epistasis is explored in Section 5. In particular, the response to selection of the mean of an additive character is derived, and additive and multiplicative fitnesses are investigated. In Section 6, the evolution of multilocus systems is studied under the assumption of weak epistasis or weak selection. Among others, the approach to quasi-linkage equilibrium and the convergence to equilibrium is examined, and asymptotic versions of Fisher's Fundamental and Robertson's Secondary Theorem of Natural Selection are derived. Finally, interpretations of the Fundamental Theorem are discussed, due to Price (1972), Ewens (1989), and Lessard (1997), who argued that Fisher considered only a certain partial change in mean fitness for which the Fundamental Theorem is exact.

1. TWO-LOCUS TWO-ALLELE MODELS

In a randomly mating diploid population with discrete generations, we consider two loci, each with two alleles. We call the alleles at the first locus $\mathcal{A}_1$ and $\mathcal{A}_2$, and those at the second $\mathcal{B}_1$ and $\mathcal{B}_2$. The recombination frequency between the two loci is denoted by r $(0 \le r \le \frac{1}{2})$. As we have seen in Chapter I.7, it is no longer possible to describe the joint dynamics at more than one locus in terms of allele frequencies, because nonrandom associations between alleles at different loci may occur. However, since we posit random mating, the evolution of systems with two or more loci can be described in terms of gametic frequencies, because gametes are randomly united to form zygotes. In the present model there are four possible gametic types, $\mathcal{A}_1\mathcal{B}_1$, $\mathcal{A}_1\mathcal{B}_2$, $\mathcal{A}_2\mathcal{B}_1$, and $\mathcal{A}_2\mathcal{B}_2$, whose relative frequencies are measured among zygotes and denoted by x_1, x_2, x_3, and x_4, respectively. Therefore, $\sum_{i=1}^4 x_i = 1$ and the state space is the simplex S_4, representable by a tetrahedron (see Figure 7.1 in Chapter I).

Selection acts only through differential viabilities. The fitness of zygotes made up of gametes i and j is designated by W_{ij} $(= W_{ji})$. These fitness values are often displayed as a 4×4 matrix. We also suppose that there is no position effect, i.e., coupling and repulsion double heterozygotes have the same fitness, $W_{14} = W_{23}$. Then the fitnesses can be expressed by the single-locus genotypes in the form of a 3×3 matrix:

$$\begin{array}{c} \\ \mathcal{A}_1\mathcal{A}_1 \\ \mathcal{A}_1\mathcal{A}_2 \\ \mathcal{A}_2\mathcal{A}_2 \end{array} \begin{array}{c} \begin{array}{ccc} \mathcal{B}_1\mathcal{B}_1 & \mathcal{B}_1\mathcal{B}_2 & \mathcal{B}_2\mathcal{B}_2 \end{array} \\ \begin{pmatrix} W_{11} & W_{12} & W_{22} \\ W_{13} & W_{14} & W_{24} \\ W_{33} & W_{34} & W_{44} \end{pmatrix} \end{array} . \tag{1.1}$$

Since fitnesses are assigned to genotypes, gametes do not have a fitness *per se*. However, each gamete i has a marginal fitness W_i, defined by averaging over all genotypes containing i, i.e.,

$$W_i = \sum_{j=1}^{4} W_{ij} x_j \ , \quad i = 1, \ldots, 4 \ . \tag{1.2}$$

Therefore, the (marginal) fitness of a gamete may change when the genetic constitution of the population changes. Further, the mean fitness of the population is

$$\overline{W} = \sum_{i,j} W_{ij} x_i x_j = \sum_i W_i x_i \ . \tag{1.3}$$

The same reasoning that led to the selection equation I(9.7) for a single locus shows that after selection the frequency of genotype ij is $p_i p_j W_{ij}/\overline{W}$. Then reproduction takes place and recombination changes the gametic frequencies according to I(7.8). Putting all this together, we find that in the next generation the gametic frequencies are given by

$$\overline{W} x_i' = x_i W_i - \eta_i r W_{14} D \ , \quad i = 1, \ldots, 4 \ , \tag{1.4}$$

where $\eta_1 = \eta_4 = 1$, $\eta_2 = \eta_3 = -1$, and $D = x_1 x_4 - x_2 x_3$ measures linkage disequilibrium. If $D = 0$, then the population is in linkage equilibrium (cf. Chapter I.7).

In this form, the equations (1.4) are due to Lewontin and Kojima (1960). Earlier, Kimura (1956) had derived the equations for the corresponding continuous-time model,

$$\dot{x}_i = x_i(m_i - \overline{m}) - \eta_i r b D \ , \quad i = 1, \ldots, 4 \tag{1.5}$$

where m_i is the marginal Malthusian fitness of gamete i, $\overline{m}$ is the mean fitness, and b is the birth rate of the double heterozygotes (cf. also Crow and Kimura 1970, pp. 195–197). These differential equations, however, should be interpreted as a weak-selection approximation to (1.4), because they are based on the assumption of Hardy–Weinberg proportions, which are not strictly satisfied in continuous time. For a precise continuous-time model and a justification of (1.5) the reader is referred to Nagylaki and Crow (1974); see also Nagylaki (1992, Chapter 8; in particular, Problem 8.15).

The analysis of two-locus models is intrinsically more involved than that of one-locus models because of the complications introduced by recombination and linkage disequilibrium. If the loci are completely linked ($r = 0$), then (1.4) is formally equivalent to a one-locus dynamics with four alleles. Therefore, Fisher's Fundamental Theorem of Natural Selection applies and mean fitness will be maximized. This is generally not true with recombination because equilibria of (1.4) may exist that are not in linkage equilibrium. Indeed, at equilibrium ($x_i' = x_i$) the equations

$$\overline{W} x_i = x_i W_i - \eta_i r W_{14} D \ , \quad i = 1, \ldots, 4 \ , \tag{1.6}$$

must be satisfied. Hence, an equilibrium point at which $D \neq 0$ must be in the interior of the simplex, and it cannot be a critical point of mean fitness because it had to satisfy $W_i - \overline{W} = 0$ for every i (cf. Chapter I.9.5). If such an equilibrium is asymptotically stable, then there must be trajectories along which fitness decreases, and the Fundamental Theorem of Natural Selection fails. Existence of equilibria with $D = 0$ requires that the fitnesses W_{ij} fulfil special relations. Therefore, the validity of the Fundamental Theorem of Natural Selection will be the exception rather than the rule. As we shall see in Section 6, however, this does not preclude the possibility that mean fitness increases during important and long periods of evolution.

We shall now briefly investigate three classes of two-locus models that have received much attention in the literature, are well understood, and display some of the typical features of multilocus systems. The first two are models without epistasis, i.e., where no interaction between loci occurs, so that the fitness values of genotypes can be defined in terms of single-locus fitness coefficients. The third is the symmetric viability model, which covers many models of stabilizing selection as studied later in this book. Additional results on two-locus two-allele models, as well as many references, may be found in Karlin (1975), Ewens (1979), and Christiansen (1999). In Section 5, the additive and the multiplicative fitness model will be extended to multiple loci.

1.1 THE ADDITIVE FITNESS MODEL

Let us suppose that fitnesses of genotypes are determined additively from fitnesses at individual loci. Then the fitness matrix (1.1) is of the form

$$\begin{pmatrix} a_{11}+b_{11} & a_{11}+b_{12} & a_{11}+b_{22} \\ a_{12}+b_{11} & a_{12}+b_{12} & a_{12}+b_{22} \\ a_{22}+b_{11} & a_{22}+b_{12} & a_{22}+b_{22} \end{pmatrix}, \tag{1.7}$$

where the fitness contributions of the one-locus genotypes are given by

$$\begin{matrix} \mathcal{A}_1\mathcal{A}_1 & \mathcal{A}_1\mathcal{A}_2 & \mathcal{A}_2\mathcal{A}_2 & \quad & \mathcal{B}_1\mathcal{B}_1 & \mathcal{B}_1\mathcal{B}_2 & \mathcal{B}_2\mathcal{B}_2 \\ a_{11} & a_{12} & a_{22} & & b_{11} & b_{12} & b_{22} \end{matrix} . \tag{1.8}$$

The behavior of this additive fitness model is rather simple, and several aspects of it can be inferred from one-locus theory. In particular, this is one of the few classes of two-locus models in which mean fitness is nondecreasing (cf. Ewens 1979). Indeed, denoting the allele frequencies of $\mathcal{A}_1$, $\mathcal{A}_2$, $\mathcal{B}_1$, $\mathcal{B}_2$ by

$$p_1 = x_1 + x_2\,,\ p_2 = x_3 + x_4\,,\ q_1 = x_1 + x_3\,,\ q_2 = x_2 + x_4\,, \tag{1.9}$$

respectively (note that $p_2 = 1 - p_1$, $q_2 = 1 - q_1$), the mean fitness becomes

$$\overline{W} = \bar{a} + \bar{b}\,, \tag{1.10a}$$

where

$$\bar{a} = a_{11}p_1^2 + 2a_{12}p_1p_2 + a_{22}p_2^2 \quad \text{and} \quad \bar{b} = b_{11}q_1^2 + 2b_{12}q_1q_2 + b_{22}q_2^2 \tag{1.10b}$$

are the single-locus mean fitnesses. Since, by (1.4) and (1.9), the gene-frequency changes Δp_1, Δp_2, Δq_1, and Δq_2 are independent of r, they are the same as in the special case $r = 0$. Therefore, the mean fitness $\overline{W}$ changes as for $r = 0$, i.e., as in a one-locus four-allele system, in which it is nondecreasing by I(9.15). It follows that in the two-locus system (1.4) with additive fitnesses the mean fitness is nondecreasing, and $\Delta\overline{W} = 0$ only at equilibria. Therefore, Fisher's Fundamental Theorem of Natural Selection is valid. This argument readily extends to multiple loci with additive fitnesses and an arbitrary number of alleles; it is due to Ewens (1969a,b).

Let us now relate the change in mean fitness, $\Delta\overline{W}$, to the additive genetic variance. For a detailed derivation, valid for multiple loci, we refer to Nagylaki (1991). Using the recursion relations (1.4) and recalling that $\Delta\overline{W}$ is independent of r, so that all terms involving r must cancel, we find by the same calculation that led to I(9.16),

$$\Delta\overline{W} = \overline{W}^{-1}\sigma^2_{\text{Gam}} + \overline{W}^{-2}\sum_{i,j} x_i x_j (W_i - \overline{W})(W_j - \overline{W})(W_{ij} - \overline{W}) \,, \tag{1.11}$$

where

$$\sigma^2_{\text{Gam}} = 2\sum_i x_i (W_i - \overline{W})^2 \tag{1.12}$$

is the *gametic variance* in fitness. Let $a_1 = a_{11}p_1 + a_{12}p_2$ and $a_2 = a_{12}p_1 + a_{22}p_2$ denote the marginal fitnesses of alleles $\mathcal{A}_1$ and $\mathcal{A}_2$, and let b_1 and b_2 be those of $\mathcal{B}_1$ and $\mathcal{B}_2$. Then $W_1 - \overline{W} = (a_1 - \bar{a}) + (b_1 - \bar{b})$, $W_2 - \overline{W} = (a_1 - \bar{a}) + (b_2 - \bar{b})$, etc., and it follows that $\sigma^2_{\text{Gam}} = \sigma^2_{\text{A}}$, where σ^2_{A} is the additive component of the total variance (cf. I(3.12) or, more generally, (3.36) below).

The second term in (1.11) vanishes if there is no dominance and, otherwise, can be estimated as in Chapter I.9.4. Therefore,

$$\Delta\overline{W} = \frac{\sigma^2_{\text{A}}}{\overline{W}}(1 + E) \,, \tag{1.13}$$

where the relative error E again satisfies I(9.21). In particular, the asymptotic form I(9.23) of the Fundamental Theorem of Natural Selection and the lower bounds I(9.22) remain valid.

Since $\Delta\overline{W} = 0$ only at equilibria, it follows from (1.12) and (1.13) (or directly by considering the case $r = 0$) that for this model at an equilibrium, $W_i = \overline{W}$ holds for every i. Therefore, (1.6) implies that all equilibria must be in linkage equilibrium. Linkage equilibrium, however, is not preserved during evolution.

Although σ^2_{Gam} equals the additive genetic variance σ^2_{A}, it is not the sum of the additive genetic variances at individual loci. Instead, by a simple calculation,

$$\sigma^2_{\text{A}} = \sigma^2_{\text{A}}(\mathcal{A}) + \sigma^2_{\text{A}}(\mathcal{B}) + 4D(a_1 - a_2)(b_1 - b_2) \tag{1.14}$$

is obtained, where in agreement with I(3.15), the single-locus marginal variance is defined by

$$\sigma^2_{\text{A}}(\mathcal{A}) = 2p_1p_2(a_1 - a_2)^2 \,. \tag{1.15}$$

For the $\mathcal{B}$ locus an analogous definition applies.

Finally, we examine existence and stability of equilibria. We have already stated that all equilibria are in linkage equilibrium. Therefore, they must be of the form

$$\hat{x}_1 = \hat{p}_1\hat{q}_1 \,, \quad \hat{x}_2 = \hat{p}_1\hat{q}_2 \,, \quad \hat{x}_3 = \hat{p}_2\hat{q}_1 \,, \quad \hat{x}_4 = \hat{p}_2\hat{q}_2 \,. \tag{1.16}$$

Such equilibria are called *product equilibria*, or *Robbins equilibria*, because Robbins (1918) proved that without selection, gamete frequencies converge to these proportions. Since $\overline{W}$ is nondecreasing, stable equilibria must be points at which $\overline{W}$ is maximized. Obviously, this is the case if and only if the single locus mean fitnesses $\bar{a}$ and $\bar{b}$ are each maximized. Suppose that $\hat{p}_1$ and $\hat{q}_1$ are stable equilibria at the $\mathcal{A}$- and $\mathcal{B}$-locus, respectively. In the case of overdominance ($a_{12} > a_{11}, a_{22}$ at the $\mathcal{A}$-locus and $b_{12} > b_{11}, b_{22}$ at the $\mathcal{B}$-locus) this implies

$$\hat{p}_1 = \frac{a_{12} - a_{22}}{2a_{12} - a_{11} - a_{22}} \quad \text{and} \quad \hat{q}_1 = \frac{b_{12} - b_{22}}{2b_{12} - b_{11} - b_{22}} \; ; \tag{1.17}$$

cf. I(9.12). Otherwise, $\hat{p}_1 = 0$ or 1, and $\hat{q}_1 = 0$ or 1. It follows that with overdominance at both loci, there is a unique polymorphic equilibrium, given by (1.16) and (1.17). It is globally asymptotically stable for all biologically feasible values of r. This result can be extended to two loci with multiple alleles (Karlin and Liberman 1978).

The second case where an interior equilibrium exists is that of underdominance at each locus. But then (1.17) is unstable and, depending on the initial condition, one of the four gametes will become fixed. If there is overdominance at only one locus (e.g. $\mathcal{A}$), then the frequency of $\mathcal{A}_1$ will converge to $\hat{p}_1$, and $\mathcal{B}_1$ or $\mathcal{B}_2$ will be lost.

1.2 THE MULTIPLICATIVE FITNESS MODEL

The multiplicative fitness model assumes that the fitness matrix (1.1) is of the form

$$\begin{pmatrix} a_{11}b_{11} & a_{11}b_{12} & a_{11}b_{22} \\ a_{12}b_{11} & a_{12}b_{12} & a_{12}b_{22} \\ a_{22}b_{11} & a_{22}b_{12} & a_{22}b_{22} \end{pmatrix} , \tag{1.18}$$

where individual loci contribute according to (1.8). Although (1.18) is again a nonepistatic fitness matrix, the model behavior is more complicated than in the additive case.

Let us begin with some simple properties. The first is that the sign of linkage disequilibrium, D, is preserved under the dynamics. This can be seen from the easily derived relation

$$\overline{W}^2 D' = D[W_1 W_4 + p_2 p_3 (a_{11}a_{22} - a_{12}^2)(b_{11}b_{22} - b_{12}^2) - r a_{12} b_{12} \overline{W}] \, , \tag{1.19}$$

because, by expanding and using $r \le \frac{1}{2}$, the expression in brackets can be shown to be nonnegative. Consequently, multiplicative selection preserves linkage equilibrium, but (1.19) does not imply that solutions converge to linkage equilibrium.

Secondly, with either overdominance or underdominance at each locus, the product equilibrium defined by (1.16) and (1.17) exists. However, even with overdominance at both loci, which we henceforth assume, it is not necessarily stable. A linear stability analysis (cf. Theorem A.4) reveals that this equilibrium is asymptotically stable if and only if

$$r > r_0 = \frac{(a_{12} - a_{11})(b_{12} - b_{11})(a_{12} - a_{22})(b_{12} - b_{22})}{(2a_{12} - a_{11} - a_{22})(2b_{12} - b_{11} - b_{22})} \tag{1.20}$$

(Bodmer and Felsenstein 1967). If selection is weak and fitness differences are of the order of s, then r_0 is of order $O(s^2)$, showing that condition (1.20) is not very stringent. Moran (1968) proved that a sufficient condition for global stability of the product equilibrium is

$$r > \min\left\{\frac{(a_{12}-a_{11})(a_{12}-a_{22})}{a_{12}(2a_{12}-a_{11}-a_{22})}, \frac{(b_{12}-b_{11})(b_{12}-b_{22})}{b_{12}(2b_{12}-b_{11}-b_{22})}\right\} . \tag{1.21}$$

It is not known how close this is to being a necessary condition.

If $r < r_0$, there exist exactly two asymptotically stable internal (completely polymorphic) equilibria, one with $D > 0$ and one with $D < 0$. Since $\overline{W}$ is a quadratic function, it can have at most one internal maximum, and $\overline{W}$ will decrease in the neighborhood of a stable equilibrium. This provides another argument that mean fitness can decrease in two-locus models. If $S = (a_{11} - a_{22})(b_{11} - b_{22}) \neq 0$, then there is a range of r-values, $r_0 < r < r^*$, for which the product equilibrium is simultaneously stable with an equilibrium with $D \neq 0$. If S is small, then $r^* \approx r_0 + S^2/(64r_0)$ (cf. Karlin and Feldman 1978, Hastings 1981a).

For general multiplicative fitnesses it can be shown that there exist at most two asymptotically stable equilibria (internal and boundary). Moreover, mean fitness can never be a maximum at the product equilibrium (1.16) (Karlin 1975). These results demonstrate that, even in the absence of epistasis, mean fitness is not necessarily increasing; linkage disequilibrium may persist forever, and completely polymorphic stable equilibria may coexist.

1.3 THE SYMMETRIC VIABILITY MODEL

Historically, the symmetric viability model has played an important role in mathematical population genetics because it arises naturally when diallelic loci are involved in the determination of a quantitative character that is under stabilizing selection. We shall take up this topic in Chapter VI.2. The symmetric viability model also exhibits epistasis yet, in several special cases, is amenable to a mathematical analysis. It assumes that the fitnesses of the genotypes are given by

$$\begin{pmatrix} 1-d & 1-b & 1-a \\ 1-c & 1 & 1-c \\ 1-a & 1-b & 1-d \end{pmatrix} , \tag{1.22}$$

where $0 \leq a, b, c, d \leq 1$, and $d \geq a$. Then the marginal fitnesses of the four gametes are

$$W_1 = 1 - dx_1 - bx_2 - cx_3 , \tag{1.23a}$$

$$W_2 = 1 - bx_1 - ax_2 - cx_4 , \tag{1.23b}$$

$$W_3 = 1 - cx_1 - ax_3 - bx_4 , \tag{1.23c}$$

$$W_4 = 1 - cx_2 - bx_3 - dx_4 , \tag{1.23d}$$

and the mean fitness of the population is

$$\overline{W} = 1 - d(x_1^2 + x_4^2) - a(x_2^2 + x_3^2) - 2b(x_1x_2 + x_3x_4) - 2c(x_1x_3 + x_2x_4). \tag{1.24}$$

Special cases of this model were investigated, for instance, by Wright (1952), Kimura (1956), Lewontin and Kojima (1960), Ewens (1968), Gale and Kearsey (1968), Hastings (1985, 1987), and Nagylaki (1989a). Bodmer and Felsenstein (1967) and, in particular, Karlin and Feldman (1970) performed extensive analyses of the general model (1.22). Our presentation is largely based on the article by Karlin and Feldman, who found all possible polymorphic equilibria and derived their stability properties for several important special cases. Since the boundary equilibria can be determined rather straightforwardly, we concentrate on the interior equilibria. There may exist two kinds of completely polymorphic equilibria.

The *symmetric equilibria* satisfy $\hat{x}_1 = \hat{x}_4$ and $\hat{x}_2 = \hat{x}_3$, and are given by

$$\hat{x}_1 = \hat{x}_4 = \tfrac{1}{4} + \hat{D}\,, \quad \hat{x}_2 = \hat{x}_3 = \tfrac{1}{4} - \hat{D}\,, \tag{1.25}$$

where $\hat{D}$ is the equilibrium value of the linkage disequilibrium function D. It is not difficult to show that the possible values of $\hat{D}$ are the solutions of the cubic equation

$$64lD^3 - 16(d-a)D^2 - 4(l-8r)D + (d-a) = 0 \tag{1.26}$$

that satisfy $|D| \le \frac{1}{4}$, where $l = 2(b+c) - (a+d)$. Equation (1.26) always has at least one solution because, if $r > 0$, the left side of (1.26) is positive at $D = \frac{1}{4}$ and negative at $D = -\frac{1}{4}$. If $b + c > \max(a, d)$, then three valid solutions exist for small r. In the special case $l = 0$, (1.26) reduces to a quadratic with the single valid solution (if $d > a$)

$$\hat{D} = \frac{r}{d-a} - \frac{1}{4(d-a)}\sqrt{(d-a)^2 + 16r^2}\,. \tag{1.27}$$

Bodmer and Felsenstein (1967) derived the conditions for local stability of the symmetric equilibria. These conditions are

$$\left|1 - r + 2l\hat{x}_1(1-2\hat{x}_1) - \tfrac{1}{2}(b+c)\right| < \overline{W}_e\,, \tag{1.28}$$

and the solutions of

$$\begin{aligned}\lambda^2\overline{W}_e^2 &- \lambda\overline{W}_e[2 - 2d\hat{x}_1 - 2a(\tfrac{1}{2} - \hat{x}_1) - \tfrac{1}{2}(b+c)]\\ &+ 1 + 4bc\hat{x}_1(\tfrac{1}{2} - \hat{x}_1) + 4ad\hat{x}_1(\tfrac{1}{2} - \hat{x}_1) - 2d\hat{x}_1 - 2a(\tfrac{1}{2} - \hat{x}_1)\\ &- \tfrac{1}{2}(b+c) + 2(b+c)[d\hat{x}_1^2 + a(\tfrac{1}{2} - \hat{x}_1)^2] = 0\end{aligned} \tag{1.29}$$

must be less in modulus than unity, where

$$\overline{W}_e = 1 - 2d\hat{x}_1^2 - 2a(\tfrac{1}{2} - \hat{x}_1)^2 - 4(b+c)\hat{x}_1(\tfrac{1}{2} - \hat{x}_1) \tag{1.30}$$

is the equilibrium mean fitness. However, as discussed by Karlin and Feldman, Bodmer and Felsenstein drew some incorrect conclusions from these equations. Karlin and Feldman (1970) investigated the stability of the symmetric equilibria in some detail and proved that for sufficiently tight linkage and any set of selection coefficients, there always exists an asymptotically stable symmetric equilibrium.

Actually, the stability of the symmetric equilibria can depend in a complicated way on r, as was first shown by Ewens (1968). Under the condition $a = d$, three symmetric equilibria may exist. They are given by

$$\text{(a)} \quad \hat{D} = 0 \quad \text{and} \quad \text{(b)} \quad \hat{D} = \pm\frac{1}{4}\left(1 - \frac{8r}{l}\right)^{1/2} \tag{1.31}$$

(Lewontin and Kojima 1960). Solution (a) always exists, and is stable if $r > \frac{1}{8}l$ and $a > |b - c|$, whereas the solutions (b) exist if and only if $r < \frac{1}{8}l$. Ewens (1968) showed that the equilibria (b) are stable if and only if

$$r^2 - r\left(\frac{bc}{b + c - a} - \tfrac{1}{2}(b + c) - a\right) + \tfrac{1}{4}a(b + c - a) > 0\,. \tag{1.32}$$

If $a < |b - c|$, then (1.32) has one solution, r^*, between 0 and $\frac{1}{8}l$, so that (1.31b) is stable for $0 \le r \le r^*$. If $a > |b - c|$, there is either no solution of (1.32) between 0 and $\frac{1}{8}l$, and (1.31b) is stable in this interval, or there are two solutions, r^* and r^{**}, and (1.31b) is stable for $0 \le r \le r^*$ and for $r^{**} \le r \le \frac{1}{8}l$ but unstable otherwise. Thus, there is a gap in the values r leading to stability of these equilibria.

In addition to the symmetric equilibria, there may be up to four *unsymmetric* interior equilibria. Unsymmetric equilibria satisfy $\hat{x}_1 \ne \hat{x}_4$, or $\hat{x}_2 \ne \hat{x}_3$, or both conditions. Karlin and Feldman (1970) proved that up to four interior unsymmetric equilibria may exist, derived explicit formulas for them in the case $b = d$, and presented formulas from which these equilibria can be calculated in the general case. These formulas are quite complicated, and we shall not reproduce them here. Karlin and Feldman also performed a local stability analysis for the case $b = c$ and proved that no more than two simultaneously stable interior equilibria are possible. Feldman and Liberman (1979) showed that as many as four boundary equilibria and two polymorphic equilibria can be simultaneously stable in this model. More recently, Hastings (1985) found that for $a = d$ but $b \ne c$, the case investigated by Lewontin and Kojima (1960) and Ewens (1968), up to four stable interior equilibria may coexist. This phenomenon can occur for arbitrarily weak selection and for arbitrarily small values of the recombination rate r, but none of the symmetric equilibria can be simultaneously stable with the unsymmetric equilibria.

Christiansen (1988) deduced several features of the dynamics of the multilocus two-allele symmetric viability model. In particular, he explored the influence of the recombination distribution and the degree of epistasis on the stability of the central equilibrium. A comprehensive treatment of the symmetric viability model and its generalization to multiple loci may be found in Christiansen (1999). Also, a generalized multiplicative model is treated there.

1.4 CYCLING

We have already seen that in two-locus systems, the Fundamental Theorem of Natural Selection may be violated, that in general fitness is not maximized at stable equilibria, and that the existence and stability of equilibrium points can depend on the recombination frequency. Furthermore, Akin (1979) proved the existence of stable limit cycles in continuous time (cf. also Akin 1982, 1983), and Hastings (1981c) did so for discrete time. The examples given by these authors are structurally stable under small perturbations of the fitness parameters and involve neither lethality nor position effect, but they do require strong epistasis. The period of these cycles is fairly long and, in some cases, almost all of the gametic frequency change occurs during a small fraction of the cycle. Thus, they may be difficult to detect in nature.

• **1.1** *The results of this section demonstrate that*

1. *In two-locus two-allele models many stable equilibria may coexist and, hence, evolution may depend sensitively on the initial conditions.*
2. *Complicated dynamical behavior is possible.*
3. *A variety of selection regimes can maintain a two-locus polymorphism.*
4. *Two-locus systems cannot be expected to be in linkage equilibrium.*
5. *Neither Fisher's Fundamental Theorem of Natural Selection nor Wright's metaphor of evolution as a hill-climbing process in an adaptive topography is of general validity in two-locus systems.*

2. THE MULTILOCUS MODEL

In this section, we formulate the general model, collect the notation, derive the basic recursion relations (2.12) for the gametic frequencies under selection and recombination, and summarize some important results about the dynamics of multilocus systems in the absence of selection. Formulation and notation are adapted from Nagylaki (1993) and Nagylaki *et al.* (1999).

2.1 DISCRETE TIME

We consider a randomly mating population with discrete and nonoverlapping generations, in which the two sexes need not be distinguished. Selection acts only through differential viabilities, which are constant, although the model can be formulated for frequency- and time-dependent fitnesses (Nagylaki 1993). The number of multiallelic loci, the linkage map, epistasis, and dominance are arbitrary. Suppose that the genetic system consists of ℓ loci and ℓ_k alleles $\mathcal{A}_{i_k}^{(k)}$ ($i_k = 1, \ldots, \ell_k$) at locus k. We use the multi-index $i = (i_1, \ldots, i_\ell)$ as an abbreviation for the gamete $\mathcal{A}_{i_1}^{(1)}\mathcal{A}_{i_2}^{(2)} \ldots \mathcal{A}_{i_\ell}^{(\ell)}$ and denote its frequency (immediately after gametogenesis) by p_i. There are $\ell_1 \cdot \ldots \cdot \ell_\ell$ different ℓ-locus gametes. Collectively, the gamete frequencies form the vector $\boldsymbol{p}$, a probability vector in the corresponding simplex. The (marginal) frequency of $\mathcal{A}_{i_k}^{(k)}$ in gametes is

$$p_{i_k}^{(k)} = \sum_{i \nmid i_k} p_i \, , \tag{2.1}$$

where the sum runs over all multi-indices i with kth component fixed as i_k. We shall repeatedly use the identity

$$\sum_{i_k} \sum_{i \nmid i_k} f_i = \sum_i f_i \, , \tag{2.2}$$

which holds for every locus k and for arbitrary functions f_i.

Let W_{ij} denote the fitness of genotype ij. We define the marginal fitness of gamete i and the mean fitness of the population by

$$W_i(\boldsymbol{p}) = \sum_j W_{ij}p_j \tag{2.3}$$

and

$$\overline{W}(\boldsymbol{p}) = \sum_{i,j} W_{ij}p_ip_j \ , \tag{2.4}$$

respectively. Usually, we shall suppress the dependence on $\boldsymbol{p}$.

To describe recombination, let $L = \{1, 2, \ldots, \ell\}$ be the set of loci and let $\{I, J\}$ be a decomposition of L, i.e., $I \cup J = L$ and $I \cap J = \emptyset$, such that I and J are each proper subsets of L and, therefore, contain at least one locus. (Clearly, the decompositions $\{I, J\}$ and $\{J, I\}$ are identified.) We designate by r_I the probability of reassociation of the genes at the loci in I, inherited from one parent, with the genes at the loci in J, inherited from the other. Further, we denote the total recombination frequency by

$$r_{\text{tot}} = \sum_I r_I \ , \tag{2.5}$$

where $\sum_I$ runs over all (different) decompositions $\{I, J\}$ of L. We designate the recombination frequency between loci k and l, such that $k < l$, by r_{kl}. This corresponds to the recombination rate r in the two-locus case. To calculate the two-locus recombination frequencies r_{kl} in terms of the linkage map, or recombination distribution, $\{r_I\}$, we define the set of sets

$$L_{kl} = \{I : k \in I \text{ and } l \in J\} \ . \tag{2.6}$$

Then we have

$$r_{kl} = \sum_{I \in L_{kl}} r_I \ . \tag{2.7}$$

▷ Let us now derive the recursion relations for the gamete frequencies. We define $R(j, k \to i)$ as the probability that a randomly chosen gamete produced by a jk individual is i. Then the frequency of gamete i in the next generation is

$$p_i' = \overline{W}^{-1} \sum_{j,k} W_{jk}p_jp_kR(j, k \to i) \ , \tag{2.8}$$

because selection acts before recombination. To express $R(j, k \to i)$ in terms of recombination frequencies r_I, we consider all decompositions $\{I, J\}$ of L and denote by i_Ij_J the vector with kth component i_k if $k \in I$ and j_k if $k \in J$. Thus, we can write $i = i_Ii_J$, $j = j_Ij_J$, $k = k_Ik_J$, and have

$$R(j, k \to i) = \sum_I R_I(j_Ij_J, k_Ik_J \to i_Ii_J) \ , \tag{2.9}$$

where $R_I(j_Ij_J, k_Ik_J \to i_Ii_J)$ is the probability that a randomly chosen gamete produced by a jk individual is i and each of the sets I and J is passed to the next generation without recombination. Hence, if we first consider drawing i_I, and then take into account recombination between I and J, we obtain

$$\begin{aligned}R_I(j_Ij_J,k_Ik_J \to i_Ii_J) = \tfrac{1}{2}\delta_{i_Ij_I}[(1-r_I)\delta_{i_Jj_J} + r_I\delta_{i_Jk_J}] \\ + \tfrac{1}{2}\delta_{i_Ik_I}[(1-r_I)\delta_{i_Jk_J} + r_I\delta_{i_Jj_J}]\,,\end{aligned} \tag{2.10}$$

where δ denotes the Kronecker delta. A simple calculation shows that

$$\begin{aligned}&\sum_{j,k} W_{jk}p_jp_kR_I(j_Ij_J,k_Ik_J \to i_Ii_J) \\ &\quad = \sum_{k_I,k_J}(1-r_I)W_{i_Ii_J,k_Ik_J}\,p_{i_Ii_J}\,p_{k_Ik_J} + \sum_{k_I,j_J} r_IW_{i_Ij_J,k_Ii_J}\,p_{i_Ij_J}\,p_{k_Ii_J}\,,\end{aligned} \tag{2.11}$$

because the sum on the left is symmetric under the simultaneous interchanges $j_I \leftrightarrow k_I$ and $j_J \leftrightarrow k_J$, so that the two terms in (2.10) contribute equally. ◁

Therefore, substitution of (2.9), (2.10), and (2.11) into (2.8) yields the recursion relations for the gametic frequencies under the combined action of selection and recombination:

$$p_i' = p_i\frac{W_i}{\overline{W}} - \Theta_i\,. \tag{2.12}$$

Here,

$$\Theta_i = \frac{1}{\overline{W}}\sum_j\sum_I r_I(W_{ij}p_ip_j - W_{i_Ij_J,j_Ii_J}\,p_{i_Ij_J}\,p_{j_Ii_J}) \tag{2.13}$$

represents a measure of linkage disequilibrium in gamete i (Nagylaki 1992, Chapter 8.2; Nagylaki 1993).

If there is no position effect (i.e., given the composition of the genotype, the arrangement of the alleles is irrelevant for the fitness of the genotype), then Θ_i simplifies to

$$\Theta_i = \frac{1}{\overline{W}}\sum_j W_{ij}\sum_I r_I(p_ip_j - p_{i_Ij_J}\,p_{j_Ii_J})\,. \tag{2.14}$$

For two diallelic loci and no position effect, we have in the notation of Section 1, $\Theta_1 = \Theta_4 = -\Theta_2 = -\Theta_3 = rW_{14}D/\overline{W}$, and (2.12) reduces to (1.4). The set of linkage disequilibria Θ_i differs from other sets of measures of linkage disequilibrium, in particular from those of Slatkin (1972); see Section 6.2 and Chapter V.4.

2.2 NO SELECTION

There are not many general conclusions about the dynamics of multilocus systems that can be drawn immediately from (2.12). Here, we shall briefly review the case of no selection, which is well understood, and relegate other cases to subsequent sections. For proofs and some further results we refer to Lyubich (1971, 1992), Nagylaki (1993), Nagylaki *et al.* (1999), Christiansen (1999), Dawson (2000), and Chapter V.4.

If all genotypes have the same fitness, then (2.12) simplifies to

$$p_i' = p_i - \Theta_i\,. \tag{2.15}$$

The *linkage-equilibrium*, or *Wright, manifold* Λ_0 is defined by

$$\Lambda_0 = \{\boldsymbol{p} : p_i = p_{i_1}^{(1)} p_{i_2}^{(2)} \dots p_{i_\ell}^{(\ell)}\} . \tag{2.16}$$

It trivially follows that $\Lambda_0 \subseteq \{\boldsymbol{p} : \Theta_i = 0 \text{ for every } i\}$. This is true also with selection, as long as there is no position effect. In the case of no selection these sets are equal, but it is unknown if this is true in general.

If all recombination frequencies r_I are positive, then it can be shown that all Θ_i tend to zero at a geometric rate. Therefore, Λ_0 is invariant and globally attracting at a uniform geometric rate. The allele frequencies $p_{i_k}^{(k)}$ remain constant during this process; cf. (4.17). All points on Λ_0 are fixed points under (2.15). The rates of approach, i.e., the eigenvalues transverse to Λ_0, can be calculated explicitly. They are given by $2^\ell - \ell - 1$ numbers λ_S, where S runs through all subsets of L with at least two elements, and λ_S is the probability that there is no recombination in S. In terms of the recombination distribution, they read

$$\lambda_S = 1 - r_{\text{tot}}^{(S)} , \quad r_{\text{tot}}^{(S)} = \sum_{I:\emptyset \neq I \cap S \neq S} r_I . \tag{2.17}$$

Here, $r_{\text{tot}}^{(S)}$ represents the total probability of recombination among the loci in S. Hence, the generic rate of approach, i.e., the largest eigenvalue, is given by $\max \lambda_S = 1 - r_{\min}$, where $r_{\min}$ is the smallest two-locus recombination frequency.

The first proof that, for an arbitrary number of loci and alleles and an arbitrary recombination distribution, the gamete frequencies converge to linkage equilibrium is due to Geiringer (1944).

2.3 CONTINUOUS TIME

A multilocus model in continuous time was developed and analyzed by Shahshahani (1979) and Akin (1979); see also Passekov (1984) and Svirezhev and Passekov (1990). It is given by the system of ordinary differential equations,

$$\dot{p}_i = p_i(m_i - \overline{m}) - \tilde{\Theta}_i , \tag{2.18}$$

where m_i is the marginal Malthusian fitness of gamete i, $\overline{m}$ is the mean fitness, and

$$\tilde{\Theta}_i = \sum_j \sum_I r_I (b_{ij} p_i p_j - b_{i_I j_J, j_I i_J} \, p_{i_I j_J} \, p_{j_I i_J}) \tag{2.19}$$

measures linkage disequilibria in gamete i. The parameters b_{ij} in (2.19) are the birth rates of the genotypes ij. (Actually, the assignment of birth rates to genotypes, instead of mating pairs, requires assumptions about fertilities of mating pairs, for instance, multiplicative fertilities.) For two loci and two alleles, (2.18) agrees with (1.5).

Although, the continuous-time model (2.18) appears to have the same qualitative behavior as the discrete-time model (2.12), no rigorous derivation of (2.18) is known that covers the parameter space of biological interest. The problem is that in a continuous-time model, Hardy–Weinberg proportions are usually not preserved and the full dynamics can only be described by genotype frequencies (cf. Nagylaki and Crow 1974, and Chapter I.10). In contrast, (2.18) is based on postulating Hardy–Weinberg proportions. However, it can be deduced as the limit of the discrete model if selection is much weaker than recombination because, then, linkage disequilibria decay rapidly (cf. Section 6.2).

3. COMPONENTS OF GENETIC VARIANCE AND COVARIANCE

For a trait controlled by a single locus, the correlation between certain relatives was calculated in Chapter I.3 from the genetic variance and its additive and dominance components. Here, we consider a quantitative trait that is determined by an arbitrary number of multiallelic loci. It will not be assumed that the population is in Hardy–Weinberg proportions, so the results are valid under arbitrary mating schemes. The aim is to derive decompositions of the total genetic variance of the trait, and of the covariance between trait and fitness, into components of evolutionary significance. This theory is a major ingredient for predictions of the response of a quantitative trait to selection and the rate of change in mean fitness. A central role in this theory is played by the concept of the average effect of an allele which is studied in detail.

For two diallelic loci this analysis was first carried out by Kimura (1965b). It was extended to two multiallelic loci by Nagylaki (1976), and to an arbitrary number of loci by Ewens and Thomson (1977), Nagylaki (1993), and Castilloux and Lessard (1995).

We retain the multilocus notation of the previous section. If P_{ij} denotes the frequency of the ordered (multilocus) genotype ij, then the frequency of gamete i is

$$p_i = \sum_j P_{ij} \tag{3.1a}$$

and that of allele $\mathcal{A}_{i_k}^{(k)}$ is

$$p_{i_k}^{(k)} = \sum_{i \nmid i_k} \sum_j P_{ij} \; . \tag{3.1b}$$

Let G_{ij} be the trait value (genotypic value) of genotype ij as measured on an appropriate scale. The marginal value G_i of gamete i, the marginal value $G_{i_k}^{(k)}$ of allele $\mathcal{A}_{i_k}^{(k)}$, and the mean (genotypic) value $\overline{G}$ of the population are defined by

$$G_i = \frac{1}{p_i} \sum_j G_{ij} P_{ij} \; , \tag{3.2a}$$

$$G_{i_k}^{(k)} = \frac{1}{p_{i_k}^{(k)}} \sum_{i \nmid i_k} \sum_j G_{ij} P_{ij} \; , \tag{3.2b}$$

$$\overline{G} = \sum_{i,j} G_{ij} P_{ij} \; , \tag{3.2c}$$

respectively. If $p_i = 0$, then G_i remains undefined (and is not needed), and if $p_{i_k}^{(k)} = 0$, then $G_{i_k}^{(k)}$ remains undefined. Henceforth, we assume $p_{i_k}^{(k)} > 0$ and $p_i > 0$.

In this chapter, deviations of quantities, such as genotypic values or fitnesses, from their mean are denoted by lower case letters. Accordingly, we define the average excess of genotype ij, of gamete i, and of allele $\mathcal{A}_{i_k}^{(k)}$ by

$$g_{ij} = G_{ij} - \overline{G} \,, \tag{3.3a}$$

$$g_i = G_i - \overline{G} = \frac{1}{p_i} \sum_j g_{ij} P_{ij} \,, \tag{3.3b}$$

$$g_{i_k}^{(k)} = G_{i_k}^{(k)} - \overline{G} = \frac{1}{p_{i_k}^{(k)}} \sum_{i \nmid i_k} \sum_j g_{ij} P_{ij} \,, \tag{3.3c}$$

respectively.

3.1 AVERAGE EFFECTS AND THE ANALYSIS OF VARIANCE

The *genotypic*, or *(total) genetic, variance* is defined by

$$\sigma_G^2 = \sum_{i,j} P_{ij} g_{ij}^2 \,. \tag{3.4}$$

However, since parents pass on their genes and not their genotypes to their progeny, the evolutionary properties of a population will depend upon the variability of measurements describing the effects of genes and not on the variability of genotypic values. In the following we define these measurements, the so-called average effects, and explore their basic properties. The concept of the average effect of an allele was introduced by Fisher (1918, 1930, 1941) in an attempt to ascribe in an optimal way a genetic value to that allele, namely such that the response to selection can be predicted as closely as possible from these average effects (cf. Chapter I.3.1)

We wish to assign to each allele $\mathcal{A}_{i_k}^{(k)}$ a parameter $\gamma_{i_k}^{(k)}$, its *average*, or *additive, effect*, so that the quantities

$$\gamma_{ij} = \gamma_i + \gamma_j \,, \tag{3.5a}$$

where

$$\gamma_i = \sum_{k=1}^{\ell} \gamma_{i_k}^{(k)} \,, \tag{3.5b}$$

approximate the genotypic excesses g_{ij} as closely as possible, i.e., in the sense of a least-squares approximation. To this end, we decompose the genotypic excess g_{ij} into an additive effect γ_{ij}, often called the *breeding value*, and a residual deviation ε_{ij},

$$g_{ij} = \gamma_{ij} + \varepsilon_{ij} \,, \tag{3.6}$$

where the γ_{ij} obey (3.5). The residual incorporates dominance and epistatic deviations. Then we define the *additive genetic* and the *residual component* of the genotypic variance by

$$\sigma_A^2 = \sum_{i,j} P_{ij} \gamma_{ij}^2 \tag{3.7}$$

and

$$\sigma_{\mathrm{R}}^2 = \sum_{i,j} P_{ij}\varepsilon_{ij}^2 \,, \tag{3.8}$$

respectively. The desired least-squares approximation is obtained by minimizing the residual variance σ_{R}^2 with respect to the $\gamma_{i_k}^{(k)}$.

▷ To carry out this minimization procedure, we substitute (3.5) and (3.6) into (3.8) and differentiate σ_{R}^2 with respect to each $\gamma_{i_k}^{(k)}$. This leads easily to the conditions

$$\sum_{i \nmid i_k} \sum_j P_{ij}\varepsilon_{ij} = 0 \quad \text{for every } k \text{ and } i_k \,. \tag{3.9}$$

Owing to (3.6), (3.3a), (3.9), and (3.5a), we obtain

$$\sum_{i,j} P_{ij}\gamma_{ij} = 2\sum_i p_i\gamma_i = 0 \,. \tag{3.10}$$

Furthermore, substitution of (3.6) into (3.3c) and using (3.9) yields

$$p_{i_k}^{(k)} g_{i_k}^{(k)} = \sum_{i \nmid i_k} \sum_j P_{ij}\gamma_{ij} \quad \text{for each } k \text{ and } i_k \,. \tag{3.11}$$

Now we show that one may assume

$$\sum_{i_k} p_{i_k}^{(k)} \gamma_{i_k}^{(k)} = 0 \quad \text{for each } k \,. \tag{3.12}$$

Indeed, if we put $\beta_k = \sum_{i_k} p_{i_k}^{(k)}\gamma_{i_k}^{(k)}$ and $\tilde{\gamma}_{i_k}^{(k)} = \gamma_{i_k}^{(k)} - \beta_k$, then $\sum_{i_k} p_{i_k}^{(k)}\tilde{\gamma}_{i_k}^{(k)} = 0$. Furthermore, we obtain from (3.10) and (3.5), by using (3.58) with $v_{ij} = 1$, that $\sum_k \beta_k = 0$; hence $\gamma_{ij} = \tilde{\gamma}_{ij}$. Therefore, the minimizing values $\gamma_{i_k}^{(k)}$ can be chosen to satisfy (3.12).

Finally, employing (3.5) and (3.1a), we deduce from (3.11)

$$\begin{aligned} p_{i_k}^{(k)} g_{i_k}^{(k)} &= \sum_{i \nmid i_k} \sum_j P_{ij} \sum_l \left(\gamma_{i_l}^{(l)} + \gamma_{j_l}^{(l)}\right) \\ &= p_{i_k}^{(k)}\gamma_{i_k}^{(k)} + \sum_{l:l \neq k} \sum_{i_l} \gamma_{i_l}^{(l)} p_{i_k i_l}^{(kl)} + \sum_l \sum_{j_l} \gamma_{j_l}^{(l)} q_{i_k j_l}^{(kl)} \,, \end{aligned} \tag{3.13}$$

for every k and i_k. Here,

$$p_{i_k i_l}^{(kl)} = \sum_{i \nmid i_k, i_l} \sum_j P_{ij} \tag{3.14}$$

denotes the frequency of gametes that carry $\mathcal{A}_{i_k}^{(k)}$ and $\mathcal{A}_{i_l}^{(l)}$, and

$$q_{i_k j_l}^{(kl)} = \sum_{i \nmid i_k} \sum_{j \nmid j_l} P_{ij} \tag{3.15}$$

is the frequency of all genotypes that carry $\mathcal{A}_{i_k}^{(k)}$ and $\mathcal{A}_{j_l}^{(l)}$ on different chromosomes, i and j, respectively. ◁

The system of linear equations (3.13), together with the constraints (3.12), provides the *defining relations of the average effects* $\gamma_{i_k}^{(k)}$ in terms of the average excesses $g_{i_k}^{(k)}$. It is important to note that, in general, the average effect of an allele depends on the genotype frequencies, and that it differs from the average excess. If all genotype frequencies are strictly positive, then the average effects are uniquely determined by (3.13), subject to the constraints (3.12), and give rise to a minimum of σ_{R}^2. The proof is relegated to Section 3.6.

We shall now recast (3.13) into a more transparent matrix form that will allow us to derive some further interesting conclusions about the average effects (cf. Ewens 1992). Let $\boldsymbol{\pi}$ denote the (column) vector of allele frequencies $p_{i_k}^{(k)}$, $\boldsymbol{g}$ the vector of average excesses $g_{i_k}^{(k)}$, and $\boldsymbol{\gamma}$ the vector of average effects $\gamma_{i_k}^{(k)}$. These vectors have length $\ell_1 + \cdots + \ell_\ell$. Naturally, but not necessarily, the components of such a vector (e.g., $\boldsymbol{\pi}$) can be ordered locus wise, i.e., such that $\boldsymbol{\pi} = (\boldsymbol{\pi}^{(1)}, \ldots, \boldsymbol{\pi}^{(\ell)})^\top$, where $\boldsymbol{\pi}^{(k)} = (p_1^{(k)}, \ldots, p_{\ell_k}^{(k)})^\top$ is the (column) vector of allele frequencies at locus k.

Then we define the entries of the matrix H by

$$\mathsf{H}(\mathcal{A}_{s_k}^{(k)}, \mathcal{A}_{r_l}^{(l)}) = H_{s_k r_l}^{(kl)} = \begin{cases} p_{s_k}^{(k)}\,, & \text{if } k = l \text{ and } s_k = r_k\,, \\ q_{s_k r_k}^{(kk)}\,, & \text{if } k = l \text{ and } s_k \neq r_k\,, \\ p_{s_k r_l}^{(kl)} + q_{s_k r_l}^{(kl)}\,, & \text{if } k \neq l\,. \end{cases} \tag{3.16}$$

This matrix may be called a genetic covariance matrix for reasons that become clear in Section 3.6. It can be written as a sum of three matrices,

$$\mathsf{H} = \mathsf{D} + \mathsf{P} + \mathsf{Q}\,, \tag{3.17}$$

where D is the diagonal matrix with the components of $\boldsymbol{\pi}$ on the diagonal, Q is the matrix with entries $q_{s_k r_l}^{(kl)}$ and P is the matrix with entries $p_{s_k r_l}^{(kl)}$ if $k \neq l$, and 0, otherwise; cf. (3.16). Designating the component-wise product of the vectors $\boldsymbol{\pi}$ and $\boldsymbol{g}$ by $\boldsymbol{\pi} \cdot \boldsymbol{g}$ (i.e., $(\boldsymbol{\pi} \cdot \boldsymbol{g})_i = \boldsymbol{\pi}_i \boldsymbol{g}_i$), we have proved the following statement:

• **3.1** *The average effects* $\gamma_{i_k}^{(k)}$ *are the components of the vector* $\boldsymbol{\gamma}$ *that is the solution of the matrix equation*

$$\boldsymbol{\pi} \cdot \boldsymbol{g} = \mathsf{H}\boldsymbol{\gamma}\,. \tag{3.18}$$

This solution is uniquely determined subject to the requirement (3.12).

In Section 3.6 it is shown that, equivalently, the average effects can be obtained from the following minimizing principle (Ewens 1992):

• **3.2** *The average effects are the components of the uniquely determined vector* $\boldsymbol{\gamma} = \boldsymbol{a}$ *that minimizes the quadratic form*

$$\boldsymbol{a}^\top \mathsf{H} \boldsymbol{a} \tag{3.19}$$

subject to the condition

$$\boldsymbol{a}^\top (\boldsymbol{\pi} \cdot \boldsymbol{g}) = \tfrac{1}{2}\sigma_{\mathrm{A}}^2\,. \tag{3.20}$$

The above analysis enables us to derive alternative representations for the additive genetic variance. Some useful identities that facilitate several calculations are derived in Section 3.6. Applying (3.60) to (3.7) yields the representation

$$\sigma_{\mathrm{A}}^2 = 2\gamma^{\top}\mathsf{H}\gamma \tag{3.21a}$$

$$= 2\sum_k \sum_{i_k} p_{i_k}^{(k)} g_{i_k}^{(k)} \gamma_{i_k}^{(k)} \,. \tag{3.21b}$$

From (3.58) and (3.9) we infer $\sum_{i,j} P_{ij}\gamma_{ij}\varepsilon_{ij} = 0$. Hence, (3.7) can be rewritten as

$$\sigma_{\mathrm{A}}^2 = \sum_{i,j} P_{ij} g_{ij} \gamma_{ij} \,. \tag{3.22}$$

Furthermore, we obtain immediately

$$\sigma_G^2 = \sum_{i,j} P_{ij}(\gamma_{ij}^2 + \varepsilon_{ij}^2 + 2\gamma_{ij}\varepsilon_{ij}) = \sigma_{\mathrm{A}}^2 + \sigma_{\mathrm{R}}^2$$

and, thus, have proved the following important result:

• **3.3** *The total genetic variance σ_G^2 of an arbitrary character can be decomposed into the additive genetic variance σ_{A}^2 and a residual component σ_{R}^2, the variance of all interaction effects,*

$$\sigma_G^2 = \sigma_{\mathrm{A}}^2 + \sigma_{\mathrm{R}}^2 \,. \tag{3.23}$$

In this generality, this result was first proved by Castilloux and Lessard (1995). A generalization to fertility selection may be found in Lessard and Castilloux (1995). Obviously, if a character is determined completely additively, i.e., without dominance or epistasis ($\varepsilon_{ij} = 0$ for every i, j; cf. Section 5.1), then $\sigma_G^2 = \sigma_{\mathrm{A}}^2$.

3.2 AVERAGE EFFECTS ON FITNESS

Of particular importance is the case when the character is fitness, i.e., $G_{ij} = W_{ij}$. Here, we define the essential quantities and state the most relevant identities. In analogy with (3.3), we define the following measures of average excess in fitness:

$$w_{ij} = W_{ij} - \overline{W} \,, \tag{3.24a}$$

$$w_i = W_i - \overline{W} = \frac{1}{p_i}\sum_j w_{ij}P_{ij} \,, \tag{3.24b}$$

$$w_{i_k}^{(k)} = W_{i_k}^{(k)} - \overline{W} = \frac{1}{p_{i_k}^{(k)}} \sum_{i \nmid i_k} \sum_j w_{ij}P_{ij} \,. \tag{3.24c}$$

As before, w_i remains undefined (and is not needed) if $p_i = 0$, and $w_{i_k}^{(k)}$ remains undefined if $p_{i_k}^{(k)} = 0$.

We decompose w_{ij} into an additive effect and a residual deviation,

$$w_{ij} = \alpha_{ij} + \nu_{ij} \; , \tag{3.25}$$

where

$$\alpha_{ij} = \alpha_i + \alpha_j \tag{3.26}$$

and

$$\alpha_i = \sum_{k=1}^{\ell} \alpha_{i_k}^{(k)} \; . \tag{3.27}$$

Here, $\alpha_{i_k}^{(k)}$ denotes the average, or additive, effect of $\mathcal{A}_{i_k}^{(k)}$ on fitness. The average effects $\alpha_{i_k}^{(k)}$ are defined implicitly by a system of linear equations analogous to (3.13) and subject to the constraints (3.12).

Let the $\boldsymbol{\alpha}$ and $\boldsymbol{w}$ denote the $\ell_1 + \cdots + \ell_\ell$-dimensional (column) vectors of average effects and average excesses of alleles, respectively. Further, let $\Delta\boldsymbol{\pi}$ denote the vector of allele-frequency changes caused by natural selection, i.e.,

$$\Delta\boldsymbol{\pi} = \boldsymbol{\pi} \cdot \boldsymbol{w}/\overline{W} \; ; \tag{3.28}$$

cf. I(9.7) and (4.17) below. Then (3.18) becomes

$$\overline{W}\Delta\boldsymbol{\pi} = \mathsf{H}\boldsymbol{\alpha} \; . \tag{3.29}$$

Equivalently, the average effects on fitness are given by

$$\boldsymbol{\alpha} = \overline{W}\mathsf{H}^{-1}\Delta\boldsymbol{\pi} \; , \tag{3.30}$$

where $\mathsf{H}^{-1}\Delta\boldsymbol{\pi}$ is well defined because the vector $\Delta\boldsymbol{\pi}$ is perpendicular to the nullspace N_{H} of H (cf. Section 3.6). It may be noted that (3.29) is a generalization of the single-locus selection dynamics I(10.14b), if $\dot{\boldsymbol{\pi}}$ and $\overline{W}\Delta\boldsymbol{\pi}$ are equated. However, in the multilocus case, (3.29) can be used only to predict the gene-frequency change for one generation; it does not describe the full dynamics of gamete frequencies.

Finally, by applying (3.21) to fitness and using (3.29), the additive genetic variance in fitness can be expressed in each of the following ways:

$$\sigma_{\mathrm{A}}^2 = 2\sum_k \sum_{i_k} p_{i_k}^{(k)} w_{i_k}^{(k)} \alpha_{i_k}^{(k)} \tag{3.31a}$$

$$= 2\boldsymbol{\alpha}^{\top}\mathsf{H}\boldsymbol{\alpha} \tag{3.31b}$$

$$= 2\overline{W}\boldsymbol{\alpha}^{\top}\Delta\boldsymbol{\pi} \; . \tag{3.31c}$$

3.3 POPULATIONS IN HARDY–WEINBERG PROPORTIONS

For populations in Hardy–Weinberg proportions, most of the above formulas can be simplifed. Notably, owing to $P_{ij} = p_i p_j$, (2.1), and (3.12), the defining equations (3.13) for the average effects reduce to

$$p_{i_k}^{(k)} g_{i_k}^{(k)} = p_{i_k}^{(k)} \gamma_{i_k}^{(k)} + \sum_{l:l\neq k} \sum_{i_l} \gamma_{i_l}^{(l)} p_{i_k i_l}^{(kl)} \; , \tag{3.32a}$$

as was first derived by Ewens and Thomson (1977); cf. also Ewens (1979, pp. 217–218). Unless all pairwise linkage disequilibria vanish (cf. Section 3.5), average effect and average excess of an allele will generally be different. In matrix form, (3.32a) becomes

$$\boldsymbol{\pi} \cdot \boldsymbol{g} = (\mathsf{D} + \mathsf{P})\boldsymbol{\gamma} , \tag{3.32b}$$

because $\mathsf{Q} = \boldsymbol{\pi}\boldsymbol{\pi}^\top$ and, by (3.12), $\boldsymbol{\pi}^\top\boldsymbol{\gamma} = 0$. If, in addition, fitness is determined by a single locus, then $\mathsf{P} = \mathbf{0}$ and

$$\mathsf{H} = \mathsf{D} + \boldsymbol{\pi}\boldsymbol{\pi}^\top = \mathsf{G}_{\boldsymbol{p}} + 2\boldsymbol{\pi}\boldsymbol{\pi}^\top , \tag{3.33}$$

where G_p is a generalization of the genetic covariance matrix defined in I(10.13). Therefore, $\boldsymbol{\pi} \cdot \boldsymbol{g} = \mathsf{D}\boldsymbol{\gamma}$, and average effects and excesses coincide, in agreement with I(3.10).

Since, under Hardy–Weinberg proportions, the analysis can be performed in terms of gamete frequencies, it is reasonable to define the *gametic variance*, measuring differences in marginal gametic values,

$$\sigma^2_{\mathrm{Gam}} = 2\sum_i p_i g_i^2 , \tag{3.34}$$

with the factor 2 because of diploidy. We decompose the gametic excess g_i into an additive effect γ_i and an epistatic deviation,

$$g_i = \gamma_i + \varepsilon_i , \tag{3.35}$$

with γ_i as in (3.5b). (But note that, in general, $\varepsilon_{ij} \neq \varepsilon_i + \varepsilon_j$.) Then, by (3.21) and (3.61), the additive genetic variance can be written as

$$\sigma^2_{\mathrm{A}} = 2\sum_i p_i \gamma_i^2 . \tag{3.36}$$

The variance of epistatic deviations, or *epistatic variance*, is defined by

$$\sigma^2_{\mathrm{Ep}} = 2\sum_i p_i \varepsilon_i^2 . \tag{3.37}$$

Like the total genetic variance, the gametic variance can be decomposed into an additive and a residual component, and the additive component is, indeed, equal to the additive genetic variance.

• **3.4** *The additive genetic variance σ^2_{A} of an arbitrary character is equal to the additive component of the gametic variance. In particular, the following decompositions obtain:*

$$\sigma^2_{\mathrm{Gam}} = \sigma^2_{\mathrm{A}} + \sigma^2_{\mathrm{Ep}} , \tag{3.38}$$

where the epistatic variance σ^2_{Ep} arises from the interaction of additive effects, and

$$\sigma^2_G = \sigma^2_{\mathrm{Gam}} + \sigma^2_{\mathrm{Gam,D}} , \tag{3.39}$$

where $\sigma^2_{\mathrm{Gam,D}}$ *represents the variance that arises from nonadditivity (dominance) of gametic effects. Comparison with* (3.23) *shows that*

$$\sigma^2_{\mathrm{R}} = \sigma^2_{\mathrm{Ep}} + \sigma^2_{\mathrm{Gam,D}} \ . \tag{3.40}$$

The proof of (3.38) is parallel to that of (3.23), whereas the proof of (3.39) is analogous to that of I(3.13) upon replacing alleles with gametes. The gametic variance σ^2_{Gam} is indeed the additive component of the genetic variance in the decomposition (3.39), because the gametes are combined in Hardy–Weinberg proportions. Therefore, σ^2_{A} is the additive component of σ^2_G. These considerations, and in particular the representation (3.21b), justify to call σ^2_{A} the *genic variance*, because it measures the variability caused by the differences among single-gene effects.

The epistatic variance σ^2_{Ep} can be decomposed into components due to interaction of additive effects. Also $\sigma^2_{\mathrm{Gam,D}}$ can be partitioned. Such interaction components enter the expressions for the correlations between relatives. We shall not pursue this topic further, but refer to Cockerham (1954), Kempthorne (1955; 1957, pp. 413–419), Ewens (1979, pp. 63–66), Bulmer (1980, pp. 46–51), and Nagylaki (1992, pp. 280–290; 1993).

Finally, we mention that, in contrast to the one-locus case (Chapter I.3), with epistasis between loci the breeding value γ_{ij} of an individual is, in general, not one half of the deviation of the expected genotypic value of its offspring from the population mean. It is instructive to prove this for the two-locus two-allele case.

3.4 ANALYSIS OF COVARIANCE

The above decomposition of variance can be extended straightforwardly to the covariance between an arbitrary character and fitness (Nagylaki 1993). For simplicity, the analysis is restricted to populations in Hardy–Weinberg proportions. It can be extended readily to the general case.

We define the *gametic*, *additive genetic*, and *epistatic covariances* of G and W by

$$\mathrm{Cov}_{\mathrm{Gam}}(G,W) = 2\sum_i p_i g_i w_i \ , \tag{3.41}$$

$$\mathrm{Cov}_{\mathrm{A}}(G,W) = 2\sum_i p_i \gamma_i \alpha_i \ , \tag{3.42}$$

$$\mathrm{Cov}_{\mathrm{Ep}}(G,W) = 2\sum_i p_i \varepsilon_i \nu_i \ , \tag{3.43}$$

where the epistatic deviation ν_i is defined by $w_i = \alpha_i + \nu_i$. Employing (3.61), we find

$$\mathrm{Cov}_{\mathrm{A}}(G,W) = 2\boldsymbol{\alpha}^{\top}\mathsf{H}\boldsymbol{\gamma} \tag{3.44a}$$

$$= 2\sum_k \sum_{i_k} p^{(k)}_{i_k} g^{(k)}_{i_k} \alpha^{(k)}_{i_k} \ . \tag{3.44b}$$

For reasons of symmetry, we also have

$$\mathrm{Cov}_{\mathrm{A}}(G,W) = 2\boldsymbol{\gamma}^{\top}\mathsf{H}\boldsymbol{\alpha} \tag{3.44c}$$

$$= 2\sum_k \sum_{i_k} p^{(k)}_{i_k} w^{(k)}_{i_k} \gamma^{(k)}_{i_k} \tag{3.44d}$$

$$= 2\overline{W}\boldsymbol{\gamma}^{\top}\Delta\boldsymbol{\pi} \ , \tag{3.44e}$$

where the last representation is a consequence of (3.44c) and (3.29). Now it is straightforward to obtain the following generalization of (3.38):

$$\mathrm{Cov}_{\mathrm{Gam}}(G,W) = \mathrm{Cov}_{\mathrm{A}}(G,W) + \mathrm{Cov}_{\mathrm{Ep}}(G,W) \ . \tag{3.45}$$

This is our desired decomposition of covariance. The component $\mathrm{Cov}_{\mathrm{A}}(G,W)$ may be called the *genic covariance*, since it is the covariance between the average effect of an allele on the character and its average excess for fitness, and between the average effect on fitness and the average excess for the character.

3.5 LINKAGE EQUILIBRIUM

For a population in linkage equilibrium further simplifications occur. In analogy with I(7.3), we define the (induced) two-locus linkage disequilibria

$$D^{(kl)}_{i_k i_l} = p^{(kl)}_{i_k i_l} - p^{(k)}_{i_k} p^{(l)}_{i_l} \ . \tag{3.46}$$

Substituting (3.46) into (3.32a), we obtain

$$p^{(k)}_{i_k} g^{(k)}_{i_k} = p^{(k)}_{i_k} \gamma^{(k)}_{i_k} + \sum_{l:l\neq k} \sum_{i_l} \gamma^{(l)}_{i_l} D^{(kl)}_{i_k i_l} \ . \tag{3.47}$$

This shows that if a population is in Hardy–Weinberg equilibrium, and if all pairwise linkage disequilibria between a locus k and the other loci vanish, then average excess and average effect of every allele at locus k coincide, i.e.,

$$g^{(k)}_{i_k} = \gamma^{(k)}_{i_k} \quad \text{for every } i_k \ . \tag{3.48}$$

From this and (3.21), we infer that in linkage equilibrium, the additive genetic variance can be written as the sum of the single-locus additive genetic variances, i.e.,

$$\sigma^2_{\mathrm{A}} = 2\sum_k \sum_{i_k} p^{(k)}_{i_k} \left(\gamma^{(k)}_{i_k}\right)^2 = 2\sum_k \sum_{i_k} p^{(k)}_{i_k} \left(g^{(k)}_{i_k}\right)^2 \ . \tag{3.49}$$

From (3.44) and (3.48), the following representations of the additive genetic covariance are obtained:

$$\mathrm{Cov}_{\mathrm{A}}(G,W) = 2\sum_k \sum_{i_k} p^{(k)}_{i_k} \gamma^{(k)}_{i_k} \alpha^{(k)}_{i_k} \tag{3.50a}$$

$$= 2\sum_k \sum_{i_k} p^{(k)}_{i_k} g^{(k)}_{i_k} w^{(k)}_{i_k} \ . \tag{3.50b}$$

*3.6 AVERAGE EFFECTS: UNIQUENESS AND OTHER PROPERTIES

First, we prove that the average effects $\gamma^{(k)}_{i_k}$ of the alleles $\mathcal{A}^{(k)}_{i_k}$ are uniquely determined for every k and i_k, if all genotype frequencies $P_{ij} > 0$. (If the population is in Hardy–Weinberg equilibrium, then it is clearly sufficient that $p_i > 0$ for every i.) The proof is a simplified version of that given by Lessard and Castilloux (1995) for fertility selection.

▷ For every genotype ij we define the $\ell_1 + \cdots + \ell_\ell$-dimensional (column) vector $\boldsymbol{z}_{ij} = (z^{(k)}_{ij,s_k})$, where $s_k = 1, \ldots, \ell_k$ and $k = 1, \ldots, \ell$, by

$$z^{(k)}_{ij,s_k} = \begin{cases} 1, & \text{if } i_k = j_k = s_k \,, \\ \frac{1}{2}, & \text{if } i_k = s_k \text{ or } j_k = s_k, \text{ and } i_k \neq j_k \,, \\ 0, & \text{otherwise} \,. \end{cases} \tag{3.51}$$

Then, for every k, i, and j, $\sum_{s_k=1}^{\ell_k} z^{(k)}_{ij,s_k} = 1$. A simple calculation shows that the $(\ell_1 + \cdots + \ell_\ell) \times (\ell_1 + \cdots + \ell_\ell)$ matrix

$$\mathsf{H} = \sum_{i,j} P_{ij} \boldsymbol{z}_{ij} \boldsymbol{z}_{ij}^\top \tag{3.52}$$

is equal to the matrix H defined in (3.16).

Let us determine the nullspace of H, i.e., the space of all $\boldsymbol{\beta}$ that satisfy

$$\mathsf{H}\boldsymbol{\beta} = \mathbf{0} \,. \tag{3.53}$$

Clearly, if $\boldsymbol{\beta}$ is a solution of (3.53), then

$$\sum_{i,j} P_{ij} (\boldsymbol{\beta}^\top \boldsymbol{z}_{ij})(\boldsymbol{z}_{ij}^\top \boldsymbol{\beta}) = \boldsymbol{\beta}^\top \mathsf{H} \boldsymbol{\beta} = \mathbf{0} \,. \tag{3.54}$$

(Actually, (3.54) is equivalent to (3.53), because H is nonnegative definite and symmetric.) This implies

$$\boldsymbol{z}_{ij}^\top \boldsymbol{\beta} = \sum_k \sum_{s_k} z^{(k)}_{ij,s_k} \beta^{(k)}_{s_k} = 0 \quad \text{if } P_{ij} > 0 \,. \tag{3.55}$$

For any pair of alleles at some locus k, $\mathcal{A}^{(k)}_{u_k}$ and $\mathcal{A}^{(k)}_{v_k}$, with $u_k \neq v_k$, choose two genotypes, ij and rs, that are homozygous for $\mathcal{A}^{(k)}_{u_k}$ and $\mathcal{A}^{(k)}_{v_k}$, respectively, but share all other alleles and occur with positive frequencies, $P_{ij} > 0$ and $P_{rs} > 0$. Formally, this means $z^{(k)}_{ij,u_k} = z^{(k)}_{rs,v_k} = 1$, $z^{(k)}_{ij,v_k} = z^{(k)}_{rs,u_k} = 0$, and $z^{(l)}_{ij,t_l} = z^{(l)}_{rs,t_l}$ for all other loci l. Then we conclude from (3.55)

$$\beta^{(k)}_{u_k} - \beta^{(k)}_{v_k} = \boldsymbol{z}_{ij}^\top \boldsymbol{\beta} - \boldsymbol{z}_{rs}^\top \boldsymbol{\beta} = 0 \,. \tag{3.56}$$

It follows that for each k a constant β_k exists such that $\beta^{(k)}_{u_k} = \beta_k$ for every u_k. Hence, we obtain from the definition of H that $\mathsf{H}\boldsymbol{\beta} = (\sum_k \beta_k)\boldsymbol{\pi}$ which, by (3.53), implies $\sum_k \beta_k = 0$. Therefore, the nullspace of H is given by

$$N_{\mathsf{H}} = \{\boldsymbol{\xi} \in \mathbf{R}^N : \boldsymbol{\xi} = (\boldsymbol{\xi}^{(1)}, \ldots, \boldsymbol{\xi}^{(\ell)})^\top, \\ \text{where } \boldsymbol{\xi}^{(k)} = \xi_k (1, \ldots, 1)^\top \text{ and } \sum\nolimits_k \xi_k = 0\} \,. \tag{3.57}$$

As a consequence, the image space of H, which is the subspace perpendicular to N_{H}, contains all vectors of allele-frequency changes $\boldsymbol{d}$, i.e., those satisfying $\sum_{i_k} d^{(k)}_{i_k} = 0$ for every k. In particular, it contains all vectors $\boldsymbol{\pi} \cdot \boldsymbol{g}$. Since H is invertible (and positive definite) on its image space, the solution of (3.18) that satsifies the constraint (3.12) is uniquely determined. ◁

In general, the weaker assumption of positive allele frequencies does not ensure uniqueness of average effects. A simple example is provided by the two-locus two-allele model with random mating and gamete frequencies $x_1 = x_4 = \frac{1}{2}$ and $x_2 = x_3 = 0$.

Next, we derive some identities that greatly facilitate several calculations involving average effects. Let $\boldsymbol{u}$ denote the $\ell_1 + \cdots + \ell_\ell$-dimensional vector with components $u_{i_k}^{(k)}$ and let $u_{ij} = \sum_k \big(u_{i_k}^{(k)} + u_{j_k}^{(k)}\big)$. Then, using (2.2), we derive for arbitrary v_{ij}

$$\sum_{i,j} P_{ij} v_{ij} u_{ij} = 2 \sum_{i,j} P_{ij} v_{ij} \sum_k u_{i_k}^{(k)} = 2 \sum_k \sum_{i_k} u_{i_k}^{(k)} \sum_{i \nmid i_k} \sum_j P_{ij} v_{ij} \,. \tag{3.58}$$

From the definition of $\boldsymbol{z}_{ij}$, we obtain $u_{ij} = \boldsymbol{z}_{ij}^\top \boldsymbol{u} = \boldsymbol{u}^\top \boldsymbol{z}_{ij}$. This, together with (3.52) implies

$$(\mathsf{H}\boldsymbol{u})_{s_k}^{(k)} = \sum_{i,j} P_{ij} z_{ij,s_k}^{(k)} u_{ij} = \sum_{i \nmid s_k} \sum_j P_{ij} u_{ij} \,. \tag{3.59}$$

Therefore, if $\boldsymbol{\gamma}$ is a vector of average effects that satisfies (3.18), we find, by substituting (3.59) into (3.58),

$$\begin{aligned} \sum_{i,j} P_{ij} \gamma_{ij} u_{ij} &= 2 \sum_k \sum_{i_k} u_{i_k}^{(k)} (\mathsf{H}\boldsymbol{\gamma})_{i_k}^{(k)} \\ &= 2\boldsymbol{u}^\top \mathsf{H} \boldsymbol{\gamma} && \text{(3.60a)} \\ &= 2 \sum_k \sum_{i_k} p_{i_k}^{(k)} g_{i_k}^{(k)} u_{i_k}^{(k)} \,. && \text{(3.60b)} \end{aligned}$$

If Hardy–Weinberg proportions are assumed and if $u_i = \sum_k u_{i_k}^{(k)}$, then (3.60) reduces to

$$\sum_i p_i \gamma_i u_i = \boldsymbol{u}^\top \mathsf{H} \boldsymbol{\gamma} = \sum_k \sum_{i_k} p_{i_k}^{(k)} g_{i_k}^{(k)} u_{i_k}^{(k)} \,. \tag{3.61}$$

In generalization of I(10.11), one can define for every allele $\mathcal{A}_{s_k}^{(k)}$ the indicator variable

$$f_{s_k}^{(k)}(i,j) = z_{ij,s_k}^{(k)} \tag{3.62}$$

that assigns to each genotype ij the frequency of a given allele $\mathcal{A}_{s_k}^{(k)}$ (with possible values 0, $\frac{1}{2}$, and 1). Then the expectation of $f_{s_k}^{(k)}$ with respect to the distribution of genotypes is

$$\bar{f}_{s_k}^{(k)} = \sum_{i,j} f_{s_k}^{(k)}(i,j) P_{ij} = p_{s_k}^{(k)} \,. \tag{3.63}$$

In particular, the entries of the matrix H can be expressed as

$$H_{s_k r_l}^{(kl)} = \mathrm{Cov}[f_{s_k}^{(k)}, f_{r_l}^{(l)}] + p_{s_k}^{(k)} p_{r_l}^{(l)} \,, \tag{3.64}$$

which generalizes I(10.13).

▷ Our final aim here is to prove the minimizing principle • **3.2** for the average effects. From (3.21) it follows immediately that

$$\boldsymbol{\gamma}^{\top}(\boldsymbol{\pi}\cdot\boldsymbol{g}) = \tfrac{1}{2}\sigma_{\mathrm{A}}^2\,. \tag{3.65}$$

(Observe that • **3.2** was not used in deriving (3.21).) Since the average effects automatically satisfy the identity (3.65), one can minimize the quadratic form

$$S(\boldsymbol{a}) = \sum_{i,j} P_{ij}\left[g_{ij} - \sum_k \left(a_{i_k}^{(k)} + a_{j_k}^{(k)}\right)\right]^2 \tag{3.66}$$

with respect to the parameters $a_{i_k}^{(k)}$ (that constitute the vector $\boldsymbol{a}$) subject to the constraints (3.12) (with $a_{i_k}^{(k)}$ instead of $\gamma_{i_k}^{(k)}$) and subject to (3.20). In this way, one obtains the same solution $\boldsymbol{a} = \boldsymbol{\gamma}$. (The reader will have noted that $S(\boldsymbol{\gamma}) = \sigma_{\mathrm{R}}^2$.) However, it is easily shown that condition (3.12) can be dropped, provided (3.20) is retained. Therefore, minimization of $S(\boldsymbol{a})$ subject to the constraint (3.20) yields the average effects $\boldsymbol{\gamma}$.

A simple calculation involving (3.4), (3.58) and (3.60) shows that the quadratic form $S(\boldsymbol{a})$ can be rewritten as

$$S(\boldsymbol{a}) = \sigma_G^2 - 4\boldsymbol{a}^{\top}(\boldsymbol{\pi}\cdot\boldsymbol{g}) + 2\boldsymbol{a}^{\top}\mathsf{H}\boldsymbol{a}\,. \tag{3.67}$$

Since σ_G^2 is constant anyway and the second term is constant under the constraint (3.20), the minimization procedure of σ_{E}^2 in Section 3.1 is equivalent to minimizing $\boldsymbol{a}^{\top}\mathsf{H}\boldsymbol{a}$ subject to the condition (3.20). This proves Result • **3.2**. ◁

4. MARGINAL SYSTEMS AND EQUILIBRIUM PROPERTIES

The fitness of an individual depends on a large number of loci, and the fitness of a single-locus genotype depends on the genetic constitution of the remainder of the genome. Typically, however, experiments and observations involve data from only one or few loci contributing to fitness. It is, therefore, of interest to assess the extent to which conclusions about the entire genetic system can be drawn from information about induced marginal fitnesses and frequencies of a subsystem involving only a small number of loci. This line of research was promoted by Ewens and Thomson (1977) and the present section is based on their work. Our main goal is to elucidate the relationship between the dynamic and equilibrium properties of the entire multilocus system, and subsystems consisting of only some of the loci. After introducing the basic notations and concepts, we derive the recursion relations for the marginal frequencies in the subsystem and explore the relation between equilibria in the entire system and equilibria apparent from observations of only some of the loci. Finally, we investigate some properties of the additive genetic variance, in particular, with regard to equilibrium properties. We derive a formula that relates the additive genetic covariance between an arbitrary trait and fitness to its single-locus marginals. The results show that the validity of conclusions about the entire genetic system, drawn from the loci investigated, depends strongly on the amount of linkage disequilibrium.

We assume random mating. Let S denote a subset of L containing at least one locus and let $T = L \setminus S$ be its complement in L. Given a multi-index i (an ℓ-locus gamete), we write i_S for the multi-index with components i_k for every k in S and call it an S-gamete. Thus, we have $i = i_S i_T$, and the (marginal) frequency of the S-gamete i_S is

$$p_{i_S}^{(S)} = \sum_{i_T} p_i \ , \tag{4.1}$$

where $\sum_{i_T}$ runs over all ℓ-locus multi-indices i with the components in S fixed as i_S. It may be noted that if S contains the single locus k, then (4.1) reduces to (2.1). If $S = L$, then $p_{i_L}^{(L)} = p_i$. We define the *induced*, or *marginal, fitness* $W_{i_S j_S}^{(S)}$ of the diploid genotype formed by the S-gametes i_S and j_S, by averaging over all genotypic combinations making up these two S-gametes, weighted appropriately by fitnesses and frequencies, i.e.,

$$p_{i_S}^{(S)} p_{j_S}^{(S)} W_{i_S j_S}^{(S)} = \sum_{i_T, j_T} W_{ij} p_i p_j \ . \tag{4.2}$$

The marginal fitness of gamete i_S is

$$W_{i_S}^{(S)} = \sum_{j_S} W_{i_S j_S}^{(S)} p_{j_S}^{(S)} \tag{4.3a}$$

$$= \frac{1}{p_{i_S}^{(S)}} \sum_{i_T} \sum_{j} W_{ij} p_i p_j \tag{4.3b}$$

$$= \frac{1}{p_{i_S}^{(S)}} \sum_{i_T} W_i p_i \tag{4.3c}$$

(provided $p_{i_S}^{(S)} > 0$, because otherwise $W_{i_S}^{(S)}$ remains undefined), and the mean fitness of the population can be written as

$$\overline{W} = \sum_{i_S} p_{i_S}^{(S)} W_{i_S}^{(S)} = \sum_{i_S, j_S} p_{i_S}^{(S)} p_{j_S}^{(S)} W_{i_S j_S}^{(S)} \ . \tag{4.4}$$

This shows that the mean fitness, as computed from marginal S-gamete frequencies and fitnesses, is identical to the true ℓ-locus mean fitness. Further, we define the average excesses in fitness of S-gametes and S-genotypes by

$$w_{i_S}^{(S)} = W_{i_S}^{(S)} - \overline{W} \ , \tag{4.5a}$$

$$w_{i_S j_S}^{(S)} = W_{i_S j_S}^{(S)} - \overline{W} \ . \tag{4.5b}$$

It is well known (cf. Section 2.1) that if fitness is determined by more than one locus, then mean fitness may decrease. Therefore, (4.4) demonstrates that the apparent one-locus mean fitness in an ℓ-locus system, being identical to the ℓ-locus mean fitness, can decrease, whereas this is impossible in a genuine one-locus system. The mean fitness can decrease even if the allele frequencies at the locus under consideration are constant. This may happen if the population is in linkage disequilibrium, but $\sigma_{\mathrm{A}}^2 = 0$.

Our first objective is to derive the recursion relations for S-gamete frequencies.

▷ For the subset S, we define the linkage disequilibria

$$\Theta_{i_S}^{(S)} = \sum_{i_T} \Theta_i \,, \tag{4.6}$$

where $T = L \setminus S$. We must also describe recombination in the subset S. For some proper, nonempty subset $Q \subset S$ we define the set of sets

$$L_Q^{(S)} = \{V : V = Q \cup U \text{ where } U \subseteq L \setminus S\} \,. \tag{4.7}$$

Then the probability of reassociation of the genes at the loci in Q, inherited from one parent, with the genes at the loci in $S \setminus Q$, inherited from the other, is

$$r_Q^{(S)} = \sum_{V : V \subset L_Q^{(S)}} r_V \,, \tag{4.8}$$

where the summation is over all decompositions. In analogy with (2.17), the total frequency of recombination among the loci in S is

$$r_{\text{tot}}^{(S)} = \sum_{Q:Q\subset S} r_Q^{(S)} \tag{4.9}$$

(cf. Nagylaki 1993). We note that if S consists of two loci, (4.7) reduces to (2.6) and (4.8) to (2.7). If there is no position effect, then a straightforward calculation shows that $\Theta_{i_k i_l}^{(kl)} = r(W_{i_k i_l}^{(kl)}/\overline{W})D_{i_k i_l}^{(kl)}$, where $D_{i_k i_l}^{(kl)}$ is defined in (3.46)

Now we are able to prove

$$\Theta_{i_S}^{(S)} = \frac{1}{\overline{W}} \sum_{j_S} \sum_{Q} r_Q^{(S)} (W_{i_S j_S}^{(S)} p_{i_S}^{(S)} p_{j_S}^{(S)} - W_{i_Q j_R, j_Q i_R}^{(S)} p_{i_Q j_R}^{(S)} p_{j_Q i_R}^{(S)}) \,, \tag{4.10}$$

where Q is as above and $R = S \setminus Q$. The reader may note the formal analogy between these induced linkage disequilibria and those defined in (2.13). Appealing successively to (4.3b) and (4.3a), we obtain from (2.13)

$$\begin{aligned}\overline{W} \sum_{i_T} \Theta_i &= \sum_I r_I [p_{i_S}^{(S)} W_{i_S}^{(S)} - \tilde{W}_{i_S}^{(S)}(I)] \\ &= \sum_I r_I \left[\sum_{j_S} p_{i_S}^{(S)} p_{j_S}^{(S)} W_{i_S j_S}^{(S)} - \tilde{W}_{i_S}^{(S)}(I) \right] \,, \end{aligned} \tag{4.11}$$

where

$$\tilde{W}_{i_S}^{(S)}(I) = \sum_{i_T} \sum_{j} W_{i_I j_J, j_I i_J}^{(S)} p_{i_I j_J}^{(S)} p_{j_I i_J}^{(S)} \,. \tag{4.12}$$

Now, we decompose I and J as $I = (I \cap S) \cup (I \cap T)$ and $J = (J \cap S) \cup (J \cap T)$, respectively. Decomposing the sums in (4.12) correspondingly and employing (4.2), we get

$$\begin{aligned}\tilde{W}_{i_S}^{(S)}(I) &= \sum_{i_{I\cap T}} \sum_{i_{J\cap T}} \sum_{j_{I\cap T}} \sum_{j_{J\cap T}} \sum_{j_{I\cap S}} \sum_{j_{J\cap S}} W_{i_I j_J, j_I i_J}^{(S)} p_{i_I j_J}^{(S)} p_{j_I i_J}^{(S)} \\ &= \sum_{j_{I\cap S}} \sum_{j_{J\cap S}} W_{i_{I\cap S}\, j_{J\cap S},\, j_{I\cap S}\, i_{J\cap S}}^{(S)} p_{i_{I\cap S}\, j_{J\cap S}}^{(S)} p_{j_{I\cap S}\, i_{J\cap S}}^{(S)} \,. \end{aligned} \tag{4.13}$$

Next, we observe that if $S \subseteq I$, then $J \cap S = \emptyset$; hence, $i_{I\cap S}j_{J\cap S} = i_S$ and $j_{I\cap S}i_{J\cap S} = j_S$. A similar argument applies if $S \subseteq J$. In each of these cases, $\tilde{W}^{(S)}_{i_S}(I)$ reduces to

$$\tilde{W}^{(S)}_{i_S}(I) = \sum_{j_S} p^{(S)}_{i_S} p^{(S)}_{j_S} W^{(S)}_{i_S j_S} \tag{4.14}$$

and, therefore, all terms on the right-hand side of (4.11) vanish that involve an $I \supseteq S$. It follows at once that the single-locus 'linkage disequilibria' $\Theta^{(k)}_{i_k}$ are zero,

$$\Theta^{(k)}_{i_k} = \sum_{i \nmid i_k} \Theta_i = 0\ . \tag{4.15}$$

To complete the proof of (4.10), we observe that every I such that $I \cap S \neq \emptyset$ and $I \cap T \neq \emptyset$ (for these I, (4.14) is not satisfied) induces a decomposition of S, i.e., $S = Q \cup R$ and $Q \cap R = \emptyset$, if $Q = I \cap S$ and $R = J \cap S$ is chosen. Since for given Q, every set I in $L^{(S)}_Q$ induces the same decomposition of S, we arrive at (4.10) by substituting (4.13) into (4.11) and observing (4.8) and (4.6). ◁

On account of (4.1), (4.3c), and (4.10), summation of (2.12) over all gametes containing i_S yields the recursion relations

$$p^{(S)\prime}_{i_S} = p^{(S)}_{i_S} \frac{W^{(S)}_{i_S}}{\overline{W}} - \Theta^{(S)}_{i_S} \tag{4.16}$$

for the marginal frequencies of S-gametes. Because of the equivalence of (2.12) and (2.8), these equations are equivalent to those of Ewens and Thomson (1977), who proved that the formal analogy of (2.8), obtained by substituting marginal fitnesses and frequencies for the original ones, indeed gives the set of recursion relations for the loci in S. If S consists of the single locus k, then (4.15) implies that (4.16) reduces to

$$p^{(k)\prime}_{i_k} = p^{(k)}_{i_k} W^{(k)}_{i_k} / \overline{W} = \sum_{i \nmid i_k} p_i W_i / \overline{W}\ . \tag{4.17}$$

The formal similarity of the full recursion relations (2.12) with the induced recursion relations (4.16) for the loci in S should not obscure the fact that normally the marginal fitnesses $W^{(S)}_{i_S j_S}$, unlike the W_{ij}, change from generation to generation and, therefore, (4.16) as well as (4.17) predict the S-gamete frequencies only one generation in advance. For long-term predictions, the full system (2.12) must be used.

We are now able to draw a number of interesting conclusions (cf. Ewens 1976, Ewens and Thomson 1977).

• **4.1** *If the entire ℓ-locus system is in equilibrium, then so is any marginal subsystem and the recursion relations of the induced system are indeed at equilibrium, i.e.,*

$$\Delta p^{(S)}_{i_S} = p^{(S)}_{i_S} w^{(S)}_{i_S} / \overline{W} = 0 \quad \textit{for every subset } S \textit{ of } L\ . \tag{4.18}$$

This result is a direct consequence of (4.1), (4.6), and the structural similarity of (2.12) and (4.16). The converse is, of course, not true. Even if all single-locus systems are at an (apparent) equilibrium, as implied by constant allele frequencies,

$$\Delta p^{(k)}_{i_k} = p^{(k)}_{i_k} w^{(k)}_{i_k} / \overline{W} = 0 \quad \text{for every } k \text{ and } i_k\ , \tag{4.19}$$

the full system is not necessarily in equilibrium because linkage disequilibria may still change. An explicit example in a two-locus two-allele system was given by Moran (1964).

Another interesting question concerns the relation of the stability properties of equilibria of the entire system to the stability properties apparent from its marginal systems. Surprisingly, no general results are achievable. For two loci with two alleles at each, Ewens and Thomson (1977) presented a symmetric fitness matrix of the form (1.22) with $a = d$ for which an unstable internal equilibrium exists, that not only displays fitness superiority of the double heterozygote, but exhibits marginal overdominance at both loci. The opposite conjecture that at a stable internal equilibrium all the reduced systems with corresponding marginal fitnesses must appear to be at a stable equilibrium was disproved by Hastings (1981b, 1982). He gave an example of a stable internal equilibrium in a two-locus two-allele model that exhibits marginal underdominance at both loci. By contrast, Karlin (1978) showed that in the presence of a stable internal equilibrium, marginal underdominance cannot occur if recombination rates are sufficiently small.

Also of evolutionary interest is the stability of boundary equilibria against the introduction of new alleles. This may occur, for instance, by mutation or immigration. A boundary equilibrium is called *externally stable* if an allele that is absent at the equilibrium is lost after being introduced at low frequencies (see Chapter I.9.5 and Appendix A.2). If it can be established, then the equilibrium is externally unstable. For completely linked loci, a boundary equilibrium is externally stable if, at this equilibrium, the marginal fitness of the new type is lower than the mean fitness of the population. This condition can be extended to the case of recombination (Liberman 1988, Christiansen 1999). The interesting aspect here is that recombination enhances external stability, i.e., the condition that allows for invasion becomes more stringent as the level of recombination increases.

For our next result, we define the (marginal) single-locus additive genetic variance in fitness for locus k by

$$\sigma_{\mathrm{A}}^{2\,(k)} = 2\sum_{i_k} p_{i_k}^{(k)} \left(\alpha_{i_k}^{(k)}\right)^2 . \tag{4.20}$$

This is in accordance with I(3.12) and (3.49).

• **4.2** *If the entire system is at an equilibrium, then the additive genetic variance in fitness is zero, $\sigma_{\mathrm{A}}^2 = 0$, and all marginal additive genetic variances, as computed from marginal fitnesses, are also zero. In contrast, if the additive genetic variance in fitness of the entire system is zero, it does not follow that the system is at an equilibrium. It follows, however, that the allele frequencies remain constant for at least one generation.*

▷ As (4.19) holds at equilibrium, it follows from (3.29) and (3.30), together with the uniqueness of average effects subject to (3.12), that (4.19) is equivalent to

$$p_{i_k}^{(k)} \alpha_{i_k}^{(k)} = 0 \text{ for every } k \text{ and } i_k . \tag{4.21}$$

(It is not assumed that $p_{i_k}^{(k)} > 0$; moreover, if the average effects are not uniquely determined, because not all gamete frequencies are positive, then one can still choose $\alpha_{i_k}^{(k)} = w_{i_k}^{(k)} = 0$.) The equivalence of (4.19) and (4.21) is remarkable, because in linkage disequilibrium average effects will generally differ from average excesses. This equivalence ensures that at equilibrium $\sigma_A^2 = 0$, and all single-locus additive variances vanish, i.e., $\sigma_A^{2\,(k)} = 0$ for every k. A similar argument applies to the marginal variances of other subsystems. Conversely, $\sigma_A^2 = 0$ implies (4.19), but not that the full system is in equilibrium because linkage disequilibria may still change. ◁

Finally, we wish to relate the true additive genetic variance σ_A^2 of the ℓ-locus system to the single-locus marginal additive genetic variances, $\sigma_A^{2\,(k)}$. More generally, as in (3.42) let $\mathrm{Cov_A}(G,W)$ denote the additive genetic covariance of a character G and fitness W, and let $\mathrm{Cov}_A^{(k)}(G,W)$ denote the corresponding single-locus marginals based on the marginal measurements of G and W, i.e.,

$$\mathrm{Cov}_A^{(k)}(G,W) = 2\sum_{i_k} p_{i_k}^{(k)} \gamma_{i_k}^{(k)} \alpha_{i_k}^{(k)} \; ; \tag{4.22}$$

cf. (3.50). Now we deduce from (4.22) and (3.47)

$$\mathrm{Cov}_A^{(k)}(G,W) = 2\sum_{i_k} p_{i_k}^{(k)} g_{i_k}^{(k)} \alpha_{i_k}^{(k)} - 2\sum_{i_k} \alpha_{i_k}^{(k)} \sum_{l \neq k} \sum_{i_l} \gamma_{i_l}^{(l)} D_{i_k i_l}^{(kl)} \,, \tag{4.23}$$

where the $D_{i_k i_l}^{(kl)}$ denote the pairwise linkage disequilibria defined in (3.46). Summing (4.23) over all loci k and using (3.44b), we obtain

$$\mathrm{Cov_A}(G,W) = \sum_k \mathrm{Cov}_A^{(k)}(G,W) + 4 \sum_{k,l:l<k} \sum_{i_k} \sum_{i_l} \alpha_{i_k}^{(k)} \gamma_{i_l}^{(l)} D_{i_k i_l}^{(kl)} \,. \tag{4.24}$$

For the variance of an arbitrary character G, this produces

$$\sigma_A^2 = \sigma_{A,LE}^2 + 4 \sum_{k,l:l<k} \sum_{i_k} \sum_{i_l} \gamma_{i_k}^{(k)} \gamma_{i_l}^{(l)} D_{i_k i_l}^{(kl)} \,, \tag{4.25}$$

where

$$\sigma_{A,LE}^2 = \sum_k \sigma_A^{2\,(k)} \tag{4.26}$$

is the sum of the single-locus additive genetic variances, or the *linkage-equilibrium variance*. Bulmer (1980) calls $\sigma_{A,LE}^2$ the genic variance. With two alleles at each locus and an additive character, so that ε_i in (3.35) vanishes, (4.25) was proved by Avery and Hill (1978). For two diallelic loci and additive fitnesses, (4.25) simplifies to (1.14). A simple consequence of (4.24) and (4.25) is that the true additive genetic (co)variance can be found by summing single-locus marginal genetic (co)variances if all pairwise linkage disequilibria (3.46) are zero. In this case, (4.25) simplifies to (3.49).

5. EVOLUTION IN THE ABSENCE OF EPISTASIS

The usage of the notion of epistasis is ambiguous in population genetics. It refers to the presence of interactions between genes at different loci. In its broadest sense, absence of epistasis means that a multilocus genotype's contribution to some character (often fitness) is representable as a function of independent single-locus genotypic effects. Otherwise, epistasis is said to occur. Multilocus systems without epistasis are easier to study than general systems and more specific results are obtainable. In this section, we first deal with the selection response of an arbitrary character that is determined additively by the single-locus genotypic values (absence of epistasis of effects on the trait). Then we consider two models in which there is no epistasis in fitness. Throughout, random mating is assumed.

5.1 ADDITIVE CHARACTERS

Here, we derive the rate of change in the mean, $\overline{G}$, of an additive character under selection. More precisely, we suppose that the loci contribute additively to the genotypic value G of the character, but admit dominance of effects, i.e.,

$$G_{ij} = \sum_{k=1}^{\ell} X^{(k)}_{i_k j_k} \ . \tag{5.1}$$

We define the single-locus means and the additive effects of single-locus genotypes and of alleles by

$$\overline{X}^{(k)} = \sum_{i_k, j_k} X^{(k)}_{i_k j_k} p^{(k)}_{i_k} p^{(k)}_{j_k} \ , \tag{5.2a}$$

$$x^{(k)}_{i_k j_k} = X^{(k)}_{i_k j_k} - \overline{X}^{(k)} \ , \tag{5.2b}$$

$$x^{(k)}_{i_k} = \sum_{j_k} x^{(k)}_{i_k j_k} p^{(k)}_{j_k} \ , \tag{5.2c}$$

respectively. It follows that $\overline{G} = \sum_k \overline{X}^{(k)}$ and

$$g_{ij} = \sum_k x^{(k)}_{i_k j_k} \ . \tag{5.3}$$

By a calculation similar (but simpler) than that leading to (3.13) and (3.32), we obtain

$$p^{(k)}_{i_k} g^{(k)}_{i_k} = p^{(k)}_{i_k} x^{(k)}_{i_k} + \sum_{l:l\neq k} \sum_{i_l} x^{(l)}_{i_l} p^{(kl)}_{i_k i_l} \ . \tag{5.4}$$

This implies that the average, or additive, effects of alleles are indeed given by

$$\gamma^{(k)}_{i_k} = x^{(k)}_{i_k} \ . \tag{5.5}$$

Therefore, (5.5), (3.35) and (3.5b) yield $g_i = \gamma_i = x_i = \sum_k x^{(k)}_{i_k}$. (Note that, because we admit dominance, $x^{(k)}_{i_k j_k} \neq x^{(k)}_{i_k} + x^{(k)}_{j_k}$.) With (5.5), the additive genetic covariance of G and W (3.44d) can be written as

$$\mathrm{Cov_A}(G, W) = 2 \sum_k \sum_{i_k} p^{(k)}_{i_k} w^{(k)}_{i_k} x^{(k)}_{i_k} \ , \tag{5.6}$$

where $w_{i_k}^{(k)}$ is the average excess in fitness of allele $\mathcal{A}_{i_k}^{(k)}$ (3.24c).

We wish to emphasize here that even for an additively determined character, the average excess $g_{i_k}^{(k)}$ and the average effect $\gamma_{i_k}^{(k)} = x_{i_k}^{(k)}$ of an allele $\mathcal{A}_{i_k}^{(k)}$ are in general different. If all pairwise linkage disequilibria between a locus k and all other loci vanish, then average excess and average effect of alleles at this locus coincide (cf. Section 3.5).

The change in the mean of the genotypic value is obtained by successively appealing to (3.2c) and Hardy–Weinberg proportions; (3.3a), $\sum_i \Delta p_i = 0$; (5.3), (2.2), (2.1); and (5.2c), (4.17), (5.6),

$$\begin{aligned}\Delta \overline{G} &= \sum_{i,j} G_{ij} \Delta p_i (2p_j + \Delta p_j) \\ &= \sum_{i,j} g_{ij} \Delta p_i (2p_j + \Delta p_j) \\ &= \sum_k \sum_{i_k, j_k} x_{i_k j_k}^{(k)} \Delta p_{i_k}^{(k)} (2p_{j_k}^{(k)} + \Delta p_{j_k}^{(k)}) \\ &= \overline{W}^{-1} \operatorname{Cov_A}(G, W) + \overline{W}^{-2} B \,, \end{aligned} \tag{5.7}$$

where

$$B = \sum_k \sum_{i_k, j_k} x_{i_k j_k}^{(k)} w_{i_k}^{(k)} w_{j_k}^{(k)} p_{i_k}^{(k)} p_{j_k}^{(k)} \,. \tag{5.8}$$

In this generality, (5.7) was first derived by Nagylaki (1989b). An important point to note is that $\overline{G}$ can change only if allele frequencies change, but not because of changes in linkage disequilibria. This is an important but special property of additively determined characters.

If there is no dominance, i.e., if

$$x_{i_k j_k}^{(k)} = x_{i_k}^{(k)} + x_{j_k}^{(k)} \,, \tag{5.9}$$

then, due to (3.12) and (5.5), $B = 0$ and the remarkably simple formula

$$\Delta \overline{G} = \operatorname{Cov_A}(G, W) / \overline{W} \tag{5.10}$$

is obtained. In this case, additive genetic (co)variance and total genetic (co)variance coincide (cf. Sections 3.3, 3.4). Robertson (1966, 1968), who found (5.10) on the basis of linear regression arguments, called this the *Secondary Theorem of Natural Selection.* For other derivations or formulations, we refer to Lande (1976), Turelli and Barton (1990), Nagylaki (1992a), and to Chapter V. A special case of (5.10) is the Li–Price covariance formula, whose continuous-time version was derived in I(10.16). Its discrete-time version is obtained from (5.10) by setting $G_{ij} = f_{s_k}^{(k)}(i,j)$; cf. (3.62), (3.63).

5.2 ADDITIVE FITNESSES

Here we generalize the results of Section 1.1 to multiple loci by assigning to each single-locus genotype $\mathcal{A}_{i_k}^{(k)} \mathcal{A}_{j_k}^{(k)}$ the fitness coefficient $A_{i_k j_k}^{(k)}$, and to the genotype ij the fitness

$$W_{ij} = \sum_{k=1}^{\ell} A_{i_k j_k}^{(k)} \,. \tag{5.11}$$

Thus, we assume additive fitness accross loci (i.e., absence of epistasis) but admit dominance. We note that in general the assumption of additive fitnesses is much more restrictive than that of an additive character, because for an additive character fitness is usually not additive (cf. Chapter V). Obviously, with additive fitnesses there is no position effect. In view of (4.4) and (5.11), the mean fitness $\overline{W}$ is the sum of its single-locus contributions, $\overline{A}^{(k)}$:

$$\overline{W} = \sum_k \overline{A}^{(k)} , \tag{5.12}$$

where

$$\overline{A}^{(k)} = \sum_{i_k, j_k} A^{(k)}_{i_k j_k} p^{(k)}_{i_k} p^{(k)}_{j_k} . \tag{5.13}$$

Denoting the marginal fitness effect of allele $\mathcal{A}^{(k)}_{i_k}$ by

$$A^{(k)}_{i_k} = \sum_{j_k} A^{(k)}_{i_k j_k} p^{(k)}_{j_k} , \tag{5.14}$$

we find from (5.5) that the average effect of allele $\mathcal{A}^{(k)}_{i_k}$ on fitness is

$$\alpha^{(k)}_{i_k} = A^{(k)}_{i_k} - \overline{A}^{(k)} . \tag{5.15}$$

The following result is due to Lyubich (1992, Theorem 9.6.13) and Nagylaki *et al.* (1999).

• **5.1** *A point* $\mathbf{p}$ *is an equilibrium of the multilocus dynamics* (2.12) *if and only if it is both a selection equilibrium for each locus* (*with fitnesses given by the* $A^{(k)}_{i_k j_k}$) *and it is in linkage equilibrium.*

In the above sense, a locus k is in equilibrium with respect to selection if

$$p^{(k)}_{i_k} \alpha^{(k)}_{i_k} = p^{(k)}_{i_k} (A^{(k)}_{i_k} - \overline{A}^{(k)}) = 0 \quad \text{for every } i_k . \tag{5.16}$$

Since in linkage equilibrium we have $\alpha^{(k)}_{i_k} = w^{(k)}_{i_k}$ [cf. (3.48)], (4.19) implies that (5.16) is equivalent to $\Delta p^{(k)}_{i_k} = 0$ (compare also • **4.2** and its proof).

Result • **5.1** shows that all equilibria are product equilibria, i.e., they are of the form

$$\hat{p}_i = \hat{p}^{(1)}_{i_1} \cdot \ldots \cdot \hat{p}^{(\ell)}_{i_\ell} , \tag{5.17}$$

where $\hat{p}^{(k)}_{i_k}$ designates the equilibrium frequency of allele $\mathcal{A}^{(k)}_{i_k}$ (compare Karlin and Liberman 1979, who called such equilibria Hardy–Weinberg equilibria, and Eq. 1.16). We now summarize some further interesting results:

• **5.2** *Suppose that fitnesses are additive. Then:*

1. *Every trajectory of the multilocus dynamics* (2.12) *converges to some equilibrium point.*
2. *The overdominance principle is valid, i.e., if each single-locus polymorphism* $\hat{\mathbf{p}}^{(k)}$ *is asymptotically stable, then the product equilibrium* (5.17) *is asymptotically stable for arbitrary but strictly positive recombination rates.*

3. *This product equilibrium is globally stable when each of the loci is diallelic or if there are only two (multiallelic) loci.*

The proofs are quite technical and we refer to Lyubich (1992) for the proof of 1., and to Karlin and Liberman (1978, 1979, 1990) for the proofs of 2. and 3.

Finally, let us investigate the validity of the Fundamental Theorem of Natural Selection in the absence of epistasis. From (5.7) we obtain that the change in mean fitness is

$$\Delta\overline{W} = \frac{\sigma_{\mathrm{A}}^2}{\overline{W}}(1+E) , \tag{5.18}$$

where σ_{A}^2 is the additive genetic variance in fitness (3.31), and

$$E = B/(\overline{W}\sigma_{\mathrm{A}}^2) \tag{5.19}$$

is the relative error with B defined as in (5.8). Therefore, $\Delta\overline{W}$ is independent of the linkage map and, assuming complete linkage, the one-locus results apply. Consequently, $\Delta\overline{W} \geq 0$ and equality holds if and only if

$$p_i w_i = p_i(W_i - \overline{W}) = 0 \quad \text{for every } i . \tag{5.20}$$

Hence, mean fitness is nondecreasing for any linkage map, as first demonstrated by Ewens (1969a,b).

If dominance is absent, then $E = 0$ and the simple and exact formula

$$\Delta\overline{W} = \sigma_{\mathrm{A}}^2/\overline{W} \tag{5.21}$$

is obtained. In general, the relative error E can be estimated by (Nagylaki 1991)

$$|E| \leq \tfrac{1}{2}s , \quad s = (\max_{i,j} W_{ij} - \min_{i,j} W_{ij})/\min_{i,j} W_{ij} . \tag{5.22}$$

Here, s signifies the greatest multilocus selection coefficient. If the genotypic fitnesses are scaled such that $\max_{i,j} W_{ij} = 1$ and if selection is weak, then (5.18) and (5.22) yield the *Asymptotic Fundamental Theorem of Natural Selection:*

$$\Delta\overline{W} = \sigma_{\mathrm{A}}^2 + O(s^3) \tag{5.23}$$

(Nagylaki 1989b, 1991). Since $\sigma_{\mathrm{A}}^2 = O(s^2)$ as $s \to 0$, (5.23) implies that σ_{A}^2 provides a good approximation to $\Delta\overline{W}$ if selection is weak (see also the discussion at the end of Section 6.4).

5.3 MULTIPLICATIVE FITNESSES

In the multiplicative fitness model, fitnesses of genotypes are given by

$$W_{ij} = \prod_{k=1}^{\ell} A_{i_k j_k}^{(k)} . \tag{5.24}$$

In Section 1.2, we have already noted that for this kind of nonepistatic selection, Fisher's Fundamental Theorem of Natural Selection fails and that multiple asymptotically stable polymorphic equilibria may coexist. However, in contrast to the additive fitness model, any population that initially is in linkage equilibrium remains in linkage equilibrium in future generations (Moran 1967; Karlin 1975; Nagylaki 1992, Problems 8.2 and 8.10).

More generally, let S and T be a decomposition of the set L of loci such that recombination in S occurs independently of recombination in T. This will be the case, for instance, if the loci in S and T do not overlap on the chromosome. If there is linkage equilibrium between S and T, i.e., if $p_i = p_{i_S}^{(S)} p_{i_T}^{(T)}$ for every gamete i, then linkage equilibrium will be preserved under multiplicative selection, i.e., $p_i' = p_{i_S}^{(S)'} p_{i_T}^{(T)'}$. The proof is straightforward by invoking (2.8), decomposing W_{jk} as

$$W_{jk} = \left(\prod_{s \in S} A_{j_s k_s}^{(s)} \right) \left(\prod_{t \in T} A_{j_t k_t}^{(t)} \right),$$

and observing that $R(j,k \to i) = R(j_S, k_S \to i_S) R(j_T, k_T \to i_T)$, which holds because interference is absent between S and T. Obviously, this property of preserving linkage equilibrium extends to any decomposition of L into sets of noninterfering loci.

Most studies of the multiplicative fitness model investigated the existence and stability properties of the polymorphic product equilibrium (5.17). Here, we summarize the main results (cf. Roux 1974; Karlin 1979; Karlin and Liberman 1979, 1982). Suppose that, for each separate locus k, a polymorphic equilibrium with equilibrium frequency vector $\hat{\boldsymbol{p}}^{(k)} = (\hat{p}_{i_k}^{(k)})$ exists. Then the product equilibrium $\hat{\boldsymbol{p}}$ defined in (5.17) is an equilibrium of the full system for any prescription of recombination frequencies. Conversely, • **4.1** implies that if $\hat{\boldsymbol{p}}$ is a polymorphic product equilibrium, then each $\hat{\boldsymbol{p}}^{(k)}$ is necessarily a polymorphism for the marginal one-locus selection model with fitnesses $A_{i_k j_k}^{(k)}$. Explicit sufficient stability conditions for the product equilibrium have been derived. Qualitatively, the product equilibrium is asymptotically stable if and only if the level of recombination is sufficiently high, but it is never stable under complete linkage.

In addition, the following reduction property holds: assume overdominance at each locus and suppose the recombination rates between adjacent loci are sufficiently high to ensure the two-locus stability of the product equilibrium, i.e., these two-locus recombination rates satisfy (1.20). Then for a large class of recombination distributions, including absence of interference, the product equilibrium is stable in the full multilocus system. Hence, local stability of the full product equilibrium can be inferred from the two-locus stability condition (1.20) applied to each adjacent pair of loci.

Stability conditions for product equilibria on the boundaries have been established by Christiansen (1990); see also Christiansen (1999).

5.4 GENERALIZED NONEPISTATIC FITNESSES

In the generalized nonepistatic selection model of Karlin and Liberman (1979), it is supposed that fitness is a linear combination of various multiplicative, additive, and neutral components. For two loci, the fitness of the genotype $\mathcal{A}_1^{(1)}\mathcal{A}_1^{(2)}/\mathcal{A}_2^{(1)}\mathcal{A}_2^{(2)}$ is assumed to be

$$W_{ij} = d_1 A_{12}^{(1)} A_{12}^{(2)} + d_2 A_{12}^{(1)} + d_3 A_{12}^{(2)} + d_4 \ , \tag{5.25}$$

where $d_1, \ldots, d_4$ are nonnegative constants and $A_{12}^{(i)}$ denotes the fitness of the single-locus genotype $\mathcal{A}_1^{(i)} \mathcal{A}_2^{(i)}$. With $d_2 = d_3 = d_4 = 0$, (5.25) reduces to the multiplicative fitness model, whereas the additive model is recovered if $d_1 = d_4 = 0$.

For this generalized nonepistatic model, the product equilibrium (5.17) exists for any level of recombination, and is polymorphic provided each locus exhibits overdominance. Sufficient conditions for stability of the product equilibrium were derived by Karlin and Liberman (1979). Qualitatively, they showed that the product equilibrium is never stable for complete linkage, but it is asymptotically stable if there is sufficient recombination, in particular, if there is free recombination. Under free recombination, it apparently is uniquely stable. Furthermore, the higher the mix of additive components in the fitness, as compared to multiplicative, the less restrictive are the requirements on recombination that entail stability. As indicated by the two-locus model with multiplicative fitnesses, many stable (polymorphic) equilibria may coexist and, in general, mean fitness is not maximized. A recent treatment of this model may be found in Christiansen (1999).

Although, these and other results demonstrate that specific problems can be solved assuming nonepistatic selection, they also show that even in the absence of epistasis, many of the one-locus results cannot be extended to multiple loci except under the additive model. Many of these complications occur because selection induces linkage disequilibria. Therefore, it is of interest to investigate to what extent deviations from linkage equilibrium occur and to what extent results such as the Fundamental Theorem of Natural Selection fail. Equivalently, we shall explore conditions guaranteeing that populations evolve in 'quasi-linkage equilibrium', and conditions implying that the Fundamental Theorem of Natural Selection holds 'approximately' or 'in most cases'. This will be the subject of the next section.

6. EVOLUTION IN MULTILOCUS SYSTEMS

The theory of two-locus models demonstrates that evolution in multilocus systems under selection may be guided by complex dynamic behavior and that no general and simple results can be expected. This, however, does not preclude the possibility that under certain biologically reasonable assumptions evolution 'mostly' occurs in a simple manner. Indeed, such simple behavior may be expected if populations evolve close to linkage equilibrium conditions, in so-called *Quasi-Linkage Equilibrium* (QLE). A central theme of this section will be Fisher's Fundamental Theorem of Natural Selection (FTNS). In particular, we shall explore assumptions ensuring the approximate validity of the FTNS and convergence to an equilibrium point. Such assumptions turn out to be 'weak epistasis', which is an extension of the concept of additive fitness that allows for small deviations from additivity, and 'weak selection', in the sense that selection coefficients are much smaller than recombination frequencies. We shall also examine the rate of evolution of the mean of a quantitative character under selection, and present a generalized version of Robertson's Secondary Theorem of Natural Selection. Finally, we shall discuss interpretations of the FTNS suggesting that Fisher

(1930, 1958) actually treated only a certain partial change in mean fitness for which his statement is true under very general assumptions. Since (almost) all of the mathematics underlying the results covered in this section is fairly advanced and involved, the exposition will be will restricted to a presentation and discussion of the basic ideas and results.

6.1 WEAK EPISTASIS

Weak epistasis means that genotypic fitnesses have the form

$$W_{ij} = \sum_{k=1}^{\ell} A^{(k)}_{i_k j_k} + \varepsilon e_{ij} \,, \tag{6.1}$$

where ε, the strength of epistasis, is sufficiently small. If $\varepsilon = 0$, then there is no epistasis and (6.1) is reduced to (5.11), i.e., fitnesses are additive between loci. It is assumed that $A^{(k)}_{i_k j_k} > 0$. Evolution of multilocus systems under weak epistasis was studied by Nagylaki *et al.* (1999), and proofs of the results stated below may be found there. Let $\mathcal{F}$ denote the set of points $\boldsymbol{p}$ satisfying (5.20). These are the points for which the selection part of (2.12) is in equilibrium and, therefore, mean fitness is the same in the next generation.

• **6.1** *If for* $\varepsilon = 0$, *each equilibrium of the multilocus dynamics* (2.12) *is hyperbolic (i.e., no eigenvalues of the linear approximation have absolute value one; cf. Appendix A.1), then for all sufficiently small* ε *the following hold:*

1. *Each trajectory of the multilocus selection equation* (2.12) *converges to an equilibrium point.*
2. *Each equilibrium of* (2.12) *is within* $O(\varepsilon)$ *of the corresponding equilibrium with* $\varepsilon = 0$ *and has the same stability properties. In particular,* $\Theta_i = O(\varepsilon)$ *holds at equilibrium for every* i, *and* $\overline{W}$ *differs by* $O(\varepsilon)$ *from its value without epistasis.*
3. *If* $\boldsymbol{p}$ *is bounded away from the set* $\mathcal{F}$, *then* $\Delta\overline{W}(\boldsymbol{p}) > 0$. *If* $\boldsymbol{p}$ *is close to* $\mathcal{F}$, *then* $\Delta\overline{W}(\boldsymbol{p}) < 0$ *can occur for arbitrarily small* $\varepsilon > 0$.

The hyperbolicity assumption in • **6.1** is satisfied generically, i.e., for almost all choices of one-locus fitness matrices $\left(A^{(k)}_{i_k j_k}\right)$. More precisely, this means that the one-locus fitness matrices can be chosen from an open and dense set of full measure of all possible one-locus fitness matrices.

6.2 WEAK SELECTION

Let us write the genotypic fitnesses in the form

$$W_{ij} = 1 + s\omega_{ij} \,, \tag{6.2}$$

where $|\omega_{ij}| \leq 1$ and s is a measure of the strength of selection. We speak of *weak selection*, or *loose linkage*, if s is small compared with all recombination frequencies r_I, i.e., if $s \ll r_{\min}$, the smallest two-locus recombination rate. No further restrictions are imposed on the parameters ω_{ij}. Under the assumption of weak selection, one may expect that linkage disequilibria are small and that, therefore, some of the features of the one-locus theory can be extended to the multilocus setting. This is indeed true, and we shall review the most important results below. These were derived by Nagylaki (1993) and Nagylaki *et al.* (1999), where the proofs may be found along with additional results. For two multiallelic loci most of this theory was developed by Nagylaki (1976, 1977, 1992).

Convergence to quasi-linkage equilibrium

In the absence of selection, the linkage-equilibrium manifold Λ_0 (2.16) is invariant and globally attracting at a uniform geometric rate, provided $r_{\min}$ and, therefore, all recombination frequencies r_I are positive (see Section 2.2). If s is small and $r_{\min} > 0$, then the theory of normally hyperbolic manifolds (see Hirsch *et al.* 1977) implies the existence of a smooth invariant manifold Λ_s close to Λ_0, which is globally attracting at a geometric rate. In the continuous-time model (2.18), this observation is due to C. Conley (see Shahshahani 1979).

For weak selection, this manifold is characterized by equations of the form

$$\Theta_i = s\psi_i(\boldsymbol{\pi}, s) \ , \tag{6.3}$$

where the ψ_i are smooth functions of the vector $\boldsymbol{\pi}$ of allele frequencies (cf. Section 3.1). Hence on Λ_s, the linkage disequilibria Θ_i are of order s. More generally, this holds for any initial values, after a sufficiently long time. Therefore, Λ_s is called the *quasi-linkage equilibrium manifold.* If the population is sufficiently close to the QLE manifold, i.e., if

$$\Theta_i = s\psi_i(\boldsymbol{\pi}, s) + O(s^2) \tag{6.4}$$

for every i, then linkage disequilibria change very slowly,

$$\Delta\Theta_i = O(s^2) \ . \tag{6.5}$$

The concept of quasi-linkage equilibrium was introduced by Kimura (1965b), who observed in the two-locus two-allele model that the function $Z = x_1x_4/x_2x_3 = 1+D/(x_2x_3)$ approaches a nearly constant value when recombination is strong relative to the amount of epistasis (cf. also Crow and Kimura 1970). We shall say that a population is in quasi-linkage equilibrium if in the current and in every subsequent generation, (6.5) holds for every i.

A direct and illuminating proof for the decay of the linkage equilibria, or the approach to QLE, was given by Nagylaki (1993). He proved that (6.5) holds after an evolutionarily short period $t_2 \approx 2(\ln s)/\ln(1-r_{\min})$. This does not necessarily imply that the more restrictive relation (6.4) is valid after the same time span, but, because of the geometric rate of approach to Λ_s, the time required until (6.4) holds will be of the same order of magnitude. We outline the main features of this proof because it provides valuable insights into the evolutionary dynamics of multilocus systems.

▷ The linkage disequilibria Θ_i are rather complicated to handle analytically. Indeed, the proofs of the results below are based on a simpler set of linkage disequilibria that do not involve the fitnesses and that may be of independent interest. Let, as in Section 4, S denote a subset of L containing at least one locus, and let $T = L \setminus S$. We define the linkage disequilibria

$$\theta_{i_S}^{(S)} = p_{i_S}^{(S)} - \tilde{p}_{i_S}^{(S)} \ , \tag{6.6a}$$

where

$$\tilde{p}_{i_S}^{(S)} = \prod_{k \in S} p_{i_k}^{(k)} . \tag{6.7a}$$

If $S = L$, we obtain

$$\tilde{p}_i = \tilde{p}_{i_L}^{(L)} = \prod_{k=1}^{\ell} p_{i_k}^{(k)} \tag{6.7b}$$

and

$$\theta_i = p_i - \tilde{p}_i . \tag{6.6b}$$

These linkage disequilibria satisfy the analog of (4.15) and are related to the Θ_i as follows:

$$\Theta_i = r_{\text{tot}} \theta_i - \sum_I r_I \theta_{i_I}^{(I)} \theta_{i_J}^{(J)} - \sum_{S:S \subset L} r_S \theta_{i_S}^{(S)} \tilde{p}_{i_T}^{(T)} + O(s) , \tag{6.8a}$$

as $s \to 0$, and

$$\Theta_i = r_{\text{tot}} \theta_i - \sum_{S:S \subset L} r_S \theta_{i_S}^{(S)} \tilde{p}_{i_T}^{(T)} + O(s^2) , \; t \geq t_1 , \tag{6.8b}$$

where the summation $\sum_{S:S \subset L}$ is over all proper subsets of L, and t_1 is defined in (6.10).

As a function of time, the linkage disequilibria $\theta_{i_S}^{(S)}$ decay according to

$$\left| \theta_{i_S}^{(S)}(t) \right| \leq \lambda_S^t a_{i_S}^{(S)} + s b_{i_S}^{(S)} , \tag{6.9a}$$

where $a_{i_S}^{(S)}$ and $b_{i_S}^{(S)}$ are constants independent of s and t, and λ_S equals 1 minus the smallest two-locus recombination rate in S. For every subset S of loci, $\lambda_S \leq \lambda_L = 1 - r_{\min}$ holds, where $r_{\min}$ denotes the smallest two-locus recombination rate in L, cf. (2.17). Therefore, λ_L determines the rate of approach to Λ_0 without selection, and to Λ_s with weak selection. Hence, for weak selection and arbitrary initial conditions, after an evolutionary short time t_1 all linkage disequilibria become small,

$$\theta_i(t) = O(s) , \quad t \geq t_1 , \tag{6.9b}$$

where

$$t_1 \sim (\ln s)/\ln(1 - r_{\min}) , \quad \text{as } s \to 0 . \tag{6.10}$$[1]

The reader may note that $-\ln s$ will only rarely exceed 5 to 10 generations. Therefore, t_1 will be considerably longer than $-\ln s$ only if $r_{\min} \ll 1$. By contrast, if the population initially is in linkage equilibrium, then $t_1 = 0$.

During the period $0 \leq t \leq t_1$, the total gene-frequency change is very small, of order st_1, because (4.17) implies

$$\Delta p_{i_k}^{(k)} = O(s) \quad \text{for all } t \geq 0 . \tag{6.11}$$

[1] The symbol $\sim$ is used for asymptotic equivalence.

As a consequence of (6.9b) and (6.11), the gamete frequencies change slowly after t_1 generations have passed, i.e.,

$$\Delta p_i = O(s) \,, \quad t \geq t_1 \,. \tag{6.12}$$

Furthermore, it can be shown for every subset $S \subseteq L$ that

$$\left|\Delta\theta_{i_S}^{(S)}(t)\right| \leq s(\lambda_S^{t-t_1} u_{i_S}^{(S)} + s v_{i_S}^{(S)}) \,, \quad t \geq t_1 \,, \tag{6.13a}$$

where $u_{i_S}^{(S)}$ and $v_{i_S}^{(S)}$ are constants independent of s and t. This implies that after a time $t_2 \approx 2t_1$, the linkage disequilibria remain almost constant,

$$\Delta\theta_i = O(s^2) \,, \quad t \geq t_2 \,. \tag{6.13b}$$

Again, the total gene-frequency change during the period $t_1 < t \leq t_2$ is very small, approximately the same as during the initial period $0 < t \leq t_1$. It is after t_2 generations that QLE has been approached because the Θ_i also satisfy (6.9) and (6.13), provided the position effect (if any) is $O(s^2)$. ◁

Therefore, (6.5) is valid for any population and arbitrary initial conditions after t_2 generations have elapsed. This is an evolutionarily short period of time because t_2 is proportional to $\ln(1/s)$. If the gametic frequencies converge to an equilibrium point, as is generically the case for weak selection (see below), then (6.11) implies that the characteristic convergence time is $t_3 \approx 1/s \gg t_2$. Therefore, most of the gene-frequency change occurs during the period $t_2 < t \leq t_3$, which, as we shall see in Section 6.4, is precisely when mean fitness increases generically.

Dynamics in the neighborhood of the linkage-equilibrium manifold

Next we describe the multilocus dynamics (2.12) for sufficiently large t, such that $\theta_i = O(s)$. Then (6.9b) and (6.11) imply that the recursion relations for the gene frequencies can be written as

$$p_{i_k}^{(k)\prime} = p_{i_k}^{(k)} \frac{\widetilde{W}_{i_k}^{(k)}}{\widetilde{W}} + O(s^2) \,, \quad t \geq t_1 \,, \tag{6.14}$$

where the leading term in (6.14), the *weak-selection approximation*, describes the dynamics on the linkage-equilibrium manifold Λ_0, i.e.,

$$\widetilde{W}_{i_k}^{(k)} = \frac{1}{p_{i_k}^{(k)}} \sum_{i|i_k} \sum_j W_{ij} \tilde{p}_i \tilde{p}_j \,, \tag{6.15a}$$

$$\widetilde{W} = \widetilde{W}(\boldsymbol{\pi}) = \sum_{i,j} W_{ij} \tilde{p}_i \tilde{p}_j \,. \tag{6.15b}$$

Now a simple calculation, parallel to that leading to I(9.14), reveals that the dynamics on Λ_0 is gradient like,

$$p_{i_k}^{(k)\prime} = p_{i_k}^{(k)} \frac{\partial \widetilde{W}}{\partial p_{i_k}^{(k)}} \Big/ \sum_j p_{j_k}^{(k)} \frac{\partial \widetilde{W}}{\partial p_{j_k}^{(k)}} \,. \tag{6.16}$$

The Theorem of Baum and Eagon (Appendix A.1) implies that the mean fitness $\widetilde{W}$ is nondecreasing for (6.16); cf. Nagylaki (1989a).

With these elements the first important result can be formulated. It shows that the true multilocus dynamics (2.12) can be approximated for a long time by the simple dynamics (6.16) on the linkage-equilibrium manifold.

• **6.2** *Choose* $\rho_{i_k}^{(k)}(t_1) = p_{i_k}^{(k)}(t_1)$ *for every* k *and* i_k. *Then*

$$p_i(t) = \rho_i(t) + O(s) , \quad t_1 \le t \le K_0/s , \tag{6.17}$$

where $\rho_i = \prod_k \rho_{i_k}^{(k)}$, *the* $\rho_{i_k}^{(k)}$ *evolve according to* (6.16), *and* K_0 *is a constant.*

If $\boldsymbol{\rho}(t)$ does not converge to some equilibrium point (which is a nongeneric case with respect to all possible fitness matrices, as we shall see below) or if $\boldsymbol{p}(t_1)$ is near the boundary of two basins of attraction, then $\boldsymbol{p}(t)$ and $\boldsymbol{\rho}(t)$ may ultimately diverge. In such a case, the restriction $t \le K_0/s$ will be necessary.

Defining $\widetilde{\omega}_{i_k}^{(k)}$ and $\widetilde{\omega}$ in analogy to (6.15), where ω_{ij} is defined in (6.2), we can rewrite (6.14) as

$$\Delta p_{i_k}^{(k)} = s p_{i_k}^{(k)} \frac{\widetilde{\omega}_{i_k}^{(k)} - \widetilde{\omega}}{1 + s\widetilde{\omega}} + O(s^2) , \quad t \ge t_1 , \tag{6.18}$$

where $\widetilde{\omega}_{i_k}^{(k)}$ and $\widetilde{\omega}$ depend only on $\boldsymbol{\pi}$ and ω_{ij} but not on s. Rescaling time t in generations as $\tau = st$, we see at least formally that as $s \to 0$, the difference equation (6.18) approaches the differential equation

$$\dot{p}_{i_k}^{(k)} = p_{i_k}^{(k)} [\widetilde{\omega}_{i_k}^{(k)}(\boldsymbol{\pi}) - \widetilde{\omega}(\boldsymbol{\pi})] \tag{6.19}$$

defined on $S_{\ell_1} \times \cdots \times S_{\ell_\ell}$. We shall refer to (6.19) as the *weak-selection limit* of (2.12), and note that the equilibria of (6.18) and (6.19) coincide.

Comparison of (6.19) with (6.16) shows that (6.19) is a Svirezhev–Shahshahani gradient of the potential function $\widetilde{\omega}$ (cf. Chapter I.10.3 and Appendix A.3). In particular, $\widetilde{\omega}$ is strictly increasing along nonconstant solutions of (6.19):

$$\dot{\widetilde{\omega}} = 2 \sum_k \sum_{i_k} p_{i_k}^{(k)} \left[\widetilde{\omega}_{i_k}^{(k)}(\boldsymbol{\pi}) - \widetilde{\omega}(\boldsymbol{\pi})\right]^2 \ge 0 . \tag{6.20}$$

Since (6.19) is a gradient system, all eigenvalues of the Jacobian matrix (see Appendix A.1, footnote 1) are real, and the eigenvalues λ of (6.19) correspond to the eigenvalues $1 + s\lambda/\hat{\widetilde{\omega}}$ of (6.18), in which $\hat{\widetilde{\omega}}$ is evaluated at equilibrium.

Convergence to equilibrium

Now we posit hyperbolicity for the weak-selection limit (6.19), i.e., we assume that all equilibria of (6.19) are hyperbolic (cf. Appendix A.2). Since all eigenvalues of the Jacobian are real, hyperbolicity of (6.19) is equivalent to the assumption that 0 must not be an eigenvalue. Hyperbolicity is a generic property, i.e., for almost all matrices (ω_{ij}), all the equilibria of the weak-selection limit are hyperbolic. This leads to the following important result, stating that under weak selection no cycling or more complicated dynamics is possible.

• **6.3** *If* $r_{\min} > 0$*, s is sufficiently small, and all equilibria of* (6.19) *are hyperbolic, then each solution* $\mathbf{p}(t)$ *of* (2.12) *converges to an equilibrium point of* (2.12) *as* $t \to \infty$.

This theorem also holds for the continuous-time model (2.18). Using the above results, in particular (6.9), the discrete-time model (2.12) can be approximated by the continuous-time model (2.18) as follows (cf. Chapter I.10.2 and Nagylaki 1992, p. 199): assume (6.2), rescale time according to $\tau = st$, set $q_i(\tau) = p_i(t)$ for $t = 0, 1, 2, \ldots$, define t_1 as in (6.10), and put $\tilde{\Theta}_i(\tau) = \Theta_i(t)/s$ for $t \geq t_1$. Then, considering $[q_i(\tau + s) - q_i(\tau)]/s$ and performing the limit $s \to 0$, one obtains $\dot{q}_i = q_i(\omega_i - \bar{\omega}) - \tilde{\Theta}_i$, which agrees with (2.18) if we set $b_{ij} = 1$ in (2.19).

Linkage equilibria and epistasis

The complexities of the multilocus dynamics arise from epistasis, because with additive fitnesses, the gametic frequencies always converge to a fixed point in linkage equilibrium (Section 5.2). Nagylaki (1993) provided explicit relations between the evolutionary change of the linkage disequilibria and the epistatic deviations in fitness. Here, we summarize without proof the main results. Since, during a short initial phase, linkage disequilibria will decay rapidly if recombination is strong relative to selection, it is of most interest to examine the long-term relation between the various measures of linkage disequilibrium and the epistatic deviations in fitness.

Let θ_i and Θ_i denote the ℓ-locus linkage disequilibria defined in (6.6b) and (2.13), respectively, and let ν_i, defined below (3.43), be the total epistatic deviation in fitness w_i. Then the following relations hold:

$$\Delta\theta_i = \overline{W}^{-1} p_i \nu_i - \Theta_i + O(s^2)\,, \quad t \geq t_1\,, \tag{6.21}$$

and

$$\Theta_i(t) = \overline{W}^{-1} p_i \nu_i + O(s^2)\,, \quad t \geq t_2\,. \tag{6.22}$$

The latter equation implies, for instance, that if epistasis is weak in the sense that $\nu_i = O(s^2)$, as is the case for multiplicative selection, then $\Theta_i = O(s^2)$ for $t \geq t_2$, which is much stronger than (6.9b). By contrast, if ν_i is comparable to s in magnitude, then $\Theta_i \approx \overline{W}^{-1} p_i \nu_i$ for $t \geq t_2$. Therefore, Θ_i has the same sign and order of magnitude as the epistatic deviation in fitness.

Both relations, (6.21) and (6.22), can be generalized to the linkage disequilibria $\theta_{i_S}^{(S)}$ (6.6) and $\Theta_{i_S}^{(S)}$ (4.6) among all loci in a subset $S \subset L$ containing at least two loci. To this end, define the epistatic deviation $\nu_{i_S}^{(S)}$, due to interactions in all subsets of S, by $p_{i_S}^{(S)} \nu_{i_S}^{(S)} = \sum_{i_T} p_i \nu_i$, where $T = L \setminus S$. Then the following relations hold:

$$\Delta\theta_{i_S}^{(S)} = \overline{W}^{-1} p_{i_S}^{(S)} \nu_{i_S}^{(S)} - \Theta_{i_S}^{(S)} + O(s^2)\,, \quad t \geq t_1\,, \tag{6.23}$$

and

$$\Theta_{i_S}^{(S)}(t) = \overline{W}^{-1} p_{i_S}^{(S)} \nu_{i_S}^{(S)} + O(s^2)\,, \quad t \geq t_2\,. \tag{6.24}$$

Equation (6.23) shows that the evolution of linkage disequilibria $\theta_{i_S}^{(S)}$ is governed primarily by the epistatic deviations $\nu_{i_S}^{(S)}$. In particular, it shows that dominance deviations and their interactions do not affect the dynamics.

For every S, the epistatic deviation $\nu_{i_S}^{(S)}$ can be expressed as $\nu_{i_S}^{(S)} = \sum_{Q\subset S}\mu_{i_Q}^{(Q)}$, where $\mu_{i_Q}^{(Q)}$ designates the effect on fitness of the (epistatic) interactions of all the loci in Q. With these much simpler epistatic parameters, Slatkin's linkage disequilibria [see V (4.22)] satisfy

$$\Delta_{i_S}^{(S)}(t) = (r_{\text{tot}}^{(S)}\overline{W})^{-1}p_{i_S}^{(S)}\mu_{i_S}^{(S)} + O(s^2)\ , \quad t \geq t_2\ . \tag{6.25}$$

For the definition and properties of $r_{\text{tot}}^{(S)}$, see (2.17) and (4.9). The remarkably simple relation (6.25) shows that epistatic interactions can be estimated from gamete frequencies and the linkage map. This does not require that the population is near equilibrium because t_2 is usually short.

6.3 THE ASYMPTOTIC SECONDARY THEOREM OF NATURAL SELECTION

On the basis of the above results, the rate of evolution of a (not necessarily additive) quantitative character can be determined. A simple calculation invoking (3.2c), (3.3a), $\sum_i \Delta p_i = 0$, and (3.3b) shows that

$$\Delta\overline{G} = 2\sum_i g_i\Delta p_i + \sum_{i,j} g_{ij}\Delta p_i\Delta p_j\ . \tag{6.26}$$

Applying (3.5b), (3.35), (6.6b), and proceeding as in (5.7), we derive

$$\begin{aligned}\sum_i g_i\Delta p_i &= \sum_i\Big(\sum_k \gamma_{i_k}^{(k)} + \varepsilon_i\Big)\Delta p_i \\ &= \tfrac{1}{2}\overline{W}^{-1}\,\text{Cov}_{\text{A}}(G,W) + \sum_i \varepsilon_i\Delta\theta_i + \sum_i \varepsilon_i\Delta\tilde{p}_i\ ,\end{aligned} \tag{6.27}$$

where the additive genetic covariance, $\text{Cov}_{\text{A}}(G,W)$, is given by (3.44).

By (6.12), the last sum in (6.26) is of order $O(s^2)$ if $t \geq t_1$. Furthermore, if $t \geq t_1$, the last sum in (6.27) can be shown to be of order $O(s^2)$ (see Nagylaki 1993, Eq. (177), for details). Therefore, the following instructive approximation for the rate of evolution of $\overline{G}$ is obtained in the limit of $s \to 0$:

$$\Delta\overline{G} = \overline{W}^{-1}\,\text{Cov}_{\text{A}}(G,W) + 2\sum_i \varepsilon_i\Delta\theta_i + O(s^2)\ , \quad t \geq t_1\ . \tag{6.28}$$

If there is no (additive) epistasis, then the sum is absent and (6.28) agrees with the exact formula (5.7). The sum in (6.28) also vanishes if the linkage disequilibria are constant. This occurs, for example, if fitnesses are multiplicative and the population is initially in linkage equilibrium (cf. Section 5.3). After quasi-linkage equilibrium has been reached (after at most t_2 generations), the linkage disequilibria θ_i change very slowly (6.13b). Therefore, (6.28) implies:

• **6.4** *The Asymptotic Secondary Theorem of Natural Selection. If selection is weak* ($s \ll r_{\min}$), *then*

$$\begin{aligned}\Delta\overline{G} &= \text{Cov}_{\text{A}}(G,W)/\overline{W} + O(s^2)\ , \quad & t \geq t_2\ , \qquad (6.29\text{a}) \\ &= 2\gamma^{\top}\Delta\pi + O(s^2)\ , & t \geq t_2\ . \qquad (6.29\text{b})\end{aligned}$$

The second equality follows from (3.44e) and directly relates the change in mean to the change of allele frequencies, with the average effects mediating this relation. It is important to note that the absolute error in (6.29) is small, but the relative error may be large. In the absence of epistasis and dominance, (6.29) simplifies to Robertson's Secondary Theorem of Natural Selection (5.10).

Applying (6.9b) and (6.13b) to (3.47), and observing $\theta_{i_k i_l}^{(kl)} = \Theta_{i_k i_l}^{(kl)}$ shows that

$$p_{i_k}^{(k)} g_{i_k}^{(k)} = p_{i_k}^{(k)} \gamma_{i_k}^{(k)} + O(s) \,, \quad t \geq t_1 \,, \tag{6.30}$$

and

$$(\Delta p_{i_k}^{(k)}) g_{i_k}^{(k)} = (\Delta p_{i_k}^{(k)}) \gamma_{i_k}^{(k)} + O(s^2) \,, \quad t \geq t_2 \,. \tag{6.31}$$

Thus, average effect and average excess differ only because of linkage disequilibrium; cf. (3.48).

If (6.30) is applied to the representation (3.44d) of $\mathrm{Cov}_{\mathrm{A}}(G, W)$, the linkage-equilibrium expression (3.50) is obtained up to an error of $O(s^2)$, because $w_{i_k}^{(k)} = O(s)$. Therefore, in (6.29a), $\mathrm{Cov}_{\mathrm{A}}(G, W)$ can be replaced by its linkage-equilibrium value (3.50). Similarly, by (6.31), $\boldsymbol{\gamma}^{\top} \Delta \boldsymbol{\pi} = \boldsymbol{g}^{\top} \Delta \boldsymbol{\pi} + O(s^2)$, which is the sum of all single-locus changes of the mean; cf. V(6.16).

The above results can be generalized to the case where genotypic values G_{ij} and fitnesses W_{ij} are time dependent. In this case, the term $\overline{\Delta G} = \sum_{ij} \Delta G_{ij} p_i' p_j'$ has to be added to the right-hand sides of (6.26) and (6.28). If, in addition, the W_{ij} and G_{ij} change sufficiently slowly, i.e., $\Delta W_{ij} = O(s^2)$ and $\Delta G_{ij} = O(s^2)$, then • **6.4** remains valid (Nagylaki 1993).

6.4 THE ASYMPTOTIC FUNDAMENTAL THEOREM OF NATURAL SELECTION

Now we derive a mathematically precise and general statement of Fisher's FTNS. We start with (6.26) which, for fitness, becomes

$$\Delta \overline{W} = 2 \sum_i w_i \Delta p_i + \sum_{i,j} w_{ij} \Delta p_i \Delta p_j \,. \tag{6.32}$$

Then, proceeding as before, and reexamining the error terms by noticing that $w_{ij} = O(s)$, we obtain the following analog of (6.28):

$$\Delta \overline{W} = \overline{W}^{-1} \sigma_{\mathrm{A}}^2 + 2 \sum_i \nu_i \Delta \theta_i + O(s^3) \,, \quad t \geq t_1 \,. \tag{6.33}$$

Here, σ_{A}^2 is the additive genetic variance in fitness, as given by (3.31), and ν_i is the epistatic deviation in fitness, defined below (3.43). In the absence of epistasis, (6.33) agrees with the exact formula (5.18). In general, we obtain:

• **6.5** *The Asymptotic Fundamental Theorem of Natural Selection. If selection is weak* $(s \ll r_{\min})$, *then*

$$\Delta \overline{W} = \sigma_{\mathrm{A}}^2 / \overline{W} + O(s^3) \,, \quad t \geq t_2 \,. \tag{6.34}$$

In analogy to (6.30) and because $w_{i_k}^{(k)} = O(s)$, we obtain for the average effects and average excesses in fitness

$$p_{i_k}^{(k)} w_{i_k}^{(k)} = p_{i_k}^{(k)} \alpha_{i_k}^{(k)} + O(s^2) , \quad t \geq t_1 . \tag{6.35}$$

Therefore, the additive genetic variance σ_{A}^2 in (6.34) can be replaced by its linkage-equilibrium approximation $\sigma_{\mathrm{A,LE}}^2$ (4.26). If fitnesses change sufficiently slowly in time, then (6.34) remains valid (Nagylaki 1993).

Equation (6.34) deceptively suggests that $\Delta\overline{W} > 0$ for large t. However, this is not necessarily the case, because σ_{A}^2 may be very small. Typically, σ_{A}^2 will be small in the neighborhood of an equilibrium. Since, even for weak selection, equilibria are, in general, not critical points of the mean fitness, $\overline{W}$ may decrease near equilibrium points. Mean fitness may also decrease during the initial phase of evolution towards QLE ($t \leq t_2$) when most change in gamete frequencies is caused by changing linkage disequilibria. For numerical examples, we refer to Moran (1964), Karlin and Carmelli (1975), Nagylaki (1977a), and Ewens (1979). Therefore, the concept of an adaptive topography is generally not applicable in multilocus systems.

However, if all equilibria of the weak-selection limit (6.19) are hyperbolic, then $\overline{W}$ increases for sufficiently small s if π is within $O(s^2)$ of Λ_s, but is not too close to an equilibrium of (6.19) on Λ_0. This will be the case during the evolutionarily long epoch $t_2 \leq t \leq t_3$, which is exactly when most of the gene-frequency change occurs (Nagylaki *et al.* 1999).

6.5 THE FUNDAMENTAL THEOREM OF NATURAL SELECTION

Fisher's (1930) formulation of the FTNS (Chapter I.9.4) gave rise to misunderstandings and controversy starting right after its publication (see Edwards (1994) for a historical review). Motivated, presumably, by Wright's (1932) concept of an adaptive topography, which associates to each gene combination an adaptive value and views evolution as a 'hill-climbing process' that increases mean fitness, Fisher's FTNS became interpreted as the statement that the increase, or rate of increase, of mean fitness of a population is approximately the current additive genetic variance in fitness (and thus nonnegative). As we have seen, this statement is not universally true and requires special assumptions. In particular, stable equilibria in general do not coincide with (local) maxima of mean fitness and evolution is not a hill-climbing process. Actually, Fisher was aware of this and strongly opposed the above interpretation, stating that:

> I have never, indeed, written about $\overline{w}$ and its relationships ... (Fisher 1958a).

See also the quotation at the end of Chapter I.10.3 from a letter to Kimura.

Fisher viewed his theorem as an exact and very general result, valid for an arbitrary number of genes and under general mating schemes. Further, in the paragraph following the statement of the theorem, he writes

> The rigor of the demonstration requires that the terms employed should be used strictly as defined; ...

Given Fisher's views about the FTNS and the fact that he later (Fisher 1941, 1958) tried to clarify some of the concepts preceding his formulation of the theorem (in particular average excess and average effect), the question remains what Fisher's original intention was.

Price (1972) argued that Fisher considered only the partial change in mean fitness that is due to gene-frequency changes. His argument was further clarified and extended by Ewens (1989, 1992, 1995). A somewhat different, but also appealing, interpretation was put forward by Lessard (1997). We shall first present the core of the Price–Ewens interpretation of the FTNS for a discrete-time model. In part, the presentation will follow Castilloux and Lessard (1995), who extended Ewens' analysis.

Since, in the spirit of Fisher, we do not assume random mating, the state of the population must be described in terms of genotype frequencies P_{ij} instead of gamete frequencies. We retain the notation of Sections 3 and 4. Assuming that selection occurs between conception of a genotype and maturity, the change in the frequency of allele $\mathcal{A}_{i_k}^{(k)}$ during this period is $p_{i_k}^{(k)} w_{i_k}^{(k)} / \overline{W}$ [see (4.17)]. Under most mating patterns, the frequency of an allele at maturity is equal to its frequency in the daughter generation at conception. Thus, the only assumption about the mating scheme will be that

$$\Delta p_{i_k}^{(k)} = p_{i_k}^{(k)} w_{i_k}^{(k)} / \overline{W} \tag{6.36}$$

holds, where Δ designates changes between generations.

Of central importance for an interpretation of the FTNS in Fisher's sense, is Fisher's concept of mean fitness and of changes in mean fitness. In his view, the substance of evolution are changes in gene frequency, and the inheritance of fitness is ascribed to the average effects. Formalizing this, we may write fitness in the form

$$W_{ij} = \overline{W} + \alpha_{ij} + \nu_{ij} \ , \tag{6.37}$$

where α_{ij} is the additive effect of genotype ij on fitness, and ν_{ij} is the residual deviation. Thus, (6.37) is equivalent to (3.25), and α_{ij} is the sum of the average allelic effects; cf. (3.26) and (3.27).

The decisive point Price and Ewens make in their interpretation is that Fisher, presumably, considered the *partial* change in mean fitness through changes of the genotype frequencies P_{ij} only, but not through changes in $\overline{W}$ and the average effects. To make this precise, we observe that the (total) change in mean fitness can be written as

$$\Delta\overline{W} = \sum_{i,j} W_{ij}(\Delta P_{ij}) + \sum_{i,j}(\Delta W_{ij})P_{ij} + \sum_{i,j}(\Delta W_{ij})(\Delta P_{ij}) \ , \tag{6.38}$$

where the second and third sum involve changes in the fitness of the genotypes, attributable to a changing (internal or external) environment. Fisher definitely did not consider such changes. Moreover, it is most likely (see Price (1972) and Ewens (1989) for discussion) that Fisher viewed the partial change in mean fitness as that due to single-locus gene-frequency changes, and disregarded nonadditive effects. Inserting (6.37) into the first sum on the right-hand side of (6.38) and omitting the term involving the epistatic interactions, $\sum_{i,j} \nu_{ij}\Delta P_{ij}$, this partial change in mean fitness, as defined by Price (1972) and Ewens (1989), is

$$\Delta_{\mathrm{PE}}\overline{W} = \sum_{i,j}(\overline{W} + \alpha_{ij})\Delta P_{ij} \ . \tag{6.39}$$

Now it is easy to show that this partial change in the mean fitness of the population is $\sigma_A^2/\overline{W}$. Indeed, using $\sum_{ij} \Delta P_{ij} = 0$, (3.58) with $v_{ij} = 1$, and (3.31c), we obtain the desired result:

$$\Delta_{PE}\overline{W} = 2\sum_k \sum_{i_k} \alpha_{i_k}^{(k)} \Delta p_{i_k}^{(k)} = \sigma_A^2/\overline{W} . \tag{6.40}$$

The reader may note that under random mating, $\Delta_{PE}\overline{W}$ is precisely the expression that yielded the leading term $\sigma_A^2/\overline{W}$ in the Asymptotic Fundamental Theorem [cf. (6.33) and (6.27)].

Fisher (1930) had used continuous-time notation in his derivation of the FTNS. However, the above reasoning also holds for continuous time and leads to the statement that the partial rate of increase in mean fitness is exactly σ_A^2 (Price 1972, Ewens 1989). Lessard and Castilloux (1995) showed that (6.40) holds true with fecundity selection if mating does not change gene frequencies in the parental generation from the time of conception to the time of reproduction, and if neither meiotic drive nor gametic selection takes place.

Although (6.40), as well as its continuous-time analog, is a very general and exact result – it holds for an arbitrary number of loci and alleles, for almost arbitrary mating patterns, and requires no restrictive assumptions about the fitness parameters, "this does not mean that Fisher would have accepted [(6.40)] as a statement of his theorem" (Price 1972).

A somewhat different interpretation of the FTNS that might even come closer to Fisher's intention, was promoted by Lessard (1997). The general model is as above. Then the frequency of a genotype ij in parents of offspring of the next generation, counted as many times as their expected number of offspring, is $P_{ij}^* = P_{ij}W_{ij}/\overline{W}$. The frequency of ij in the offspring of the next generation, denoted P'_{ij}, is generally different from P_{ij}^* because of sexual reproduction. Nevertheless, it can be written in the form

$$P'_{ij} = P_{ij}V_{ij}/\overline{V} , \tag{6.41}$$

where V_{ij} will be a complicated function of the state of the population involving parameters for survival, mating, fecundity, recombination, and segregation. However, the identity $\overline{V} = \sum_{i,j} P_{ij}V_{ij} = \overline{W}$ will hold.

Importantly, the average excess $w_{i_k}^{(k)}$ of an allele will be the same, irrespective of whether the fitness of genotype ij is defined according to the original definition W_{ij}, or by the quantity V_{ij}, defined implicitly in (6.41). As a consequence, if, in analogy to (6.37), we write

$$V_{ij} = \overline{V} + \alpha_{ij} + \beta_{ij} , \tag{6.42}$$

and minimize the variance of the β_{ij}, the average allelic effects $\alpha_{i_k}^{(k)}$ are the same as in (3.27). Therefore, we can decompose the change of genotype frequencies from one generation to the next as

$$\Delta P_{ij} = P_{ij}(V_{ij} - \overline{V})/\overline{V} = (\Delta P_{ij})_A + (\Delta P_{ij})_E , \tag{6.43}$$

where $(\Delta P_{ij})_A = P_{ij}\alpha_{ij}/\overline{V}$ and $(\Delta P_{ij})_E = P_{ij}\beta_{ij}/\overline{V}$. It should be emphasized that the term $(\Delta P_{ij})_A$ represents the change ascribable only to changes in gene frequencies.

The decomposition (6.43) leads to the following one for the first term on the right-hand side of (6.38):

$$\sum_{i,j} W_{ij}(\Delta P_{ij}) = \sum_{i,j} W_{ij}(\Delta P_{ij})_{\mathrm{A}} + \sum_{i,j} W_{ij}(\Delta P_{ij})_{\mathrm{E}} \, .$$

Lessard (1997) suggested that $\Delta_{\mathrm{L}}\overline{W} = \sum_{i,j} W_{ij}(\Delta P_{ij})_{\mathrm{A}}$ be the change in mean fitness considered by Fisher (1930, 1958) in his Fundamental Theorem of Natural Selection, because this interpretation of change in mean fitness appears to be consistent with Fisher's writings, and because it can be shown that

$$\Delta_{\mathrm{L}}\overline{W} = \sigma_{\mathrm{A}}^2/\overline{W} \, . \tag{6.44}$$

Although these interpretations lead to a 'universally' valid theorem, their biological significance has yet to be established.

*6.6 *OPTIMIZING PRINCIPLES*

Here, we generalize Kimura's Maximum Principle to arbitrary multilocus systems and show that it is equivalent to the minimizing procedures of Section 3.1 that yielded the average effects. This observation is due to Ewens (1992, 1995). Let $\Delta\boldsymbol{\pi}$ denote the (column) vector of allele-frequency changes caused by natural selection. Applied to fitness, Result • **3.2** can be formulated as follows:

• **6.6** *The vector of average effects is the unique solution* $\boldsymbol{\alpha} = \boldsymbol{a}$ *which minimizes the quadratic form* $\boldsymbol{a}^\top \mathsf{H}\boldsymbol{a}$ *subject to the constraint* $\boldsymbol{a}^\top \Delta\boldsymbol{\pi} = \sigma_{\mathrm{A}}^2/(2\overline{W})$.

Let us introduce the linear transformation

$$\overline{W}\boldsymbol{d} = \mathsf{H}\boldsymbol{a} \, . \tag{6.45}$$

Then $\boldsymbol{d}$ is a vector of allele-frequency changes, i.e., it satisfies $\sum_{i_k=1}^{\ell_k} d_{i_k}^{(k)} = 0$ for every $k = 1, \dots, \ell$. Actually, as shown in the uniqueness proof of the average effects in Section 3.6, the transformation (6.45) establishes a one-to-one correspondence between all allele-frequency vectors $\boldsymbol{d}$ and all vectors $\boldsymbol{a}$ that satisfy $\sum_{i_k=1}^{\ell_k} a_{i_k}^{(k)} p_{i_k}^{(k)} = 0$ for every k. Now, a simple calculation using (6.45) establishes that minimizing $\boldsymbol{a}^\top \mathsf{H}\boldsymbol{a}$ subject to $\boldsymbol{a}^\top \Delta\boldsymbol{\pi} = \sigma_{\mathrm{A}}^2/(2\overline{W})$ is equivalent to *minimizing*

$$\boldsymbol{d}^\top \mathsf{H}^{-1}\boldsymbol{d} \tag{6.46}$$

subject to the constraint

$$\boldsymbol{\alpha}^\top \boldsymbol{d} = \frac{\sigma_{\mathrm{A}}^2}{2\overline{W}} \, . \tag{6.47}$$

Comparison of (6.45) with (3.29) shows that the unique solution must be $\boldsymbol{d} = \Delta\boldsymbol{\pi}$.

By expanding $(\boldsymbol{d} - \Delta\boldsymbol{\pi})^\top \mathsf{H}^{-1}(\boldsymbol{d} - \Delta\boldsymbol{\pi})$, we immediately infer that minimizing (6.46) subject to (6.47) is equivalent to *maximizing*

$$\boldsymbol{d}^\top \mathsf{H}^{-1}\Delta\boldsymbol{\pi} \tag{6.48}$$

subject to the constraint

$$d^\top \mathsf{H}^{-1} d = \frac{\sigma_A^2}{2\overline{W}^2}\,. \tag{6.49}$$

The unique solution of this procedure is $d = \Delta\pi$.

A glance at (6.40) reveals that the left-hand side of (6.47), $\alpha^\top d$, is one half of the partial change in mean fitness, $\Delta_{PE}\overline{W}$, considered by Price (1972) and Ewens (1989), provided the allele-frequency change is d. Of course, we also have $\alpha^\top d = \frac{1}{2}\Delta_L\overline{W}$; cf. (6.46). In addition, by (3.30), the quadratic form (6.48) is proportional to $\Delta_{PE}\overline{W}$ and to $\Delta_L\overline{W}$. Therefore, the maximizing principle defined by (6.48) and (6.49) is a direct multilocus generalization of Kimura's Maximum Principle.

In generalization of the Svirezhev–Shahshahani metric I(10.15), a genetic distance can be introduced on the vector space of all allele-frequency changes. If p and π denote the vectors of gamete and allele frequencies in the parental generation, respectively, and $\pi + d$ the vector of allele frequencies in the offspring generation, then we define the genetic distance by

$$\|d\|_p = \sqrt{d^\top \mathsf{H}^{-1} d}\,. \tag{6.50}$$

Recall from Sections 3.2 and 3.6 that the matrix H^{-1} depends on the allele frequencies, all pairwise linkage disequilibria and, unless Hardy–Weinberg proportions obtain, on pairwise associations between alleles on different chromosomes; it is well defined and positive definite on the vector space of all gene-frequency changes. From (3.30) and (3.31), we infer that the genetic distance between parent and offspring caused by natural selection is

$$\|\Delta\pi\|_p = \frac{1}{\overline{W}}\sqrt{\tfrac{1}{2}\sigma_A^2}\,. \tag{6.51}$$

If fitness depends only on one locus and the population mates at random, then (3.33) holds and H^{-1} is calculated to be

$$\mathsf{H}^{-1} = \mathsf{D}^{-1} - \tfrac{1}{2}\mathsf{U}\,, \tag{6.52}$$

where U is a matrix of 1s. In this case, the distance (6.50) reduces to the Svirezhev–Shahshahani metric I(10.15).

Therefore, the optimizing principles derived above can be stated as follows (Ewens 1992):

• **6.7** *Natural selection acts in such a way as to minimize the (squared) genetic distance* $d^\top \mathsf{H}^{-1} d$ *between parental and daughter gene frequencies, subject to the requirement that the partial increase in mean fitness is the natural selection value* $\sigma_A^2/\overline{W}$.

The converse maximizing principle is Kimura's Maximum Principle:

• **6.8** *Among all allele-frequency changes* d *of the same length* $\|d\|_p = \sigma_A^2/(2\overline{W}^2)$, *the vector* d *that maximizes the partial increase in mean fitness,* $2\alpha^\top d$, *is the natural selection vector* $\Delta\pi$.

These principles are equivalent to the minimization procedure • **6.6** which yields the average effects. The main results of this subsection can be generalized to fertility selection (Lessard and Castilloux 1995).

III

Classical Mutation-Selection Models

Natural selection and mutation are two central factors guiding biological evolution: mutation generates the genetic variability upon which selection can act. This was clearly recognized by the pioneers of population genetics, Fisher, Haldane, and Wright, who developed mathematical models quantifying the relative importance of selection and mutation in maintaining genetic variation. In traditional models, two alleles per locus, the wild type and a mutant, are considered, and the equilibrium frequencies of the alleles can be calculated under recurrent mutation and various assumptions on the selective values of the genotypes. In many instances, however, more than two alleles per locus may occur.

Prior to 1970, only few general results about mutation-selection models with multiple alleles per locus were available (see Crow and Kimura 1970). Moran (1976) demonstrated existence, uniqueness, and global stability of an equilibrium for a haploid mutation-selection model. Previously, Thompson and McBride (1974) had derived the solution of a system of differential equations occurring in the theory of the evolution of macromolecules that is formally equivalent to the haploid mutation-selection equation. At the same time, the stepwise-mutation model, describing variation caused by electrophoretically detectable alleles, was introduced (Ohta and Kimura 1973) and analyzed (Moran 1976, 1977; Kingman 1977). A quite different model, the so-called continuum-of-alleles model, was suggested by Crow and Kimura (1964) and Kimura (1965a). It is based on the assumption that at a locus, an infinite sequence of different alleles can be generated by mutation, and that every mutation may produce a new allele whose effect is drawn from a continuous distribution. Under the assumption of weak selection and small mutational effects, Crow and Kimura derived the equilibrium distribution and calculated the mutation load and the genetic variance of a quantitative trait under mutation-selection balance. Such models are particularly useful in quantitative genetics and will be treated separately in Chapter IV.

This chapter is devoted to the basic properties of classical models incorporating mutation and selection at a single gene locus. By classical models, we mean that only a finite number of alleles (mutant types) per locus is considered. In addition, the stepwise-mutation, or ladder, model is included, in which a possibly infinite sequence of mutants occurs.

In Section 1, the basic mutation-selection model is introduced for asexual haploid populations. It is shown that under very general conditions a unique, completely polymorphic equilibrium exists which is globally asymptotically stable. This is not true in diploid populations, where stringent conditions on the fitnesses and the mutation distribution are needed to establish a unique stable equilibrium. It can be shown that when at least three alleles segregate, complicated dynamical behavior such as limit cycles may occur. This does not happen if mutation is weak relative to selection. General dynamical and equilibrium properties of the haploid and diploid mutation-selection model with an arbitrary number of alleles at a single locus are explored in Section 2. In the limit of weak mutation, simple approximations for the equilibrium distribution and its mean fitness are derived in Section 3. It is shown that Haldane's principle, stating that at mutation-selection balance the equilibrium mean fitness depends only on the mutation rate, but not on the fitnesses, is valid under very general conditions on the mutation pattern and the selection regime. The error term is proved to be of order $O(\mu^2/s)$. The related concept of the genetic load is also introduced. In Section 4, a generalized version of the haploid stepwise-mutation model is examined.

Throughout this chapter, populations are assumed to be sufficiently large that random genetic drift can be ignored, to mate at random (if sexual), and to have constant genotypic fitnesses. Both discrete- and continuous-time models are treated.

1. DYNAMICS IN HAPLOID POPULATIONS

1.1 DISCRETE GENERATIONS

We consider a haploid, asexually reproducing population, in which k types (alleles), labeled as $\mathcal{A}_1, \ldots, \mathcal{A}_k$, may occur. Let the fitness of $\mathcal{A}_i$ be W_i, let its relative frequency in generation t be $p_i = p_i(t)$, so that $\sum_i p_i = 1$, and let $\boldsymbol{p} = (p_1, \ldots, p_k)^\top$. As previously, allele frequencies in successive generations are denoted by p_i and p_i'. Frequencies are measured in offspring before selection. Thus, the life cycle begins with selection, which is followed by reproduction during which mutation occurs. We designate the probability that an $\mathcal{A}_i$ individual has an $\mathcal{A}_j$ offspring (where $j \neq i$) by the mutation rate μ_{ij}, and use the convention $\mu_{ii} = 0$ for every i.

After selection, the frequency of $\mathcal{A}_i$ is $p_i^* = p_i W_i / \overline{W}$; see I(9.1). Then reproduction and mutation occur, and change the allele frequencies p_i^* according to I(6.1). Therefore, substituting p_i^* for p_i in I(6.1), we obtain the *mutation-selection equation*

$$p_i' = p_i \frac{W_i}{\overline{W}} + \frac{1}{\overline{W}} \sum_j (p_j W_j \mu_{ji} - p_i W_i \mu_{ij}) \ . \tag{1.1}$$

For our purposes, it is convenient to cast (1.1) into matrix form. Let us define the $k \times k$ mutation matrix $\tilde{\mathsf{U}} = (\tilde{u}_{ij})$ by

$$\tilde{u}_{ij} = \begin{cases} 1 - \sum_l \mu_{il} \ , & i = j \ , \\ \mu_{ji} \ , & i \neq j \ , \end{cases} \tag{1.2}$$

and the mutation-selection matrix $\mathsf{C} = (c_{ij})$ by

$$c_{ij} = \tilde{u}_{ij} W_j \ . \tag{1.3}$$

Then a simple calculation shows that

$$\bar{c} = \sum_i (\mathsf{C}\boldsymbol{p})_i = \overline{W} \ , \tag{1.4}$$

and (1.1) can be rewritten as

$$p' = \frac{1}{\bar{c}} \mathrm{C} p \,. \tag{1.5}$$

The state space is the simplex S_k defined in I(9.24). Observing that $n(t) = \mathrm{C}^t n(0)$ is the solution of $n' = \mathrm{C}n$, and $p(t) = n(t) / \sum_i n_i(t)$, it follows immediately that (1.1) has the explicit solution

$$p(t) = \mathrm{C}^t p_0 \Big/ \sum_i (\mathrm{C}^t p_0)_i \,, \tag{1.6}$$

where $p(0) = p_0 \in S_k$ designates the initial frequency distribution.

It is our aim to show that all solutions $p(t)$ of (1.1), or (1.5), starting from an arbitrary initial value in S_k, converge to a unique equilibrium distribution $\hat{p}$ that is completely polymorphic, i.e., $\hat{p}_i > 0$ for every i. Some positivity condition on the mutation rates and the fitnesses will be required to achieve such a result, because otherwise equilibria may exist with $\hat{p}_i = 0$ for one or several i.

Actually, more complicated behavior can occur, as the following example shows. Consider two alleles, $\mathcal{A}_1$ and $\mathcal{A}_2$, and assume $W_1 = W_2 = 1$ and $\mu_{12} = \mu_{21} = 1$. Inserting these values into (1.1) gives $p_1' = 1 - p_1$ and, furthermore, $p_1'' = p_1$. Thus, periodic orbits exist in this simple model. (In the corresponding continuous-time model (1.8), all solutions converge to $p_1 = p_2 = \frac{1}{2}$.) The appropriate condition for achieving uniqueness and stability is that a positive integer n exists so that through a series of n steps, every allele $\mathcal{A}_i$ gives rise to descendants of every type $\mathcal{A}_j$ with positive probability. Mathematically, this means that for some integer $n \geq 1$ all entries of the matrix C^n must be positive (> 0). Such a matrix is called *primitive*. A matrix C^n can be positive[1] only if $W_i > 0$ for every i. In this case, C is primitive if and only if $\tilde{\mathrm{U}}$ is primitive. The following result was proved by Moran (1976):

• **1.1** *If the matrix* C *defined in* (1.3) *is primitive, then the mutation-selection dynamics* (1.5) *admits a unique equilibrium,* $\hat{p}$, *that satisfies* $\hat{p}_i > 0$ *for every* i. *This equilibrium is the unique solution of*

$$\hat{\overline{W}}\, \hat{p} = \mathrm{C}\hat{p} \,, \tag{1.7}$$

where $\hat{\overline{W}} = \sum_i W_i \hat{p}_i$ *is the equilibrium mean fitness, and it is globally asymptotically stable.*

▷ By the Perron–Frobenius Theorem (see Appendix B.1), the primitive matrix C has an eigenvalue $r > 0$ with a positive eigenvector $\hat{q}$, i.e., $\mathrm{C}\hat{q} = r\hat{q}$, and $\hat{q}$ is unique except for multiplication by positive constants. In addition, r is the unique eigenvalue with this property. Assuming that $\hat{q}$ is normalized, $\sum_i \hat{q}_i = 1$, we observe from this eigenvalue equation that $\hat{\bar{c}} = \sum_i (\mathrm{C}\hat{q})_i = r$. From (1.4) we infer that the equilibrium mean fitness is $\hat{\overline{W}} = r$. Since the equilibrium solution $\hat{p}$ must satisfy (1.7), and since $\hat{q}$ is unique, we conclude that $\hat{p} = \hat{q}$, which proves the assertion about existence, uniqueness, and positivity of $\hat{p}$. Global stability also follows from Perron–Frobenius theory, because for any initial vector q_0, $r^{-t}\mathrm{C}^t q_0$ converges to some positive multiple

[1] Throughout, positive matrix or positive vector means that all components are > 0; nonnegative means ≥ 0. See also Appendix B.

of $\hat{\boldsymbol{p}}$. This observation, together with the fact that a solution $\boldsymbol{p}(t)$ of (1.5) is normalized and given by (1.6), shows that any solution of (1.5), and hence of (1.1), converges to $\hat{\boldsymbol{p}}$. ◁

1.2 OVERLAPPING GENERATIONS

The notation is the same as above, except that m_i stands for the Malthusian parameter of allele $\mathcal{A}_i$ (cf. Chapter I.8.2). In continuous time, there is more than one way to model the evolution under selection and mutation. The most frequently used differential equation is

$$\dot{p}_i = p_i(m_i - \overline{m}) + \sum_j (p_j \mu_{ji} - p_i \mu_{ij}) \,, \tag{1.8}$$

in which selection and mutation are decoupled. It is based on the assumption that mutation occurs continuously, and that the nonadditive interaction of selectional and mutational forces in a small time interval Δt is of order $(\Delta t)^2$, and thus negligible (see Crow and Kimura 1970, Nagylaki 1992). Given that mutation rates μ_{ij} (defined as probability of mutation per unit time) are typically very small, this is a reasonable assumption. The reader may also note that mutation rates in continuous time may differ from those in discrete time because, for the latter, a time unit is one generation.

Another way to obtain a continuous-time model is to replace $\overline{W}(p_i' - p_i)$ in the discrete-time equation (1.1) by $\dot{p}_i$. The resulting differential equation has the advantage of having exactly the same equilibria as (1.1) (see Hadeler 1981). Under the assumption of weak selection and weak mutation, this differential equation, as well as the discrete-time equation (1.1), simplifies to (1.8), as is shown by replacing μ_{ij} by $h\mu_{ij}$, setting $W_i = a + hm_i$, rescaling time $t \to t/h$, and letting $h \to 0$ (cf. Chapter I.10.2).

A third way to derive a mutation-selection equation in continuous time is to consider birth and death rates, b_i and d_i, as in Chapter I.10.1, and to assume that mutation occurs during reproduction. Then, in a small time interval Δt, a fraction $b_i \Delta t$ of $\mathcal{A}_i$ individuals produce an offspring. Among these offspring, $b_i \mu_{ij} \Delta t$ are of type $\mathcal{A}_j$ and $b_i(1 - \sum_j \mu_{ij})\Delta t$ are of type $\mathcal{A}_i$. Defining $m_i = b_i - d_i$, this reasoning leads to

$$\dot{p}_i = p_i(m_i - \overline{m}) + \sum_j (p_j b_j \mu_{ji} - p_i b_i \mu_{ij}) \,. \tag{1.9}$$

Equations (1.9) and (1.8) are formally equivalent if the μ_{ij} are redefined.

We shall always use (1.8) as our continuous-time mutation-selection model. An existence and stability result can be proved that is analogous to that for the discrete-time dynamics. To this end, (1.8) is rewritten as

$$\dot{\boldsymbol{p}} = \mathsf{A}\boldsymbol{p} - \bar{a}\boldsymbol{p} \,, \tag{1.10}$$

where $\mathsf{A} = (a_{ij})$ is the matrix defined by

$$a_{ij} = m_i \delta_{ij} + \mu_{ji} \tag{1.11}$$

and $\bar{a} = \sum_{i,j} a_{ij} p_i p_j$. The differential equation (1.10) can be solved explicitly by an argument analogous to the one that led to (1.6). For given initial condition $\boldsymbol{p}(0) = \boldsymbol{p}_0$, the solution is

$$\boldsymbol{p}(t) = \frac{e^{\mathsf{A}t}\boldsymbol{p}_0}{\sum_i (e^{\mathsf{A}t}\boldsymbol{p}_0)_i} \,, \tag{1.12}$$

because the matrix exponential $e^{\mathsf{A}t} = \sum_{n=0}^{\infty} \frac{1}{n!}(\mathsf{A}t)^n$ generates the solution of the associated linear differential equation $\dot{\boldsymbol{n}} = \mathsf{A}\boldsymbol{n}$ for the nonnormalized frequencies. This observation is due to Thompson and McBride (1974), who investigated a special case of (1.9) as a model for the evolution of self-reproducing macromolecules when replication errors occur (see Eigen 1971).

Existence, uniqueness, and strict positivity of an equilibrium solution $\hat{\boldsymbol{p}}$ follows from (1.12), with $\hat{\boldsymbol{p}} = \boldsymbol{p}(t) = \boldsymbol{p}_0$, and the discrete-time case if the matrix A is assumed to be irreducible (which slightly weaker than primitivity; see Appendix B.1), because then $e^{\mathsf{A}t}$ is positive and, hence, primitive. Similarly, global asymptotic stability of $\hat{\boldsymbol{p}}$ in continuous time, follows from the proof in discrete time. An analogous result holds for the differential equation (1.9) by redefining A appropriately.

2. DYNAMICS IN DIPLOID POPULATIONS

For diploid populations subject to mutation and selection, equilibria are not necessarily uniquely determined. Even periodic orbits and stable limit cycles may occur. Of particular interest are, therefore, conditions on the mutation and fitness parameters that ensure existence, uniqueness, and stability of polymorphic equilibria.

2.1 THE MUTATION-SELECTION EQUATIONS

For the discrete-time model we use the notation of Chapter I.9.2; for the continuous-time model we use that of Chapter I.10.1. In both cases, μ_{ij} denotes the mutation rate from $\mathcal{A}_i$ to $\mathcal{A}_j$ and $\mu_{ii} = 0$. Allele frequencies are measured among zygotes before selection, and the life cycle begins with selection, which is followed by the production of germ cells, during which mutation occurs, and the formation of zygotes. The same argument that led to the haploid mutation-selection equation (1.1), but now by substituting I(9.7) into I(6.1), yields the diploid mutation-selection dynamics for a population with discrete generations,

$$p_i' = p_i \frac{W_i}{\overline{W}} + \frac{1}{\overline{W}} \sum_j (p_j W_j \mu_{ji} - p_i W_i \mu_{ij}) \ , \tag{2.1}$$

where W_i is the marginal fitness of the allele $\mathcal{A}_i$, defined I(9.6).

For a population with overlapping generations, we shall use the differential equation

$$\dot{p}_i = p_i(m_i - \overline{m}) + \sum_j (p_j \mu_{ji} - p_i \mu_{ij}) \ , \tag{2.2}$$

where m_i is the marginal Malthusian fitness of $\mathcal{A}_i$; see I(10.2). Both models are classical; for continuous-time, see Crow and Kimura (1970, Chapter 6.4); for discrete time, Nagylaki (1977, Chapter 4.9). For a derivation of the continuous-time model (2.2) consult Nagylaki (1974). It can also be obtained as the weak-selection weak-mutation limit of (2.1).

Obviously, the dynamics of (2.1) remains unchanged if all fitnesses W_{ij} are multiplied by the same positive constant, and (2.2) remains unchanged if the same constant is added to every m_{ij}. For multiplicative fitnesses, $W_{ij} = W_i W_j$, the discrete-time recursion (2.1) reduces to the haploid recursion (1.1), and for additive fitnesses, $m_{ij} = m_i + m_j$, the continuous-time equation (2.2) reduces to the haploid equation (1.8).

2.2 THE CASE OF TWO ALLELES

For the present purpose it is convenient to parameterize the fitness values of the genotypes $\mathcal{A}_1\mathcal{A}_1$, $\mathcal{A}_1\mathcal{A}_2$, $\mathcal{A}_2\mathcal{A}_2$ as in Chapter I.9.3: $W_{11} = 1$, $W_{12} = 1-hs$, $W_{22} = 1-s$. Instead of p_1 and p_2, we write p and $q = 1 - p$. Then the marginal fitnesses and the mean fitness are given by I(9.9) and I(9.10), respectively. For the mutation rates we write $\mu = \mu_{12}$ and $\nu = \mu_{21}$ and require $\mu+\nu < 1$. A straightforward calculation shows that the equilibria of the mutation-selection equation (2.1) are the solutions p of

$$\begin{aligned} &p^3 s(2h-1) + p^2 s[2 - 3h + \mu h + \nu(1-h)] \\ &\qquad + p[-s(1-h) + \mu(1-hs) + \nu(1-2s+hs)] - \nu(1-s) = 0 \end{aligned} \tag{2.3}$$

in the interval $[0, 1]$. As we shall see below, there may be one, two, or three such solutions, depending on the parameters. Some elementary, but lengthy, algebra shows the following (Norman 1974, Nagylaki 1992):

• **2.1** *If* $0 < s < 1$ *and* $h \le \frac{1}{2}$, *or* $s < 0$ *and* $h \ge \frac{1}{2}$, *then* (2.3) *has a unique solution in* $[0, 1]$. *Because* $\mu+\nu < 1$, *this equilibrium is globally asymptotically stable. Convergence is monotonic.*

This result includes a number of interesting special cases, such as no dominance ($h = \frac{1}{2}$), complete dominance of $\mathcal{A}_1$ ($h = 0$), and overdominance ($h < 0$) (in all these cases $s > 0$ is assumed).

The equilibrium solutions are simple only in special cases. We restrict our attention to the case $\nu = 0$, in which back mutations from the deleterious (and thus rare) allele $\mathcal{A}_2$ to $\mathcal{A}_1$ are ignored. It will be convenient to give the precise formulas in terms of $q = 1 - p$. Obviously, $\hat{q}^{(0)} = 1$ is always an equilibrium, because if $\mathcal{A}_1$ is initially not present in the population, it will not arise by mutation. Since $\nu = 0$, (2.3) reduces to a quadratic equation which, if $4\mu/s \le 1$, has the following solutions in $[0, 1]$ (cf. Bürger 1983a):

$$\hat{q}^{(1)} = \frac{h(1+\mu)}{2(2h-1)}\left[1 - \sqrt{1 - \frac{4\mu(2h-1)}{(1+\mu)^2 h^2 s}}\right] \qquad \text{if } h \ne \tfrac{1}{2}, \tag{2.4a}$$

$$\hat{q}^{(1)} = \frac{2\mu}{s(1+\mu)} \qquad \text{if } h = \tfrac{1}{2}, \tag{2.4b}$$

and

$$\hat{q}^{(2)} = \frac{h(1+\mu)}{2(2h-1)}\left[1 + \sqrt{1 - \frac{4\mu(2h-1)}{(1+\mu)^2 h^2 s}}\right] \qquad \text{if } h > h_c, \tag{2.5}$$

where

$$h_c = \frac{1-\mu/s}{1-\mu}. \tag{2.6}$$

Note that the case $h > h_c$ includes underdominance, i.e., $h > 1$. If $h < h_c$, then $\hat{q}^{(2)}$ is biologically not meaningful because $\hat{q}^{(2)} > 1$. If this holds, then the equilibrium $\hat{q}^{(1)}$

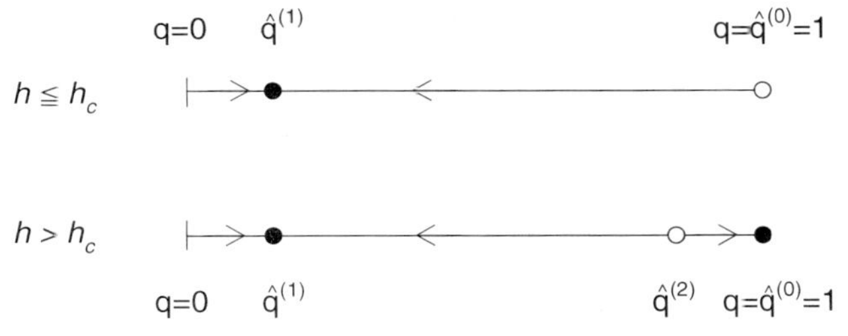

Figure 2.1 Equilibria and dynamics for the diallelic mutation-selection equation with one-way mutation. The drawing on top displays the case $h \le h_c$, that on bottom is for $h > h_c$. Stable equilibria are indicated by $\bullet$, unstable ones by $\circ$.

is globally asymptotically stable. If $h > h_c$, then three equilibria coexist. They satisfy $0 < \hat{q}^{(1)} < \hat{q}^{(2)} < \hat{q}^{(0)} = 1$, and $\hat{q}^{(1)}$ and $\hat{q}^{(0)}$ are asymptotically stable, whereas $\hat{q}^{(2)}$ is unstable (see Figure 2.1). Thus, for one-way mutation, a simple and explicit classification of the stability of equilibria is available. It can be shown that this is also valid for the case $0 < \nu \ll \mu$. Then, of course, $\hat{q}^{(0)} < 1$ and $\hat{q}^{(0)} \approx 1$. Analogous results hold for the differential equation (2.2).

We point out that for $h_c \le h \le 1$ the pure selection model has one globally asymptotically stable boundary equilibrium ($\hat{q} = 0$), but the introduction of mutation, however weak, leads to two stable and one unstable equilibria. Thus, already with two alleles, the diploid mutation-selection dynamics may be qualitatively different from the haploid dynamics.

Assuming that μ is of smaller order than s, simple approximations for the equilibrium frequencies can be derived in the following cases:

If $h = 0$, then

$$\hat{q}^{(1)} = \sqrt{\frac{\mu}{s}}\,. \tag{2.7a}$$

If $h \gg \sqrt{\mu/s}$, then

$$\hat{q}^{(1)} \approx \frac{\mu}{hs}\,. \tag{2.7b}$$

If $h > h_c$, then the unstable interior equilibrium is admissible and satisfies

$$\hat{q}^{(2)} \approx \frac{h}{2h-1} - \frac{\mu}{hs}\,. \tag{2.7c}$$

For multiplicative selection coefficients, $W_{11} = 1$, $W_{12} = 1 - t$, $W_{22} = (1-t)^2$, one obtains (exactly)

$$\hat{q}^{(1)} = \frac{\mu}{t}\,. \tag{2.7d}$$

The equilibrium frequency of a recessive deleterious mutant ($h = 0$) is much higher than that of an intermediate or dominant deleterious mutant ($h \gg \sqrt{\mu/s}$), because it occurs mostly in heterozygotes, against which selection is ineffective.

The case of weak mutation can also be treated by perturbation arguments. If $s > 0$ and $h < 1$, then introduction of sufficiently weak mutation [such that $h < (1-\mu/s)/(1-\mu)$; cf. (2.6)] leads only to very small disturbances of the equilibria which, in particular, maintain their stability properties. Then a stable equilibrium at the boundary will move inwards, and only the situation displayed in the upper drawing of Figure 2.1 can occur. The case $h = 1$ is nonregular (see Section 3.1) and this simple perturbation argument does not apply. It is generally held that the case of weak mutation is biologically the most important, and we shall return to it in Section 3.1.

2.3 STABILITY AND CYCLING

The analysis of the diallelic model demonstrates that equilibria under mutation-selection balance in diploid populations are, in general, not uniquely determined. Indeed, even more complicated behavior, such as stable periodic orbits, can occur. Only a few sufficient conditions on the selection regime and/or mutation matrix are known that imply uniqueness and global stability. The first is a simple consequence of what has been shown for haploid populations.

• **2.2** *If, in discrete time, fitness is multiplicative and the matrix defined by* (1.3) *is primitive, then the mutation-selection dynamics* (2.1) *admits a unique, completely polymorphic equilibrium* $\hat{\boldsymbol{p}}$. *This equilibrium is globally asymptotically stable. In continuous time, an analogous result holds for the dynamics* (2.2) *if Malthusian fitness is additive and the matrix defined by* (1.11) *is irreducible. These statements apply in particular to the neutral mutation model* I(6.1).

Further progress can be achieved if the mutation rates are assumed to satisfy

$$\mu_{ij} = \mu_j \quad \text{for } j \neq i \;, \tag{2.8}$$

i.e., if the mutation rates depend only on the target genes. Following Kingman (1977, 1978), this is called the *house-of-cards (HC) mutation model.* (Actually, Wright (1949, 1969) used this assumption earlier to get stationary distributions in a stochastic model; cf. Appendix E.3.) As we shall see in Chapter IV, this is a reasonable approximation in situations, in which most mutants are deleterious and the variance of the distribution of mutational effects is large compared with the existing variance in the population. An important special case of the HC-mutation model is obtained if all mutation rates are equal, i.e., if $\mu_{ij} = \mu$ for every $i \neq j$. Besides, (2.8) is automatically satisfied in the case of two alleles per locus.

Let us denote the total mutation rate by

$$\mu = \sum_{i=1}^{k} \mu_i \;, \tag{2.9}$$

and assume $\mu \leq 1$. Adapting an argument of Nagylaki (1977; 1992, Chapter 6.3), we will show that

$$\tilde{V}(\boldsymbol{p}) = \overline{W}(\boldsymbol{p})^{1-\mu} \prod_i p_i^{2\mu_i} \tag{2.10}$$

is a Lyapunov function for (2.1) and, consequently, plays a role generalizing that of mean fitness in the pure selection model.

▷ A straightforward calculation shows that with the HC-mutation rates (2.8), the recursion (2.1) can be written as

$$p_i' = p_i\omega_i/\bar{\omega} = p_i\frac{\partial\tilde{V}}{\partial p_i}\bigg/\sum_j p_j\frac{\partial\tilde{V}}{\partial p_j}\,, \tag{2.11}$$

where

$$\omega_i = (1-\mu)W_i + \frac{\mu_i\overline{W}}{p_i} \quad\text{and}\quad \bar{\omega} = \sum_i \omega_i p_i = \overline{W}\,. \tag{2.12}$$

We can approximate $\tilde{V}$ as closely as desired by replacing the real numbers μ_i and μ by rational numbers ν_i and ν, and call this function V_*. Then there exists a positive integer l such that $F = V_*^l$ is a homogeneous polynomial in $\boldsymbol{p}$ with nonnegative coefficients. Therefore, with arbitrarily high accuracy, (2.11) holds for F instead of $\tilde{V}$. Now it follows directly from the inequality of Baum and Eagon (Appendix A.1) that F is nondecreasing along orbits and the change in F is zero only at equilibrium. This implies that $\tilde{V}$ is a Lyapunov function that plays the same role as mean fitness $\overline{W}$ in the pure selection model. ◁

With HC-mutation rates (2.8), the differential equation (2.2) can be written as

$$\dot{p}_i = p_i(f_i - \bar{f})\,, \tag{2.13}$$

where

$$f_i = m_i + \frac{\mu_i}{p_i} \quad\text{and}\quad \bar{f} = \sum_i f_i p_i = \overline{m} + \mu\,. \tag{2.14}$$

Then, as proved by Passekov (1978) and Hofbauer (1985),

$$V(\boldsymbol{p}) = \tfrac{1}{2}\overline{m} + \sum_i \mu_i \ln p_i \tag{2.15}$$

is a global Lyapunov function, because

$$\dot{V}(\boldsymbol{p}) = \sum_i \frac{\partial V}{\partial p_i}\dot{p}_i = \sum_i f_i p_i (f_i - \bar{f}) = \sum_i p_i (f_i - \bar{f})^2 \geq 0\,. \tag{2.16}$$

Obviously, the above derivations do not require strict positivity of the mutation rates μ_i. In mathematical terms, (2.13) is a gradient system, a so-called Svirezhev–Shahshahani gradient (Appendix A.3), in which V plays the role of a potential. In general, the mutation-selection equation (2.2) is not a gradient system, unless (2.8) is satisfied. Summarizing, we have shown the following:

• **2.3** *If mutation rates satisfy the HC-condition* (2.8), *then all orbits of the mutation-selection equations* (2.1) *and* (2.2) *converge to the set of equilibria. These are given by the solutions of the equations* $\omega_i(\boldsymbol{p}) = \text{const.}$ $(i = 1, \ldots, k)$ *for* (2.1), *and of* $f_i(\boldsymbol{p}) = \text{const.}$ $(i = 1, \ldots, k)$ *for* (2.2).

An interesting question is, under which assumptions an equilibrium that is stable in the pure selection model remains stable after the introduction of mutation. The following result was derived by Hofbauer (1985) and generalizes a theorem of Hadeler (1981).

• **2.4** *Suppose the pure selection model admits an asymptotically stable, completely polymorphic equilibrium. Then, for every choice of mutation rates satisfying the HC-condition* (2.8) *with* $\mu \leq 1$, *the mutation-selection equations* (2.1) *and* (2.2) *have exactly one equilibrium, which is globally asymptotically stable. The conclusion remains valid if the pure selection equilibrium is not isolated, but* $\mu_i > 0$ *for every* i.

▷ In view of the result • **2.3**, it is sufficient to show that the mutation-selection equilibrium is unique. We already know from Chapters I.9.5 and I.10.3 that a stable, completely polymorphic equilibrium is a strict interior maximum of the mean fitness function $\overline{W} = \overline{W}(\boldsymbol{p})$, or $\overline{m} = \overline{m}(\boldsymbol{p})$. Hence, the quadratic forms $\overline{W}$ and $\overline{m}$ are strictly concave on S_k, and so is $\ln \overline{W}$. Also $\sum_i \mu_i \ln p_i$ is concave. Therefore, the Lyapunov function $V(\boldsymbol{p})$ for the continuous-time model is strictly concave on S_k and can have only one critical point, which is the global maximum. If the pure selection equilibrium is not isolated, but $\mu_i > 0$ for every i, then $\sum_i \mu_i \ln p_i$ is strictly concave, and the assertion follows. In the case of discrete time, observe that the differential equation resulting from (2.1) by replacing $p_i' - p_i$ by $\dot{p}_i$ has the same equilibria as (2.1) (cf. Section 1.2). For this differential equation, a calculation parallel to (2.16) shows that

$$\ln \tilde{V}(\boldsymbol{p}) = (1-\mu) \ln \overline{W}(\boldsymbol{p}) + 2 \sum_i \mu_i \ln p_i$$

is a Lyapunov function that is strictly concave under the given assumptions (Hofbauer 1985). Hence, it has a unique critical point, and this is the globally asymptotically stable equilibrium of (2.1). ◁

The conclusion of the above result does not hold if selection alone produces a globally asymptotically stable equilibrium at the *boundary* of S_k. This follows from the diallelic case considered previously (take, e.g., $h = 1$ and $0 < \nu \ll \mu \leq s/4$, so that the equilibrium corresponding to $\hat{q}^{(2)}$ is in $[0, 1]$). The result • **2.4** cannot be extended to general mutation rates, because a theorem of Akin (1979) implies that for every mutation matrix (μ_{ij}) not satisfying (2.8), a fitness matrix (m_{ij}) can be chosen such that periodic orbits occur in the continuous-time mutation-selection dynamics (2.2). This is also true for the discrete-time dynamics.

Hofbauer (1985) gave an explicit example of stable limit cycles in the continuous-time mutation-selection equation by assuming that all heterozygotes have equal fitness, all homozygotes have equal but higher fitness, and that the mutation rates are cyclically symmetric, i.e., $\mu_{ij} = \mu_{j-i}$. A different class of examples was provided by Baake and Wiehe (1997). They showed that stable limit cycles can exist in three-allele systems in which homozygotes may have different fitnesses and heterozygotes are intermediate, though with some degree of dominance. All these examples require that the strengths of selection and mutation are comparable, in a sense. The biological relevance of these cycling results is not easy to judge because, on the one hand, mutation rates are often much smaller than selection coefficients. On the other hand, at the molecular level, many mutations are likely to be neutral or nearly neutral, but then stochastic events might dominate evolution.

For sufficiently small mutation rates, the combined action of selection and mutation can be treated as a perturbation of the pure selection equation. Then (externally) stable boundary equilibria will move inwards and remain asymptotically stable (but they need not remain globally asymptotically stable, even if they are so without mutation), and some unstable boundary equilibria will leave the simplex. This perturbation-theoretic approach will be applied in the following section; but no quantitative results exist on how large such perturbations can be without destroying the stability properties.

3. EQUILIBRIUM MEAN FITNESS AND GENETIC LOADS

Haldane (1937) investigated the effect of recurrent deleterious mutation on the equilibrium mean fitness. For sufficiently large population size he found that

> ... the loss of fitness to the species depends entirely on the mutation rate and not at all on the effect of the gene upon fitness of the individual carrying it ....

This is often called Haldane's principle.

The investigation of the influence of deleterious mutations on the mean fitness of a population is of considerable biological interest. In a sufficiently large population, deleterious mutants are kept at low frequency by selection, and their main effect is to reduce the mean fitness. In small populations, however, they exert additional effects, because they may become fixed by random genetic drift, and they may lead to a process, known as Muller's ratchet, during which deleterious mutants accumulate on chromosomes. Chromosomes with zero or few deleterious mutations will be lost by random genetic drift and, if there is little or no recombination (as in asexuals), chromosomes with fewer mutations will not be re-established. Both of these processes result in a decrease of mean fitness, and may lead to eventual extinction of a population. Other interesting applications concern the evolutionary significance of sex and recombination (see Chapter VII.5).

Therefore, recurrent deleterious mutations impose a 'load' on a population by reducing their mean fitness below the maximum possible level. The word load was coined by Muller (1950), who rediscovered Haldane's principle. The genetic load is usually defined as the proportion by which the fitness of the average genotype in a population is reduced in comparison with the best genotype, i.e.,

$$L = \frac{W_{\max} - \overline{W}}{W_{\max}} \,. \tag{3.1}$$

The concept of genetic load has often been criticized on the grounds that a best genotype does not exist in a real population, because fitnesses change in time and may be frequency dependent; or if recombination is taken into account and the best type exists in theory, it is extremely unlikely to ever be produced. Of course, the idealizations on which every model is based have to be kept in mind when applying it. Here, the word genetic load will be used in a technical sense as a synonym for the reduction in mean fitness, caused by one or several genetic mechanisms, relative to the idealized situation when these mechanisms are absent or neglected.

We shall commence with the analysis of the mutation-selection model with multiple alleles under the assumption that mutation is much weaker than selection, and derive some properties of the equilibrium distribution and approximations for the resulting mutation load. Then we shall discuss other kinds of genetic loads.

3.1 THE MUTATION LOAD

Haldane's (1937, 1957) work on the mutation load has been extended in several directions and applied to various subjects (e.g. Muller 1950, Crow and Kimura 1964, King 1966, Kimura and Maruyama 1966, Crow 1970, Fraser and Mayo 1974, Kondrashov 1982, Kondrashov and Crow 1988, Crow 1992). In all these investigations, mutations were assumed to be unconditionally deleterious, and back mutations to the optimal type were ignored. As is well known (see Kimura and Maruyama 1966, and Crow 1970), the mutation load is essentially the same in asexual and sexual populations if a single locus is considered. Under the assumption that mutation is weaker than selection, we shall extend Haldane's principle to 'arbitrary' mutation patterns among alleles. As a warm up, we start with the simple asexual case.

Asexual populations

We consider the discrete-time mutation-selection equation (1.1) and assume that $\mathcal{A}_1$ is the fittest allele, i.e., $W_1 > W_i$ for every $i = 2, \ldots, k$. Then $\hat{\boldsymbol{p}}^o = (1, 0, \ldots, 0)$ is the globally asymptotically stable equilibrium of (1.1) without mutation. Hence, the perturbed system (with μ_{ij} small) has an asymptotically stable equilibrium nearby, $\hat{\boldsymbol{p}}(\mu)$, which satisfies $\hat{\boldsymbol{p}}^o - \hat{\boldsymbol{p}}(\mu) = O(\mu)$ as $\mu \to 0$, because mutation is a linear perturbation. Therefore, to first order we have

$$\hat{\bar{W}} \hat{p}_i(\mu) = \hat{p}_i(\mu) W_i + W_1 \mu_{1i} \,, \quad i \geq 2 \,,$$

or

$$\hat{p}_i(\mu) \approx \frac{W_1 \mu_{1i}}{W_1 - W_i} \,, \quad i \geq 2 \,, \tag{3.2}$$

because replacing $\mu_{1i}/(\hat{\bar{W}} - W_i)$ by $\mu_{1i}/(W_1 - W_i)$ introduces an error smaller than $O(\mu)$. It follows that

$$\begin{aligned} \hat{\bar{W}} &= W_1 \left(1 - \sum_{i=2}^{k} \hat{p}_i(\mu)\right) + \sum_{i=2}^{k} W_i \hat{p}_i(\mu) \\ &= W_1 + \sum_{i=2}^{k} (W_i - W_1) \hat{p}_i(\mu) \\ &\approx W_1 - W_1 \sum_{i=2}^{k} \mu_{1i} = W_1 \left(1 - \sum_{i=2}^{k} \mu_{1i}\right) . \end{aligned} \tag{3.3}$$

Hence, in the first approximation, the mutation load is

$$L = \sum_{i=2}^{k} \mu_{1i} \,, \tag{3.4}$$

which is the total mutation rate of the fittest allele to all others.

If there is no back mutation to the fittest allele, i.e., if $\mu_{j1} = 0$ for $j > 1$, then the formula (3.4) is even exact: take $i = 1$ in (1.1) and cancel through p_1. This is a well known result (for continuous time, see Crow 1970, p. 148). If there are two or more alleles with maximal fitness, the mutation load may strongly depend on the pattern of mutation between optimal alleles.

Example. Consider three alleles, $\mathcal{A}_1$, $\mathcal{A}_2$, $\mathcal{A}_3$, with fitness values 1, 1, $1-s$, respectively, and mutation rates $\mu_{ij} = \mu u_{ij}$ such that $u_{32} = u_{31} = 0$, i.e., there is no back mutation from $\mathcal{A}_3$ to $\mathcal{A}_1$ and $\mathcal{A}_2$. Moreover, we assume that $u_{12}+u_{13} = 1$, $u_{21}+u_{23} = 1$, and recall that $u_{ii} = 0$. Then the equilibrium mean fitness is $\overline{W} = 1-\mu\left(1-\sqrt{u_{12}u_{21}}\right)$. Therefore, the load can attain any value between 0 ($u_{12} = u_{21} = 1$) and μ ($u_{12} = 0$ or $u_{21} = 0$), but it is still independent of the selection coefficient s (because there is no back mutation). ◇

Diploid populations

We shall derive a first-order approximation for the allele frequencies and the mean fitness at mutation-selection balance, and we estimate the order of magnitude of the error made as $\mu/s \to 0$.

▷ Instead of (2.1), let us first consider the more general difference equation

$$\Delta p_i^o = (p_i^o)' - p_i^o = p_i^o f_i(\boldsymbol{p}^o)\ , \quad i = 1,\dots,k\ , \tag{3.5}$$

and its perturbation

$$\Delta p_i = F_i(\boldsymbol{p},\varepsilon) = p_i f_i(\boldsymbol{p}) + \varepsilon v_i(\boldsymbol{p})\ , \quad i = 1,\dots,k\ . \tag{3.6}$$

Suppose, $\hat{\boldsymbol{p}}^o$ is a regular[2] and saturated, or externally stable (see Appendix A.2), equilibrium point of the unperturbed equation (3.5). After renumbering the indices, this means

$$\hat{p}_i^o > 0 \text{ and } f_i(\hat{\boldsymbol{p}}^o) = 0\ , \quad 1 \le i \le l\ , \tag{3.7a}$$

and

$$\hat{p}_i^o = 0 \text{ and } f_i(\hat{\boldsymbol{p}}^o) < 0\ , \quad l < i \le k\ , \tag{3.7b}$$

because in (3.7b) the case $f_i(\hat{\boldsymbol{p}}^o) = 0$ is excluded by the regularity assumption on $\hat{\boldsymbol{p}}^o$; see also Remark 3 below. Note that (3.7b) states that the growth rates $\Delta p_i^o/p_i^o = f_i(\hat{\boldsymbol{p}}^o)$ at $\hat{\boldsymbol{p}}^o$ are negative, which means that none of the absent alleles can invade the equilibrium state $\hat{\boldsymbol{p}}^o$. If only one allele is present at $\hat{\boldsymbol{p}}^o$, then $l = 1$ and (3.7b) implies stability of $\hat{\boldsymbol{p}}^o$. Furthermore, every (stable or unstable) interior fixed point $\hat{\boldsymbol{p}}^o$ is externally stable, because then $l = k$, and (3.7b) is an empty condition.

By the implicit function theorem, there is a smooth curve $\hat{\boldsymbol{p}}(\varepsilon)$ of equilibrium points of (3.6) that is defined for small ε, such that $\hat{\boldsymbol{p}}(0) = \hat{\boldsymbol{p}}^o$ and

$$\hat{\boldsymbol{p}}(\varepsilon) = \hat{\boldsymbol{p}}^o + \varepsilon \boldsymbol{q} + O(\varepsilon^2)\ , \tag{3.8}$$

where $\boldsymbol{q}$ is the solution of the linearized equation

$$D_{\boldsymbol{p}}F(\hat{\boldsymbol{p}}^o,0)\boldsymbol{q} + D_\varepsilon F(\hat{\boldsymbol{p}}^o,0) = 0\ .$$

[2] The equilibrium $\hat{\boldsymbol{p}}^o$ is regular if the Jacobian matrix $D_{\boldsymbol{p}}F(\hat{\boldsymbol{p}}^o,0)$ of first-order partial derivatives of $F = (F_1,\dots,F_k)$ evaluated at $(\hat{\boldsymbol{p}}^o,0)$ is invertible; cf. Appendix A 1.

For the components with $\hat{p}_i = 0$, this simplifies to

$$f_i(\hat{\boldsymbol{p}}^o)q_i + v_i(\hat{\boldsymbol{p}}^o) = 0\,, \quad l < i \leq k\,,$$

so that (3.8) turns into

$$\hat{p}_i(\varepsilon) = -\varepsilon\frac{v_i(\hat{\boldsymbol{p}}^o)}{f_i(\hat{\boldsymbol{p}}^o)} + O(\varepsilon^2)\,, \quad l < i \leq k\,. \tag{3.9}$$

Equation (3.9) reflects the well known fact that a regular, externally stable fixed point of (3.5) moves into the interior of the state space after introduction of mutation terms, whereas an externally unstable fixed point (with at least one of the inequalities in (3.7b) reversed) will move out for $\varepsilon > 0$, thus becoming unacceptable. Hence, the calculation of the mutation load makes sense only at externally stable equilibria.

Let us now return from (3.6) to the mutation-selection equation (2.1) and set

$$f_i(\boldsymbol{p}) = \frac{W_i(\boldsymbol{p}) - \overline{W}(\boldsymbol{p})}{\overline{W}(\boldsymbol{p})}\,. \tag{3.10}$$

Then we obtain from the definition of mean fitness, using $\boldsymbol{p} = \hat{\boldsymbol{p}}^o + (\boldsymbol{p} - \hat{\boldsymbol{p}}^o)$, (3.10), and (3.7),

$$\overline{W}(\boldsymbol{p}) = \overline{W}(\hat{\boldsymbol{p}}^o) + 2\overline{W}(\hat{\boldsymbol{p}}^o)\sum_{i=l+1}^{k} p_i f_i(\hat{\boldsymbol{p}}^o) + \overline{W}(\boldsymbol{p} - \hat{\boldsymbol{p}}^o)\,. \tag{3.11}$$

Taking $\boldsymbol{p} = \hat{\boldsymbol{p}}(\varepsilon)$, substituting (3.9), and using (3.8) together with the homogeneity of $\overline{W}$, we get

$$\overline{W}(\hat{\boldsymbol{p}}(\varepsilon)) - \overline{W}(\hat{\boldsymbol{p}}^o) = -2\varepsilon\overline{W}(\hat{\boldsymbol{p}}^o)\sum_{i=l+1}^{k} v_i(\hat{\boldsymbol{p}}^o) + O(\varepsilon^2)\,. \tag{3.12}$$

With $\mu_{ij} = \varepsilon u_{ij}$ and

$$v_i(\boldsymbol{p}) = \frac{1}{\overline{W}}\sum_{j=1}^{k}(p_j W_j u_{ji} - p_i W_i u_{ij})\,, \tag{3.13}$$

we arrive at the desired approximation

$$\frac{\overline{W}(\hat{\boldsymbol{p}}^o) - \overline{W}(\hat{\boldsymbol{p}}(\varepsilon))}{\overline{W}(\hat{\boldsymbol{p}}^o)} = 2\varepsilon\sum_{i=l+1}^{k}\sum_{j=1}^{l}\hat{p}_j^o u_{ji} + O(\varepsilon^2) \tag{3.14}$$

as $\varepsilon \to 0$. ◁

Therefore, the *first-order approximation* of the mutation load,

$$L = 2\sum_{i=l+1}^{k}\sum_{j=1}^{l}\hat{p}_j^o \mu_{ji}\,, \tag{3.15}$$

has the same biological meaning as the total rate of mutations occurring at the equilibrium $\hat{\boldsymbol{p}}^o$ and leading to new (less fit) alleles, with a factor 2 because of diploidy.

Remarks and Examples. 1. If one homozygote (the wild type) has higher fitness than all other genotypes, i.e., $W_{11} > W_{ij}$ for all $(i,j) \neq (1,1)$, then the above analysis applies with $\hat{\boldsymbol{p}}^o = (1,0,\dots,0)$ and $\mu_{ij} = \mu u_{ij}$. Therefore, (3.14) yields

$$\overline{W}(\hat{\boldsymbol{p}}(\mu)) \approx W_{11}\left(1 - 2\sum_{j=2}^{k} \mu_{1j}\right), \tag{3.16}$$

and the error made in (3.16) is of order $O(\mu^2)$ as $\mu \to 0$. Thus, to first order, the mutation load is equal to the total mutation rate to deleterious alleles,

$$L = 2\sum_{j=2}^{k} \mu_{1j} . \tag{3.17}$$

In the absence of back mutations, (3.16) and (3.17) are exact, as has long been known (for continuous time, see Crow 1970, pp. 136–138; for discrete time, Nagylaki 1977, 1992, Eq. 4.102).

2. Equations (3.9) and (3.7a) provide simple approximations for the allele frequencies at mutation-selection balance. In the above example, set $W_{11} = 1$ and $W_{ij} = 1 - s_{ij}$ with $s_{ij} > 0$ if $(i,j) \neq (1,1)$. Then it follows that at mutation-selection equilibrium, $\hat{\boldsymbol{p}} = \hat{\boldsymbol{p}}(\mu)$, the frequencies of the deleterious alleles are

$$\hat{p}_i \approx \frac{\mu_{1i}}{s_{1j}} , \quad i \geq 2 . \tag{3.18a}$$

Defining the average selection coefficient of the deleterious $\mathcal{A}_1\mathcal{A}_i$ heterozygotes by $\tilde{s} = (\sum_{i\geq 2} \hat{p}_i s_{1i})/(1-\hat{p}_1)$, we find that the total frequency of deleterious mutations is

$$1 - \hat{p}_1 \approx \frac{1}{\tilde{s}} \sum_{i\geq 2} u_{1i} . \tag{3.18b}$$

This is the ratio of the total forward mutation rate to the mean wild-type/mutant heterozygous selection coefficient. For continuous time and in the absence of back mutation, (3.18) may be found in Crow and Kimura (1970, Chapter 6.4); for discrete time in Nagylaki (1977; 1992, Chapter 4.9).

3. Let us assume in (2.1) that selection is weak, but mutation is much weaker, i.e., $W_{ij} = 1 + s\omega_{ij}$, where the ω_{ij} are of order 1 and s is small, and $\mu_{ij} = \mu u_{ij}$ with $\mu \ll s$. If we set $f_i = (\omega_i - \bar{\omega})/\overline{W}$ in (3.10), then we obtain from (3.12), where now the left-hand side is $\bar{\omega}(\hat{\boldsymbol{p}}(\varepsilon)) - \bar{\omega}(\hat{\boldsymbol{p}}^o)$, and with $\varepsilon = \mu/s$ and v_i as in (3.13),

$$\overline{W}(\hat{\boldsymbol{p}}(\mu/s)) = \overline{W}(\hat{\boldsymbol{p}}^o)\left(1 - 2\sum_{i=l+1}^{k}\sum_{j=1}^{l} \hat{p}_j^o \mu_{ji}\right) + O(\mu^2/s) \tag{3.19}$$

as $\mu \to 0$, $s \to 0$, and $\mu/s \to 0$. Therefore, the mutation load is approximately given by (3.15), and the error is of order μ^2/s. (Clearly, this requires that our general assumption is satisfied, i.e., that $\hat{\boldsymbol{p}}^o$ is a regular and externally stable equilibrium of the pure selection equation.)

4. If, in the diallelic case, the mutant $\mathcal{A}_2$ is completely recessive, i.e., the fitness of $\mathcal{A}_1\mathcal{A}_2$ equals the fitness of $\mathcal{A}_1\mathcal{A}_1$, then the above argument does not apply because the equilibrium $\hat{\boldsymbol{p}}^o = (1,0)$ is not regular, since $f_i(\hat{\boldsymbol{p}}^o) = 0$ and the fitness has zero slope at $\hat{\boldsymbol{p}}^o$. Direct calculation shows that

$$\hat{p}_2(\varepsilon) = \varepsilon^{1/2} + O(\varepsilon) \quad (\text{with } \varepsilon = \mu/s) \quad \text{and} \quad L = \mu + O(\mu\sqrt{\mu/s}) \tag{3.20}$$

instead of (3.9) and (3.17) (cf. (2.7) and Crow 1970, p. 137).

5. If $l = k$, that is, if $\hat{\boldsymbol{p}}$ is an interior equilibrium, then $L = 0$ in (3.15). This means that the mutation load in (3.14) is actually of smaller order $O(\varepsilon^2)$, or $O(\mu^2/s)$ following Remark 3. As an illustration, consider again the case of two alleles $\mathcal{A}_1$ and $\mathcal{A}_2$, with fitnesses of $\mathcal{A}_1\mathcal{A}_1$, $\mathcal{A}_1\mathcal{A}_2$, $\mathcal{A}_2\mathcal{A}_2$ being $1 - s_1$, 1, $1 - s_2$, respectively, and mutation rates $\mu_{12} = \mu_1$, $\mu_{21} = \mu_2$. Then the second-order approximation of the mutation load can be calculated to be:

$$L = (s_1 + s_2)\left(\frac{\mu_2}{s_2} - \frac{\mu_1}{s_1}\right)^2 = \frac{(\mu_2/p_1 - \mu_1/p_2)^2}{s_1 + s_2}. \tag{3.21}$$

Here, $\boldsymbol{p} = (p_1, p_2) = (s_2, s_1)/(s_1 + s_2)$ is the overdominant selection equilibrium. If mutation alone leads to the same equilibrium as selection, then $L = 0$ of course. Otherwise, L is of order μ^2/s. ◇

For the continuous-time model (2.2), the load has to be defined as the absolute reduction in Malthusian fitness, i.e., $L = m_{\max} - \bar{m}$. This case was investigated by Bürger and Hofbauer (1994), whose proofs and results are analogous to those above.

Let us summarize the main results:

• **3.1** *The first-order approximation of the mutation load is the total rate of mutations leading to new (less fit) alleles; cf.* (3.4) *and* (3.15). *The error term is of order* $O(\mu^2/s)$ *if the fitness differences between genotypes are of order* $O(s)$. *This holds for arbitrary mutation patterns, for haploid and diploid populations, for an arbitrary number of alleles, and in discrete and continuous time, provided the equilibrium without mutation is regular (i.e., not degenerate as for completely recessive mutants) and externally stable (i.e., if the equilibrium is at the boundary of the simplex, mutation pushes it into the simplex).*

In Chapter IV, we shall extend Haldane's principle to models such as the stepwise-mutation model and to quantitative traits. In Chapter VI, we shall see that this principle is closely related to the question of how much genetic variation can be maintained by a balance between stabilizing selection and mutation.

3.2 OTHER GENETIC LOADS

Besides the mutation load, many other kinds of genetic loads have been considered in the literature. For example, the segregation load (Haldane 1937) caused by inferior homozygotes that arise by segregation from better adapted heterozygotes, or the load caused by recombination if the chromosomes $\mathcal{A}_1\mathcal{B}_1$ and $\mathcal{A}_2\mathcal{B}_2$ have higher fitness than $\mathcal{A}_1\mathcal{B}_2$ and $\mathcal{A}_2\mathcal{B}_1$. Closely related to the genetic load is another concept introduced by Haldane (1957), namely 'the cost of natural selection' which is now called the substitutional load. Haldane considered an initially rare but favorable gene, and computed the amount of selective death, or cost, needed to carry out such a substitution. He defined the cost as the sum of all expressions of the form (3.1), but with $W_{\max}$ replaced by the average fitness of the favorable allele, and the summation being over all generations required for the substitution. The surprising result is that this substitutional load is roughly independent of the selection coefficient and depends only on the initial frequency of the favorable mutant. The substitutional load has played a central role in the development of the neutral theory of evolution (cf. Kimura 1983). A comprehensive treatment of the theory of genetic loads may be found in Crow (1970, 1992). For well-taken criticisms, in particular, concerning the substitutional load, see Ewens (1979).

Let us take a closer look at the segregation load, and assume that the fitnesses of the genotypes $\mathcal{A}_1\mathcal{A}_1$, $\mathcal{A}_1\mathcal{A}_2$, and $\mathcal{A}_2\mathcal{A}_2$ are $1-s_1$, 1, and $1-s_2$, respectively. Let p and $q=1-p$ denote the frequencies of $\mathcal{A}_1$ and $\mathcal{A}_2$. Then, at equilibrium (see Chapter I.9.3), the allele frequencies are $\hat{p}=s_2/(s_1+s_2)$ and $\hat{q}=s_1/(s_1+s_2)$, and the mean fitness is $1-s_1s_2/(s_1+s_2)$. Hence, the segregation load is

$$L=\frac{s_1s_2}{s_1+s_2}\,, \tag{3.22}$$

and, therefore, half the harmonic mean of s_1 and s_2. It can also be written as

$$L=s_1\hat{p}^2+s_2\hat{q}^2 \tag{3.23a}$$

$$=s_1\hat{p}=s_2\hat{q}\,. \tag{3.23b}$$

The first representation of L shows that the allele with the greatest selective disadvantage contributes the least to the segregation load, and the second shows that L can be computed from only one allele.

These results carry over to the multi-allelic case. Let the fitness of the genotype $\mathcal{A}_i\mathcal{A}_j$ be $W_{ij}=1-s_{ij}$, and let us consider a completely polymorphic equilibrium with $\hat{p}_i>0$ for $i=1,\dots,k$. Then, all marginal fitnesses satisfy $\hat{W}_i=\hat{\bar{W}}$. It follows that the load is

$$L=\sum\nolimits_j s_{ij}\hat{p}_j \qquad \text{for every } i\,. \tag{3.24}$$

This implies $L/s_{ii}\geq\hat{p}_i$ and, by summing over all i, we find $L\sum_i s_{ii}^{-1}\geq 1$. Denoting the harmonic mean of the s_{ii} by $\tilde{s}$, we obtain

$$L\geq\tilde{s}/k\,. \tag{3.25}$$

This extends the principles mentioned in the diallelic case that (i) the segregation load can be estimated from the frequency and the homozygous effect of one gene, and that (ii) genes which are less harmful contribute more to the load than highly disadvantageous ones (see Crow 1970). The Fundamental Theorem of Natural Selection implies that the segregational load is minimized by selection. It is arguable that the segregation load is not always a load for a population, because heterozygote advantage often occurs by hybridization of homozygous individuals from different populations, and this confers an advantage, not a load.

We shall encounter Haldane's principle on the mutation load several times, but also recombination, segregation, and other loads.

4. THE STEPWISE-MUTATION MODEL

The *stepwise-mutation*, or *ladder, model* was introduced by Ohta and Kimura (1973, 1975) to estimate the number of electrophoretically detectable alleles. An electrophoretic analysis measures the electrophoretic mobility of proteins and detects mutations that cause a change in the electric charge and, hence, in the electrophoretic mobility. Therefore, this model assumes that allelic states are expressed by integers, and that if an allele changes its state (charge) by a single-step mutation, the change occurs so that it moves either one step in the positive direction or one step in the negative direction. Recently, this model has been applied to describe mutation at microsatellite loci (e.g., Kimmel and Chakraborty 1996). Stochastic and deterministic variants of this model have been investigated. In the former type of model, there is no selection and the distribution of allele frequencies is due to a balance between mutation and random genetic drift (see Kimura 1983 for a review). In the latter type, there is selection but the population is assumed to be infinitely large.

Here we deal with a generalized version of the deterministic model that ignores random genetic drift (see Moran 1976, 1977, and Kingman 1977, 1980). The population is assumed to consist of haploid individuals which reproduce asexually. Let I be a countable set, e.g. $I = \{\dots, -2, -1, 0, 1, 2, \dots\}$, as we shall assume here without loss of generality. We consider a single locus at which the possible alleles are $\mathcal{A}_i$, $i \in I$. The relative frequency of $\mathcal{A}_i$ in generation t is $p_i(t)$, so that $\sum_i p_i(t) = 1$, where $\sum_i$ always means $\sum_{i \in I}$. Let $\mathcal{A}_i$ have fitness W_i with

$$0 < W_i \leq 1 \,. \tag{4.1}$$

The mutation rate from $\mathcal{A}_i$ to $\mathcal{A}_j$ is μ_{ij} ($\mu_{ii} = 0$), and it is assumed that

$$\sum_j \mu_{ij} < 1 \quad \text{for every } i \,. \tag{4.2}$$

We define the infinite-dimensional mutation matrix $\tilde{\mathsf{U}} = (\tilde{u}_{ij})$ by

$$\tilde{u}_{ij} = \begin{cases} 1 - \sum\limits_l \mu_{il} \,, & i = j \,, \\ \mu_{ji} \,, & i \neq j \,. \end{cases} \tag{4.3}$$

This matrix is *stochastic*, i.e., it satisfies $\sum_j \tilde{u}_{ji} = 1$ for every i. We assume that $\tilde{\mathsf{U}}$ is *irreducible* and *aperiodic* which, for a finite matrix, is the same as primitive (see Appendix B.2 for definitions). Furthermore, we define the matrix $\mathsf{C} = (c_{ij})$ by

$$c_{ij} = \tilde{u}_{ij} W_j \,, \tag{4.4}$$

and note that, due to (4.1) and (4.2), the matrix $\mathsf{C}^t = (c_{ij}^{(t)})$ has finite entries for every integer t, and the spectral radius $r = r(\mathsf{C})$ satisfies

$$0 < r \leq 1 \,. \tag{4.5}$$

Then, as in Section 1.1, the mutation-selection dynamics is given by

$$\boldsymbol{p}' = \frac{1}{\bar{c}}\,\mathsf{C}\boldsymbol{p}\,, \tag{4.6}$$

where

$$\bar{c} = \sum_i (\mathsf{C}\boldsymbol{p})_i = \overline{W} \tag{4.7}$$

is the mean fitness. Equation (4.6) also applies to diploid populations if fitnesses are multiplicative.

The following example shows that in an infinite state space I, additional assumptions are necessary to obtain convergence to an equilibrium distribution. Let $W_i = 1$ for every i and assume $p_o(0) = 1$, for some $o \in I$, and $p_i(0) = 0$, otherwise. Then the solution of (4.6) is

$$p_i(t) = c_{io}^{(t)} = \tilde{u}_{io}^{(t)}\,. \tag{4.8}$$

If mutation occurs only in single steps so that every allele $\mathcal{A}_i$ mutates to its nearest neighbours, $\mathcal{A}_{i-1}$ and $\mathcal{A}_{i+1}$, with probabilities ν and μ, respectively, and to no other alleles, then the sequence $p_i(t)$ tends to zero for every i because there is no selection. (In mathematical terms, this mutation matrix is not positive recurrent; see Appendix B.2.) In practice, the index set I could never be infinite, but when I is very large, this dissipative behavior corresponds to an equilibrium distribution that is very thinly spread over I. Therefore, conditions are needed which ensure that a few alleles are sufficiently advantageous to prevent the population spreading itself too thinly over many different alleles.

Without proof, we state the following theorem of Kingman (1977):[3]

• **4.1** *If*

$$\limsup_{i\in I} W_i < r, \tag{4.9}$$

then there is a unique probability distribution $\hat{\mathbf{p}} = (\hat{p}_i)_{i\in I}$, *satisfying*

$$\mathsf{C}\hat{\boldsymbol{p}} = r\hat{\boldsymbol{p}}, \tag{4.10}$$

and every solution of the mutation-selection equation (4.6) *converges to* $\hat{\mathbf{p}}$, *provided the initial distribution* $\mathbf{p}(0)$ *has a finite support* (*i.e.*, $p_i(0) > 0$ *only for a finite number of indices*).

We note that assumptions (4.1) and (4.2) together with the well known estimate $r \le \sup_j \sum_i |c_{ij}|$ for the spectral radius yield the inequality $r \le \sup_i W_i$. Therefore, condition (4.9) implies that each given set of fitness values $\{W_i\}$ assumes a maximum. It also implies that only finitely many W_i are greater or equal than r. Obviously, (4.9) is satisfied if the W_i decay to zero as $|i| \to \infty$. Summation over all components in the left- and right-hand sides of (4.10) leads, in conjunction with (4.7), to

$$r = \hat{\bar{c}} = \hat{\overline{W}}\,. \tag{4.11}$$

---

[3] $\limsup W_i$ designates the limit superior of the sequence W_i, i.e., the largest of its accumulation points.

As r may be difficult to compute, condition (4.9) is not very useful. Therefore, let J be some finite subset of I, let C_J denote the corresponding submatrix, and r_J its spectral radius. Then we have $r = \sup_J r_J$ (Seneta 1981, Theorem 6.8), and (4.9) holds if and only if

$$\limsup_{i \in I} W_i < r_J \tag{4.12}$$

for some finite J. Taking $J = \{j\}$, $j \in I$, it is sufficient that

$$\limsup_{i \in I} W_i < W_j \big(1 - \sum_l \mu_{jl}\big) . \tag{4.13}$$

Condition (4.12) has a simple genetic interpretation (Kingman 1977): because r_J is the mean equilibrium fitness of the population when all alleles outside J are rendered lethal, (4.12) simply means that the equilibrium mean fitness of such a population is higher than all but a finite number of the original fitness values W_i. In the single-step mutation model, in which $\mu_{ij} = \frac{1}{2}\mu$ if $|i - j| = 1$ and $\mu_{ij} = 0$ otherwise, the assumptions $W_0 = 1$ and $W_i = W_{-i} < 1$ for every $i \neq 0$ lead to the sufficient condition

$$\limsup_{i \in I} W_i < 1 - \mu , \tag{4.14}$$

a result obtained by Moran (1977).

Moran (1977) and Kingman (1977) give examples showing that (4.14) and (4.9) are not necessary conditions for the existence of a unique equilibrium distribution. For instance, if $\mu_{ij} = \mu_j$ for every i, j and $\sum_j \mu_j = 1$, as in the HC-mutation model (2.8), then $\hat{p}_i = \mu_i$ holds independently of the fitness values. (In a diploid model with HC-mutation rates, the equilibrium distribution is, of course, dependent on the fitness values.)

Moran (1977) presented a nice example, showing that it is possible to turn a model with a stable equilibrium distribution into a model with no solution by *reducing* the value of one of the W_i. He considered the single-step mutation model with $\mu < 1$, $W_0 = 1$, $W_i = W_{-i}$, $1 - \mu < W_i = W < 1$ for every $i > 1$, and $W_1 \leq W$. Then (4.14) is not satisfied, but Moran could show that, nevertheless, a unique, globally asymptotically stable equilibrium distribution exists if $W_c < W_1 \leq W$ for some critical value $W_c > 0$, whereas such an equilibrium does not exist if $W_1 < W_c$. Thus, the alleles $\mathcal{A}_1$ and $\mathcal{A}_{-1}$ act as a buffer between the most advantageous allele $\mathcal{A}_0$ and the rest, and selection becomes too weak relative to mutation to counteract the dissipation of the solutions under the mutation-selection dynamics.

Equation (4.14) defines a critical mutation rate above which stable equilibria do not necessarily exist. This phenomenon is equivalent to the error threshold found in models for polynucleotide replication (cf. Eigen and Schuster 1977, Swetina and Schuster 1982, Wagner and Krall 1993, Higgs 1994, Baake and Wiehe 1997, Wiehe 1997, Baake and Gabriel 1999).

In Chapters IV.2 and IV.3, we shall derive results closely related to • **4.1** in a more general framework. In particular, we shall prove that, if instead of (4.9) the stronger assumption that the W_i vanish at infinity (i.e., $\lim_{|i|\to\infty} W_i = 0$) is imposed, existence and uniqueness follows for arbitrary mutation rates, and convergence to equilibrium occurs for any initial probability distribution. In Chapter IV.5.3, it will be shown that

$$r = \hat{\overline{W}} = 1 - \mu + O(\mu^2) \quad \text{as } \mu \to 0 , \tag{4.15}$$

thus, identifying (4.14) as a good approximation to (4.9).

IV

Mutation-Selection Models for Quantitative Traits

Quantitative traits such as body weight, brain volume, growth rate, or milk production exhibit continuous or almost continuous variation. Typically, such traits are controlled by many gene loci and, therefore, are also called *polygenic*. In addition, these traits are influenced by environmental effects, to varying degrees. Many morphological, physiological, or behavioral characters, as well as characters of economic importance, fall into this category. Since selection – natural and artificial – occurs to a considerable extent at the level of phenotypes, fitness is a continuous quantitative trait. It is natural and convenient to describe the composition of a population with respect to a polygenic character, in terms of a continuous probability distribution defined on an appropriate scale of measurement. The same applies to many meristic characters, i.e., characters for which the phenotype is expressed in discrete, integral classes (e.g., litter size or number of abdominal bristles on a fruit fly) if the number of possible phenotypes is sufficiently large.

Crow and Kimura (1964) and Kimura (1965a) introduced a model that has been widely used to study the evolution of polygenic characters subject to selection. It is based on the assumption that at each locus, in principle, infinitely many different alleles can be generated by mutation and every mutation produces a new allele. This idea is substantiated by the consideration of DNA nucleotide sequences. Assume an average length of the amino acid coding part of a gene of approximately 900 nucleotides. Then, with four possible nucleotides at each site, the number of possible alleles is about $4^{900} \approx 10^{542}$. Data indeed indicate that many different mutants occur, in higher organisms, on the order of 0.1 or more per genome per generation, and the range of their phenotypic effects may be large, though a large fraction of mutants has small or no measurable effects. Recent experiments also provide information about the distribution of mutational effects (see Chapter VII.1). In the model, to each allele an effect x, measured on the character's scale, is assigned. The effect y of a new allele is assumed to differ only slightly from the parent allele x, according to a (conditional) probability density $u(y - x)$. This has been called the *continuum-of-alleles model*, because at each locus, allelic effects are potentially continuously distributed. It was used by Crow and Kimura (loc. cit.) to calculate the mutation load and the genetic variance of a polygenic trait under mutation-selection balance. Compared with the classical diallelic model, the continuum-of-alleles model is at the other extreme on a scale of models enumerated with regard to allele number, and it may be nearer to reality for quantitative traits. As we shall see, in some situations the continuum-of-alleles model also leads to simpler, and perhaps biologically more realistic, results than diallelic models with their inherent symmetries.

The purpose of this chapter is to develop the mathematical theory of mutation and selection for the continuum-of-alleles model and, more generally, for models allowing for an arbitrary number of alleles per locus, finite or infinite. As a by-product, we shall obtain some of the results of Chapter III as special cases. Throughout this chapter, we concentrate on the dynamics and equilibrium behavior of gene-frequency distributions in an asexually reproducing population of effectively infinite size, so that random genetic drift can be ignored. Also, environmental influences are neglected. Otherwise, the theory is developed in a rather general setting, and applies to wide classes of fitness functions and mutation distributions. The efforts to achieve this generality appear to be justified in the light of recent and ongoing experimental approaches to uncover the distribution of mutational effects on quantitative characters. In the next chapter, a method involving cumulants and generating functions is developed that facilitates the derivation of the selection dynamics of the phenotypic distribution, as determined by multiple loci, from the present single-locus haploid results, provided effects are additive and recombination is not too weak. The mathematics underlying this chapter is much deeper than that in most other chapters. Therefore, the most general form of the model as well as the proofs are separated from the basic properties, ideas and results, and are relegated to separate sections. Mathematical background material is summarized in Appendix C.

Section 1 is mainly of introductory and motivational character. The haploid continuum-of-alleles model of Crow and Kimura is introduced, and two well known, but controversial, approximations for the equilibrium variance maintained under a balance between mutation and stabilizing selection are presented. In Section 2, a general model is formulated that contains as special cases the continuum-of-alleles model, the stepwise-mutation model, and the classical model with a finite number of alleles. Sufficient conditions for existence, uniqueness, and global stability of an equilibrium distribution are stated. The mathematical theory and proofs of (more general versions of) these results are given in Section 3. A further generalization, in which frequency distributions of allelic effects are described by probability measures, not necessarily possessing a density, is treated in Section 4. In the final section, first- and second-order approximations to the equilibrium mean fitness are derived, and Haldane's principle on the mutation load is generalized to models with an infinite number or a continuum of alleles. Applications and extensions to multilocus models are treated in subsequent chapters.

1. THE CONTINUUM-OF-ALLELES MODEL

Since its introduction (Crow and Kimura 1964, Kimura 1965a), the continuum-of-alleles model has played an important role in several analyses aimed at resolving the problem of how much genetic variation can be maintained at a balance between mutation and stabilizing selection. Here, we shall formulate this model in its original form and present two qualitatively different approximations for the equilibrium variance that are based on different assumptions about the biological parameters. These are the Gaussian allelic approximation of Kimura (1965a) and the house-of-cards approximation of Turelli (1984).

Consider a gene locus contributing to a quantitative character. In the model, it is assumed that every mutation produces a new allele different from the pre-existing ones, and that the effect of the mutant allele differs only slightly from that of its parent allele. The effect of an allele on the character is denoted by (a real number) x. The mutation rate is denoted by μ. Let the probability density be $u(\xi)$ so that an allele with effect x mutates to an allele with effect $x+\xi$, conditional on the assumption that a mutation occurs. This is called the random-walk mutation model (cf. Section 2.1). A typical example is the Gaussian mutation distribution

$$u(\xi) = \frac{1}{\sqrt{2\pi}\gamma} \exp\left(-\frac{\xi^2}{2\gamma^2}\right) , \tag{1.1}$$

where γ^2 is called the variance of mutational effects. This model leads to a continuous distribution of allelic effects, x, whose density at time t is designated by $p(x,t)$. Under the assumption of overlapping generations, fitness is measured in terms of the Malthusian parameter $m(x)$. We shall be concerned with *stabilizing selection*, i.e., selection for an intermediate optimum (cf. Chapters VI and VII). A typical and simple example for a fitness function modeling stabilizing selection is

$$m(x) = -sx^2 . \tag{1.2}$$

This yields the so-called quadratic optimum model (cf. Chapter VI). The reader may recall from Chapter I.10.1 that the selection dynamics is unaltered by adding a constant to $m(x)$.

We consider a haploid locus in isolation, and a mutation distribution u and fitness function m that may be more general than in (1.1) and (1.2), respectively (cf. Section 2.4). Under the assumption that the interaction of selectional and mutational forces is negligible in small time intervals Δt (cf. Chapter III.1.2), the evolution of the density of allelic effects is determined by the integro-differential equation (Kimura 1965a)

$$\frac{\partial p(x,t)}{\partial t} = [m(x) - \overline{m}(t)]p(x,t) + \mu \left[\int u(x-y)p(y,t)\,\mathrm{d}y - p(x,t) \right] , \tag{1.3}$$

where $\overline{m}(t) = \int m(x)p(x,t)\,\mathrm{d}x$ is the mean fitness and $\int = \int_{-\infty}^{\infty}$. For a derivation, see Section 2.4. The structure of this equation is completely analogous to that of the classical haploid mutation-selection equation III(1.8), the first term on the right-hand side describing the change caused by selection, the second term that by mutation. It is intuitively obvious (and can be easily proved) that the differential equation (1.3) preserves the property of being a probability density.

Our main interest is in equilibrium solutions of this differential equation, i.e., in probability densities $p = p(x)$ that satisfy

$$[\mu + \overline{m} - m(x)]p(x) = \mu \int u(x-y)p(y)\,\mathrm{d}y . \tag{1.4}$$

We shall prove in Section 3 in a much more general context that the solution is uniquely determined and positive.[1] However, it cannot be calculated explicitly. Therefore, approximation techniques are necessary to obtain information about its shape.

For simplicity, let us now assume that u and m are given by (1.1) and (1.2), respectively. Since they are symmetric and the equilibrium density is uniquely determined, it follows that the equilibrium density is symmetric, and its mean coincides with the optimum zero. A simple, but important, observation is that the equilibrium mean fitness, $\hat{\bar{m}}$, satisfies the inequalities

$$-\mu \leq \hat{\bar{m}} < 0 \,. \tag{1.5a}$$

The right inequality is a consequence of the fact that $m(x) \leq 0$, and the left inequality follows from this, together with the nonnegativity of the right-hand side of (1.4), in particular for $x = 0$. The most important quantity is the variance $\hat{\sigma}^2$ of the equilibrium distribution, because it measures the amount of variability that can be maintained in such a model. For the present fitness function, we have $\hat{\bar{m}} = -s\hat{\sigma}^2$. Therefore, (1.5a) is equivalent to

$$0 < \hat{\sigma}^2 < \frac{\mu}{s} \,, \tag{1.5b}$$

and we have obtained a simple upper bound for the equilibrium variance. This is one of the central observations on which part of the analysis in this chapter is based. However, let us first review some 'classical' results.

1.1 THE GAUSSIAN ALLELIC APPROXIMATION

For the quadratic fitness function (1.2), Kimura (1965a) proposed the following procedure to approximate the solution of (1.4). By performing the transformation $y \to y+x$ and then expanding $p(x+y)$ in a Taylor series about x, the integral expression on the right-hand side of (1.4) can be represented as

$$\begin{aligned} \int u(x-y)p(y)\,\mathrm{d}y = {} & p(x) + \frac{\mathrm{d}p(x)}{\mathrm{d}x}\int yu(-y)\,\mathrm{d}y \\ & + \tfrac{1}{2}\frac{\mathrm{d}^2p(x)}{\mathrm{d}x^2}\int y^2u(-y)\,\mathrm{d}y + \dots \,. \end{aligned} \tag{1.6}$$

The integral terms are the moments of the mutation distribution u. Under the assumption that its mean is zero, its variance is γ^2, and that the higher-order terms in (1.6) can be neglected (as for a Gaussian with small γ^2), substitution into (1.4) yields

$$s\left(x^2 - \int y^2 p(y)\,\mathrm{d}y\right)p(x) = \tfrac{1}{2}\mu\gamma^2\frac{\mathrm{d}^2p(x)}{\mathrm{d}x^2} \,. \tag{1.7}$$

As noted by Kimura, this is a Weber equation, whose solution is uniquely determined subject to the constraints $p(x) \geq 0$, $\lim_{|x|\to\infty} p(x) = 0$, and $\int p(x)\,\mathrm{d}x = 1$. It is

[1] Throughout, a function f is called positive if $f(x) > 0$ for all x. If $f(x) \geq 0$ for all x, then f is called nonnegative.

straightforward to check that (1.7) is satisfied by the Gaussian density with mean zero and variance

$$\hat{\sigma}^2(\mathrm{G}) = \sqrt{\frac{\mu}{s}\frac{\gamma^2}{2}} \,. \tag{1.8}$$

Numerical calculations (Turelli 1984, Bürger 1986c, and Figure 6.1 in Chapter VI) show that the *Gaussian allelic approximation*[2] (1.8) for the genetic variance is accurate if

$$2\gamma^2 < \frac{\mu}{s} \,. \tag{1.9}$$

This approximation has significantly influenced the development of evolutionary quantitative genetics. It was generalized to multiple unlinked or linked loci by Latter (1970) and Lande (1975). Fleming (1979) analyzed a generalized discrete-time, multilocus version of Crow and Kimura's model, and determined second-order approximations under scaling assumptions about the parameters that lead to the Gaussian allelic approximation in first order (see also Nagylaki 1984, and Chapters VI.6, VI.7). It was Lande's work that greatly popularized this topic, because he combined his mathematical analysis with a review of available data and concluded that mutation-selection balance is an important factor in maintaining high levels of genetic variation despite relatively strong stabilizing selection.

1.2 THE HOUSE-OF-CARDS APPROXIMATION

A completely different approximation was proposed by Turelli (1984). It is based on the assumption that the variance γ^2 of the mutation distribution u is much larger than the equilibrium genetic variance $\hat{\sigma}^2$ of $\hat{p}$. Then the integral on the right-hand side of (1.4) can be approximated as

$$\int u(x-y)\hat{p}(y)\,\mathrm{d}y \approx u(x) \,, \tag{1.10}$$

where equality holds if $\hat{p}$ is the point measure located at zero. Using (1.10), we infer from (1.4) that the equilibrium density $\hat{p}$ has the explicit form

$$\hat{p}(x) = \frac{\mu u(x)}{\mu + \hat{\bar{m}} - m(x)} \,, \tag{1.11}$$

where the equilibrium mean fitness $\hat{\bar{m}}$ is determined by the normalization condition

$$\int \hat{p}(x)\,\mathrm{d}x = 1 \,. \tag{1.12}$$

For the quadratic fitness function (1.2), we obtain from (1.11) that (1.12) can be written as

$$\frac{\mu}{s}\int \frac{u(x)}{\alpha_{\mu/s} + x^2}\,\mathrm{d}x = 1 \,, \tag{1.13}$$

[2] We call this the Gaussian allelic approximation, to distinguish it from other approximations involving Gaussian densities.

where

$$\alpha_{\mu/s} = \frac{1}{s}(\mu + \hat{\bar{m}}) = \frac{\mu}{s} - \hat{\sigma}^2 \,. \tag{1.14}$$

From (1.5b), we infer that $\alpha_{\mu/s} \to 0$ as $\mu/s \to 0$.

If u is Gaussian, the left-hand side of (1.13) can be integrated explicitly (e.g. with *Mathematica*, Wolfram 1996) and becomes

$$\sqrt{\frac{\pi}{2}}\,\frac{\mu}{s\gamma^2}\,\frac{\exp\left(\frac{\alpha_{\mu/s}}{2\gamma^2}\right)}{\sqrt{\frac{\alpha_{\mu/s}}{\gamma^2}}}\left[1 - \mathrm{erf}\sqrt{\frac{\alpha_{\mu/s}}{2\gamma^2}}\right] = 1\,,$$

where $\mathrm{erf}(x) = 2\pi^{-1/2}\int_0^x \exp(-t^2)\,dt$ denotes the error function. For $\mu/(s\gamma^2) \to 0$ (hence $\alpha_{\mu/s}/\gamma^2 \to 0$), the left-hand side is, to leading order in $\mu/(s\gamma^2)$,

$$\sqrt{\frac{\pi}{2\gamma^2}}\,\frac{\mu}{s}\,\frac{1}{\sqrt{\alpha_{\mu/s}}}\,.$$

Equating this to one, we obtain from (1.14) the asymptotic equivalence

$$\hat{\sigma}^2 \sim \frac{\mu}{s} - \frac{\pi}{2\gamma^2}\left(\frac{\mu}{s}\right)^2 \quad \text{as } \mu/(s\gamma^2) \to 0\,. \tag{1.15}$$

Actually, the right-hand side of (1.15) provides a lower bound for the equilibrium variance. The approximation

$$\hat{\sigma}^2(\mathrm{HC}) = \frac{\mu}{s} \tag{1.16}$$

has been called the *house-of-cards* approximation (henceforth abbreviated as HC-approximation) and was derived by Turelli (1984) in a different way for the corresponding discrete-time model. Equation (1.5b) shows that the HC-approximation is an upper bound for the true equilibrium genetic variance. The second-order expansion (1.15) was derived by Bürger and Hofbauer (1994). In Chapter VI.6, these approximations will be extended to discrete-time models and to much more general mutation distributions, and in Chapter VI.7 to multiple loci.

Turelli (1984) called (1.16) the house-of-cards approximation because it is based on the approximation (1.10). The latter is equivalent to the assumption that the mutation distribution depends only on the effect of the new mutant and is independent of the original allelic state of the gene. Thus, the mutant is chosen at random from a fixed distribution; in the present case from (1.1). We have already used this assumption in Chapter III.2.3. This kind of mutation distribution, for which the mutant's effect is independent of its original state, was used in a slightly different context by Kingman (1977, 1978), because it reflects the tendency for most mutations to be deleterious. He argued that such mutations will presumably 'upset the evolutionary house of cards built up by the careful improvements of many generations'. We shall refer to Kingman's assumption about the mutation distribution (see also (2.4) below) as the house-of-cards (HC) mutation model. This should not be confused with the HC-approximation (1.16) for the genetic variance because, as we shall see in Chapter VI.6, the latter is valid under almost any kind of mutation distribution in the limit of small mutation rates.

The qualitative difference between the Gaussian approximation (1.8) and the HC-approximation (1.16) is evident, and is displayed in Figure 6.1 in Chapter VI. Based on a review of available data, this discrepancy between the Gaussian and the HC-approximation, which persists in multilocus models, caused Turelli to question Lande's hypothesis that large amounts of genetic variation can be maintained through a balance between mutation and stabilizing selection. We shall examine this hypothesis in detail in Chapter VII.

Here, we give a very brief summary of empirical results to indicate the range of parameter values that is of biological interest. A more detailed review and the relevant references may be found in Chapter VII.1. Despite substantial recent progress, current knowledge about the actual size of the parameters s, μ, γ^2, the number of loci (ℓ), and the form of the fitness function and the mutation distribution is still limited. One of the reasons is that in typical experiments only compound quantities such as the gametic mutation rate ($\ell\mu$) or the total input of new variance produced by mutation per generation and zygote ($2\ell\mu\gamma^2$) are measured. Fitness functions are usually determined by regression methods. With a quadratic regression, the resulting fitness function is quadratic by definition. More complex regressions are rarely statistically justified. Selection intensities on genotypic values, s, may range from 0 to $\frac{1}{2}$ if stabilizing selection acts on the phenotype and environmental effects are taken into account and normalized ($s = (2V_s)^{-1} = [2(\omega^2 + \sigma_E^2)]^{-1}$, $\sigma_E^2 = 1$; see Chapter V.1.2). Moderately strong selection means that s ranges from $s = 0.01$ to $s = 0.05$. Typical mutation rates per locus are on the order of $\mu = 10^{-5}$–10^{-6}, though there is some evidence from *Drosophila* that loci influencing bristle number exist with μ between 10^{-3} and 10^{-4}. The variance of mutational effects, γ^2, may be on the order of 0.05, with values smaller or larger by a factor 10 also possible. Little is known about the form of the mutation distribution, but recent data have suggested a highly leptokurtic (L-shaped) mutation distributions. The random-walk mutation model appears to be plausible when dealing with quantitative traits. For an application of the above model to characters related to metabolism when mutations alter enzyme activities, we refer to Clark (1991).

2. THE GENERAL MODEL AND ITS BASIC PROPERTIES

Motivated by the structural similarity of the mutation-selection equations, (1.3) for a continuum of possible alleles, III(1.8) for a finite number of alleles, and III(4.6) for the stepwise-mutation model, we develop a general model that includes all these cases, and many more. We formulate the model in discrete and continuous time, and give simple sufficient conditions for the mutation distribution and the fitness function to guarantee a unique equilibrium distribution, which is globally asymptotically stable. For the HC-mutation model, a simple necessary and sufficient condition is found. The proof of the general case involves deeper mathematical methods and is given in Section 3.

2.1 THE MODEL

We consider an effectively infinite, haploid population, in which individuals are characterized by their type x, where x is an element in the state space $\mathcal{X}$ of all admissible types. The theory we shall develop is applicable to a broad range of possible choices of $\mathcal{X}$, though our presentation is mainly tailored to fit models with a continuum of possible types. Here are some examples:

Examples. 1. If the types are identified with a finite number of alleles $\mathcal{A}_1, \ldots, \mathcal{A}_k$ at a given gene locus, it is appropriate to choose $\mathcal{X} = \{1, 2, \ldots, k\}$.

2. The stepwise-mutation model of Chapter III.4 is obtained if $\mathcal{X} = \{0, \pm 1, \pm 2, \ldots\}$.

3. If x is interpreted as the contribution of a gene to a given quantitative trait, as in the continuum-of-alleles model, one may choose $\mathcal{X} = \mathbf{R}$, the set of real numbers, or if the trait values are constrained, $\mathcal{X}$ may be some interval in $\mathbf{R}$. For instance, the choice $\mathcal{X} = [0, 1]$ will be the natural one if relative fitness is the trait under consideration.

4. More generally, if the given locus contributes to n characters, then $\mathcal{X}$ may be any multidimensional interval in $\mathbf{R}^n$ or the whole of $\mathbf{R}^n$. ◇

In mathematical terms, $\mathcal{X}$ is a locally compact space. Type densities and mutation distributions are taken with respect to a given measure on $\mathcal{X}$. If $\mathcal{X} = \mathbf{R}$ or if $\mathcal{X}$ is a (multidimensional) interval, we shall take the ordinary (Lebesgue) measure and $\int_{\mathcal{X}} f(x)\,\mathrm{d}x$ is to be understood in the usual way. If $\mathcal{X}$ is a discrete set, finite or infinite, we take the counting measure and $\int_{\mathcal{X}} f(x)\,\mathrm{d}x$ has to be interpreted as $\sum_{x \in \mathcal{X}} f(x)$.

The mutation rate of a type x is denoted by $\mu(x)$, $0 \le \mu(x) \le 1$. Presently, it is not assumed that all types have the same mutation rate, though in most applications $\mu = \mu(x)$ will be constant. Conditional on the assumption of a mutation event, the probability density of mutations from type x to type y is designated by $u(x, y)$. Therefore, we have $u(x, y) \ge 0$,

$$\int_{\mathcal{X}} u(x, y)\,\mathrm{d}y = 1 \quad \text{for all } x \text{ in } \mathcal{X}\ , \tag{2.1}$$

and $\mu(x)u(x, y)$ is the fraction of individuals of type y that originate through mutation from individuals of type x during one generation. Here are some examples of mutation distributions that have appeared in the literature and will be used in the sequel.

Examples. 5. Let $\mathcal{X}$ be a finite or countably infinite discrete set. To cover this case in our general theory, we set $\mu(x) = 1$ for all x. Therefore, in contrast to models with a continuum of possible types, $1 - u(x, x) = \sum_{y: y \ne x} u(x, y)$ has to be interpreted as the mutation probability of type x. Then the mutation model of Chapter III is obtained by defining

$$\mu_{ij} = \begin{cases} 0\ , & i = j\ , \\ u(i, j)\ , & i \ne j\ . \end{cases} \tag{2.2}$$

6. *The random-walk mutation model.* This is the model used by Crow and Kimura (1964). It is supposed that x mutates to $x + \xi$ with (conditional) probability $u(\xi)$. Formally, this is expressed as

$$u(x, y) = u_{\mathrm{RW}}(y - x)\ , \tag{2.3}$$

where u_{RW} is a probability density. This choice is possible only if $\mathcal{X}$ forms a group with respect to addition, for example, if $\mathcal{X} = \mathbf{R}^n$ or $\mathcal{X} = \{0, \pm 1, \pm 2, \ldots\}$.

7. *The house-of-cards mutation model*, introduced by Kingman (1977, 1978), assumes that the mutation distribution is independent of the parent gene, i.e., $\mu(x) \equiv \mu$ and

$$u(x, y) = u_{\mathrm{HC}}(y) \qquad \text{for all } x \text{ and } y\ , \tag{2.4}$$

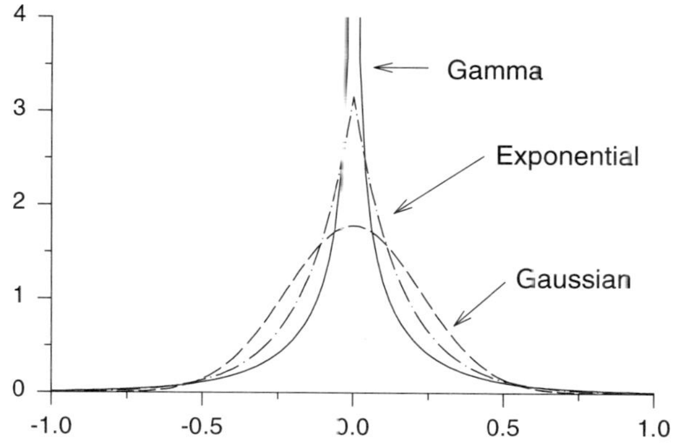

Figure 2.1 Three different mutation distributions serving as candidates for u_{RW}, u_{HC}, and u_{ZC}. All have variance $\gamma^2 = 0.05$. The dahsed line represents a Gaussian distribution (kurtosis = 0), the dash-dotted line represents an exponential distribution reflected about zero (kurtosis = 3), and the solid line a Γ-distribution reflected about zero with $\theta = \frac{1}{2}$ (kurtosis = 26/3).

where u_{HC} is a probability density. This assumption about the mutation distribution is of considerable mathematical convenience, because it significantly simplifies the mathematical analysis. We recall from Chapter III.2.3 that the HC-mutation model arises automatically in the case of only two alleles per locus.

8. The following mutation model was introduced by Zeng and Cockerham (1993) and interpolates between (2.3) and (2.4). It is defined for $\mathcal{X} = \mathbf{R}^n$, and assumes

$$u(x, y) = u_{ZC}(y - cx) \tag{2.5}$$

for some probability density u_{ZC} and a constant c. ◇

If $\mathcal{X} = \mathbf{R}$, then in Examples 6, 7, and 8, typical mutation distributions include the Gaussian distribution (1.1), as used by Turelli (1984) and others, or the family of Γ-distributions reflected about zero,

$$u(x) = \frac{d^\theta}{2\Gamma(\theta)} |x|^{\theta-1} \exp(-d\,|x|) , \tag{2.6}$$

as used by Hill (1982b), for instance. The latter has variance $\theta(\theta+1)/d^2$ and kurtosis $2(3+\theta-\theta^2)/\theta(\theta+1)$ (cf. Appendix D.3). This distribution is unimodal if $\theta \leq 1$ and, for $\theta = 1$, includes the exponential distribution as a special case (see Figure 2.1). Some experiments (e.g., Mackay *et al.* 1992, López and López-Fanjul 1993a,b) indicate that distributions of mutational effects tend to be highly leptokurtic, as is the reflected Γ-distribution for large θ.

Let $p(x, t)$ denote the type density (with respect to dx) at time t. For a population with discrete generations, the fitness (viability) of a type x is denoted by $W(x)$, where $W(x) \geq 0$. The mean fitness of the population is $\overline{W}(t) = \int_{\mathcal{X}} W(x)p(x,t)\,dx$. Then, provided $0 < \overline{W}(t) < \infty$, the density of types after selection becomes

$$p_s(x, t) = p(x, t)W(x)/\overline{W}(t) . \tag{2.7}$$

Assuming that mutation occurs during gametogenesis, which follows selection, the type densities evolve according to

$$p(x,t+1) = [1-\mu(x)]p_s(x,t) + \int_{\mathcal{X}} p_s(y,t)\mu(y)u(y,x)\,\mathrm{d}y\ . \tag{2.8}$$

This equation describes the mutation-selection dynamics in discrete time for our general model. For the continuum-of-alleles, it was first used by Fleming (1979). If $\mathcal{X} = \{1,2,\ldots,k\}$, then (2.8) reduces to III(1.1) by observing (2.2). If $\mathcal{X} = \{0,\pm1,\pm2,\ldots\}$, then (2.8) reduces to III(4.6). It is easily verified that (2.8) preserves the property of being a density, i.e., if $p(x,t) \geq 0$ and $\int_{\mathcal{X}} p(x,t)\,\mathrm{d}x = 1$, then $p(x,t+1)$ has these properties.

2.2 EXISTENCE AND STABILITY OF EQUILIBRIA

In Chapter III.1, existence, uniqueness, and global stability of an equilibrium was proved in the haploid mutation-selection model with a finite number of alleles. Since the present model is formulated in a rather general framework, it will not come as a surprise that several technical conditions have to be imposed on the fitness function and the mutation distribution for an equilibrium of (2.8) to exist and be uniquely determined.

It may be recalled that in the stepwise-mutation model of Chapter III.4, such conditions were needed, because with an infinite number of possible types, the type density may wander to infinity, due to directional selection, for instance, or it may become thinner and thinner because selection cannot prevent the dissipation caused by mutation. A simple example of the latter behavior is obtained in the continuum-of-alleles model with random-walk mutation and no selection, because then the variance of the type density satisfies the recurrence relation $\sigma^2(t+1) = \sigma^2(t) + \mu\gamma^2$, where γ^2 denotes the variance of the mutation distribution. A much more sophisticated (and surprising) example was provided by Moran (1977) in the stepwise-mutation model (see Chapter III.4). Wagner and Krall (1993) showed for $\mathcal{X} = \{0,1,2,\ldots\}$ and mutation occuring at rate μ from type k to $k+1$ that an equilibrium solution for all $0 < \mu < 1$ can exist only if there is a sequence of types whose fitnesses decrease to zero. If $W_i \geq c > 0$ for every i, then no equilibrium distribution can exist if $\mu > 1-c$.

Under directional selection with an unbounded fitness function $W(x)$, solutions of (2.8) exist only as long as $\overline{W}(t) < \infty$. For given $W(x)$, the mean fitness $\overline{W}(t)$ is finite for all t only if the initial distribution $p(x,0)$ decays sufficiently rapidly as $x \to \infty$ (cf. Bürger 1993 and Chapter VII.7.2).

In the continuum-of-alleles model, equilibria are given by probability distributions. Such a distribution may have a density or it may have singularities, i.e., atoms of probability, at which a positive probability is associated to a single point. This happens if mutation is either absent or too weak relative to selection, in the sense that not enough mutants close to the optimal type are produced. The phenomenon also occurs with pleiotropy, when each allele affects several characters, and is investigated in Section 4.

At first, we are interested only in equilibrium distributions that possess a density. Two general mathematical theorems about existence, uniqueness, and global stability of an equilibrium distribution for the mutation-selection equation (2.8) and its continuous-time analogue (2.20) are proved in the next section. Here, we list a number of simple but important special cases for which these theorems are valid. Two general assumptions need to be imposed on the fitness function that are related to stabilizing selection, and ensure that $0 < \overline{W}(t) \leq 1$ for all $t \geq 0$. Assuming discrete generations, we require the following

(W1) $W(x)$ is continuous, positive, bounded, and normalized such that $0 < W(x) \leq 1$.
(W2) There exists at least one x_O, the optimum type, with $W(x_O) = 1$.

The archetypical fitness function describing stabilizing selection in discrete-time models is the Gaussian $W(x) = \exp[-s(x - x_O)^2]$.

The following four cases cover most of the situations that have been investigated in the literature:

(A) $\mathcal{X} = \{1, 2, \dots, k\}$ and the mutation-selection matrix C defined in III(1.3), with μ_{ij} as in (2.2), is primitive (cf. Appendix B).
(B) $\mathcal{X}$ is a countably infinite set, e.g., the set $\mathcal{X} = \{0, \pm 1, \pm 2 \dots\}$ in the stepwise-mutation model. The mutation matrix is positive, i.e., $u(x, y) > 0$ for all $x, y \in \mathcal{X}$, and $W(x)$ vanishes at infinity, i.e., $\lim_{|x|\to\infty} W(x) = 0$.
(C) $\mathcal{X} = \mathbf{R}$ or $\mathcal{X} = \{0, \pm 1, \pm 2, \dots\}$ and posit the random-walk mutation model (2.3) with $\mu(x) = \mu > 0$ for all x and u_{RW} nonnegative. Furthermore, $W(x)$ vanishes at infinity. If $\mathcal{X} = \mathbf{R}$, it is also required that $u_{\text{RW}}(x) \geq u_0$ for some constant $u_0 > 0$ in an interval containing $x = 0$, and that $W(x) \geq 1 - c|x - x_O|$ for all x in a small interval around x_O and some $c > 0$. Thus, mutations to neighbouring states occur with positive probability, and the fitness function must not decrease too rapidly near the optimum x_O.
(D) $\mathcal{X}$ is arbitrary and the HC-mutation model (2.4) is assumed with u_{HC} positive and

$$\mu \int_{\mathcal{X}} \frac{u_{\text{HC}}(y)}{1 - W(y)} \, dy \geq 1 \,, \tag{2.9}$$

including divergence of the integral.

In (C) the conditions $u_{\text{RW}}(x) \geq u_0 > 0$ and $W(x) \geq 1 - c|x - x_O|$ prevent the build up of an atom of probability at x_O. In (D) this is prevented by (2.9). These special cases, and the more general mutation distributions considered in Theorem 3.1, presumably cover most biologically relevant situations, because fitness functions on the level of genotypic values can be expected to be smooth due to the smoothing effects of environmental contributions (cf. Nagylaki 1989a, and Chapter V.1.2).

Suppose that the general assumptions (2.1), (W1) and (W2) are satisfied, as well as one of the cases (A), (B), (C), or (D). Then the following results hold:

• **2.1** *There exists a unique equilibrium density* $\hat{p}$ *of* (2.8). *It is positive. Thus, the equilibrium is totally polymorphic. If the mutation rate* μ *is constant, the equilibrium mean fitness satisfies*

$$1 - \mu \leq \hat{\overline{W}} < 1 \,. \tag{2.10}$$

• **2.2** *For the HC-mutation model,* (2.9) *is a necessary and sufficient condition for the existence of an equilibrium distribution that has a density. Otherwise, the equilibrium distribution may contain atoms of probability.*

• **2.3** *The unique equilibrium density $\hat{p}$ is globally asymptotically stable, i.e., for arbitrary initial distributions with density $p(x,0) = p_0(x)$, convergence to $\hat{p}$ occurs in the following sense*

$$\lim_{t\to\infty} \int_{\mathcal{X}} |\hat{p}(x) - p(x,t)| \,\mathrm{d}x = 0 \,. \tag{2.11}$$

These results were first proved in Bürger (1988a,b), except for the second one, which is essentially due to Kingman (1978). Moran (1977) and Kingman (1977) proved existence and uniqueness of an equilibrium solution in the stepwise-mutation model, with mutation occurring only to next neighbours, under weaker assumptions about $W(x)$ than that given in (B) or (C) (cf. Chapter III.4). However, under their assumptions on the fitness function, the equilibrium exists only for sufficiently small mutation rates, and their stability results are weaker than the present one. It is the assumption that $W(x)$ vanishes at infinity that guarantees existence and stability for arbitrary mutation rates.

2.3 THE HC-MUTATION MODEL

It is instructive to prove existence and uniqueness separately for the HC-mutation model, because the proof is elementary, reflects the basic idea on which the general case is based, and condition (2.9) is necessary and sufficient. Let us recall from case (D) that u_{HC} is assumed to be positive (to avoid degeneracies similar to the ones that may occur in the pure selection model). Furthermore, we assume that $W(x)$ is not identical to one, i.e., there is selection. Then the equilibrium density, if it exists, can be calculated easily from (2.8). It is given by

$$\hat{p}(x) = \frac{\mu u_{\mathrm{HC}}(x)}{1 - \frac{1-\mu}{\hat{\overline{W}}} W(x)} \,, \tag{2.12}$$

where $\hat{\overline{W}}$ denotes the equilibrium mean fitness. necessary conditions for $\hat{p}$ to be a probability density are $1 - \mu \le \hat{\overline{W}} < 1$ and $\int \hat{p}(x)\,\mathrm{d}x = 1$.

To prove existence and uniqueness of such a solution, we introduce a parameter β, define

$$p_\beta(x) = \frac{\mu u_{\mathrm{HC}}(x)}{1 - \beta W(x)} \,, \tag{2.13}$$

and observe, by decomposing the numerator in (2.12) as $\mu u_{\mathrm{HC}}(1-\beta W) + \mu u_{\mathrm{HC}} \beta W$, that $\int_{\mathcal{X}} p_\beta(x)\,\mathrm{d}x = 1$ if and only if $\beta = (1-\mu)/\overline{W}_\beta$, where $\overline{W}_\beta$ is the mean fitness of p_β. We show that an equilibrium density exists, and is of the form (2.12), if and only if a unique b with $1 - \mu \le b < 1$ exists such that $\int_{\mathcal{X}} p_b(x)\,\mathrm{d}x = 1$. For the proof, we define

$$I(\beta) = \mu \int_{\mathcal{X}} \frac{u_{\mathrm{HC}}(x)}{1 - \beta W(x)} \,\mathrm{d}x \,, \tag{2.14}$$

and examine this function in the interval $[1-\mu, 1]$. Then, because $W(x) \le 1$, but not identical to one, it follows that $I(1-\mu) < 1$. Furthermore, $I(\beta)$ is strictly monotone increasing. Therefore, by B. Levi's monotone-convergence theorem (e.g. Hewitt and

Stromberg 1965), we have

$$\lim_{\beta\uparrow 1} I(\beta) = \mu \int_{\mathcal{X}} \frac{u_{\mathrm{HC}}(x)}{1 - W(x)}\, \mathrm{d}x \tag{2.15}$$

if the right-hand side is finite, and $\lim_{\beta\uparrow 1} I(\beta) = \infty$, otherwise. Hence, the desired b exists if and only if (2.9) holds. This proves that a unique equilibrium density exists if (2.9) is satisfied. As we shall see in Section 4, the equilibrium distribution has atoms of probability if (2.9) is not fulfilled.

To summarize, the equilibrium density is of the form (2.12). If equality holds in (2.9), then $\hat{\overline{W}} = 1 - \mu$; otherwise, we have $1 - \mu < \hat{\overline{W}} < 1$. Thus, to order μ, the equilibrium mean fitness is independent of the strength of selection, a fact well known from simpler models as Haldane's principle (cf. Chapter III.3 and Section 5).

2.4 CONTINUOUS TIME

For a population with overlapping generations, a Malthusian fitness value $m(x)$ is assigned to each type x, and we have to replace the assumptions (W1) and (W2) by

(m1) $m(x)$ is continuous, bounded from above, and normalized such that $m(x) \le 0$.
(m2) There exists at least one x_O with $m(x_O) = 0$.

The assumptions (A) – (D) have to be replaced by

(A') $\mathcal{X} = \{1, 2, \ldots, k\}$ and the mutation-selection matrix A defined in III(1.11) is irreducible.

(B') $\mathcal{X}$ is a countably infinite set, e.g., $\mathcal{X} = \{0, \pm 1, \pm 2, \ldots\}$ as in the stepwise-mutation model. The mutation matrix $u(x, y)$ is irreducible and $m(x)$ satisfies $\lim_{|x|\to\infty} m(x) = -\infty$.

(C') $\mathcal{X} = \mathbf{R}$ or $\mathcal{X} = \{0, \pm 1, \pm 2, \ldots\}$ and posit the random-walk mutation model (2.3) with $\mu(x) = \mu > 0$ for all x. Assume $\lim_{|x|\to\infty} m(x) = -\infty$. If $\mathcal{X} = \mathbf{R}$, then it is additionally required that $u_{\mathrm{RW}}(x) \ge u_0$ for some constant $u_0 > 0$ in an interval containing $x = 0$, and $m(x) \ge -c\,|x - x_O|$ for all x and some $c > 0$ in a small interval around x_O.

(D') $\mathcal{X}$ is arbitrary and the HC-mutation model (2.4) is assumed with u_{HC} positive and

$$\mu \int_{\mathcal{X}} \frac{u_{\mathrm{HC}}(y)}{-m(y)}\, \mathrm{d}y \ge 1\,, \tag{2.16}$$

including divergence of the integral.

Let us derive the integro-differential equation for the evolution of type densities under mutation and selection from the recursion (2.8) by performing a weak-selection approximation. For this purpose, we require that $\overline{m}(t) = \int_{\mathcal{X}} m(x)p(x,t)\, \mathrm{d}x$, the mean Malthusian fitness, is finite (see below). If we set

$$W(x) = e^{sm(x)}\,, \tag{2.17}$$

we obtain by expansions $W(x) = 1 + sm(x) + s^2 h(x)$ and $\overline{W}(t) = 1 + s\overline{m}(t) + s^2 \bar{h}(t)$, where $h(x)$ is the remainder term and $\bar{h}(t) = \int h(x)p(x,t)\, \mathrm{d}x$ is finite because $\overline{W}(t)$

and $\overline{m}(t)$ are. Rescaling time according to $\tau = st$, where s may be interpreted as generation length, and setting $q(x,\tau) = p(x,t)$, we obtain

$$\frac{\partial q(x,\tau)}{\partial \tau} = \lim_{s\to 0} \frac{1}{s}\left[q(x,\tau+s) - q(x,\tau)\right] = \lim_{s\to 0} \frac{1}{s}\left[p(x,t+1) - p(x,t)\right] . \tag{2.18}$$

Further, after rescaling mutation rates according to $\mu(x) = s\tilde{\mu}(x)$, a simple calculation shows that for every x,

$$\begin{aligned} p(x,t+1) - p(x,t) = {}& \overline{W}(t)^{-1} s\Big\{[m(x) - \overline{m}(t)] + s[h(x) - \bar{h}(t)]\Big\}p(x,t) \\ & + \overline{W}(t)^{-1} s\Big\{\int [1 + sm(y) + s^2h(y)]p(y,t)\tilde{\mu}(y)u(y,x)\,\mathrm{d}y \\ & - [1 + sm(x) + s^2h(x)]\tilde{\mu}(x)p(x,t)\Big\} . \end{aligned} \tag{2.19}$$

Substituting (2.19) into (2.18), performing the limit, and writing t for τ, we obtain the desired integro-differential equation that describes the dynamics of type densities in continous time:

$$\frac{\partial q(x,t)}{\partial t} = [m(x) - \overline{m}(t)]q(x,t) + \int_{\mathcal{X}} q(y,t)\tilde{\mu}(y)u(y,x)\,\mathrm{d}y - \tilde{\mu}(x)q(x,t) . \tag{2.20}$$

Therefore, we have the approximation

$$\begin{aligned} \Delta p(x,t) &\approx s\frac{\partial q(x,t)}{\partial t} \\ &= s[m(x) - \overline{m}(t)]p(x,t) + \int_{\mathcal{X}} p(y,t)\mu(y)u(y,x)\,\mathrm{d}y - \mu(x)p(x,t) . \end{aligned} \tag{2.21}$$

This approximation requires that $m(x)$ is independent of t and s, but it may depend on $p(x)$. For constant μ and the random-walk mutation model, (2.20) reduces to Kimura's equation (1.3). The reader may recall from Chapter III.1 that $\mu(x)$ is a probability in (2.8), whereas it is a rate in (2.20). Abstract arguments show that for any given initial condition p_0 that is a density, the solution $p(x,t)$ of (2.20) with $p(x,0) = p_0(x)$ exists for all $t > 0$ and satisfies $-\infty < \overline{m}(t) < 0$ (cf. Bürger 1988a). Conditions and proofs for existence and uniqueness of solutions of (2.20) for the case in which the fitness function is not bounded above, as in models of directional selection, may be found in Bürger (1991b,c).

It may be of interest to point out that there is a formal relationship between the equilibrium equation (1.4) with $m(x) = -sx^2$ and Schrödinger equations. Applying the Fourier transformation to (1.4), we obtain

$$-\frac{\partial^2\varphi(z)}{\partial z^2} - \frac{\mu}{s}V(z)\varphi(z) = \psi\varphi(z) , \tag{2.22}$$

where $\varphi(z)$ and $V(z)$ are the Fourier transforms of $p(x)$ and $u(x)$, respectively, and $\psi = -(\mu + \overline{m})/s < 0$. Here, $-V(z)$ plays the role of the potential. This is an eigenvalue problem, and the equilibrium solution of Kimura's equation (1.4) corresponds to the ground state of (2.22), which is the (unique) eigenfunction corresponding to the lowest eigenvalue ψ of (2.22). This relationship motivated the present author's mathematical approach to this subject (cf. Bürger 1986c).

Under the above-stated assumptions, the existence, uniqueness, and stability results stated in Section 2.2 for the discrete-time model carry over to the continuous-time case if (2.10) is replaced by (1.5a). Proofs may be found in Bürger (1986c, 1988a,b) and Bürger and Bomze (1996).

*3. A RIGOROUS MATHEMATICAL TREATMENT

The proofs of the above-stated results require methods from functional analysis and operator theory, which are summarized in Appendix C. The book of Hewitt and Stromberg (1965) may serve as a general reference for measure theory. In this section, we shall state more general conditions under which the results • **2.1** – **2.3** hold, and provide rigorous derivations of the corresponding mathematical theorems for the discrete-time model. The presentation essentially follows Bürger (1988b), though with some generalizations and simplifications.

We begin by settling the notation and the general assumptions. Let $\mathcal{X}$ denote a locally compact space endowed with a positive σ-finite measure λ. For notational simpicity, we will always write $\int \ldots \mathrm{d}x$ instead of $\int_{\mathcal{X}} \ldots \lambda(\mathrm{d}x)$. $L^1(\lambda)$ is the Banach space of absolutely λ-integrable complex-valued functions on $\mathcal{X}$, i.e., $f \in L^1(\lambda)$ if $\|f\|_1 = \int |f(x)| \,\mathrm{d}x < \infty$. For an essentially bounded measurable function h, we define $\|h\|_\infty = \operatorname{ess\,sup}_{x\in\mathcal{X}} |h(x)|$. The fitness function $W : \mathcal{X} \to \mathbf{R}_+$ is supposed to be measurable, positive, bounded, and normalized such that $0 < W(x) \leq 1$ and $\|W\|_\infty = 1$. The mutation kernel $u : \mathcal{X} \times \mathcal{X} \to \mathbf{R}_+$ is assumed to be measurable and such that $\int u(y,x)\,\mathrm{d}x = 1$ for all $y \in \mathcal{X}$. Furthermore, $\mu(x)$ is measurable and $0 \leq \mu(x) \leq 1$. If $\mathcal{X}$ is discrete, we define $\mu(x) \equiv 1$ (cf. Section 2.1, Example 5). The density of types, $x \mapsto p(x,t)$, determines the function $p(t) \in L^1(\lambda)$.

We define the following operators on $L^1(\lambda)$:

$$Tf(x) = [1-\mu(x)]W(x)f(x) \,, \tag{3.1}$$

$$Cf(x) = \int f(y)W(y)\mu(y)u(y,x)\,\mathrm{d}y \,, \tag{3.2}$$

$$Af = Tf + Cf \,. \tag{3.3}$$

For discrete $\mathcal{X}$, we observe that $A = C$ because $\mu(x) = 1$. For the haploid model with finitely many alleles, C coincides with the matrix defined in II(1.3), for the stepwise-mutation model with that in III(4.4). The operators T, C, and A are bounded and nonnegative because of the assumptions imposed on W, μ, and u. Then (2.8) can be rewritten as

$$p(t+1) = Ap(t) \Big/ \int Ap(x,t)\,\mathrm{d}x \,, \tag{3.4}$$

and the solution $p(t)$ exists (and is a density) for all $t \geq 0$. The equilibrium solutions of (3.4) are precisely the functions in $L^1(\lambda)$ that satisfy

$$\alpha f_\alpha = Af_\alpha \tag{3.5}$$

for some complex number α, since (3.5) implies $\alpha = \int A f_\alpha \, dx / \int f_\alpha \, dx$. In the present context, we are only interested in the existence of nonnegative solutions of (3.5), normalized such that $\int f_\alpha \, dx = 1$. For notational convenience, we set

$$\tau = \|(1-\mu)w\|_\infty \, .$$

Theorem 3.1 *Assume that the following conditions are satisfied.*
1. *For every subset $S \subseteq \mathcal{X}$ with $\lambda(S) > 0$ and $\lambda(\mathcal{X} \setminus S) > 0$, we have:*

$$\int_{\mathcal{X}\setminus S} \int_S \mu(y) u(y,x) \, dy \, dx > 0 \, . \tag{3.6}$$

2. *There exists an $n \geq 1$ such that C^n is compact.*
3. *Let J be a set of positive measure in which $(1-\mu)W$ attains the value τ, i.e., $\operatorname{ess\,sup}_{x\in J}[1-\mu(x)]W(x) = \tau$. Assume there is a constant $u_0 > 0$, such that*

$$\mu(y)u(y,x) \geq u_0 \qquad \textit{for all } (y,x) \in J \times J \tag{3.7}$$

and

$$u_0 \int_J \frac{W(y)}{\tau - [1-\mu(y)]W(y)} \, dy > 1 \, , \tag{3.8}$$

where divergence of the integral is admitted.

Then a unique nonnegative equilibrium density $\hat{p}$ of (3.4) *exists. It is positive (almost everywhere) and the corresponding eigenvalue a satisfies*

$$\tau < a = \int A\hat{p}(x) \, dx = \hat{\overline{W}} < 1 \, . \tag{3.9}$$

Remarks. a) Condition (3.6) is equivalent to irreducibility (cf. Appendix C) of the integral operator C, because $W(x) > 0$ for all x. It implies that $\mu(x) > 0$ λ-*a.e.*[3] If $\mu(y)u(y,x) > 0$ for all y, x, then (3.6) is clearly fulfilled. Without the assumption of irreducibility, equilibrium solutions are in general not uniquely determined.

b) We state some conditions implying power compactness of C (cf. Theorem C.1). Suppose $\mathcal{X}$ is a group, $u(y,x) = u_{\mathrm{RW}}(x-y)$, and u_{RW} is a probability density. If W vanishes at infinity, then C is compact. If, for general $u(y,x)$,

$$\int \operatorname*{ess\,sup}_{y\in\mathcal{X}} W(y)\mu(y)u(y,x) \, dx < \infty$$

holds, then C^2 is compact. Of course, if $\mathcal{X}$ is compact, in particular if it is a finite set, the latter condition is always satisfied if u is bounded, because W is assumed to be bounded. In the HC-case (2.4), the operator C is always compact. If $\mathcal{X}$ is a countably-infinite discrete set, then C is compact if $W(x)$ vanishes at inifinity. Power compactness of C prevents solutions of (3.4) to wander to infinity or to become thinner and thinner.

[3] λ-*a.e.* is short-hand for λ-almost everywhere, and means 'for all x except of a set of λ-measure zero'.

c) Inequality (3.8) guarantees absolute continuity of the equilibrium distribution Otherwise, atoms of probability may occur (see Section 4). The reader may note that (3.7) and (3.8) are much more stringent conditions than (2.9) in the HC-case. Obviously, if $W(x) \geq 1 - c\,|x - x_O|$, as in cases (C) and (C') of Section 2, then (3.8) is satisfied. If $\mathcal{X}$ is discrete and $u(x_O, x_O) > 0$, then condition 3 is trivially satisfied. The smaller the set J is chosen, the larger will u_0 be.

d) If $\mu(x) = \mu$ for all x, then condition **3** can be simplified as follows:

3'. Let J be a set of positive measure in which W attains its essential supremum, i.e., $\operatorname{ess\,sup}_{x\in J} W(x) = 1$, and assume there is a constant $u_0 > 0$ such that (3.7) and

$$u_0 \int_J \frac{W(y)}{1 - W(y)}\,\mathrm{d}y > 1 - \mu \tag{3.8'}$$

hold. The inequality (3.8') is satisfied, for instance, if there is a constant $c > 0$ such that $W(x) \geq 1 - c\,|x - x_O|$ for $x \in J$, because then the left-hand side of (3.8') is infinite. ◇

Proof of Theorem 3.1. For all real α satisfying $\alpha > \tau$, we define the family of operators

$$(C_\alpha f)(x) = \int f(y)\,\frac{W(y)}{\alpha - [1 - \mu(y)]W(y)}\,\mu(y)u(y, x)\,\mathrm{d}y \tag{3.10}$$

on $L^1(\lambda)$. (Note that for discrete $\mathcal{X}$ we simply have $C_\alpha = \frac{1}{\alpha}A$.) It follows that each C_α is nonnegative, bounded, and satisfies $\|C_\alpha\| \leq \|\mu W\|_\infty/(\alpha - \tau)$. It is trivial to verify that f_α is an eigenfunction of A corresponding to the eigenvalue α if and only if

$$C_\alpha g_\alpha = g_\alpha\,, \quad \text{where } g_\alpha(x) = [\alpha - (1 - \mu)W(x)]f_\alpha(x)\,. \tag{3.11}$$

Hence, (3.5) and (3.11) are equivalent, and to prove Theorem 3.1 it has to be shown that a unique $a \in (\tau, 1)$ exists, such that C_a has 1 as an algebraically simple eigenvalue.

Strict positivity of W, together with assumption (3.6), implies

$$\int_{\mathcal{X}\setminus S} \int_S f(y)\,\frac{W(y)}{\alpha - [1 - \mu(y)]W(y)}\,\mu(y)u(y, x)\,\mathrm{d}y > 0\,. \tag{3.12}$$

Therefore, each C_α is irreducible (Appendix C). Since C^n is compact and

$$C_\alpha \leq \frac{1}{\alpha - \tau}\,C\,,$$

it follows that C_α^{2n} is compact (Lemma C.1). Hence, every C_α satisfies the hypothesis of Theorem C.2. Therefore, the spectral radius $r(C_\alpha) > 0$ is an eigenvalue of C_α with a unique normalized eigenfunction that is positive λ-*a.e.* Moreover, no such operator C_α has another nonnegative eigenfunction. Therefore, to prove Theorem 3.1 it remains to show that a unique a exists with $r(C_a) = 1$ (cf. Figure 3.1).

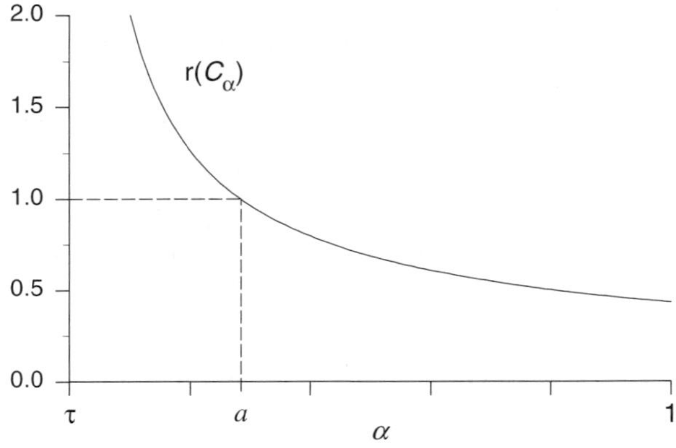

Figure 3.1 The spectral radius $r(C_\alpha)$ as a function of $\alpha \in [\tau, 1]$. The unique value a, where $r(C_a) = 1$, is equal to the equilibrium mean fitness $\hat{\bar{W}}$.

For this purpose, we first observe that

$$r(C_\alpha) \leq ||C_\alpha|| \leq \frac{||\mu W||_\infty}{\alpha - \tau} ,$$

whence $r(C_\alpha) \leq 1$ follows if $\alpha \geq 1$. Lemma C.2 implies that $\alpha \mapsto r(C_\alpha)$ is continuous, and Lemma C.3 shows that $\alpha \mapsto r(C_\alpha)$ is strictly monotone decreasing. If we can show that $\lim\limits_{\alpha \downarrow \tau} r(C_\alpha) > 1$, then a unique a exists such that $r(C_a) = 1$ and $\tau < a < 1$. This, however, is a consequence of assumption 3 of Theorem 3.1, as we show subsequently.

Let φ_J denote the characteristic function of an interval J chosen so that (3.7) holds. Then

$$\begin{aligned}(C_\alpha \varphi_J)(x) &= \int_J \frac{W(y)}{\alpha - [1-\mu(y)]W(y)} \, \mu(y) u(y,x) \, \mathrm{d}y \\ &\geq u_0 \varphi_J(x) \int_J \frac{W(y)}{\alpha - [1-\mu(y)]W(y)} \, \mathrm{d}y \, .\end{aligned}$$

It follows that

$$||C_\alpha^n||^{1/n} \geq u_0 \int_J \frac{W(y)}{\alpha - [1-\mu(y)]W(y)} \, \mathrm{d}y \quad \text{for every } n \geq 1$$

and, by performing the limit $n \to \infty$,

$$r(C_\alpha) \geq u_0 \int_J \frac{W(y)}{\alpha - [1-\mu(y)]W(y)} \, \mathrm{d}y \, . \tag{3.13}$$

The right-hand side is a strictly monotone decreasing function of α. Therefore, by B. Levi's monotone-convergence theorem (Hewitt and Stromberg 1965), and by condition (3.8), we obtain

$$\lim_{\alpha \downarrow \tau} r(C_\alpha) \geq u_0 \int_J \frac{W(y)}{\tau - [1-\mu(y)]W(y)} \, \mathrm{d}y > 1 \, .$$

If the normalized eigenfunction of C_a corresponding to the eigenvalue 1 is denoted by $\hat{q}$, then

$$\hat{p}(x) = \frac{[a - [1 - \mu(x)]W(x)]^{-1}\hat{q}(x)}{\int [a - [1 - \mu(x)]W(x)]^{-1}\hat{q}(x)\,\mathrm{d}x} \tag{3.14}$$

is the uniquely determined equilibrium density of (2.8) and (3.4). Note that $a > \tau$, together with $0 \le W(x) \le 1$ and $\hat{q} \in L^1(\lambda)$, implies that the integral on the right-hand side of (3.14) is finite and positive. This finishes the proof of Theorem 3.1. ◁

Theorem 3.2 *Suppose that conditions 1 – 3 of Theorem 3.1 are satisfied. In addition, assume that all eigenvalues $\psi \neq a$ of A satisfy $|\psi| < a$. Then the unique positive equilibrium solution $\hat{p}$ of* (3.4) *is globally asymptotically stable in the sense that for any $p_0 \ge 0$ with $\int p_0(x)\,\mathrm{d}x = 1$, the uniquely determined solution $p(x,t)$ with $p(x,0) = p_0(x)$ satisfies*

$$\lim_{t\to\infty} \int |p(x,t) - \hat{p}(x)|\,\mathrm{d}x = 0\,. \tag{3.15}$$

Remark. The additional assumption that a is the only eigenvalue in the peripheral spectrum is fulfilled, for example, if for every $f \ge 0$, $f \in L^1(\lambda)$, there is some $n \in \mathbb{N}$ such that $A^n f$ is positive λ-*a.e.* (cf. Theorem C.2). This is the case if $C^n f$ is positive. This additional assumption is a peculiarity of the discrete-time model, since otherwise periodic orbits may occur (cf. Chapter III.1.1). ◇

Proof of Theorem 3.2. Let an initial distribution $p_0 \in L^1(\lambda)$ be given such that $p_0 \ge 0$ and $\int p_0 = 1$. According to Theorem C.3 (with $K = A$ and $r(K) = a$) and adapting its notation, there is a unique decomposition

$$p_0 = c\hat{p} + v_0\,,$$

where c is a complex number and $v_0 \in E_2$. It follows that $A^n p_0 = ca^n\hat{p} + A_2^n v_0$. Since $p(\,.\,,n) = A^n p_0 / \int A^n p_0$, we obtain

$$\begin{aligned}
||p(\,.\,,n) - \hat{p}||_1 &= \left\| \frac{c\hat{p} + a^{-n}A_2^n v_0}{c + a^{-n}\int A_2^n v_0} - \hat{p} \right\|_1 \\
&\le \left\| \frac{c\hat{p}}{c + a^{-n}\int A_2^n v_0} - \hat{p} \right\|_1 + a^{-n} \left\| \frac{A_2^n v_0}{c + a^{-n}\int A_2^n v_0} \right\|_1 \\
&= \left| \frac{c}{c + a^{-n}\int A_2^n v_0} - 1 \right| ||\hat{p}||_1 + \frac{||A_2^n v_0||_1}{a^n} \frac{1}{c + a^{-n}\int A_2^n v_0}\,.
\end{aligned}$$

Since $r(A_2) < a$, it follows that $\left|a^{-n}\int A_2^n v_0\right| \le a^{-n}\,||A_2^n v_0||_1 \to 0$ as $n \to \infty$. This proves that $||p(\,.\,,n) - \hat{p}||_1 \to 0$ as $n \to \infty$, unless $c = 0$. However, $c = 0$ is impossible, because of the following argument. Suppose that $c = 0$. Then we would have $v_0 = p_0 \ge 0$. For the adjoint operator A^* on $L^\infty(\mathcal{X})$, the number a is again the unique eigenvalue with a positive eigenfunction and has a one-dimensional eigenspace. It follows that $v_0 \in E_2$ is orthogonal to this eigenspace, which is impossible because of the positivity. Hence, the assumption $c = 0$ leads to a contradiction. This finishes the proof of Theorem 3.2. ◁

For the continuous-time equation (2.20), analogous theorems can be proved with a few obvious modifications (cf. Bürger 1988a). Condition 3 in Theorem 3.1 has to be replaced by

3". Let J be a set of positive measure in which $m-\mu$ attains its essential supremum. Assume that there is a constant $u_0 > 0$ such that (3.7) holds and

$$u_0 \int_J [\operatorname{ess\,sup}(m-\mu) + \mu(x) - m(x)]^{-1}\, \mathrm{d}x > 1\,. \tag{3.16}$$

For constant $\mu(x)$, (3.16) reduces to

$$u_0 \int_J \frac{1}{-m(x)}\, \mathrm{d}x > 1\,. \tag{3.17}$$

This simplifies and slightly generalizes Proposition 3.4 in Bürger (1988a).

*4. A MODEL IN TERMS OF PROBABILITY MEASURES

In the previous sections, certain conditions on the local behavior of the mutation distribution and the fitness function near the fitness optimum were needed to prove existence of an equilibrium density. Here we show that without these conditions, probability mass may accumulate at the fitness optimum. In such a case, no equilibrium density exists, but an atom of probability builds up. Suppose, for example, that mutation is absent. Then (2.8) can be solved explicitly, the solution being

$$p(x,t+1) = W(x)^n p(x,0) \Big/ \int_{\mathcal{X}} W(x)^n p(x,0)\, \mathrm{d}x\,.$$

If $\mathcal{X} = \mathbf{R}$ (or some multidimensional interval), and if $W(x)$ is bounded and attains its maximum value at x_0, then $p(x,t)$ tends to infinity as $x \to x_O$. For all other values of x, $p(x,t)$ tends to zero. Kingman (1978) and Bürger and Bomze (1996) explored situations of mutation-selection balance which lead to an equilibrium distribution that has no density. A simple example is the following (cf. Bürger 1988b): let $\mathcal{X} = [0,1]$, $u(x,y) = 1$ for all x, y, and let $W(x) = 1 - \sqrt{x}$. Then, for any mutation rate μ with $0 < \mu < 1$, the equilibrium distribution consists of a point measure at $x = 0$ and a 'density' with mass less than one; cf. (4.21). The reason is that $W(x)$ decreases too rapidly near the optimum $x = 0$ in order that mutation can prevent the accumulation of types near $x = 0$. Another example will be discussed at the end of this section.

Sometimes, it may also be desirable to start with monomorphic initial populations, described by a point measure, and not by a density. Therefore, we shall develop a haploid mutation-selection model that generalizes that of Section 2, by describing the population in terms of probability measures instead of probability densities. From a mathematical point of view, this appears to be the natural setting for such models. Related approaches, partially in different contexts, were developed by Eshel (1971, 1972), Kingman (1978), and Karlin (1988), who considered special fitness functions that describe directional selection, such as $W(x) = sx$ and $W(x) = e^{sx}$, but admitted more general mutation distributions than we do.

In the present model, the dynamics of the type distribution under mutation and selection is represented by an integro-difference equation for Borel probability measures. The purpose is to characterize the structure of possible equilibrium distributions, and to give sufficient and necessary conditions for existence and uniqueness. The structure of equilibrium distributions is examined with the help of Lebesgue's Decomposition Theorem, according to which every measure can be represented as the sum of an absolutely continuous (with a density) and a singular one. For the HC-mutation model, a complete characterization is obtained, extending that of Kingman (1978). To a large extent, the presentation follows Bürger and Bomze (1996), who developed and analyzed the corresponding continuous-time model.

4.1 THE MODEL

The basic model is that of Section 3, except that type densities are described in a more general way. We denote the total frequency of types at time t by the positive finite measure Q_t. Thus, for a Borel set $Y \subseteq \mathcal{X}$ the total number of individuals of type $y \in Y$ is $Q_t(Y)$. According to the definition of fitness (viability), the type distribution after selection is

$$Q_t^s(Y) = \int_Y W(y)\, Q_t(\mathrm{d}y)\ .$$

Since we are only interested in the evolution of relative frequencies, let $P_t = Q_t/Q_t(\mathcal{X})$ denote the probability measure corresponding to Q_t. Then, the relative frequency distribution P_t^s after selection is

$$P_t^s(Y) = \frac{1}{\overline{W}(t)} \int_Y W(y)\, P_t(\mathrm{d}y)\ , \tag{4.1}$$

where $\overline{W}(t) = \int_{\mathcal{X}} W(x)\, P_t(\mathrm{d}x)$ is the mean fitness, which satisfies $0 < \overline{W}(t) \leq 1$.

If P_t is the distribution before mutation, then the distribution after mutation is

$$P_t^\mu(Y) = \int_Y \left(\int_{\mathcal{X}} \mu(x)u(x,y)\, P_t(\mathrm{d}x) \right) \lambda(\mathrm{d}y) + \int_Y [1-\mu(y)]\, P_t(\mathrm{d}y)\ , \tag{4.2}$$

where $\mu(y)$ is the mutation probability of type y, and the mutation kernel $u(x,y)$ satisfies (2.1). (The reader may note that for notational clarity the roles of x and y have been interchanged compared with the previous sections.)

Combining (4.1) with (4.2), and assuming that mutation occurs after selection, the basic mutation-selection equation for the evolution of the frequency distribution P_t is obtained,

$$\begin{aligned} \overline{W} P_{t+1}(Y) = & \int_Y [1-\mu(y)] W(y)\, P_t(\mathrm{d}y) \\ & + \int_Y \left(\int_{\mathcal{X}} W(x)\mu(x)u(x,y)\, P_t(\mathrm{d}x) \right) \lambda(\mathrm{d}y)\ , \end{aligned} \tag{4.3}$$

where $t \geq 0$ and $Y \subseteq \mathcal{X}$ may be any Borel set in $\mathcal{X}$. If P_t has a density $p(t)$ (with respect to λ), i.e., if $P_t = p(t)\lambda$, then (4.3) simplifies to (2.8).

The investigation of the solutions of (4.3) is simplified by interpreting (4.3) as a difference equation on the Banach space $\mathcal{M} = \mathcal{M}(\mathcal{X})$ of complex-valued, finite Borel measures endowed with the variational norm $\|\,.\,\|$, i.e., $\|P\| = |P|\,(\mathcal{X})$. To this end, we define in generalization of (3.1), (3.2), and (3.3) the following linear operators on $\mathcal{M}$

$$(TQ)(Y) = \int_Y [1-\mu(y)]W(y)\,Q(\mathrm{d}y)\,, \tag{4.4}$$

$$(CQ)(Y) = \int_Y \left(\int_{\mathcal{X}} W(x)\mu(x)u(x,y)\,Q(\mathrm{d}x)\right)\lambda(\mathrm{d}y), \tag{4.5}$$

$$A = T + C\,. \tag{4.6}$$

We shall use the same notation for the operators T, C, and A as operators on $\mathcal{M}$ and on $L^1(\lambda)$, because we always identify absolutely continuous measures with respect to (w.r.t.) λ with their densities.

The first important observation is that C is a nonnegative, bounded operator on $\mathcal{M}$ whose image is contained in $L^1(\lambda)$. Hence, we may write

$$CQ(y) = \int_{\mathcal{X}} W(x)\mu(x)u(x,y)\,Q(\mathrm{d}x)\,, \tag{4.7}$$

thus, identifying CQ with its density w.r.t. λ. To prove this, we have to show that for $Q \in \mathcal{M}$ the measure CQ is absolutely continuous. Indeed, $CQ \in L^1(\lambda)$ because, by Fubini's theorem and by our general assumptions (2.1), $\mu(x) \le 1$, and $W(x) \le 1$, we have

$$\begin{aligned}\|CQ\|_1 &\le \int_{\mathcal{X}} \left(\int_{\mathcal{X}} W(x)\mu(x)u(x,y)\,|Q|\,(\mathrm{d}x)\right)\lambda(\mathrm{d}y)\\ &= \int_{\mathcal{X}} \left(\int_{\mathcal{X}} W(x)\mu(x)u(x,y)\,\lambda(\mathrm{d}y)\right)|Q|\,(\mathrm{d}x)\\ &= \int_{\mathcal{X}} W(x)\mu(x)\,|Q|\,(\mathrm{d}x)\\ &\le \|Q\|\,.\end{aligned} \tag{4.8}$$

Therefore, C is bounded with image in $L^1(\lambda)$. Clearly, C is nonnegative.

Since T is also bounded and nonnegative, A is a bounded and nonnegative operator on $\mathcal{M}$. Therefore, the mutation-selection dynamics (4.3) can be rewritten as a difference equation on $\mathcal{M}$,

$$P_{t+1} = AP_t/AP_t(\mathcal{X}) \tag{4.9}$$

[cf. (3.4)], where

$$AP_t(\mathcal{X}) = \int_{\mathcal{X}} AP_t(\mathrm{d}x) = \overline{W}(t)\,. \tag{4.10}$$

As A is bounded and nonnegative on $\mathcal{M}$, and because $P_t(\mathcal{X}) = 1$ implies $P_{t+1}(\mathcal{X}) = 1$, solutions of (4.9) exist for all $t \ge 0$ and for any initial probability measure P_0.

4.2 STRUCTURE AND EXISTENCE OF EQUILIBRIUM DISTRIBUTIONS

We denote the subset of probability measures of $\mathcal{M}$ by $\mathcal{P}$. The equilibria of (4.9) are the solutions P of the eigenvalue problem

$$AP = aP \tag{4.11}$$

on $\mathcal{M}$ with $P \in \mathcal{P}$. Since the eigenvalue a satisfies $a = AP(\mathcal{X}) = \hat{\overline{W}}$, and the fitness function satisfies $0 < W(x) \leq 1$, the inequalities $0 < a \leq 1$ are obtained. From now on, we suppose that C is irreducible as an operator on $L^1(\lambda)$. This is equivalent to the validity of (3.6), because W is positive by assumption. However, we do not yet assume conditions 2 and 3 of Theorem 3.1.

As discussed in Section 2, and more fully in Bürger and Bomze (1996), there are cases where no equilibrium solution exists. We first investigate the structure of possible solutions of (4.11) and deal with existence thereafter. For notational convenience we set, as in Section 3,

$$\tau = \|(1-\mu)W\|_\infty \ ,$$

and we introduce the level sets

$$W_\alpha = \{x \in \mathcal{X} : [1-\mu(x)]W(x) = \alpha\} \ , \quad 0 \leq \alpha \leq \tau \ ,$$

and

$$N_0 = \{x \in \mathcal{X} : \mu(x) = 0\} \ .$$

Irreducibility of C implies $\lambda(N_0) = 0$.

Given that a solution $P \in \mathcal{P}$ of (4.11) exists, Lebesgue's Decomposition Theorem states that there are probability measures $R, S \in \mathcal{P}$ and a number $\rho \in [0,1]$, such that

$$P = (1-\rho)R + \rho S \ , \tag{4.12}$$

where R possesses a density p with respect to λ and S is singular, i.e., S is concentrated on a λ-null set. We already know from (4.7) that CS is absolutely continuous. Substituting (4.12) into (4.11) and decomposing the left- and the right-hand sides into an absolutely continuous and a singular part, we obtain the following equations, which jointly are equivalent to (4.11):

$$(1-\rho)AR + \rho CS = a(1-\rho)R \ , \tag{4.13a}$$

$$\rho TS = a\rho S \ . \tag{4.13b}$$

First, we settle the case $\rho = 1$ of a singular solution $P = S$, i.e., $AS = aS$.

Theorem 4.1 *A singular solution of (4.11) exists in $\mathcal{P}$ if and only if $N_0 \neq \emptyset$. More precisely, any number a, such that $a = W(x)$ for some $x \in N_0$, may be an eigenvalue, and any (singular) measure S with $S(N_0 \cap W_a) = 1$ solves (4.11). Every eigenvalue corresponding to a singular equilibrium solution satisfies $0 < a \leq \tau$.*

Proof. First suppose $S(N_0 \cap W_a) = 1$ for some $S \in \mathcal{P}$ and $0 \leq a \leq \tau$. Since S is a probability measure, we have $S(N_0) = 1$. This implies singularity of S, because

$\lambda(N_0) = 0$. By using Fubini's theorem, we derive from $S(N_0) = 1$:

$$\begin{aligned} CS(Y) &= \int_Y \left[\int_{\mathcal{X}} W(x)\mu(x)u(x,y)\, S(\mathrm{d}x)\right] \lambda(\mathrm{d}y) \\ &\leq \int_{\mathcal{X}} \left[\int_{\mathcal{X}} W(x)\mu(x)u(x,y)\, S(\mathrm{d}x)\right] \lambda(\mathrm{d}y) \\ &= \int_{\mathcal{X}} W(x)\mu(x)\, S(\mathrm{d}x) \\ &= 0, \quad \text{for all Borel sets } Y \in \mathcal{X}\ . \end{aligned}$$

Therefore, we have $AS = TS = aS$, because $S(W_a) = 1$ means $[1-\mu(x)]W(x) = W(x) = a$ S-*a.e.*

Conversely, if $\rho = 1$, then (4.13a) implies $\int_{\mathcal{X}} \mu(x)W(x)\, S(\mathrm{d}x) = CS(\mathcal{X}) = 0$, and hence $S(N_0) = 1$ because $W(x) > 0$ for all x. In addition, (4.13b) tells us that $TS = aS$, which entails $S(W_a) = 1$. To summarize, we have $AS = aS$ if and only if $S(N_0 \cap W_a) = 1$. The assumption $W(x) > 0$ implies $a > 0$. ◁

The above theorem has the following biological interpretation: if S describes a population consisting only of individuals that do not mutate (i.e., $\mu(x) = 0$ S-*a.e.*) and have the same fitness ($W(x) = a$), then it will remain unchanged.

Now we treat the more interesting case $0 \leq \rho < 1$.

Lemma 4.2 *If there is a nonnegative solution $P \in \mathcal{P}$ of (4.11) with $0 \leq \rho < 1$, then $a \geq \tau$ and the density p of R satisfies $p > 0$ λ-a.e. If $a = \tau$, then $\lambda(W_\tau) = 0$.*

Proof. Let $P \geq 0$, $P \neq 0$, be a solution of (4.11). Then, using the positivity of C and (4.13a), we obtain

$$\begin{aligned} 0 &\leq (1-\rho)Cp(x) \\ &\leq (1-\rho)Cp(x) + \rho CS(x) \\ &= (1-\rho)Cp(x) + (1-\rho)(ap - Ap)(x) \\ &= (1-\rho)(ap - Tp)(x) \\ &= (1-\rho)\big[a - [1-\mu(x)]W(x)\big]p(x) \qquad \lambda\text{-}a.e. \end{aligned} \tag{4.14}$$

Suppose that $a < \tau$, and let $I = \{x \in \mathcal{X} : [1-\mu(x)]W(x) > a\}$. Then $\lambda(I) > 0$, and (4.14) implies that $p(x) = 0$ for almost all $x \in I$ and $Cp(x) = 0$ λ-*a.e.* on I. This is impossible on a set of positive measure because C is irreducible. Therefore, $a \geq \tau$ holds. The same argument shows that irreducibility of C yields $p > 0$ λ-*a.e.* This proves the first statement. Similarly, (4.14) implies the second if $a = \tau$. ◁

The following theorem classifies the structure of non-singular equilibrium solutions of (4.11).

Theorem 4.3

1. *Suppose that a solution $P \in \mathcal{P}$ of (4.11) exists and is given in the form (4.12) with $R = p\lambda$ and $0 \leq \rho < 1$. Then the eigenvalue a corresponding to P satisfies $\tau \leq a \leq 1$, and precisely the following cases can occur:*

(A) *If $a > \tau$, then $\rho = 0$, i.e., P is absolutely continuous. Its λ-density p is a nonnegative solution of the eigenvalue problem*

$$Ap = ap\ , \qquad p \in L^1(\lambda)\ . \tag{4.15}$$

(B) *If $a = \tau$ and $\rho = 0$, then the λ-density of P is given by*

$$Ap = \tau p \,. \tag{4.16}$$

(C) *If $a = \tau$ and $0 < \rho < 1$, then $W_\tau \neq \emptyset$, $\lambda(W_\tau) = 0$, and S is a singular measure with $S(W_\tau) = 1$. The density p of R satisfies*

$$(A - \tau)p = -\frac{\rho}{1-\rho} CS \,, \tag{4.17}$$

where the parameter ρ is determined by $\int_{\mathcal{X}} p(x)\lambda(\mathrm{d}x) = R(\mathcal{X}) = 1$.

2. *Conversely, if $p \in L^1_+(\lambda)$, $\int_{\mathcal{X}} p(x)\lambda(\mathrm{d}x) = 1$, is a solution of* (4.15) *or* (4.16), *then $P = p\lambda$ is a solution of* (4.11) *with the same a. If S is as in case* (C), *and p and ρ satisfy* (4.17), *then $P = (1-\rho)p\lambda + \rho S$ is a solution of* (4.11) *with $a = \tau$.*

Proof. 1. (A) Suppose that $\rho > 0$. Then (4.13b) implies that $TS = aS$, i.e., $[1 - \mu(x)]W(x) = a$ holds S-*a.e.*, and consequently $a = \tau$, a contradiction. This proves the first assertion. That the λ-density p has to satisfy (4.15) follows from (4.13a).

(B) This is obvious from (4.13a) and (4.13b).

(C) The proof of (A) shows that the assumptions $a = \tau$ and $\rho > 0$ imply $[1 - \mu(x)]W(x) = \tau$, S-*a.e.*, which can be expressed as $S(W_\tau) = 1$. In particular, the level set W_τ cannot be void. For $a = \tau$ we get, by applying (4 13a) to W_τ,

$$\begin{aligned}
&\int_{W_\tau} \left[(1-\rho)Cp(y) + \rho \int_{\mathcal{X}} W(x)\mu(x)u(x,y)\, S(\mathrm{d}x)\right] \lambda(\mathrm{d}y) \\
&\quad = (1-\rho) \int_{W_\tau} \{[(1-\mu(y))W(y) - \tau]\, p(y) + Cp(y)\}\lambda(\mathrm{d}y) + \rho CS(W_\tau) \\
&\quad = [(1-\rho)(A-\tau)R + \rho CS]\,(W_\tau) = 0 \,.
\end{aligned}$$

Since C is irreducible, $0 < \rho < 1$, and $p > 0$ holds λ-*a.e.* (Lemma 4.2), the left-hand side can be zero only if $\lambda(W_\tau) = 0$. Hence, S is singular with $S(W_\tau) = 1$, and (4.17) follows from (4.13a).

2. This is a simple consequence of (4.13a) and (4.13b). ◁

It follows that if $W_\tau = \emptyset$ or $\lambda(W_\tau) > 0$, then $\rho = 0$ or a nonnegative solution P of (4.11) does not exist. Another simple consequence of Lemma 4.2 and Theorem 4.3 is

Corollary 4.4 *Let P be an equilibrium solution that is not purely singular. Then the equilibrium mean fitness satisfies $\|(1-\mu)W\|_\infty \leq \hat{\overline{W}} = a \leq 1$. If P has a singular part, i.e., if $\rho > 0$, then $\hat{\overline{W}} = \|(1-\mu)W\|_\infty$.*

By Theorem 4.3, we have reduced the problem of existence and uniqueness of equilibrium solutions of (4.9) to the solution of (4.15), (4.16), or (4.17) in $L^1(\lambda)$. The eigenvalue problem (4.15) is identical to (3.5), and Theorem 3.1 provides sufficient conditions for its solution. In what follows, the operators T, C, and A are always considered as operators on $L^1(\lambda)$. The operators C_α are defined as in (3.10). Actually, the following characterization of case (A) of Theorem 4.3 is a simple consequence of Theorem 3.1.

Theorem 4.5 *Let C be irreducible and power compact. Then a solution $P \in \mathcal{P}$ of* (4.11) *corresponding to an eigenvalue $a > \tau$ exists if and only if*

$$\text{there is some } \alpha > \tau \text{ such that } r(C_\alpha) > 1 \, . \tag{4.18}$$

In this case, P is the unique nonnegative, normalized solution of (4.11) *satisfying $0 \leq \rho < 1$. In fact, P is absolutely continuous, i.e. $\rho = 0$. Its density p is positive, and is the unique nonnegative, normalized solution of the eigenvalue problem* (4.15).

Proof. Assertion (4.18) follows from (the proof of) Theorem 3.1. It remains to show that it is impossible to have a solution p of (4.15) with $a > \tau$, simultaneously with a solution $\tilde{p}$ of (4.17). If this were the case, then, by Lemma 4.8 in Schaefer (1974, Chapter V), there existed $p' \in L^\infty(\lambda)$ such that

$$A'p' = ap' \, , \quad p' \geq 0, \quad \text{and } \langle p, p' \rangle > 0 \, , \tag{4.19}$$

where A' is the dual operator of A on $L^\infty(\lambda)$. This, together with the strict positivity of $\tilde{p}$ (Lemma 4.2), yields

$$0 < a\langle p', \tilde{p} \rangle = \langle A'p', \tilde{p} \rangle = \langle p', A\tilde{p} \rangle = \tau \langle p', \tilde{p} \rangle - \left\langle p', \frac{\rho}{1-\rho} CS \right\rangle ,$$

which implies $a \leq \tau$, a contradiction. ◁

Under a number of technical assumptions about the operator C_τ on $L^1(\lambda)$, it can be proved that case (A) of Theorem 4.3 occurs if and only if $r(C_\tau) > 1$, case (B) occurs if and only if $r(C_\tau) = 1$, and case (C) occurs if and only if $r(C_\tau) < 1$ (cf. Bürger and Bomze 1996). For the HC-mutation model, Theorem 4.5 provides a simple classification of all possible equilibrium states.

4.3 THE HC-MUTATION MODEL

For the HC-mutation model (2.4), every equilibrium distribution P satisfies $CP = \hat{\overline{W}} \mu u_{\mathrm{HC}} = a \mu u_{\mathrm{HC}}$. Writing P in the form $P = (1-\rho)p\lambda + \rho S$, where $\rho \in (0,1)$, and noting that the left-hand side of (4.13a) is equal to $(1-\rho)TR + CP$, the equilibrium equations (4.13) for the absolutely continuous and the singular part become

$$(1-\rho)(1-\mu)W(x)p(x) + a\mu u_{\mathrm{HC}}(x) = a(1-\rho)p(x) \, , \quad \lambda\text{-}a.e., \tag{4.20a}$$

$$\rho(1-\mu)WS = a\rho S \, . \tag{4.20b}$$

Since $\mu(x) \equiv \mu > 0$, no purely singular solutions exist (Theorem 4.1) and $\tau = 1 - \mu$.

Theorem 4.6 *Assume that $\{x \in \mathcal{X} : W(x) = 1\} \neq \emptyset$, hence $W_\tau \neq \emptyset$. Then a unique $\rho \in [0,1)$ exists such that* (4.20) *admits at least one solution. In particular, the following can be shown:*

(A) *If* (2.9) *holds, i.e., if*

$$\mu \int_{\mathcal{X}} \frac{u_{\mathrm{HC}}(x)}{1 - W(x)} \, \lambda(\mathrm{d}x) \geq 1 \, ,$$

then a unique equilibrium solution P exists. It is absolutely continuous, i.e., $\rho = 0$, and its λ-density is given by (2.12), where $1 - \mu \leq \hat{\overline{W}} < 1$. The equilibrium mean fitness is $\hat{\overline{W}} = 1 - \mu$ if and only if equality holds in (2.9).

(B) *Otherwise, we have $\rho = 1 - \mu \displaystyle\int_{\mathcal{X}} \frac{u_{\mathrm{HC}}(x)}{1 - W(x)}\, \lambda(\mathrm{d}x) > 0$, $a = \tau = 1 - \mu$, and*

$$p(x) = \frac{\mu u_{\mathrm{HC}}(x)}{(1-\rho)[1 - W(x)]}\,. \tag{4.21}$$

It follows that $\lambda(W_\tau) = 0$, and the singular part S of the solution may be any singular probability measure concentrated on W_τ.

Proof. (A) was proved in Section 2.3 and (B) follows from what we have shown above. ◁

Part (B) shows that atoms of probability occur if (2.9) is not satisfied. In this case, the equilibrium distribution is uniquely determined only if a single fitness optimum exists.

Waxman and Peck (1998) observed that if alleles contribute to three traits or more, i.e., $\mathcal{X} = \mathbf{R}^n$ with $n \geq 3$, then the equilibrium distribution contains an atom of probability if the mutation distribution is multivariate Gaussian and the fitness function is $W(x) = \prod_{l=1}^{n} \exp(-x_l^2/2V_{\mathrm{s}})$ and, hence, is smooth. Coppersmith *et al.* (1999) extended and refined the model of Waxman and Peck, and mapped it onto problems in quantum mechanics. However, a simple exercise in integration shows that for this model the integral occuring in the condition of Theorem 4.6 (A) is infinite if $k = 1$ or 2, but finite if $k \geq 3$. Therefore, condition (2.9) is satisfied for $k \geq 3$ only if μ is unrealistically large. Hence, case (B) of Theorem 4.6 applies and the equilibrium distribution contains an atom of probability at the position of the optimum. It is obvious from the structure of condition (2.9), and its application to the above fitness function, that for $k \geq 3$, (2.9) will fail to hold for a wide class of smooth fitness functions and mutation distributions, unless μ is very large. If such mutation distributions are biologically reasonable is another question. As indicated by Theorem 4.3 and condition (3.8), this will extend to mutation models different from the HC model. For a (speculative) discussion of the possible biological implications of this phenomenon, we refer to Waxman and Peck (1998) and Wagner (1998).

5. EQUILIBRIUM MEAN FITNESS AND MUTATION LOAD

In Chapter III.3, Haldane's principle that the mutation load, $L = (W_{\max} - \overline{W})/W_{\max}$, is independent of the selection coefficient was extended to include a finite number of alleles and (almost) arbitrary mutation patterns between different alleles. It is the purpose of this section to prove that this is also the case for the general model treated above, in particular for the continuum-of-alleles model and the stepwise-mutation model. Thus, we shall refine result (2.10) that the equilibrium mean fitness is between $1 - \mu$ and 1, and show that it is asymptotically equivalent to $1 - \mu$ as $\mu \to 0$. For a broad class of fitness functions and mutation distributions also the second-order term of the equilibrium mean fitness is determined. In contrast to the first-order term, the second-order term depends on the decay of the fitness function near the maximum fitness value and on the form of the mutation distribution. Throughout this section, the fitness function $W(x)$ and the mutation distribution $u(x, y)$ are considered as fixed, whereas the mutation rate μ varies. All asymptotic results apply in the limit $\mu \to 0$. The main results are (5.5) and (5.6), (5.21) and (5.22), and (5.31) and (5.33).

5.1 A CONTINUUM OF ALLELES

For notational simplicity, we only consider models with a continuum of possible types, e.g., $\mathcal{X} = \mathbf{R}^k$ or $\mathcal{X}$ is some interval in $\mathbf{R}^k$, $k \geq 1$. Discrete allelic models will be treated separately. We assume discrete generations, a constant mutation rate μ, and a fitness function $W(x)$ satisfying conditions (W1) and (W2) of Section 2.2. We are only interested in the case that the equilibrium distribution has a density. Otherwise, we already know from Corollary 4.4 that the equilibrium mean fitness is equal to $1 - \mu$. Therefore, let us assume that $W(x)$ and the mutation distribution $u(x, y)$ satisfy condition 2 of Theorem 3.1 with $n = 1$, and condition 3' of Remark d) following Theorem 3.1. These hold in the special cases (C) and (D) of Section 2.2. However, we do not require condition 1, i.e., irreducibility of the mutation distribution. Thus, we include cases where, for example, mutation occurs only to non-optimal types. Without the assumption of irreducibility, the equilibrium distribution is not necessarily unique. Throughout this section we consider *some* equilibrium distribution with a density p_μ and equilibrium mean fitness $\overline{W}_\mu$, the subscript μ signifying the dependence on the mutation rate. We denote the set of optimal types by

$$W_{\mathrm{Opt}} = \{x \in \mathcal{X} : W(x) = 1\} , \tag{5.1}$$

and note that it is not empty because, by (W2), $x_O \in W_{\mathrm{Opt}}$. The reader may note that (3.8') is fulfilled whenever $\lambda(W_{\mathrm{Opt}}) > 0$.

With the present assumptions, it follows directly from (2.8) that the equilibrium mean fitness satisfies

$$1 - \mu \leq \overline{W}_\mu \leq 1 . \tag{5.2}$$

The case $\overline{W}_\mu = 1$ can occur only if the equilibrium density is confined to the set W_{Opt}, which is possible only if no mutation to suboptimal types occurs. We exclude this trivial case from our consideration.

Let us define the mutation operator

$$Uf(x) = \int_{\mathcal{X}} f(y)W(y)u(y, x)\,\mathrm{d}y , \tag{5.3}$$

and observe that $\mu U = C$, with C as in (3.2). By our assumptions, U is a compact operator on $L^1(\lambda)$. The following asymptotic result for the equilibrium mean fitness will be proved in Section 5.2:

$$\lim_{\mu \to 0} \frac{1}{\mu} \left[\overline{W}_\mu - \left(1 - \mu + \mu \int_{W_{\mathrm{Opt}}} Up_\mu(x)\,\mathrm{d}x \right) \right] = 0 . \tag{5.4}$$

As the integral does not necessarily vanish, the equilibrium mean fitness, and thus the mutation load, may depend on details of the equilibrium density. A simple example was given below III(3.4).

However, under additional assumptions, the equilibrium mean fitness is independent of the fitness function W and of the specific form of the mutation distribution u. Suppose, for instance, that there exists a *unique optimal type* x_O or, more generally, let $\lambda(W_{\mathrm{Opt}}) = 0$. Then (5.4) reduces to

$$\lim_{\mu \to 0} \frac{\overline{W}_\mu - (1-\mu)}{\mu} = 0 , \tag{5.5a}$$

or

$$\overline{W}_\mu \sim 1 - \mu \quad \text{as } \mu \to 0 . \tag{5.5b}$$

This can be reformulated as

• **5.1** *For a continuum of alleles and any number of discrete optimal types, the equilibrium mean fitness is, to first order in μ, independent of the fitness function and the mutation distribution. The first-order approximation of the mutation load is*

$$L = \mu . \tag{5.6}$$

If the assumptions stated above are not satisfied, so that the equilibrium solution has a singular part, then $L = \mu$ holds exactly (Corollary 4.4).

An analogous result is valid for the corresponding continuous-time model (Bürger and Hofbauer 1994). In Section 5.4 we shall calculate second-order terms for some specific fitness functions and mutation distributions. The approximation (5.6) is in sharp contrast to Crow and Kimura's (1964) original result,

$$L = \frac{\mu \bar{u}_{\mathrm{RW}}}{\bar{u}_{\mathrm{RW}}^2 + \gamma^2} + \sqrt{\tfrac{1}{2}\mu s(\gamma^2 + \bar{u}_{\mathrm{RW}}^2)} , \tag{5.7}$$

which was based on the Gaussian allelic approximation (1.8). In (5.7), $\bar{u}_{\mathrm{RW}}$ and γ^2 are the mean and variance, respectively, of the mutation distribution u_{RW}; cf. (2.3). As we shall show in Chapter VI.6, the HC-approximation (1.16) is a consequence of the generalization (5.6) of Haldane's principle.

For the *HC-mutation model* (2.4), we have $Up_\mu = u_{\mathrm{HC}}$, so that (5.4) simplifies to

$$\lim_{\mu \to 0} \frac{1}{\mu} \left[\overline{W}_\mu - \left(1 - \mu \int_{\mathcal{X} \setminus W_{\mathrm{Opt}}} u_{\mathrm{HC}}(x)\, \mathrm{d}x \right) \right] = 0 . \tag{5.8}$$

Therefore, the mutation load is, to first order in μ, the total mutation rate toward suboptimal alleles.

$$L = \mu \int_{\mathcal{X} \setminus W_{\mathrm{Opt}}} u_{\mathrm{HC}}(x)\, \mathrm{d}x . \tag{5.9}$$

*5.2 PROOF OF (5.4)

We rewrite the equilibrium equation (3.5) as

$$a_\mu p_\mu = (1-\mu) W p_\mu + \mu U p_\mu , \tag{5.10}$$

where p_μ denotes the equilibrium density depending on the parameter μ, and $a_\mu = \overline{W}_\mu$ is the equilibrium mean fitness. From (5.2) and the assumption that mutations occur (also) to suboptimal types, we already know that

$$1 - \mu \leq a_\mu = \overline{W}_\mu < 1 \,. \tag{5.11}$$

Let us assume $a_\mu > 1 - \mu$, because otherwise nothing has to be proved, and define ε_μ by

$$a_\mu = (1-\mu)(1+\varepsilon_\mu) \,. \tag{5.12}$$

Then,

$$0 < \varepsilon_\mu < \frac{\mu}{1-\mu} \,, \tag{5.13}$$

and (5.10) is equivalent to

$$p_\mu(x) = \frac{\mu}{1-\mu} \frac{Up_\mu(x)}{\varepsilon_\mu + 1 - W(x)} \,. \tag{5.14}$$

Using this identity, we obtain

$$\begin{aligned}
\overline{W}_\mu - 1 &= \frac{\mu}{1-\mu} \int \frac{W(x) - 1}{\varepsilon_\mu + 1 - W(x)} Up_\mu(x)\,\mathrm{d}x \\
&= \frac{\mu}{1-\mu} \int -Up_\mu(x)\,\mathrm{d}x + \frac{\mu}{1-\mu} \int \frac{\varepsilon_\mu}{\varepsilon_\mu + 1 - W(x)} Up_\mu(x)\,\mathrm{d}x \\
&= -\frac{\mu}{1-\mu}\overline{W}_\mu + \frac{\mu}{1-\mu} \int\limits_{W_{\mathrm{Opt}}} Up_\mu(x)\,\mathrm{d}x \\
&\quad + \frac{\mu}{1-\mu} \int\limits_{\mathcal{X}\setminus W_{\mathrm{Opt}}} \frac{\varepsilon_\mu}{\varepsilon_\mu + 1 - W(x)} Up_\mu(x)\,\mathrm{d}x \,,
\end{aligned}$$

where $\overline{W}_\mu$ is obtained from (5.3) by recalling (2.1). It follows that

$$\frac{1}{\mu}\left[\overline{W}_\mu - \left(1 - \mu + \mu \int_{W_{\mathrm{Opt}}} Up_\mu(x)\,\mathrm{d}x\right)\right] = \int\limits_{\mathcal{X}\setminus W_{\mathrm{Opt}}} \frac{\varepsilon_\mu}{\varepsilon_\mu + 1 - W(x)} Up_\mu(x)\,\mathrm{d}x \,. \tag{5.15}$$

To finish the proof of (5.4), it has to be shown that the right-hand side tends to zero as $\mu \to 0$. Using (5.13) together with $1 - W(x) \geq 0$ and $Up_\mu(x) \geq 0$, we obtain

$$\frac{\varepsilon_\mu}{\varepsilon_\mu + 1 - W(x)} Up_\mu(x) \leq \frac{\mu}{\mu + (1-\mu)[1 - W(x)]} Up_\mu(x) \,.$$

Therefore, it suffices to prove

$$\int\limits_{\mathcal{X}\setminus W_{\mathrm{Opt}}} \frac{\mu_n}{\mu_n + (1-\mu_n)[1 - W(x)]} Up_{\mu_n}(x)\,\mathrm{d}x \to 0 \tag{5.16}$$

for every sequence $\mu_n \to 0$. This is a consequence of the following argument.

Since $\int p_\mu(x)\,dx = 1$ and U is a compact operator on $L^1(\lambda)$, every subsequence of Up_μ is contained in a compact subset of $L^1(\lambda)$. Now we use the well known fact that for an L^1-convergent sequence $(f_n) \subseteq L^1(\lambda)$, i.e., $\|f_n - f\|_1 \to 0$, and a uniformly bounded sequence (g_n), i.e., $g_n(x) \le$ const. λ-*a.e.* for every n, such that $g_n(x) \to 0$ λ-*a.e.*, the sequence $(g_n f_n)$ converges to 0 in $L^1(\lambda)$. It is easy to show that the same assertion holds for any sequence (f_n) that is contained in a compact subset of $L^1(\lambda)$. Thus, if we choose $f_n = Up_{\mu_n}$, $g_n(x) = \mu_n[\mu_n + (1-\mu_n)(1-W(x))]^{-1}$ if $x \in \mathcal{X} \setminus W_{\text{Opt}}$, and $g_n(x) = 0$ otherwise, we obtain $\|g_n f_n\|_1 \to 0$ which proves (5.16).

5.3 DISCRETE ALLELES

For a finite number of alleles, it was shown in Chapter III.3.1 that the first-order approximation of the mutation load is the total mutation rate of the fittest allele to all others. An example was given, showing that this is not true if there are two or more alleles with maximal fitness. Let us consider a general countable discrete state space $\mathcal{X}$, e.g. the stepwise-mutation model, and assume that a unique optimal type x_O exists, i.e. $W(x_O) = 1$ and $W(x) < 1$ for all other $x \in \mathcal{X}$. For simplicity, let all alleles have the same mutation rate μ. Then $u(x,x) = 1 - \mu$, and we define $v(y,x)$ by $u(y,x) = \mu v(y,x)$ if $y \ne x$, and $v(x,x) = 0$. Furthermore, if in analogy to (5.3) we define

$$Uf(x) = \sum_{y \in \mathcal{X}} f(y)W(y)v(y,x) , \tag{5.17}$$

then we obtain (5.14). As $\mathcal{X}$ is discrete, and because a unique optimal type is assumed, there exists a constant c such that

$$c = \sup_{\mu>0} \sup_{x \ne x_O} \frac{1}{1 + \varepsilon_\mu - W(x)} < \infty , \tag{5.18}$$

where ε_μ is defined in (5.12). Summing (5.14) over all $x \ne x_O$ yields

$$\sum_{x \ne x_O} p_\mu(x) \le \frac{\mu}{1-\mu} c , \tag{5.19}$$

because $\sum_{x \in \mathcal{X}} Up_\mu(x) \le 1$. Evaluating the equilibrium equation (5.10) at the optimal type x_O and using (5.19), we obtain

$$\overline{W}_\mu p_\mu(x_O) = (1-\mu)p_\mu(x_O) + O(\mu^2) \qquad \text{as } \mu \to 0 . \tag{5.20}$$

Again by (5.19), it follows that $p_\mu(x_O)$ is of order one. Hence, (5.20) yields the desired result:

$$\overline{W}_\mu = 1 - \mu + O(\mu^2) . \tag{5.21}$$

Thus we have proved:

• **5.2** *For a finite or countably infinite number of possible alleles, all with the same mutation rate μ, and a unique optimal type, the mutation load satisfies*

$$L = \mu + O(\mu^2) \qquad \textit{as } \mu \to 0 . \tag{5.22}$$

If different types have different mutation rates, an analogous result can be shown along the lines of the above proof if μ is taken to be the mutation rate of the type with the maximum value of $[1 - u(x,x)]W(x)$. Unless differences in mutation rates, $1 - u(x,x)$, between types are of the same order of magnitude as selective differences, μ will be the mutation rate of the type with maximal fitness, i.e., of x_O.

As a special case of (5.22), a well known result of Kimura and Maruyama (1966) is obtained. Suppose that fitness is a function of the number of mutant genes in an individual, and characterize each genotype by the number of mutations it carries. Thus, an appropriate state space is $\mathcal{X} = \{0, 1, 2, \dots\}$. It is assumed that the mutation-free type, 0, has maximal fitness $W_0 = 1$, and all other types have fitnesses $W_i < 1$ such that W_i decreases to zero as $i \to \infty$. Assuming that the number of new mutations follows a Poisson distribution with mean $\tilde{\mu}$, (5.22) implies that the mutation load is approximately $L = 1 - e^{-\tilde{\mu}}$, which is exactly the probability that a mutation occurs. Thus, as already stressed by Kimura and Maruyama, in asexuals non-multiplicative interaction of fitness effects of mutants (epistasis) has no effect on the first-order approximation of the equilibrium mean fitness (cf. Chapter VII.5).

*5.4 SECOND-ORDER TERMS

The accuracy of the first-order approximation, $1 - \mu$, for the equilibrium mean fitness is determined by the second-order term. For a broad class of fitness functions and mutation distributions the second-order term can be calculated. By (5.22), this second-order term is $O(\mu^2)$ for discrete allelic models. This is not necessarily the case with a continuum of alleles. The results derived below, in particular (5.31), (5.32), (5.33), and (5.34), apply to $\mathcal{X} = \mathbf{R}$, but also, with trivial modifications, to any $\mathcal{X}$ being an interval in $\mathbf{R}$.

a) We first deal with the *HC-mutation model*, (2.4), which allows a somewhat more general treatment. The proof is elementary but relegated to Section 5.5. Let the fitness function $W(x)$ satisfy the general assumptions (W1) and (W2) of Section 2.2, and assume that $x_O = 0$ is the unique fitness maximum, $W(x_O) = 1$. Moreover, suppose that $W(x)$ behaves locally near $x_O = 0$ like $1 - s\,|x|^q$, $q > 0$, i.e.,

$$W(x) = 1 - s\,|x|^q + f(x)\ , \tag{5.23}$$

where $f(x) = o(|x|^q)$. These assumptions imply that there are intervals J_1 and J_2, containing $x_O = 0$ in their interior, such that

$$W(x) \geq 1 - 2s\,|x|^q\ , \qquad x \in J_1\ , \tag{5.24a}$$
$$W(x) \leq 1 - \tfrac{1}{2}s\,|x|^q\ , \qquad x \in J_2\ , \tag{5.24b}$$

and there is a constant such that

$$W(x) \leq \text{const.} < 1, \qquad x \notin J_2\ . \tag{5.24c}$$

The mutation distribution is assumed to be of the form

$$u_{\mathrm{HC}}(x) = |x|^{-1/n}\, g(x)\ , \tag{5.25}$$

where

(i) g is an arbitrary, bounded, nonnegative function such that $\int u_{\text{HC}} = 1$ and $g(x) \geq g_0 > 0$ for $x \in [-\eta, \eta]$ (with $g_0 > 0$, $\eta > 0$ fixed but arbitrary), and

(ii) $n \geq 1$ (including ∞) and $q > \theta = 1 - 1/n$.

Thus, if $n = \infty$, we have $u_{\text{HC}} = g$ and u_{HC} may, for instance, be a Gaussian or an exponential distribution reflected about 0. A Γ-distribution reflected about 0 (2.6) is obtained for

$$g(x) = \frac{d^\theta}{2\Gamma(\theta)} \exp(-d\,|x|) \ .$$

▷ As in (2.14), let us define

$$I(\beta) = \mu \int_{\mathcal{X}} \frac{u_{\text{HC}}(x)}{1 - \beta W(x)} \,\mathrm{d}x \ . \tag{5.26}$$

We have to approximate the value $\beta = b_\mu$ that satisfies

$$I(b_\mu) = 1 \ , \tag{5.27}$$

because the equilibrium mean fitness is given by

$$\overline{W}_\mu = (1 - \mu)/b_\mu \ , \tag{5.28}$$

where $b_\mu > 1 - \mu$ (cf. Section 2.3). In the next subsection it is proved that

$$c_1' \,\mu\, s^{-\theta/q} \,(1 - \beta)^{-1+\theta/q} \leq I(\beta) \leq c_2' \,\mu\, s^{-\theta/q} \,(1 - \beta)^{-1+\theta/q} \tag{5.29}$$

for positive constants c_1', c_2', and small enough μ. Together with (5.27) and (5.28), these inequalities, applied to $\beta = b_\mu$, yield after some rearrangement

$$1 - \mu + c_1\, \mu \left(\frac{\mu}{s}\right)^{\frac{\theta}{q-\theta}} \leq \overline{W}_\mu \leq 1 - \mu + c_2\, \mu \left(\frac{\mu}{s}\right)^{\frac{\theta}{q-\theta}} \ , \tag{5.30}$$

where c_1 and c_2 are appropriate constants. Note that $\theta/(q - \theta) > 0$ holds under our assumptions, and this is actually necessary in order that condition (2.9) is satisfied. ◁

Thus, we have the asymptotic equality

$$\frac{\overline{W}_\mu - (1 - \mu)}{\mu} \sim (\text{const.}) \left(\frac{\mu}{s}\right)^{\frac{\theta}{q-\theta}} \qquad \text{as } \mu \to 0 \ . \tag{5.31}$$

It is interesting to note that the error term of the load L is $O(\mu^2/s)$, as in (5.22) for discrete alleles, if and only if $q = 2\theta$. If $u_{\text{HC}} = g$ is bounded near the optimum $x_O = 0$, as for a Gaussian or a reflected exponential distribution, and if $W(x) \sim 1 - sx^2$ for x near x_O, then $\theta = 1$, $q = 2$, and we obtain

$$\frac{\overline{W}_\mu - (1 - \mu)}{\mu} \sim (\text{const.})\, \frac{\mu}{s} \qquad \text{as } \mu \to 0 \ . \tag{5.32}$$

This is valid, for example, if $W(x)$ is a Gaussian fitness function. If, in addition, u_{HC} is Gaussian with variance γ^2, as in (1.1), the constant in (5.32) can be shown to be $\pi/(2\gamma^2)$ [see (1.15) and Turelli 1984, Eq. (3.9)]. It is easy to show that (5.31) and (5.32) also hold under the weaker condition $\mu/s \to 0$.

These results show that the second-order term of the equilibrium mean fitness depends on the *local* behavior of the fitness function and of the mutation distribution near the optimum. If $\theta \geq q$, the estimates leading to (5.30) do not apply because the integrals do not necessarily exist and the equilibrium distribution may have an atom of probability at x_O. Then, by Corollary 4.4, the equilibrium mean fitness equals $1-\mu$.

b) Let $u(x,y)$ be a *bounded mutation distribution*, i.e., $u(x,y) \leq \|u\|_\infty < \infty$ for all x and y, that satisfies conditions 1 and 2 of Theorem 3.1, and 3' in the following Remark d. If, in addition, the fitness function $W(x)$ satisfies (W1), (W2), and (5.23), then the following asymptotic equality can be proved

$$\frac{\overline{W}_\mu - (1-\mu)}{\mu} \sim (\mathrm{const.}) \left(\frac{\mu}{s}\right)^{\frac{1}{q-1}} , \quad \text{as } \mu \to 0 . \tag{5.33}$$

In particular, (5.32) also holds for any bounded mutation distribution and all fitness functions behaving like $1 - sx^2$ near the optimum. The proof is rather technical and is given in Section 5.6. Actually, slightly more can be shown: if μ and s are such that $u(x,y) > 0$ for all x and y with $|x_O - \max(x,y)| \leq \mu/[s(1-\mu)]$, then

$$\frac{\overline{W}_\mu - (1-\mu)}{\mu} \geq (\mathrm{const.}) \left(\frac{\mu}{s}\right)^{\frac{1}{q-1}} . \tag{5.34}$$

Thus, (5.34) is valid for all μ and s if the mutation distribution is positive.

Asymptotic estimates completely analogous to (5.31), (5.32), (5.33), and (5.34) hold in the continuous-time model, and were first derived by Bürger and Hofbauer (1994).

*5.5 PROOF OF (5.29)

First, we prove the right inequality of (5.29). The inequalities (5.24b,c) yield

$$\frac{u_{\mathrm{HC}}(x)}{1-\beta W(x)} \leq \begin{cases} \dfrac{u_{\mathrm{HC}}(x)}{1-\beta(1-\frac{1}{2}s\,|x|^q)} , & \text{if } x \in J_2 , \\ (\mathrm{const.})\, u_{\mathrm{HC}}(x) , & \text{if } x \notin J_2 . \end{cases} \tag{5.35}$$

Therefore, we obtain from (5.26) and $\int u_{\mathrm{HC}} = 1$,

$$\begin{aligned} I(\beta) &\leq \mu \int_{J_2} \frac{u_{\mathrm{HC}}(x)}{1-\beta(1-\frac{1}{2}s\,|x|^q)}\, \mathrm{d}x + (\mathrm{const.})\, \mu \int_{R\setminus J_2} u_{\mathrm{HC}}(x)\, \mathrm{d}x \\ &\leq \mu \int_{\infty}^{\infty} \frac{u_{\mathrm{HC}}(x)}{1-\beta(1-\frac{1}{2}s\,|x|^q)}\, \mathrm{d}x + (\mathrm{const.})\, \mu \\ &\leq 2\mu \|g\|_\infty \int_0^\infty \frac{|x|^{-1/n}}{1-\beta+\frac{1}{2}\beta s\,|x|^q}\, \mathrm{d}x + (\mathrm{const.})\, \mu , \end{aligned} \tag{5.36}$$

where $\|g\|_\infty = \sup_x g(x) < \infty$ by (i).

This integral is calculated to be (e.g., by *Mathematica*, Wolfram 1996)

$$\int_0^\infty \frac{|x|^{-1/n}}{1-\beta+\frac{1}{2}\beta s\,|x|^q}\,\mathrm{d}x = \frac{2^{\theta/q}\,\pi}{\beta^{\theta/q}\,q\,\sin(\pi\theta/q)}\,s^{-\theta/q}\,(1-\beta)^{-1+\theta/q}\,. \tag{5.37}$$

Inserting (5.37) into (5.36) implies

$$I(\beta) \le \mu\left(c_1''\,s^{-\theta/q}\,(1-\beta)^{-1+\theta/q} + c_2''\right) \le (\text{const.})\,\mu\,s^{-\theta/q}\,(1-\beta)^{-1+\theta/q}\,, \tag{5.38}$$

where the second inequaltiy requires sufficiently small μ and $\beta > 1-\mu$.

Now we prove the left inequality of (5.29). By (5.24a) and assumption (i) on g, with η chosen such that $[-\eta,\eta] \subseteq J_1$, we obtain

$$I(\beta) \ge 2g_0\mu\int_0^\eta \frac{|x|^{-1/n}}{1-\beta(1-2s\,|x|^q)}\,\mathrm{d}x = \frac{2g_0\mu}{\beta s}\int_0^\eta \frac{|x|^{-1/n}}{\zeta+|x|^q}\,\mathrm{d}x\,, \tag{5.39}$$

where $\zeta = (1/\beta - 1)/s$. Since $1-\mu \le \beta \le 1$, ζ satisfies $0 \le \zeta \le \mu/(s-s\mu)$. Hence, $\zeta \to 0$ as $\mu \to 0$, and we can restrict our attention to $\zeta \in (0, \eta^q)$. Thus, using (5.39) and the transformation $y = x^\theta$, we obtain

$$\begin{aligned} I(\beta) &\ge \frac{2g_0\mu}{\beta s}\int_0^{\zeta^{1/q}} \frac{|x|^{-1/n}}{\zeta+|x|^q}\,\mathrm{d}x \\ &= \frac{2g_0\mu}{\beta s\theta}\int_0^{\zeta^{\theta/q}} \frac{\mathrm{d}y}{\zeta+y^{q/\theta}} \\ &\ge \frac{g_0\mu}{\beta s\theta\zeta}\int_0^{\zeta^{\theta/q}} \mathrm{d}y \\ &= \frac{g_0\mu}{\beta s\theta}\,\zeta^{\frac{\theta}{q}-1} \\ &\ge (\text{const.})\,\mu\,s^{-\theta/q}\,(1-\beta)^{-1+\theta/q}\,, \end{aligned} \tag{5.40}$$

which proves (5.29). If $W(x)$ is given explicitly, then the constants in (5.38) and (5.40) can be calculated explicitly by using Chebyshev's inequality (or the explicit form of u_{HC}) in (5.36).

*5.6 *PROOF OF (5.33) AND (5.34)*

We define the following operators on $L^1(\lambda)$:

$$(C_\alpha f)(x) = \mu\int f(y)\,\frac{W(y)}{\alpha-(1-\mu)W(y)}\,u(y,x)\,\mathrm{d}y \tag{5.41}$$

and

$$(\tilde{C}_\alpha f)(x) = \frac{\mu}{\alpha-(1-\mu)W(x)}\int f(y)W(y)u(y,x)\,\mathrm{d}y\,, \tag{5.42}$$

where $\alpha > 1 - \mu$; cf. (3.10). Both operators are nonnegative and bounded because of the general assumptions, and C_α is power compact by assumption 2 of Theorem 3.1. (An estimate of $\|\tilde{C}_\alpha\|$ is derived below.)

In the proof of Theorem 3.1 we showed the existence of a unique value $\alpha = a_\mu$ such that $r(C_{a_\mu}) = 1$, r denoting the spectral radius. This value a_μ equals the mean equilibrium fitness; cf. (3.9). By deriving upper and lower bounds for $r(C_\alpha)$, it is then possible to obtain such bounds for a_μ, and thus for the mean fitness $\overline{W}_\mu$.

The derivation of a lower bound is very similar to the HC-case. Actually, we only have to combine (3.13), the estimates in (5.39) and (5.40) for the case $\theta = 1 - 1/n = 1$, and the trivial fact that $W(y)$ is bounded away from zero near the optimum, to obtain

$$r(C_\alpha) \geq c_1' \frac{\mu}{s} \zeta^{\frac{1}{q}-1} , \tag{5.43}$$

where $\zeta = [\alpha/(1-\mu) - 1]/s$ and c_1' is a positive constant. In fact, (5.43) holds for all μ and s such that $\mu/[s(1-\mu)] \in J$ with J as in (3.7). Since $r(C_{a_\mu}) = 1$, this implies

$$\overline{W}_\mu = a_\mu \geq 1 - \mu + c_1 \mu \left(\frac{\mu}{s}\right)^{\frac{1}{q-1}} , \tag{5.44}$$

which holds for all μ and s such that $\mu/[s(1-\mu)] \in J$ or, equivalently, for all μ and s such that $u(x,y) > 0$ if $|x_O - x, y| \leq \mu/[s(1-\mu)]$. This proves (5.34).

The derivation of an upper bound is more complicated. It is based on the following inequality for the spectral radii of C_α and $\tilde{C}_\alpha$:

$$r(C_\alpha) \leq r(\tilde{C}_\alpha) \leq \|\tilde{C}_\alpha\| . \tag{5.45}$$

Only the first inequality has to be proved; the second is well known [Appendix, (C.2)]. For $\alpha > 1 - \mu$, the equation

$$C_\alpha f = \psi f , \qquad f \in L^1(\lambda) ,$$

is equivalent to

$$\tilde{C}_\alpha h = \psi h , \qquad h = f[\alpha - (1-\mu)W]^{-1} \in L^1(\lambda) .$$

Therefore, the operators C_α and $\tilde{C}_\alpha$ have the same eigenvalues (it is important to note that $f \in L^1$ if and only if $h \in L^1$, because $[\alpha - (1-\mu)W]^{-1}$ is bounded). By assumption, C_α is power compact, hence its spectrum consists only of eigenvalues and $r(C_\alpha)$ is an eigenvalue (Theorem C.2). It follows that the spectrum of C_α is a subset of that of $\tilde{C}_\alpha$, and this implies (5.45).

An upper bound for $r(C_\alpha)$ is obtained from the following estimates (for simplicity it is assumed that (3.7) holds for $J = J_2$), some being parallel to those in (5.36) and

(5.37),

$$
\begin{aligned}
\|\tilde{C}_\alpha\| &= \sup_{\|f\|_1=1} \|\tilde{C}_\alpha f\|_1 \\
&= \mu \sup_{\|f\|_1=1} \left[\int_{-\infty}^{\infty} \frac{1}{\alpha - (1-\mu)W(x)} \int_{-\infty}^{\infty} u(y,x) W(y)\, |f(y)| \, \mathrm{d}y \, \mathrm{d}x \right] \\
&\leq \mu \sup_{\|f\|_1=1} \left[\int_{J_2} \frac{1}{\alpha - (1-\mu)(1 - \frac{1}{2} s\, |x|^q)} \int_{-\infty}^{\infty} u(y,x) W(y)\, |f(y)| \, \mathrm{d}y \, \mathrm{d}x \right] \\
&\quad + \mu\, (\mathrm{const.}) \sup_{\|f\|_1=1} \left[\int_{R \setminus J_2} \int_{-\infty}^{\infty} u(y,x) W(y)\, |f(y)| \, \mathrm{d}y \, \mathrm{d}x \right] \\
&\leq \frac{4}{1-\mu} \frac{\mu}{s} \sup_{\|f\|_1=1} \left[\int_0^{\infty} \frac{1}{\zeta + |x|^q} \int_{-\infty}^{\infty} u(y,x) W(y)\, |f(y)| \, \mathrm{d}y \, \mathrm{d}x \right] \\
&\quad + \mu\, (\mathrm{const.}) \sup_{\|f\|_1=1} \left[\int_{-\infty}^{\infty} \left(\int_{-\infty}^{\infty} u(y,x) \, \mathrm{d}x \right) W(y)\, |f(y)| \, \mathrm{d}y \, \mathrm{d}x \right] \\
&\leq \frac{4}{1-\mu} \frac{\mu}{s} \sup_{\|f\|_1=1} \left[\int_0^{\infty} \frac{1}{\zeta + |x|^q} \mathrm{d}x \, \|u\|_\infty \, \|f\|_1 \right] + \mu\, (\mathrm{const.}) \sup_{\|f\|_1=1} \|f\|_1 \\
&= c_1' \frac{\mu}{s} \frac{\pi}{q \sin(\pi/q)} \zeta^{\frac{1}{q}-1} + c_2' \mu \, ,
\end{aligned}
\tag{5.46}
$$

where $\zeta = 2[\alpha/(1-\mu) - 1]/s$. This, together with (5.45) and the condition $r(C_{a_\mu}) = 1$, yields

$$
\overline{W}_\mu = a_\mu \leq 1 - \mu + c_2\, \mu \left(\frac{\mu}{s} \right)^{\frac{1}{q-1}} \, , \tag{5.47}
$$

and finishes the proof of (5.33).

V

Dynamical Equations for Quantitative Traits under Selection

Quantitative characters are determined by a potentially large number of gene loci which often have individually small effects. Since, as noted previously, many traits of evolutionary or economic relevance are of this kind, it is of substantial interest to understand how their distribution is shaped by the basic genetic and evolutionary mechanisms, and how it responds to selection.

In this chapter we set out a general theory for analyzing the dynamics of multi-locus systems under selection, recombination, and mutation that is developed for applications to quantitative genetics. It uses generating functions and multivariate cumulants for representing the frequency distribution of genotypes and phenotypes in a population. Cumulants are a convenient set of parameters, including the mean, the variance, and the third central moment, for analyzing the evolutionary forces of selection, recombination, and mutation. Their advantages are most pronounced if genes act additively to produce a quantitative trait. We assume a randomly mating, diploid population with no sex differences. The population is supposed to be sufficiently large so that random genetic drift can be ignored. The number of alleles per locus is arbitrary, their labels or effects may be discrete or continuous. The general model neither assumes weak selection, nor additivity of effects, nor linkage equilibrium. It is also not based on any assumptions about the distribution of phenotypic or breeding values. Under the additive model of quantitative genetics, explicit recursion relations for the dynamics of the (multivariate) cumulants of the distribution of genotypes and of the distribution of phenotypic values can be derived. Various applications and special cases are investigated in detail.

Pioneering analyses to elucidate the genetic basis of inheritance and the response to selection of the mean of a quantitative character are due to Galton (1889), Pearson (1903), Fisher (1918) and Wright (1921), and their students Smith (1936) and Lush (1937). Whereas the analyses of Galton and Pearson were of a purely statistical nature, Fisher reconciled their biometric description with Mendelian genetics by assuming that a large number of unlinked loci with small additive effects determine the character. The work of Fisher and Wright forms the basis of classical quantitative genetic theory and its applications to animal and plant breeding (see, e.g., Bulmer 1980, Mather and Jinks 1982, Mayo 1987, Falconer and Mackay 1996, Lynch and Walsh 1998). Lande (1975, 1979, and later) extended these approaches, derived equations for the change of the mean phenotype of a set of quantitative characters in terms of the additive genetic covariance matrix and the so-called selection gradient, and applied these to numerous problems of evolutionary biology (see Lande 1988 for a review, and Chapter VII). His, as well as the previous, approaches were based on the empirically testable assumption of a (multivariate) normal distribution of the characters and assumed that the variances and covariances change on a much slower time scale than the mean.

More recently, several attempts were undertaken to establish the dynamics of the distribution of a polygenic character under selection from *genetic* principles (e.g., Barton and Turelli 1987, 1991, Turelli and Barton 1990, Frank and Slatkin 1990; cf. also Bulmer 1980). These analyses concentrated on the derivation of equations for the change of the mean, the variance, and the higher moments of the allele-frequency distribution at each locus and subsequent extension to the multilocus situation. This extension faces some severe difficulties, because the moments of the phenotypic distribution are complicated functions of the moments of the distribution of genotypic frequencies, and because linkage disequilibria have to be taken into account. The basic idea of the present approach is to describe the change caused by selection in terms of the cumulant generating function of the distribution of genotypic frequencies and to use (multivariate) cumulants instead of moments. This is most successful if genes contribute additively to the phenotypic character, because the cumulants have a very convenient additivity property. Compared with the moments, they have several other advantages; for instance, they are the most natural set of parameters for describing deviations from normality (Appendix D contains a concise review). However, since the change of linkage disequilibria under recombination is best described by central moments, a relatively complex, but readily automated, algorithm is needed to calculate the recursions of the cumulants from one generation to the next. This approach was first carried out in full generality by Turelli and Barton (1994). For weak selection and under the assumption of linkage equilibrium it was introduced by Bürger (1991a).

In the first section notations and basic facts on the additive model of quantitative genetics are collected, and formulas for the fitness of genotypes from the fitness of phenotypes are derived. Under the assumption that selection is sufficiently weak to ignore deviations from linkage equilibrium, in Section 2 differential equations are derived for the dynamics of the cumulants of the allele-frequency distribution and the distribution of phenotypes. The principal method is the same as that in the general case with linkage disequilibria, but technically much simpler. In Section 3, the response to selection is studied in the presence of linkage disequilibria. The frequency distribution of genotypes and, in particular, of linkage disequilibria is described by multivariate cumulants. Based on two different, but equivalent, approaches, the dynamical equations for the response to selection of the multivariate cumulants are derived. The first (Section 3.2) is based on the assumption that the fitness function can be approximated by a polynomial or Taylor series, whereas the second (Section 3.4) is based on selection gradients. Section 4 is devoted to recombination and linkage. The recursion equations for the multivariate cumulants across generations under selection and recombination are derived, the relations between various measures of linkage disequilibrium are highlighted, and the dynamics to and in quasi-linkage equilibrium is explored. The change caused by mutation is treated in Section 5. In Section 6, these results are applied to find the recursion relations for the change between generations of the cumulants of the distribution of phenotypic values. As an immediate application, the classical case of a normal distribution of breeding values is treated.

For numerical investigations of multilocus systems, the dynamical equations derived in this chapter may not be well suited because, in general, the systems can only be closed under assumptions about the distribution of gamete frequencies. However, they provide useful structural insights into the way in which means, variances, covariances, and higher-order cumulants of allele-frequency distributions, as well as the linkage disequilibria, evolve under selection, recombination, and mutation. Moreover, these recursion relations show explicitly how the between-generation change of the mean, the variance, and the higher cumulants of a quantitative trait depends on the underlying genetics and the shape of the fitness function. Thus, they reveal which contributions by genetic mechanisms are missing in certain approximations, and when they are likely to be negligible. Some simple and useful results that emerge for stabilizing and directional selection will be treated in the following chapters.

1. PRELIMINARIES

In a randomly mating diploid population, we consider ℓ loci and describe the state of the newly produced zygotic ℓ-locus genotype by the random vector $(\boldsymbol{X}, \boldsymbol{X}^*)$ of length 2ℓ, where the components X_i and X_i^* describe the allelic state at locus i on the paternally and maternally inherited chromosome, respectively. In this chapter, we are mainly interested in the case where these ℓ loci contribute to a quantitative character. Then the value of X_i may be taken as a real number representing the contribution of the allele to the character, as is natural for a continuum of alleles, as well as for diallelic and multiallelic loci. (More generally, the values of the X_i might be elements of a state space such as admitted in Chapter IV and denoted there by $\mathcal{X}$.) Let the values of $\boldsymbol{X}$ and $\boldsymbol{X}^*$, i.e., of specific paternally and maternally inherited gametes, be denoted by $\boldsymbol{x} = (x_1, \dots, x_\ell)$ and $\boldsymbol{x}^*$, respectively.[1] Zygotes are designated by $\boldsymbol{z} = (\boldsymbol{x}, \boldsymbol{x}^*) = (x_1, \dots, x_\ell, x_1^*, \dots, x_\ell^*)$.

We will be concerned with the evolutionary change of the distribution $P(\boldsymbol{z}, t)$ of diploid ℓ-locus genotypes caused by selection, recombination, and mutation. Unless needed, we shall suppress the time variable t. Under random mating, $\boldsymbol{x}$ and $\boldsymbol{x}^*$ are independent in zygotes, so that $P(\boldsymbol{z}) = p(\boldsymbol{x})p(\boldsymbol{x}^*)$, where $p(\boldsymbol{x})$ is the frequency (density) of $\boldsymbol{x}$, and the frequency of gametes is the same in both sexes. However, since selection introduces deviations from Hardy–Weinberg proportions, in general the joint distribution $P(\boldsymbol{x}, \boldsymbol{x}^*)$ must be considered when studying the response to selection. As an introduction, the next section deals with the relatively simple case of populations in linkage equilibrium in which some of the fundamental ideas can be highlighted. Before, we collect the basic model assumptions and facts used throughout this chapter.

1.1 THE ADDITIVE MODEL OF QUANTITATIVE GENETICS

In general, little is known about how genes interact to produce a quantitative character. However, with the advent of powerful molecular techniques, data are beginning to accumulate. A simple, but very useful, model is the so-called additive genetic model devised by Fisher (1918) and Wright (1921). In this model, the phenotypic value,

[1] In this chapter vectors and multi-indices are always denoted by bold letters.

P, of a single polygenic character is assumed to be determined by a component, G, attributable to the influence of the genotype, and an 'environmental' component, E, attributable to all non-genetic circumstances that influence the phenotype, such that

$$P = G + E \, . \tag{1.1}$$

The genotypic value G (or breeding value) is determined additively by all paternal and maternal allelic effects, i.e.,

$$G = \sum_{i=1}^{\ell} (X_i + X_i^*) \, . \tag{1.2}$$

This, of course, rests on the assumption that the genotypic value can be closely approximated by the sum of the additive, or average, effects of the contributing genes (cf. Chapters II.3, II.5, II.6). The additivity assumption (1.2) will be a good approximation if dominance and epistatic deviations are small. It is a special case of II(5.1), in which dominance was admitted.

We assume that the environmental component E is a normally distributed random variable with mean 0 and variance σ_E^2, which is independent of G. It is often called the environmental effect, or environmental deviation, and may be considered as resulting from micro-environmental variation, developmental noise, and cytoplasmatic effects unique to individuals.

Hence, denoting the means of the distributions of phenotypic values and genotypic values by $\overline{P}$ and $\overline{G}$, respectively, and their respective variances by σ_P^2 and σ_G^2, we have

$$\overline{P} = \overline{G} \quad \text{and} \quad \sigma_P^2 = \sigma_G^2 + \sigma_E^2 \, . \tag{1.3}$$

We recall from the end of Chapter II.3.1 that under the additive model, additive genetic and total genetic variance coincide, i.e., $\sigma_{\mathrm{A}}^2 = \sigma_G^2$. In this case we shall simply speak of genetic variance and use the notation σ_G^2.

In practice, it is often necessary, and also possible, to choose an appropriate scale on which the character is determined additively. For comprehensive treatments of the biological and statistical aspects of the additive model as well as of its applications, the reader is referred to Falconer and Mackay (1996) and Lynch and Walsh (1998).

1.2 FITNESS OF GENOTYPES FROM PHENOTYPES

Selection acts on phenotypes, whereas the recurrence relations for gamete and genotype frequencies involve genotypic fitnesses. Therefore, the fitness of genotypes, $W(\boldsymbol{z})$, has to be calculated from the phenotypic fitness $W_P(P)$. In this chapter, we do not posit that fitnesses are constant in time. Since environmental effects are assumed to be Gaussian with mean zero and variance σ_E^2, the mean fitness of individuals with genotypic value G is

$$W(G) = (2\pi\sigma_E^2)^{-\frac{1}{2}} \int_{-\infty}^{\infty} W_P(G + E) \exp[-E^2/(2\sigma_E^2)] \, \mathrm{d}E \tag{1.4}$$

and, by (1.2),

$$W(\boldsymbol{z}) = W(\boldsymbol{x}, \boldsymbol{x}^*) = W(G) \tag{1.5}$$

for all $\boldsymbol{x}$ and $\boldsymbol{x}^*$ such that $\sum_i (x_i + x_i^*) = G$.

An important point to note is that environmental contributions have a smoothing effect on the genotypic fitness $W(G)$ (Nagylaki 1989a). More precisely, if W_P is any bounded Lebesgue-integrable function, and if the environmental effects can be described by a smooth (i.e., n-times differentiable) function, for example a Gaussian, then, being a convolution, $W(G)$ is smooth (Apostol 1974, pp. 283 and 328). Thus, even if the fitness landscape of phenotypes is rugged and bizarre, the fitness landscape of genotypic values will be smooth (though it may have multiple peaks).

If the phenotypic fitness function has a Taylor series expansion,

$$W_P(P) = \sum_{k=0}^{K} a_k P^k \ , \tag{1.6}$$

then a straightforward calculation (see, e.g., Gröbner and Hofreiter 1975) shows that

$$W(G) = \sum_{k=0}^{K} a_k G^k \left[\sum_{\nu=0}^{[k/2]} \frac{k!}{\nu!(k-2\nu)!} \left(\frac{\sigma_E^2}{2G^2} \right)^{\nu} \right] \ , \tag{1.7}$$

where $[k/2]$ denotes the largest integer $\leq k/2$. A simple rearrangement yields a Taylor series of the form

$$W(G) = \sum_{k=0}^{K} s_k G^k \ , \tag{1.8}$$

where $K = \infty$ is admissible. If $a_k = 0$ for $k \geq 5$, then

$$\begin{aligned} s_0 &= a_0 + \sigma_E^2 a_2 + 3\sigma_E^2 a_4 \ , \\ s_1 &= a_1 + 3\sigma_E^2 a_3 \ , \\ s_2 &= a_2 + 6\sigma_E^2 a_4 \ , \\ s_3 &= a_3 \ , \\ s_4 &= a_4 \ . \end{aligned} \tag{1.9}$$

From (1.8) it is obvious that the sequence of selection coefficients s_k has to decay to zero at geometric rate (or faster) if genotypic values with $|G| > 1$ are admitted.

The following are important fitness functions for modeling stabilizing selection (cf. Chapters VI and VII). If phenotypic fitness is a quadratic function,

$$W_P(P) = 1 - s(P - P_O)^2 \ , \tag{1.10}$$

as in Wright's (1935a,b) *quadratic optimum model*, then the genotypic fitness function is again quadratic:

$$W(G) = 1 - s\sigma_E^2 - s(G - P_O)^2 \ . \tag{1.11}$$

In a discrete-time model, the choice (1.10) requires that selection is sufficiently weak that the frequencies of types with phenotypic value P such that $|P - P_O| \geq s^{-1/2}$ can be neglected.

This problem does not occur for *Gaussian stabilizing selection* (Haldane 1954), i.e., if

$$W_P(P) = \exp\left\{-\frac{(P-P_O)^2}{2\omega^2}\right\}. \tag{1.12}$$

Then, using (1.4), the genotypic fitness function is again Gaussian,

$$W(G) = \exp\left\{-\frac{(G-P_O)^2}{2V_s}\right\}, \tag{1.13}$$

where $V_s = \omega^2 + \sigma_E^2$ is a reciprocal measure of the strength of stabilizing selection. Thus, large V_s (indicating a wide fitness function) corresponds to weak stabilizing selection. Denoting $s = 1/(2V_s)$, the coefficients s_k in the series expansion (1.8) of $W(G)$ become

$$s_0 = 1 - s(P_O^2 + \sigma_E^2) + O(s^2)\,, \quad s_1 = 2sP_O + O(s^2)\,, \quad s_2 = -s + O(s^2)\,, \tag{1.14}$$

and $s_k = O(s^2)$ if $k \geq 3$. This is in agreement with (1.11). Often, the quadratic fitness function (1.10), or (1.11), is used in continuous-time models, whereas the Gaussian (1.12), or (1.13), is used in discrete time (cf. Chapter IV.2.4).

In general, employing (1.8) and the polynomial theorem [Appendix, (D.1)], we obtain from (1.5) the representation

$$W(\boldsymbol{z}) = \sum_{k=0}^{K} s_k \left(\sum_{i=1}^{2\ell} z_i\right)^k = \sum_{k=0}^{K} s_k \sum_{\boldsymbol{k}:|\boldsymbol{k}|=k} \frac{k!}{\boldsymbol{k}!}\,\boldsymbol{z}^{\boldsymbol{k}} \tag{1.15}$$

for the genotypic fitness function. Here, $\boldsymbol{k} = (k_1, \ldots, k_{2\ell})$ is a multi-index of length 2ℓ such that $|\boldsymbol{k}| = \sum_i k_i = k$ (see Appendix D for basic facts on multi-indices which are always assumed to be nonnegative). The mean fitness of the population is

$$\overline{W} = \int W_P(P) f_P(P)\, \mathrm{d}P = \int W(G) f(G)\, \mathrm{d}G = \int W(\boldsymbol{z}) P(\boldsymbol{z})\, \mathrm{d}\boldsymbol{z}\,, \tag{1.16}$$

where f_P and f denote the densities of the distributions of phenotypic and genotypic values, respectively, and the integrals are to be replaced by sums over all possible phenotypes/genotypes if the underlying state space is discrete. In this chapter, integrals of the form $\int \ldots \mathrm{d}\boldsymbol{z}$ indicate integration over the whole (2ℓ-dimensional) state space.

2. POPULATIONS IN LINKAGE EQUILIBRIUM

In this section, we outline the essential features of the analysis of polygenic traits subject to selection, recombination, and mutation under the simplifying assumption of global linkage equilibrium. This appears to be justified, because (i) in the general case, technical complications may obscure the basic ideas for the mathematically less educated reader and (ii), as we have seen in Chapter II.6, the evolution of a population may be approximated by its linkage-equilibrium dynamics if selection is sufficiently weak relative to recombination. That the assumption of linkage equilibrium yields good approximations for several selection models will be further substantiated in Section 4.3 and, in particular, for some models of stabilizing selection, in Chapter VI; see also Chapters VII.2 and VII.7.

We shall represent the distributions of phenotypic and genotypic values, and of the allele frequencies by their mean, their variance, and their higher-order cumulants. Cumulants are a set of parameters that describe the shape of a probability distribution, such as moments. Compared with moments, however, they have properties that make them a much more convenient tool for investigating additive polygenic traits, and we refer to Appendix D for a concise summary of the properties of cumulants. Our aim here is to derive the differential equations for the change of the cumulants of the allele-frequency distribution at a particular locus, and for the cumulants of the distributions of genotypic and phenotypic values. The derivation in this section is based on the continuum-of-alleles model and on the assumptions of random mating, no sex differences, additive gene action, linkage equilibrium, Hardy–Weinberg proportions, and absence of random drift. It follows and extends the approach in Bürger (1991a) and Bürger and Hofbauer (1994).

The subscripts i and j always refer to specific loci; for instance, $p_i(x,t)$ denotes the (haploid) allelic distribution at locus i in generation t, μ_i the mutation rate at locus i, etc. We shall omit the time variable t, unless we wish to emphasize time dependence. Sums and products without limits, such as $\sum_i$ or $\prod_i$, are taken over all ℓ loci. Then, assuming linkage equilibrium, the distribution of ℓ-locus diploid genotypes is

$$P(\boldsymbol{x},\boldsymbol{x}^*) = \prod_i p_i(x_i) \prod_i p_i(x_i^*) \,. \tag{2.1}$$

Let

$$\psi_i(\xi,t) = \int e^{\xi x} p_i(x,t)\,\mathrm{d}x \tag{2.2}$$

denote the moment generating function of p_i. Then the cumulant generating function is defined by

$$\Psi_i(\xi,t) = \ln \psi_i(\xi,t) = \sum_{n=1}^{\infty} c_n^{(i)}(t)\frac{\xi^n}{n!} \,, \tag{2.3}$$

where $c_n^{(i)} = c_n^{(i)}(t)$ is the *nth cumulant* of p_i. Denoting an nth partial derivative with respect to ξ by D^n, we have

$$c_n^{(i)}(t) = (D^n\Psi_i)(0,t) \,. \tag{2.4}$$

Similarly, we define the nth cumulant of the distribution $f(G)$ of genotypic values as the coefficient of ξ^n in the Taylor series expansion of its cumulant generating function and designate it by C_n. Then, because of the assumptions of linkage equilibrium (2.1), Hardy–Weinberg equilibrium, additive gene action (1.2), and equivalent sexes, we obtain from (D.17)

$$C_n = 2\sum_i c_n^{(i)} \,. \tag{2.5}$$

It is this additivity property of the cumulants that makes them the appropriate tool for analyzing the dynamics of additive polygenic traits. In many instances, we shall consider a fixed locus and write c_n instead of $c_n^{(i)}$. The derivation of the dynamics of cumulants rests on some properties of marginal systems that we summarize below.

2.1 MARGINAL PROPERTIES

In Chapter II.4, marginal properties of classical multilocus models were investigated. With trivial changes these can be extended to the continuum-of-alleles case.

Let $y_i = \sum_{j:j\neq i} x_j$ denote the sum of paternal contributions to G from all loci except i, and let $y^* = \sum_j x_j^*$. By (2.1), the kth moment about zero of the distribution of $G - x_i = y_i + y^*$ is

$$B_{k,i} = \int \cdots \int (y_i + y^*)^k \prod_{j\neq i} p_j(x_j)\,\mathrm{d}x_j \prod_j p_j(x_j^*)\,\mathrm{d}x_j^* \,. \tag{2.6}$$

In terms of cumulants and omitting the index i, the first four of them can be written as

$$\begin{aligned} B_0 &= 1\,,\\ B_1 &= C_1 - c_1\,,\\ B_2 &= C_2 - c_2 + (C_1 - c_1)^2\,,\\ B_3 &= C_3 - c_3 + 3(C_2 - c_2)(C_1 - c_1) + (C_1 - c_1)^3 \end{aligned} \tag{2.7}$$

(cf. Appendix D). Then a simple application of the binomial theorem, using the representation (1.8) of the fitness function, together with the decomposition $G = (y_i + y^*) + x_i$, yields the following expression for the marginal fitness of an allele at locus i with effect x_i:

$$\begin{aligned} W_i(x_i) &= \int \cdots \int W(G) \prod_{j\neq i} p_j(x_j)\,\mathrm{d}x_j \prod_j p_j(x_j^*)\,\mathrm{d}x_j^* \\ &= \sum_{k=0}^{K} s_k \sum_{l=0}^{k} \binom{k}{l} x_i^l B_{k-l,i}\,. \end{aligned} \tag{2.8}$$

It should be noted that the marginal fitness of an allele depends on the current composition $p(\boldsymbol{x})$ of the population and, thus, implicitly on time.

By series expansion of $W(G)$, the mean fitness (1.16) may be written as

$$\overline{W} = \sum_{k=0}^{K} s_k M_k^0\,, \tag{2.9}$$

where $M_k^0 = M_k^0(t)$ denotes the kth moment about zero of $f(G)$.

The recursion relations II(4.17) for the marginal allele frequencies under selection can be straightforwardly extended to the continuum-of-alleles model. The inclusion of mutation is also straightforward, because mutation acts independently at different loci. Let us fix some locus i and omit the index i throughout, i.e., $p(x) = p_i(x_i)$, $W(x) = W_i(x_i)$, and $c_n = c_n^{(i)}$. Then, in analogy to IV(2.8), the mutation-selection dynamics for the marginal distribution $p(x)$ is given by

$$\overline{W}\, p'(x) = p(x)W(x) + \mu \int p(y)W(y)u(y-x)\,\mathrm{d}y - \mu p(x)W(x)\,, \tag{2.10}$$

where $\mu = \mu_i$ denotes the mutation rate and $u(y - x) = u_i(y_i - x_i)$ the conditional transition probability between alleles under the random-walk mutation model (cf. Chapter IV.2.1). To simplify the exposition, and because linkage disequilibria will be ignored, we assume weak selection, i.e., $W(G) = W(0) + sm(G) + s^2h(G)$ for appropriate functions m and h. Putting $w(G) = sm(G)$, $\overline{w} = \int w(G)f(G)\,\mathrm{d}G$, and following the procedure in Chapter IV.2.4, we approximate (2.10) by the differential equation

$$\frac{\partial}{\partial t}p(x,t) = p(x)[w(x) - \overline{w}] + \mu \int p(y)u(y-x)\,\mathrm{d}y - \mu p(x) \, , \tag{2.11}$$

where $w(x) = w_i(x_i)$ is defined according to (2.8), with $w(G)$ instead of $W(G)$, and hence depends on $p(\boldsymbol{x})$. For multilocus selection models, it was shown in Chapter II.4 that any marginal subsystem satisfies the equilibrium conditions if the full system is at equilibrium. This result has a straightforward extension to mutation-selection models (Hastings 1989). Obviously, it also applies to the present setting.

2.2 THE CUMULANT EQUATIONS FOR ALLELE FREQUENCIES

We continue to consider a fixed but arbitrary locus i, and omit this index. The decisive step in the present approach is the description of the evolution of allele frequencies by a differential equation for the cumulant generating function. Let

$$\varphi(\xi) = \int e^{\xi x}u(x)\,\mathrm{d}x$$

denote the moment generating function of the conditional mutation distribution u. Multiplication of both sides of (2.11) by $e^{\xi x}$, integration with respect to x, and the trivial observation that $\partial\Psi/\partial t = (\partial\psi/\partial t)/\psi$ yield the desired differential equation

$$\frac{\partial}{\partial t}\Psi(\xi,t) = \frac{1}{\psi(\xi,t)}\int e^{\xi x}w(x)p(x,t)\,\mathrm{d}x - \overline{w}(t) + \mu[\varphi(\xi) - 1] \, . \tag{2.12}$$

As in (1.8), let us denote the coefficients of the Taylor series of $w(G)$ by s_k. The following results are most useful if $w(G)$ is a polynomial of low degree. Then, using (2.4), substituting in (2.12) the representation (2.8) of w, and appealing to the obvious identity $D^l\psi(\xi,t) = \int e^{\xi x}x^l p(x,t)\,\mathrm{d}x$, we obtain the differential equation describing the evolution of the cumulants of the allele-frequency distribution under selection and mutation,

$$\dot{c}_n = \sum_{k=1}^{K} s_k \sum_{l=1}^{k} \binom{k}{l} B_{k-l}S_{nl} + \mu v_n \qquad \text{for every } n \geq 1 \, , \tag{2.13}$$

where v_n denotes the nth moment about zero of the mutation distribution u and

$$S_{nl} = D^n\left(\frac{D^l\psi}{\psi}\right)(0) \, . \tag{2.14}$$

Straightforward l-fold differentiation of $\psi = \exp \Psi$ yields the following expressions for the quantities S_{nl}:

$$S_{n1} = c_{n+1} , \tag{2.15a}$$

$$S_{n2} = c_{n+2} + \sum_{k=0}^{n} \binom{n}{k} c_{k+1} c_{n+1-k} , \tag{2.15b}$$

$$S_{n3} = c_{n+3} + 3 \sum_{k_1+k_2=n} \frac{n!}{k_1!\, k_2!} c_{k_1+1}\, c_{k_2+2}$$

$$+ \sum_{k_1+k_2+k_3=n} \frac{n!}{k_1!\, k_2!\, k_3!} c_{k_1+1}\, c_{k_2+1}\, c_{k_3+1} , \tag{2.15c}$$

$$S_{n4} = c_{n+4} + \sum_{k_1+k_2=n} \frac{n!}{k_1!\, k_2!} [4c_{k_1+1}\, c_{k_2+3} + 3c_{k_1+2}\, c_{k_2+2}]$$

$$+ 6 \sum_{k_1+k_2+k_3=n} \frac{n!}{k_1!\, k_2!\, k_3!} c_{k_1+1}\, c_{k_2+1}\, c_{k_3+2}$$

$$+ \sum_{k_1+k_2+k_3+k_4=n} \frac{n!}{k_1!\, k_2!\, k_3!\, k_4!} c_{k_1+1}\, c_{k_2+1}\, c_{k_3+1}\, c_{k_4+1} , \tag{2.15d}$$

where the sums are over all $k_i \geq 0$ that satisfy the indicated constraints. Higher-order terms can be calculated easily using a formula-manipulation program such as *Mathematica* (Wolfram 1996), but this will seldom be necessary. It is also easy to see that, in general, S_{nl} depends on the cumulants $c_1, \dots , c_{n+l}$.

From (2.13) and (2.15), the recursion relations under mutation and selection for the first four cumulants are given by

$$\begin{aligned}\dot{c}_1 = {} & s_1 c_2 + s_2[c_3 + 2C_1 c_2] + s_3[c_4 + 3C_1 c_3 + 3(C_1^2 + C_2)c_2] \\ & + s_4[c_5 + 4C_1 c_4 + 6(C_1^2 + C_2)c_3 \\ & \quad + 4(C_1^3 + 3C_1 C_2 + C_3)c_2] + \dots + \mu v_1 ,\end{aligned} \tag{2.16a}$$

$$\begin{aligned}\dot{c}_2 = {} & s_1 c_3 + s_2[c_4 + 2c_2^2 + 2C_1 c_3] \\ & + s_3[c_5 + 6c_2 c_3 + 3C_1(c_4 + 2c_2^2) + 3(C_1^2 + C_2)c_3] \\ & + s_4[c_6 + 8c_2 c_4 + 6c_3^2 + 4C_1(c_5 + 6c_2 c_3) + 6(C_1^2 + C_2)(c_4 + 2c_2^2) \\ & \quad + 4(C_1^3 + 3C_1 C_2 + C_3)c_3] + \dots + \mu v_2 ,\end{aligned} \tag{2.16b}$$

$$\begin{aligned}\dot{c}_3 = {} & s_1 c_4 + s_2[c_5 + 6c_2 c_3 + 2C_1 c_4] + s_3[c_6 + 6(c_2^3 + c_3^2 + c_2 c_4) \\ & \quad + 3C_1(c_5 + 6c_2 c_3) + 3(C_1^2 + C_2)c_4] + \dots + \mu v_3 ,\end{aligned} \tag{2.16c}$$

$$\begin{aligned}\dot{c}_4 = {} & s_1 c_5 + s_2[c_6 + 8c_2 c_4 + 6c_3^2 + 2C_1 c_5] + s_3[c_7 + 12c_2 c_5 + 30c_3 c_4 + 36c_2^2 c_3 \\ & \quad + 3C_1(c_6 + 6c_3^2 + 8c_2 c_4) + 3(C_1^2 + C_2)c_5] + \dots + \mu v_4 .\end{aligned} \tag{2.16d}$$

These equations cannot be truncated to give a closed finite system, unless information about the distribution of allele frequencies is available, because the rate of change of cumulants of given order depends on higher-order cumulants. In addition, these equations show that the allele-frequency dynamics at a particular locus depends on the distribution of all other loci through the phenotypic cumulants C_1, C_2, etc., except when fitness is linear. In this case, we simply have $\dot{c}_n = s_1 c_{n+1}$.

For a general fitness function w, the following covariance formula is easily derived:

$$\dot{c}_1 = \mathrm{Cov}(X, w) \ , \tag{2.17}$$

where $X(= X_i)$ denotes the allelic effect at locus i. In fact, it holds in a much more general context, because its validity neither requires additive gene action nor the assumption of linkage equilibrium; cf. (3.22a) below.

2.3 THE CUMULANT EQUATIONS FOR PHENOTYPES

Owing to the additivity property (2.5) of the cumulants, the dynamics of the cumulants of the distribution of genotypic values is simply

$$\dot{C}_n = 2 \sum_{i=1}^{\ell} \dot{c}_n^{(i)} \ , \quad n \geq 1 \ , \tag{2.18}$$

with $\dot{c}_n^{(i)}$ as in (2.13). The dynamics of the phenotypic distribution also follows immediately, because the distribution of E is assumed to be Gaussian with mean zero and variance σ_E^2, and independent of the distribution of G. Therefore, the phenotypic cumulants satisfy

$$C_2^{(P)} = C_2 + \sigma_E^2 \quad \text{and} \quad C_n^{(P)} = C_n \text{ if } n \neq 2 \ , \tag{2.19}$$

and we have $\dot{C}_n^{(p)} = \dot{C}_n$ for every $n \geq 1$.

With a method that is slightly different from that above, the differential equation for the genotypic and phenotypic mean, $\overline{P} = \overline{G}$, can be calculated easily for arbitrary fitness functions (see Section 6 for the derivation in a more general setting). It is given by

$$\dot{\overline{G}} = \sum_{k=1}^{K} s_k \left(M_{k+1}^0 - \overline{G} M_k^0 \right) \ . \tag{2.20}$$

It may be observed that the definition (2.14) implies that $S_{1k}(C_1, \ldots, C_{k+1}) = M_{k+1}^0 - \overline{G} M_k^0$, and the moments M_k^0 can be expressed in terms of cumulants (see Appendix D). Writing σ_G^2 for C_2, (2.20) reads

$$\begin{aligned} \dot{\overline{G}} = {} & s_1 \sigma_G^2 + s_2 (C_3 + 2\overline{G}\sigma_G^2) + s_3 (C_4 + 3\overline{G} C_3 + 3\sigma_G^4 + 3\overline{G}^2 \sigma_G^2) \\ & + s_4 (C_5 + 4\overline{G} C_4 + 10\sigma_G^2 C_3 + 6\overline{G}^2 C_3 + 4\overline{G}^3 \sigma_G^2 + 12\overline{G}\sigma_G^4) + \ldots \ . \end{aligned} \tag{2.21}$$

If $w(G)$ is the quadratic fitness function (1.11), then the rate of change of the mean is

$$\dot{\overline{G}} = 2s\sigma_G^2 (P_O - \overline{G}) - sC_3 \ . \tag{2.22}$$

Summing (2.17) over all loci and paternal and maternal gametes, and noting that for an additive character the total genetic covariance, $\mathrm{Cov}_G(G, w)$, equals the additive genetic covariance, $\text{Cov}_A(G, w)$ (cf. Chapters II.3.4 and II.5.1), we obtain that (2.20) is equivalent to the Secondary Theorem of Natural Selection II(5.10), i.e.,

$$\dot{\bar{G}} = \text{Cov}_G(G, w) \; . \tag{2.23}$$

For an *asexually reproducing population*, the differential equations (2.13) can be simplified, because a constant fitness is assigned to each allele, and we may assume that $w(x) = \sum_{k=1}^{K} s_k x^k$. Then we have $C_i = c_i$ and $B_k = 0$ if $k > 0$. Therefore, the dynamics of cumulants is given by

$$\dot{c}_n = \sum_{k=1}^{K} s_k S_{nk} + \mu v_n . \tag{2.24}$$

With the equations derived in this section, the evolution of the phenotype can be predicted in terms of the (empirically measurable) cumulants if the cumulants of the initial population are given. However, since in general not all the cumulants are known and the rate of change $\dot{C}_n$ depends on higher-order cumulants, the above equations can be used only to predict a few generations with accuracy. For applications, we refer to Section 6 and Chapters VI and VII. The reader may also note that the rate of change of the (additive) genetic variance σ_G^2 cannot be expressed solely in terms of cumulants of genotypic, or phenotypic, values, but depends on the allelic distribution at individual loci. This is obvious from (2.16b) by observing that $\dot{c}_2$ is a nonlinear function of the cumulants.

*3. SELECTION RESPONSE AND LINKAGE DISEQUILIBRIUM

In this section, we derive formulas for the response to selection of the frequency distribution $P(\boldsymbol{z})$ of multilocus genotypes without ignoring linkage disequilibrium. It is supposed that the population is in Hardy–Weinberg proportions before selection. These will be distorted by selection. Thus, instead of the independent haploid one-locus distributions $p_i(x_i)$, we have to consider the joint distribution $P(\boldsymbol{z})$. However, the assumption of Hardy–Weinberg proportions is not used before (3.16). After settling the basic notations and facts, we shall represent the selection response of the multivariate cumulants in terms of their weak-selection responses. Then we calculate these weak selection responses, and close with the consideration of some special cases.

3.1 BASIC DEFINITIONS AND IDENTITIES

Until the end of this subsection, we shall not impose any assumptions (such as additivity) about the mapping of genotypes $\boldsymbol{z} = (\boldsymbol{x}, \boldsymbol{x}^*)$ to genotypic values G. Selection changes the genotype frequencies according to

$$\Delta_s P(\boldsymbol{z}) = P(\boldsymbol{z}) \, \frac{W(\boldsymbol{z}) - \overline{W}}{\overline{W}} \; . \tag{3.1}$$

The fundamental idea for studying the multilocus dynamics is to describe the distribution of genotypes in terms of, now multivariate, generating functions. Let $\psi = \psi(\boldsymbol{\zeta})$ denote the moment generating function of $P(\boldsymbol{z})$, and let $\Psi = \ln \psi$ be its cumulant generating function.[2] (We tacitly assume that these exist, in particular, that moments and cumulants of all order exist.) Then, by (3.1), selection changes ψ according to

$$\Delta_s \psi(\boldsymbol{\zeta}) = \frac{1}{\overline{W}} \int \exp(\boldsymbol{\zeta} \bullet \boldsymbol{z}) W(\boldsymbol{z}) P(\boldsymbol{z}) \,\mathrm{d}\boldsymbol{z} - \psi(\boldsymbol{\zeta}) , \tag{3.2}$$

where $\boldsymbol{\zeta} \bullet \boldsymbol{z} = \sum_{i=1}^{2\ell} \zeta_i z_i$, and the change of Ψ is

$$\begin{aligned} \Delta_s \Psi &= \ln(\psi + \Delta_s \psi) - \ln(\psi) = \ln[1 + (\Delta_s \psi)/\psi] \\ &= \sum_{\nu=1}^{\infty} (-1)^{\nu-1} \frac{1}{\nu} \left(\frac{\Delta_s \psi}{\psi} \right)^{\nu} . \end{aligned} \tag{3.3}$$

Let $\boldsymbol{n} = (\boldsymbol{j}, \boldsymbol{j}^*)$ denote a multi-index of length 2ℓ, where the comma separates the 'paternal' multi-index $\boldsymbol{j}$ from the 'maternal' index $\boldsymbol{j}^*$. All considered multi-indices are nonnegative. We write $c_{\boldsymbol{n}}$ for the *multivariate* cumulant of order $\boldsymbol{n}$ of the distribution $P(\boldsymbol{x}, \boldsymbol{x}^*)$ of diploid genotypes. On some occasions, to emphasize that we are dealing with cross-gamete cumulants, we shall use the notation $c_{\boldsymbol{j},\boldsymbol{j}^*}$ instead of $c_{\boldsymbol{n}}$. Because we assume no sex differences, $c_{\boldsymbol{j},\boldsymbol{k}} = c_{\boldsymbol{k},\boldsymbol{j}}$ for all multi-indices $\boldsymbol{j}$ and $\boldsymbol{k}$ of length ℓ.

Property (D.15) of the cumulants implies that a population is in Hardy–Weinberg equilibrium with respect to all loci if and only if $c_{\boldsymbol{j},\boldsymbol{j}^*} = 0$ for every $\boldsymbol{j}$ and $\boldsymbol{j}^*$ such that $\boldsymbol{j} > \boldsymbol{0}$ *and* $\boldsymbol{j}^* > \boldsymbol{0}$. Consequently, in a randomly mating population all cross-gamete cumulants are zero before selection. This is one of the main advantages of cumulants compared with moments. Selection, however, changes cumulants both within and between gametes.

For low-order cumulants we shall use the following more convenient notation. Let $\boldsymbol{e}_i$ denote the vector $(0, \ldots, 1, \ldots, 0)$ of length ℓ, where the 1 is at position i. Then we write

$$\kappa_i = c_{\boldsymbol{e}_i,\boldsymbol{0}} = \mathrm{E}[X_i] , \tag{3.4a}$$

$$\kappa_{ij} = c_{\boldsymbol{e}_i+\boldsymbol{e}_j,\boldsymbol{0}} = \mathrm{Cov}[X_i, X_j] , \tag{3.4b}$$

$$\kappa_{ijk} = c_{\boldsymbol{e}_i+\boldsymbol{e}_j+\boldsymbol{e}_k,\boldsymbol{0}} = \mathrm{E}[(X_i - \kappa_i)(X_j - \kappa_j)(X_k - \kappa_k)] , \tag{3.4c}$$

etc., for the paternal and, because we ignore sex differences, also for the maternal within-gamete cumulants. (Expectations are, of course, taken with respect to the full distribution $P(\boldsymbol{z})$.) Thus, in the notation of Section 2, we have for a population in Hardy–Weinberg and in linkage equilibrium $\kappa_i = c_1^{(i)}$, $\kappa_{ii} = c_2^{(i)}$, etc. For the cross-gamete cumulants, we use the notation

$$\kappa_{i,j} = c_{\boldsymbol{e}_i,\boldsymbol{e}_j} = \mathrm{Cov}[X_i, X_j^*] , \tag{3.4d}$$

$$\kappa_{ij,k} = c_{\boldsymbol{e}_i+\boldsymbol{e}_j,\boldsymbol{e}_k} = \mathrm{E}[(X_i - \kappa_i)(X_j - \kappa_j)(X_k^* - \kappa_k)] . \tag{3.4e}$$

[2] The reader is referred to Appendix D for the definitions and basic properties of cumulants and moments, and for a concise summary of the multivariate notation used here and below.

Additional subscripts are added if more loci are involved. Recall that before selection all the $\kappa_{i,j}$, $\kappa_{ij,k}$, etc. are zero.

Linkage disequilibria are quantified by cumulants $c_{\boldsymbol{j},\mathbf{0}}$, where $\boldsymbol{j}$ is a multi-index of length ℓ with two or more non-zero entries. By property (D.15), two loci, i and j, are in linkage equilibrium, if all cumulants κ_{ij}, κ_{iij}, κ_{ijj}, etc. vanish. Three loci are in linkage equilibrium if cumulants of any order involving at least two of the loci vanish. Global linkage equilibrium obtains if and only if $c_{\boldsymbol{j},\mathbf{0}} = 0$ for every multi-index $\boldsymbol{j}$ of length ℓ with more than one positive entry. Except for special cases, these measures of linkage disequilibrium are different from all those introduced previously (see Section 4). However, for diallelic loci with allelic effects $x_i = 0, 1$ at each locus, the standard measures of linkage disequilibrium (cf. Chapter I.7) are obtained, i.e., $\kappa_{ij} = p(x_i x_j) - p_i p_j$ and $\kappa_{ijk} = p(x_i x_j x_k) - p_i \kappa_{jk} - p_j \kappa_{ki} - p_k \kappa_{ij}$, where p_i denotes the frequency of the 1-allele at locus i.

From (3.3) the exact selection response of the cumulants of arbitrary order is obtained by differentiation:

$$\begin{aligned} \Delta_{\mathrm{s}} c_{\boldsymbol{n}} &= (D^{\boldsymbol{n}} \Delta_{\mathrm{s}} \Psi)(\mathbf{0}) \\ &= \sum_{\nu=1}^{|\boldsymbol{n}|} (-1)^{\nu-1} \frac{1}{\nu} D^{\boldsymbol{n}} \left(\frac{\Delta_{\mathrm{s}} \psi}{\psi} \right)^{\nu} (\mathbf{0}) , \end{aligned} \tag{3.5}$$

where $D^{\boldsymbol{n}} = D^{\boldsymbol{n}}_{\boldsymbol{\zeta}}$ denotes the partial derivative with respect to $\boldsymbol{\zeta}$ (see Appendix D.1). Next, let us introduce the notation

$$\tilde{\Delta}_{\mathrm{s}} c_{\boldsymbol{n}} = \overline{W} D^{\boldsymbol{n}} \left(\frac{\Delta_{\mathrm{s}} \psi}{\psi} \right) (\mathbf{0}) . \tag{3.6}$$

This is a multiple of the $\boldsymbol{n}$th derivative of the first-order term in the Taylor-series expansion (3.3) of $\Delta_{\mathrm{s}} \Psi$. We call $\tilde{\Delta}_{\mathrm{s}} c_{\boldsymbol{n}}$ the *weak-selection response* of $c_{\boldsymbol{n}}$ for reasons that will become clear in Section 3.2. Using (3.6) and applying the differentiation formula (D.3) to $\Delta_{\mathrm{s}} \psi / \psi$, we can express (3.5) as

$$\Delta_{\mathrm{s}} c_{\boldsymbol{n}} = \sum_{\nu=1}^{|\boldsymbol{n}|} (-1)^{\nu-1} \frac{1}{\nu} \overline{W}^{-\nu} \sum_{\boldsymbol{k}_1 + \ldots + \boldsymbol{k}_\nu = \boldsymbol{n}} \frac{\boldsymbol{n}!}{\boldsymbol{k}_1! \cdot \ldots \cdot \boldsymbol{k}_\nu!} \tilde{\Delta}_{\mathrm{s}} c_{\boldsymbol{k}_1} \cdot \ldots \cdot \tilde{\Delta}_{\mathrm{s}} c_{\boldsymbol{k}_\nu} , \tag{3.7}$$

where the inner sum is over all nonzero (because $c_{\mathbf{0}} = 0$) multi-indices $\boldsymbol{k}_i$ of length 2ℓ that satisfy the indicated constraint. Thus, the strong-selection response $\Delta_{\mathrm{s}} c_{\boldsymbol{n}}$ can be expressed in terms of the weak-selection responses $\tilde{\Delta}_{\mathrm{s}} c_{\boldsymbol{k}_i}$, $\boldsymbol{k}_i \le \boldsymbol{n}$. The latter have the advantage that they can be calculated more easily.

For low-order cumulants, (3.7) simplifies considerably. With the notation introduced above, one obtains

$$\begin{aligned} \overline{W} \Delta_{\mathrm{s}} \kappa_i &= \tilde{\Delta}_{\mathrm{s}} \kappa_i , \\ \overline{W} \Delta_{\mathrm{s}} \kappa_{ij} &= \tilde{\Delta}_{\mathrm{s}} \kappa_{ij} - \overline{W} \Delta_{\mathrm{s}} \kappa_i \Delta_{\mathrm{s}} \kappa_j , \\ \overline{W} \Delta_{\mathrm{s}} \kappa_{ijk} &= \tilde{\Delta}_{\mathrm{s}} \kappa_{ijk} - (\tilde{\Delta}_{\mathrm{s}} \kappa_{ij} \Delta_{\mathrm{s}} \kappa_k + \tilde{\Delta}_{\mathrm{s}} \kappa_{ik} \Delta_{\mathrm{s}} \kappa_j + \tilde{\Delta}_{\mathrm{s}} \kappa_{jk} \Delta_{\mathrm{s}} \kappa_i) \\ &\quad + 2\overline{W} \Delta_{\mathrm{s}} \kappa_i \Delta_{\mathrm{s}} \kappa_j \Delta_{\mathrm{s}} \kappa_k , \end{aligned} \tag{3.8}$$

and structurally identical equations hold for the change of the cross-gamete cumulants.

By virtue of (3.7) and (3.8), we have reduced the calculation of the selection response to the calculation of the weak-selection response $\tilde{\Delta}_{\mathrm{S}} c_{\boldsymbol{n}}$ for $\boldsymbol{n} \geq \boldsymbol{0}$. More explicit expressions for the selection response can be obtained under more specific assumptions about the gene interaction or the fitness function.

Denoting the nth cumulant of the distribution of breeding values $f(G)$ by C_n (hence $C_1 = \overline{G}$ and $C_2 = \sigma_G^2$), and assuming an additive trait, (1.2), the additivity property (D.13) of the cumulants informs us that

$$C_n = \sum_{\boldsymbol{n}:|\boldsymbol{n}|=n} \frac{n!}{\boldsymbol{n}!} c_{\boldsymbol{n}} . \tag{3.9}$$

However, under random mating only the ℓ-variate (within-gamete) cumulants of $p(\boldsymbol{x})$ contribute to C_n if measured among zygotes. In linkage equilibrium, (3.9) is reduced to (2.5).

3.2 WEAK SELECTION

From now on we assume the additive model of Section 1.1. Then the weak-selection response $\tilde{\Delta}_{\mathrm{S}} c_{\boldsymbol{n}}$ of the cumulants can be expressed explicitly in terms of the cumulants before selection and the coefficients s_k of the Taylor series expansion of $W(G)$. Here we summarize the results. The derivations are relegated to Section 3.3. If $|\boldsymbol{n}| \geq 1$, then the weak-selection response is

$$\tilde{\Delta}_{\mathrm{S}} c_{\boldsymbol{n}} = \sum_{k=1}^{K} s_k S_k(\boldsymbol{n}) , \tag{3.10}$$

where

$$S_k(\boldsymbol{n}) = \sum_{\boldsymbol{k}:|\boldsymbol{k}|=k} \frac{k!}{\boldsymbol{k}!} D^{\boldsymbol{n}} \left(\frac{1}{\psi} D^{\boldsymbol{k}} \psi \right) (\boldsymbol{0}) . \tag{3.11}$$

Thus, we see from (3.7) that the weak-selection response is obtained from the strong-selection response by omitting all terms involving products of the s_k. It may be noted, however, that weak selection in the present sense does not necessarily imply weak selection in the sense of Chapter II.6.2, unless $\sum_{k=1}^{\infty} |s_k| = O(s)$ as $s \to 0$. If this condition is satisfied, then we have by (3.10), $\tilde{\Delta}_{\mathrm{S}} c_{\boldsymbol{n}} = O(s)$ for every $\boldsymbol{n}$, and by (3.7) and because $\overline{W} = 1 + O(s)$,

$$\Delta_{\mathrm{S}} c_{\boldsymbol{n}} = \tilde{\Delta}_{\mathrm{S}} c_{\boldsymbol{n}} + O(s^2) \quad \text{as } s \to 0 . \tag{3.12}$$

Another way to interpret the weak-selection response is to approximate, in analogy to Section 2.1, the discrete-time recursion (3.1) by the differential equation

$$\frac{\partial}{\partial t} P(\boldsymbol{z}, t) = P(\boldsymbol{z}, t)[W(\boldsymbol{z}) - \overline{W}] , \tag{3.13}$$

which is a good approximation if $W(\boldsymbol{z}) = 1 + sw(\boldsymbol{z})$ and s is small. Passing over to the cumulant generating function and using (3.2), we get

$$\frac{\partial}{\partial t} \Psi(\boldsymbol{\zeta}, t) = \overline{W} \, \frac{\Delta_{\mathrm{S}} \psi(\boldsymbol{\zeta}, t)}{\psi(\boldsymbol{\zeta}, t)} . \tag{3.14}$$

Therefore, setting $\tilde{\Delta}_s \Psi = \frac{\partial}{\partial t}\Psi$, differentiating with respect to $\boldsymbol{\zeta}$, and recalling (3.6), we obtain $\frac{\mathrm{d}}{\mathrm{d}t} c_{\boldsymbol{n}} = \tilde{\Delta}_s c_{\boldsymbol{n}}$. If one is interested only in the weak-selection response, but not in using (3.7), then all terms of order less than $O(s)$ can be omitted on the right-hand side of (3.13).

The expressions $S_k(\boldsymbol{n})$ can be computed explicitly in terms of the multivariate cumulants. With the notation $c_{\boldsymbol{n}\bullet} = \sum_{|\boldsymbol{j}|=1} c_{\boldsymbol{n}+\boldsymbol{j}}$, $c_{\boldsymbol{n}\bullet\bullet} = \sum_{|\boldsymbol{j}|=2} c_{\boldsymbol{n}+\boldsymbol{j}}$, etc., where the multi-index $\boldsymbol{j}$ in the summation has length 2ℓ, one finds

$$S_1(\boldsymbol{n}) = c_{\boldsymbol{n}\bullet}\ , \tag{3.15a}$$

$$S_2(\boldsymbol{n}) = c_{\boldsymbol{n}\bullet\bullet} + \sum_{\boldsymbol{k}=\boldsymbol{0}}^{\boldsymbol{n}} \binom{\boldsymbol{n}}{\boldsymbol{k}} c_{\boldsymbol{k}\bullet} c_{\boldsymbol{n}-\boldsymbol{k}\bullet}\ , \tag{3.15b}$$

$$\begin{aligned} S_3(\boldsymbol{n}) = c_{\boldsymbol{n}\bullet\bullet\bullet} &+ 3 \sum_{\boldsymbol{k}_1+\boldsymbol{k}_2=\boldsymbol{n}} \frac{\boldsymbol{n}!}{\boldsymbol{k}_1!\,\boldsymbol{k}_2!}\, c_{\boldsymbol{k}_1\bullet}\, c_{\boldsymbol{k}_2\bullet\bullet} \\ &+ \sum_{\boldsymbol{k}_1+\boldsymbol{k}_2+\boldsymbol{k}_3=\boldsymbol{n}} \frac{\boldsymbol{n}!}{\boldsymbol{k}_1!\,\boldsymbol{k}_2!\,\boldsymbol{k}_3!}\, c_{\boldsymbol{k}_1\bullet}\, c_{\boldsymbol{k}_2\bullet}\, c_{\boldsymbol{k}_3\bullet}\ , \end{aligned} \tag{3.15c}$$

$$\begin{aligned} S_4(\boldsymbol{n}) = c_{\boldsymbol{n}\bullet\bullet\bullet\bullet} &+ \sum_{\boldsymbol{k}_1+\boldsymbol{k}_2=\boldsymbol{n}} \frac{\boldsymbol{n}!}{\boldsymbol{k}_1!\,\boldsymbol{k}_2!} \left[4 c_{\boldsymbol{k}_1\bullet}\, c_{\boldsymbol{k}_2\bullet\bullet\bullet} + 3 c_{\boldsymbol{k}_1\bullet\bullet}\, c_{\boldsymbol{k}_2\bullet\bullet}\right] \\ &+ 6 \sum_{\boldsymbol{k}_1+\boldsymbol{k}_2+\boldsymbol{k}_3=\boldsymbol{n}} \frac{\boldsymbol{n}!}{\boldsymbol{k}_1!\,\boldsymbol{k}_2!\,\boldsymbol{k}_3!}\, c_{\boldsymbol{k}_1\bullet}\, c_{\boldsymbol{k}_2\bullet}\, c_{\boldsymbol{k}_3\bullet\bullet} \\ &+ \sum_{\boldsymbol{k}_1+\boldsymbol{k}_2+\boldsymbol{k}_3+\boldsymbol{k}_4=\boldsymbol{n}} \frac{\boldsymbol{n}!}{\boldsymbol{k}_1!\,\boldsymbol{k}_2!\,\boldsymbol{k}_3!\,\boldsymbol{k}_4!}\, c_{\boldsymbol{k}_1\bullet}\, c_{\boldsymbol{k}_2\bullet}\, c_{\boldsymbol{k}_3\bullet}\, c_{\boldsymbol{k}_4\bullet}\ . \end{aligned} \tag{3.15d}$$

Thus, by combining (3.7) with (3.10), (3.11), and (3.15), we have found explicit expressions for the selection response of multivariate cumulants of arbitrary order if the fitness function can be approximated by a quartic polynomial. We also note that we have not yet used that the population is in Hardy–Weinberg proportions before selection. A simple procedure for deriving the terms with $k \geq 5$ is outlined below (3.21). We shall treat applications with specific fitness functions in Sections 3.6 and 3.7, and in Chapters VI and VII. The above formulas generalize those in Section 2 for the univariate within-locus cumulants.

Let $\kappa_{i\bullet} = \sum_{j=1}^{\ell} \kappa_{ij} = c_{e_i\bullet}$, $\kappa_{i\bullet\bullet} = \sum_{j,k=1}^{\ell} \kappa_{ijk} = c_{e_i\bullet\bullet}$, etc. Using, for the first time, the assumption that the population is in Hardy–Weinberg proportions, we observe that $C_1 = 2\kappa_\bullet = c_{\boldsymbol{0}\bullet}$ and $C_2 = 2\kappa_{\bullet\bullet} = c_{\boldsymbol{0}\bullet\bullet}$, because cross-gamete cumulants are zero before selection. Then, for the low-order cumulants, the weak-selection response (3.10) becomes, to given order in s_k,

$$\begin{aligned} \tilde{\Delta}_s \kappa_i = {} & s_1 \kappa_{i\bullet} + s_2[\kappa_{i\bullet\bullet} + 2C_1\kappa_{i\bullet}] \\ & + s_3[\kappa_{i\bullet\bullet\bullet} + 3C_1\kappa_{i\bullet\bullet} + 3(C_1^2 + C_2)\kappa_{i\bullet}] \\ & + s_4[\kappa_{i\bullet\bullet\bullet\bullet} + 4C_1\kappa_{i\bullet\bullet\bullet} + 6(C_1^2 + C_2)\kappa_{i\bullet\bullet} + 4(C_1^3 + 3C_1C_2 + C_3)\kappa_{i\bullet}] + \dots\ , \end{aligned} \tag{3.16a}$$

$$\begin{aligned} \tilde{\Delta}_s \kappa_{ij} = {} & s_1 \kappa_{ij\bullet} + s_2[\kappa_{ij\bullet\bullet} + 2C_1\kappa_{ij\bullet} + 2\kappa_{i\bullet}\kappa_{j\bullet}] \\ & + s_3[\kappa_{ij\bullet\bullet\bullet} + 3(\kappa_{i\bullet\bullet}\kappa_{j\bullet} + \kappa_{j\bullet\bullet}\kappa_{i\bullet}) + 3C_1(\kappa_{ij\bullet\bullet} + 2\kappa_{i\bullet}\kappa_{j\bullet}) + 3(C_1^2 + C_2)\kappa_{ij\bullet}] \end{aligned}$$

$$
\begin{aligned}
&+ s_4[\kappa_{ij\ldots.} + 4(\kappa_{i.}\kappa_{j\ldots} + \kappa_{j.}\kappa_{i\ldots}) + 6\kappa_{i..}\kappa_{j..} \\
&\qquad + 4C_1(\kappa_{ij\ldots} + 3\kappa_{i.}\kappa_{j..} + 3\kappa_{j.}\kappa_{i..}) + 6(C_1^2 + C_2)(\kappa_{ij..} + 2\kappa_{i.}\kappa_{j.}) \\
&\qquad + 4(C_1^3 + 3C_1C_2 + C_3)\kappa_{ij.}] + \ldots\,, \qquad (3.16b)
\end{aligned}
$$

$$
\begin{aligned}
\tilde{\Delta}_s\kappa_{ijk} &= s_1\kappa_{ijk.} + s_2[\kappa_{ijk..} + 2C_1\kappa_{ijk.} + 2(\kappa_{ij.}\kappa_{k.} + **)] \\
&\quad + s_3\{\kappa_{ijk\ldots} + 6\kappa_{i.}\kappa_{j.}\kappa_{k.} + 3(\kappa_{ij.}\kappa_{k..} + \kappa_{ij..}\kappa_{k.} + **) \\
&\qquad + 3C_1[\kappa_{ijk..} + 2(\kappa_{ij.}\kappa_{k.} + **)] + 3(C_1^2 + C_2)\kappa_{ijk.}\} + \ldots\,, \qquad (3.16c)
\end{aligned}
$$

$$
\begin{aligned}
\tilde{\Delta}_s\kappa_{ijkl} &= s_1\kappa_{ijkl.} + s_2[\kappa_{ijkl..} + 2C_1\kappa_{ijkl.} + 2(\kappa_{ijk.}\kappa_{l.} + ***) + 2(\kappa_{ij.}\kappa_{kl.} + **)] \\
&\quad + s_3\{\kappa_{ijkl\ldots} + 3(\kappa_{ij..}\kappa_{kl.} + *****) + 6(\kappa_{ij.}\kappa_{k.}\kappa_{l.} + *****) \\
&\qquad + 3[(\kappa_{ijk..}\kappa_{l.} + \kappa_{ijk.}\kappa_{l..}) + ***] \\
&\qquad + 3C_1[\kappa_{ijkl..} + 2(\kappa_{ij.}\kappa_{kl.} + **) + 2(\kappa_{ijk.}\kappa_{l.} + ***)] \\
&\qquad + 3(C_1^2 + C_2)\kappa_{ijkl.}\} + \ldots\,, \qquad (3.16d)
\end{aligned}
$$

where $**$ ($***$, $*****$) denotes two (three, five) additional terms, each obtained by permuting the subscripts. These formulas for the selection response agree with those derived by Turelli and Barton (1994), which are presented in Section 3.4. In linkage equilibrium, when all cross-locus cumulants vanish, (3.16) simplifies to (2.16) if $\tilde{\Delta}_s\kappa_i$ is replaced by $\dot{c}_1$, etc.

For cross-gamete cumulants $\boldsymbol{n} = (\boldsymbol{j}, \boldsymbol{j}^*)$, most terms in (3.15) vanish, because only those terms in a sum are nonzero for which *all* $\boldsymbol{k}_i$ are of the form $(\boldsymbol{l}, \boldsymbol{0})$ or $(\boldsymbol{0}, \boldsymbol{l})$ with $\boldsymbol{n} \geq \boldsymbol{l} \geq \boldsymbol{0}$. In particular, the weak-selection response of the cross-gamete cumulants up to order four is found to be

$$
\tilde{\Delta}_s\kappa_{i,j} = 2s_2\kappa_{i.}\kappa_{j.} + 3s_3[\kappa_{i..}\kappa_{j.} + \kappa_{j..}\kappa_{i.} + 2C_1\kappa_{i.}\kappa_{j.}] + \ldots\,, \qquad (3.17a)
$$

$$
\begin{aligned}
\tilde{\Delta}_s\kappa_{i,jk} &= 2s_2\kappa_{i.}\kappa_{jk.} + 3s_3[2\kappa_{i.}\kappa_{j.}\kappa_{k.} \\
&\qquad + 2C_1\kappa_{jk.}\kappa_{i.} + \kappa_{i..}\kappa_{jk.} + \kappa_{i.}\kappa_{jk..}] + \ldots\,, \qquad (3.17b)
\end{aligned}
$$

$$
\begin{aligned}
\tilde{\Delta}_s\kappa_{i,jkl} &= 2s_2\kappa_{i.}\kappa_{jkl.} + 3s_3[2\kappa_{i.}(\kappa_{j.}\kappa_{kl.} + **) \\
&\qquad + \kappa_{i.}\kappa_{jkl..} + \kappa_{i..}\kappa_{jkl.} + 2C_1\kappa_{i.}\kappa_{jkl.}] + \ldots\,, \qquad (3.17c)
\end{aligned}
$$

$$
\begin{aligned}
\tilde{\Delta}_s\kappa_{ij,kl} &= 2s_2\kappa_{ij.}\kappa_{kl.} + 3s_3[\kappa_{ij..}\kappa_{kl.} + \kappa_{ij.}\kappa_{kl..} \\
&\qquad + 2\kappa_{ij.}\kappa_{k.}\kappa_{l.} + 2\kappa_{kl.}\kappa_{i.}\kappa_{j.} + 2C_1\kappa_{ij.}\kappa_{kl.}] + \ldots\,. \qquad (3.17d)
\end{aligned}
$$

These equations can be obtained formally from (3.16) by inserting the comma at the appropriate position, and recalling that all cross-gamete cumulants are zero before selection. They quantify the deviations from Hardy–Weinberg proportions caused by selection. The evaluation of (3.15) that led to (3.16) is readily automated with the help of a computer language for symbolic computations such as *Mathematica* (Wolfram 1996). Then higher-order terms can be quickly derived (*Mathematica* notebooks may be obtained from the author on request).

The above equations form the basis for the derivation of the change of the multivariate cumulants and for the cumulants of the phenotypic distribution across generations. It is important to note that, in general, the recursions for the cumulants are not a closed system of equations, because the change of a cumulant always depends on the value of cumulants of higher order before selection. Thus, in concrete applications additional assumptions are needed for closing up the system of recursions.

*3.3 DERIVATION OF THE WEAK-SELECTION RESPONSE

The representation (3.10) of the weak-selection response follows from (3.6), (3.2), and (1.15), because

$$\begin{aligned}
\tilde{\Delta}_{\mathrm{s}} c_{\boldsymbol{n}} &= \overline{W} D^{\boldsymbol{n}} \left(\frac{\Delta_{\mathrm{s}} \psi}{\psi} \right) (\mathbf{0}) \\
&= D^{\boldsymbol{n}} \left[\frac{1}{\psi(\boldsymbol{\zeta})} \int \exp(\boldsymbol{\zeta} \cdot \boldsymbol{z}) P(\boldsymbol{z}) W(\boldsymbol{z}) \, \mathrm{d}\boldsymbol{z} \right] \Bigg|_{\boldsymbol{\zeta}=\mathbf{0}} \\
&= \sum_{k=0}^{K} s_k \sum_{\boldsymbol{k}:|\boldsymbol{k}|=k} \frac{k!}{\boldsymbol{k}!} D^{\boldsymbol{n}} \left[\frac{1}{\psi(\boldsymbol{\zeta})} \int \exp(\boldsymbol{\zeta} \cdot \boldsymbol{z}) P(\boldsymbol{z}) \boldsymbol{z}^{\boldsymbol{k}} \, \mathrm{d}\boldsymbol{z} \right] \Bigg|_{\boldsymbol{\zeta}=\mathbf{0}} \\
&= \sum_{k=0}^{K} s_k \sum_{\boldsymbol{k}:|\boldsymbol{k}|=k} \frac{k!}{\boldsymbol{k}!} D^{\boldsymbol{n}} \left(\frac{1}{\psi(\boldsymbol{\zeta})} D^{\boldsymbol{k}} \psi(\boldsymbol{\zeta}) \right) \Bigg|_{\boldsymbol{\zeta}=\mathbf{0}} .
\end{aligned} \tag{3.18}$$

To derive (3.15), we introduce the notation

$$S(\boldsymbol{n}, \boldsymbol{k}) = D^{\boldsymbol{n}} \left(\frac{1}{\psi} D^{\boldsymbol{k}} \psi \right) (\mathbf{0}) . \tag{3.19}$$

These expressions are readily evaluated in terms of cumulants by the following method. Recalling $\psi = \exp \Psi$, straightforward differentiation yields

$$\begin{aligned}
D^{i} \psi / \psi &= D^{i} \Psi , \\
D^{ij} \psi / \psi &= D^{ij} \Psi + D^{i} \Psi D^{j} \Psi , \\
D^{ijk} \psi / \psi &= D^{ijk} \Psi + D^{ij} \Psi D^{k} \Psi + D^{ik} \Psi D^{j} \Psi + D^{jk} \Psi D^{i} \Psi + D^{i} \Psi D^{j} \Psi D^{k} \Psi ,
\end{aligned} \tag{3.20}$$

with the obvious short-hand notation D^i for $D^{\boldsymbol{e}_i}$, etc. Higher-order derivatives can be efficiently calculated with *Mathematica.* Applying the differential operator $D^{\boldsymbol{n}}$ to these expressions and using Leibniz' product rule (D.2), evaluating at $\mathbf{0}$, and inserting the definition of cumulants, one obtains

$$S(\boldsymbol{n}, \boldsymbol{e}_i) = c_{\boldsymbol{n}+\boldsymbol{e}_i} , \tag{3.21a}$$

$$S(\boldsymbol{n}, \boldsymbol{e}_i + \boldsymbol{e}_j) = c_{\boldsymbol{n}+\boldsymbol{e}_i+\boldsymbol{e}_j} + \sum_{\boldsymbol{l}=\mathbf{0}}^{\boldsymbol{n}} \binom{\boldsymbol{n}}{\boldsymbol{l}} c_{\boldsymbol{l}+\boldsymbol{e}_i} c_{\boldsymbol{n}-\boldsymbol{l}+\boldsymbol{e}_j} , \tag{3.21b}$$

$$\begin{aligned}
S(\boldsymbol{n}, \boldsymbol{e}_i + \boldsymbol{e}_j + \boldsymbol{e}_k) = c_{\boldsymbol{n}+\boldsymbol{e}_i+\boldsymbol{e}_j+\boldsymbol{e}_k} &+ \sum_{\boldsymbol{l}=\mathbf{0}}^{\boldsymbol{n}} \binom{\boldsymbol{n}}{\boldsymbol{l}} c_{\boldsymbol{l}+\boldsymbol{e}_i+\boldsymbol{e}_j} c_{\boldsymbol{n}-\boldsymbol{l}+\boldsymbol{e}_k} \\
&+ 2 \text{ similar terms with } i, j, k \text{ permuted} \\
&+ \sum_{\boldsymbol{l}=\mathbf{0}}^{\boldsymbol{n}} \sum_{\boldsymbol{m}=\mathbf{0}}^{\boldsymbol{l}} \binom{\boldsymbol{n}}{\boldsymbol{l}} \binom{\boldsymbol{l}}{\boldsymbol{m}} c_{\boldsymbol{m}+\boldsymbol{e}_i} c_{\boldsymbol{l}-\boldsymbol{m}+\boldsymbol{e}_j} c_{\boldsymbol{n}-\boldsymbol{l}+\boldsymbol{e}_k} .
\end{aligned} \tag{3.21c}$$

Now, appropriate summation yields (3.15a,b,c).

Actually, there is a simpler method for deriving (3.15) which we illustrate for (3.15d). Let $g = g(x)$ be a function of one variable and denote by D^n the nth derivative with respect to x. Then

$$(D^4 \exp g)/\exp g = D^4 g + 4Dg\, D^3 g + 3(D^2 g)^2 + 6(Dg)^2 D^2 g + (Dg)^4 ,$$

and (3.15d) is obtained by applying D^n to the right-hand side, using the univariate version of the generalized product rule (D.3), and substituting $c_{k_i\cdot}$ for $D^{k_i}(Dg)$, $c_{k_i\cdot\cdot}$ for $D^{k_i}(D^2 g)$, etc. This method is easily extended to terms $S_k(\boldsymbol{n})$ with $k \geq 5$.

With a slightly different method, the change of the cumulants κ_i and κ_{ij} can be derived for any fitness functions W of the form (1.8). By (3.6) and (3.2), we have

$$\begin{aligned}\tilde{\Delta}_{\mathrm{s}}\kappa_i &= \overline{W} D^{\boldsymbol{e}_i}\left(\frac{\Delta_{\mathrm{s}}\psi}{\psi}\right)(\boldsymbol{0}) \\ &= \overline{W} D^{\boldsymbol{e}_i}\Delta_{\mathrm{s}}\psi(\boldsymbol{0}) \\ &= \int x_i W(\boldsymbol{z})P(\boldsymbol{z})\,\mathrm{d}\boldsymbol{z} - \kappa_i\overline{W} \\ &= \mathrm{Cov}(X_i, W)\ . \end{aligned} \tag{3.22a}$$

This covariance formula is closely related to the Secondary Theorem of Natural Selection; see II(5.10) and (6.15).

From (3.22a), a simple calculation employing (1.15) yields

$$\begin{aligned}\tilde{\Delta}_{\mathrm{s}}\kappa_i &= \sum_{k=0}^{K} s_k \sum_{\boldsymbol{k}:|\boldsymbol{k}|=k} \frac{k!}{\boldsymbol{k}!}\int z_i \boldsymbol{z}^{\boldsymbol{k}} P(\boldsymbol{z})\,\mathrm{d}\boldsymbol{z} - \kappa_i\overline{W} \\ &= \sum_{k=0}^{K} s_k \sum_{\boldsymbol{k}:|\boldsymbol{k}|=k} \frac{k!}{\boldsymbol{k}!} m^0_{\boldsymbol{k}+\boldsymbol{e}_i} - \kappa_i\overline{W}\ , \end{aligned} \tag{3.22b}$$

where $m^0_{\boldsymbol{k}} = \int \boldsymbol{z}^{\boldsymbol{k}} P(\boldsymbol{z})\,\mathrm{d}\boldsymbol{z}$ denotes the moment about zero of order $\boldsymbol{k}$. In a similar way, one obtains

$$\tilde{\Delta}_{\mathrm{s}}\kappa_{ij} = \mathrm{Cov}(X_i X_j - \kappa_j X_i - \kappa_i X_j, W) \tag{3.23a}$$

$$= \sum_{k=0}^{K} s_k \sum_{\boldsymbol{k}:|\boldsymbol{k}|=k} \frac{k!}{\boldsymbol{k}!}\Big[m^0_{\boldsymbol{k}+\boldsymbol{e}_i+\boldsymbol{e}_j} - (\kappa_i m^0_{\boldsymbol{k}+\boldsymbol{e}_j} + \kappa_j m^0_{\boldsymbol{k}+\boldsymbol{e}_i}) + m^0_{\boldsymbol{k}}(\kappa_i\kappa_j - \kappa_{ij})\Big]\ . \tag{3.23b}$$

In both cases, the moments can be expressed in terms of cumulants. Using (D.10) or (D.11), it is straightforward to show that for $k \leq 4$, (3.22b) and (3.23b) coincide with (3.16a) and (3.16b), respectively. We note that (3.22a) and (3.23a) are not based on the additivity assumption (1.2), but are generally valid.

3.4 THE SELECTION-GRADIENT APPROACH

Turelli and Barton (1994) used a somewhat different method to derive the selection response of the multivariate cumulants. Here we outline their approach without proofs, but retain our notation from above. In Chapter I.10, it was shown that at a single locus the change of the genotype frequencies between generations can be written as a generalized gradient system. This was extended by Turelli and Barton (1990) to the multilocus setting, who showed that the change in genotypic frequencies caused by selection is (exactly)

$$\Delta_{\mathrm{s}} P(\boldsymbol{z}) = \int G(\boldsymbol{z},\boldsymbol{y}) \frac{\partial \ln \overline{W}}{\partial P(\boldsymbol{y})}\, \mathrm{d}\boldsymbol{y} \,, \tag{3.24a}$$

where

$$G(\boldsymbol{z},\boldsymbol{y}) = P(\boldsymbol{z})[\delta(\boldsymbol{z}-\boldsymbol{y}) - P(\boldsymbol{y})] \,. \tag{3.24b}$$

Here, $\boldsymbol{y}$ and $\boldsymbol{z}$ denote diploid ℓ-locus genotypes, δ is the Dirac delta function if there is a continuum of possible types, and it is the Kronecker delta in models with discrete alleles. In the latter case, the integral is replaced by a sum, and (3.24) is equivalent to I(10.14a) and I(10.13). Then the change of the moment generating function of $P(\boldsymbol{z})$ caused by selection can be expressed as

$$\Delta_{\mathrm{s}} \psi(\boldsymbol{\zeta}) = \int \Gamma_*(\boldsymbol{\zeta},\boldsymbol{\eta}) \frac{\partial \ln \overline{W}}{\partial \Psi(\boldsymbol{\eta})}\, \mathrm{d}\boldsymbol{\eta} \,, \tag{3.25a}$$

where

$$\Gamma_*(\boldsymbol{\zeta},\boldsymbol{\eta}) = \exp[\Psi(\boldsymbol{\zeta}+\boldsymbol{\eta}) - \Psi(\boldsymbol{\eta})] - \exp[\Psi(\boldsymbol{\zeta})] \tag{3.25b}$$

and Ψ is the cumulant generating function. By differentiation, it can then be shown that

$$\Delta_{\mathrm{s}} m^0_{\boldsymbol{n}} = \sum_{\boldsymbol{k}\geq \boldsymbol{0}} \Gamma(\boldsymbol{n},\boldsymbol{k}) \mathcal{L}_{\boldsymbol{k}} \,, \tag{3.26a}$$

where

$$\mathcal{L}_{\boldsymbol{k}} = \frac{\partial \ln \overline{W}}{\partial c_{\boldsymbol{k}}} \tag{3.26b}$$

is the so-called *selection gradient* with respect to the cumulant $c_{\boldsymbol{k}}$, and

$$\Gamma(\boldsymbol{n},\boldsymbol{k}) = (D^{\boldsymbol{n}}_{\boldsymbol{\zeta}} D^{\boldsymbol{k}}_{\boldsymbol{\eta}} \Gamma_*)(\boldsymbol{0},\boldsymbol{0}) \,. \tag{3.26c}$$

This result is rather general because no assumptions about gene interaction have been imposed, and because it applies to very general fitness functions. For more detailed results, however, the genetic model has to be specified.

Under the additive model, the selection gradients can be simplified by using (3.9), and become

$$\mathcal{L}_{\boldsymbol{k}} = \frac{k!}{\boldsymbol{k}!} \mathcal{L}_k = \frac{k!}{\boldsymbol{k}!} \frac{\partial \ln \overline{W}}{\partial C_k} \,, \tag{3.27}$$

where $k = |\boldsymbol{k}|$. Then, in terms of the distribution of breeding values $f(G)$, the following expressions can be found:

$$\mathcal{L}_k = \frac{1}{\overline{W}} \frac{(-1)^k}{k!} \int_{-\infty}^{\infty} W(G) D^k_G f(G)\, \mathrm{d}G \tag{3.28a}$$

$$= \frac{1}{\overline{W}} \frac{1}{k!} \int_{-\infty}^{\infty} f(G) D^k_G W(G)\, \mathrm{d}G \,, \tag{3.28b}$$

where D_G^k denotes the k-fold derivative with respect to G. These selection gradients are generalizations of the 'directional' and 'stabilizing selection gradients' defined by Lande (1979) and Lande and Arnold (1983); cf. Chapter VII.7.5. With the formulas (3.28), the selection gradients can be calculated if, for example, $f(G)$ can be approximated by a simple probability density or if the fitness function $W(G)$ can be approximated by a polynomial. If $W(G)$ is as in (1.8), then from (3.28b),

$$\mathcal{L}_k = \frac{1}{\overline{W}} \sum_{j\geq 0} \binom{k+j}{k} s_{k+j} M_{k+j}^0 \ . \tag{3.29}$$

In particular, $\mathcal{L}_0 = 1$; cf. (2.9). If the fitness function is a quartic, i.e., $s_k = 0$ for $k \geq 5$, then

$$\overline{W}\mathcal{L}_1 = s_1 + 2s_2C_1 + 3s_3(C_1^2 + C_2) + 4s_4(C_1^3 + 3C_1C_2 + C_3) \ , \tag{3.30a}$$
$$\overline{W}\mathcal{L}_2 = s_2 + 3s_3C_1 + 6s_4(C_1^2 + C_2) \ , \tag{3.30b}$$
$$\overline{W}\mathcal{L}_3 = s_3 + 4s_4C_1 \ , \tag{3.30c}$$
$$\overline{W}\mathcal{L}_4 = s_4 \ . \tag{3.30d}$$

Finally, the weak-selection response of the cumulants up to order three can be expressed in terms of the selection gradients $\mathcal{L}_k$:

$$\overline{W}^{-1}\tilde{\Delta}_s\kappa_i = \kappa_{i.}\mathcal{L}_1 + \kappa_{i..}\mathcal{L}_2 + \kappa_{i...}\mathcal{L}_3 + \kappa_{i....}\mathcal{L}_4 \ , \tag{3.31}$$

$$\begin{aligned}\overline{W}^{-1}\tilde{\Delta}_s\kappa_{ij} &= \kappa_{ij.}\mathcal{L}_1 + [2\kappa_{i.}\kappa_{j.} + \kappa_{ij..}]\mathcal{L}_2 \\ &\quad + [3(\kappa_{i.}\kappa_{j..} + \kappa_{j.}\kappa_{i..}) + \kappa_{ij...}]\mathcal{L}_3 \ ,\end{aligned} \tag{3.32}$$

$$\begin{aligned}\overline{W}^{-1}\tilde{\Delta}_s\kappa_{ijk} &= \kappa_{ijk.}\mathcal{L}_1 + [2(\kappa_{i.}\kappa_{jk.} + **) + \kappa_{ijk..}]\mathcal{L}_2 \\ &\quad + [6\kappa_{i.}\kappa_{j.}\kappa_{k.} + 3(\kappa_{ij.}\kappa_{k..} + \kappa_{k.}\kappa_{ij..} + **) + \kappa_{ijk...}]\mathcal{L}_3 \ .\end{aligned} \tag{3.33}$$

By substituting (3.29) into (3.31)–(3.33), it follows that these equations coincide with (3.16a)–(3.16c), respectively. Turelli and Barton (1994) provided additional higher-order terms. The two approaches (Taylor series expansion *vs.* selection gradients) are equivalent, and it depends on the kind of application which one is the more convenient. It appears, however, that the equations (3.15) are much easier and faster to compute using a symbolic formula manipulation program such as *Mathematica* (Wolfram 1996).

*3.5 LINKAGE EQUILIBRIUM REVISITED

Equations (2.16) display the rate of change of the univariate cumulants for a population in global linkage equilibrium. However, in general, selection induces linkage disequilibria. Since for a population in linkage equilibrium the generating function ψ can be decomposed into the product $\prod_i \psi_i$, we find from (3.11) that

$$S_k(\boldsymbol{n}) = \begin{cases} 0 \ , & \text{if } k < |N| \ , \\ \displaystyle\sum_{\substack{\boldsymbol{k}:|\boldsymbol{k}|=k, \\ k_i>0 \text{ if } i\in N}} \frac{k!}{\boldsymbol{k}!} \prod_{i\in N} S_{n_i k_i} \ , & \text{otherwise} \ , \end{cases} \tag{3.34}$$

where $N = \{i : n_i > 0\}$, $|N|$ is the number of elements in N, and $S_{n_i k_i}$ is defined by (2.14). For the change of the univariate cumulants (κ_i, κ_{ii}, etc.), substitution of (3.34) into (3.10) yields (2.16); for the cross-locus and cross-gamete cumulants of order two and three, we obtain

$$\begin{aligned}\tilde{\Delta}_{\mathrm{s,LE}}\kappa_{ij} &= \tilde{\Delta}_{\mathrm{s,LE}}\kappa_{i,j} = 2s_2\kappa_{ii}\kappa_{jj} + 3s_3(\kappa_{iii}\kappa_{jj} + \kappa_{ii}\kappa_{jjj} + 2C_1\kappa_{ii}\kappa_{jj}) \\ &\quad + s_4[4(\kappa_{iiii}\kappa_{jj} + \kappa_{ii}\kappa_{jjjj}) + 6\kappa_{iii}\kappa_{jjj} + 12(C_1^2 + C_2)\kappa_{ii}\kappa_{jj} \\ &\qquad + 12C_1(\kappa_{iii}\kappa_{jj} + \kappa_{ii}\kappa_{jjj})] + \dots , \end{aligned} \tag{3.35}$$

$$\tilde{\Delta}_{\mathrm{s,LE}}\kappa_{ijk} = \tilde{\Delta}_{\mathrm{s,LE}}\kappa_{i,jk} = 6s_3\kappa_{ii}\kappa_{jj}\kappa_{kk} + \dots , \tag{3.36}$$

$$\begin{aligned}\tilde{\Delta}_{\mathrm{s,LE}}\kappa_{iij} &= \tilde{\Delta}_{\mathrm{s,LE}}\kappa_{j,ii} = 2s_2\kappa_{iii}\kappa_{jj} \\ &\quad + 3s_3[2\kappa_{ii}^2\kappa_{jj} + \kappa_{iii}\kappa_{jjj} + \kappa_{iiii}\kappa_{jj} + 2C_1\kappa_{iii}\kappa_{jj}] + \dots , \end{aligned} \tag{3.37}$$

$$\tilde{\Delta}_{\mathrm{s,LE}}\kappa_{i,ij} = 6s_3\kappa_{ii}^2\kappa_{jj} + \dots , \tag{3.38}$$

$$\tilde{\Delta}_{\mathrm{s,LE}}\kappa_{ijkl} = \tilde{\Delta}_{\mathrm{s,LE}}\kappa_{i,jkl} = \tilde{\Delta}_{\mathrm{s,LE}}\kappa_{ij,kl} = 24s_4\kappa_{ii}\kappa_{jj}\kappa_{kk}\kappa_{ll} + \dots , \tag{3.39}$$

where i, j, k, l are pairwise distinct indices. It may be noted, that these equations follow from (3.16b,c,d) or (3.17) by omitting all cross-locus and cross-gamete cumulants, as is appropriate for a population in Hardy–Weinberg and linkage equilibrium.

The selection response of cross-locus and cross-gamete cumulants involving n different loci depends only on the selection coefficients $s_n, s_{n+1}, \dots$. However, unless the fitness function in (3.13) is linear, covariances will be changed by selection to nonzero values. Consequently, in successive generations higher-order cross cumulants will be driven away from zero.

3.6 A MULTIVARIATE GAUSSIAN ALLELIC DISTRIBUTION

As a simple application, we derive the weak-selection response of a Gaussian allelic distribution (for which $c_{\boldsymbol{n}} = 0$ if $|\boldsymbol{n}| > 2$) under linear and quadratic selection. Let $W(G) = s_0 + s_1 G + s_2 G^2$. For a Gaussian distribution, we have from (3.15a,b) that $S_1(\boldsymbol{n}) = 0$ if $|\boldsymbol{n}| \geq 2$, and $S_2(\boldsymbol{n}) = 0$ if $|\boldsymbol{n}| \geq 3$. Since, by (3.10), $\tilde{\Delta}_{\mathrm{s}} c_{\boldsymbol{n}} = s_1 S_1(\boldsymbol{n}) + s_2 S_2(\boldsymbol{n})$, it follows that $\tilde{\Delta}_{\mathrm{s}} c_{\boldsymbol{n}} = 0$ if $|\boldsymbol{n}| \geq 3$. Therefore, a Gaussian distribution remains Gaussian under this sort of selection.

For a quadratic fitness function, it follows from (3.16a,b) that

$$\tilde{\Delta}_{\mathrm{s}}\kappa_i = \kappa_{i\boldsymbol{.}}(s_1 + 2s_2 C_1) , \tag{3.40a}$$

$$\tilde{\Delta}_{\mathrm{s}}\kappa_{ij} = 2s_2\kappa_{i\boldsymbol{.}}\kappa_{j\boldsymbol{.}} . \tag{3.40b}$$

For linear fitness, this simplifies to

$$\tilde{\Delta}_{\mathrm{s}}\kappa_i = s_1\kappa_{i\boldsymbol{.}} , \tag{3.41a}$$

$$\tilde{\Delta}_{\mathrm{s}}\kappa_{ij} = 0 , \tag{3.41b}$$

i.e., all variances and covariances remain unchanged. However, recombination leads to departures from the Gaussian shape, unless fitness is linear and the population is in linkage equilibrium, when all covariances vanish [see (4.8)]. It may also be noted from (3.40b) that quadratic selection induces linkage disequilibria.

For a normally distributed trait in a haploid asexually reproducing population, the above univariate versions of (3.40) and (3.41) (with $\kappa_i = C_1$, $\kappa_{i.} = \kappa_{ii} = C_2$, and $\kappa_{ij} = 0$ if $i \neq j$) provide the exact weak-selection recursions for linear and quadratic fitness functions.

3.7 EXPONENTIAL DIRECTIONAL SELECTION

For exponential directional selection, i.e., for the fitness function

$$W(G) = e^{sG} , \tag{3.42}$$

where s is a measure of the strength of directional selection, simple and explicit recursions can be derived for arbitrarily strong selection. If the initial distribution is in linkage equilibrium, the dynamics can be solved explicitly. In the additive model, exponential selection leads to multiplicative fitnesses of genotypes. Because of the specific form of the fitness function, the following direct method can be applied.

Let Ψ and Ψ_s denote the multivariate cumulant generating functions of the distribution $P(\boldsymbol{z})$ of genotypes before and after selection, respectively, $c_{\boldsymbol{n}}$ and $c_{\boldsymbol{n},s}$ the corresponding cumulants, and $\mathbf{1}$ a vector of 1's. Then we obtain from (3.2)

$$\Psi_s(\boldsymbol{\zeta}) = \Psi(\boldsymbol{\zeta} + s\mathbf{1}) - \ln \overline{W} , \tag{3.43}$$

where $\overline{W} = \int e^{s\boldsymbol{z}} P(\boldsymbol{z})\, \mathrm{d}\boldsymbol{z}$ is the mean fitness. From the identity $D_{\zeta}^{\boldsymbol{n}}(\boldsymbol{\zeta} + s\mathbf{1})^{\boldsymbol{k}} = \frac{\boldsymbol{k}!}{(\boldsymbol{k}-\boldsymbol{n})!}(\boldsymbol{\zeta} + s\mathbf{1})^{\boldsymbol{k}-\boldsymbol{n}}$ and the definition of the cumulants, we derive

$$\begin{aligned} c_{\boldsymbol{n},s} &= (D_{\zeta}^{\boldsymbol{n}} \Psi_s)(\mathbf{0}) = (D_{\zeta}^{\boldsymbol{n}} \Psi)(s\mathbf{1}) \\ &= D_{\zeta}^{\boldsymbol{n}} \left[\sum_{\boldsymbol{k}:|\boldsymbol{k}|\geq 1} c_{\boldsymbol{k}} \frac{(\boldsymbol{\zeta} + s\mathbf{1})^{\boldsymbol{k}}}{\boldsymbol{k}!} \right] \Bigg|_{\boldsymbol{\zeta}=\mathbf{0}} \\ &= \sum_{\boldsymbol{k}:\boldsymbol{k}\geq \boldsymbol{n}} c_{\boldsymbol{k}} \frac{s^{k-n}}{(\boldsymbol{k}-\boldsymbol{n})!} \\ &= \sum_{\boldsymbol{k}\geq \mathbf{0}} c_{\boldsymbol{n}+\boldsymbol{k}} \frac{s^k}{\boldsymbol{k}!} , \end{aligned} \tag{3.44}$$

where $k = |\boldsymbol{k}|$ and $n = |\boldsymbol{n}|$. It follows immediately that exponential selection preserves Hardy–Weinberg proportions and linkage equilibrium, because if all cross-gamete or cross-locus cumulants are zero, they remain unaltered. This is well known from multiplicative fitness models (Chapters II.1.2 and II.5.3). The shape of a multivariate Gaussian allelic distribution is preserved, too, because in this case $c_{\boldsymbol{n}} = 0$ if $|\boldsymbol{n}| > 2$. This is in analogy to the weak-selection response (3.41) under a linear fitness function.

The selection response of the cumulants of the distribution $f(G)$ of breeding values is derived analogously by applying the above arguments to the univariate distribution $f(G)$. Instead of (3.44), we obtain

$$\Delta_s C_n = \sum_{k>0} \frac{s^k}{k!} C_{n+k} . \tag{3.45}$$

If the population is in Hardy–Weinberg and linkage equilibrium, then the change between generations is simply $\Delta C_n = \Delta_s C_n$, because exponential selection and recombination maintain Hardy–Weinberg and linkage equilibrium. A straightforward calculation yields the following explicit solution in terms of the initial distribution:

$$C_n(t) = (D^n \Psi_0)(st) = \sum_{k>0} \frac{(st)^k}{k!} C_{n+k}(0) , \qquad (3.46)$$

where Ψ_0 denotes the cumulant generating function of the initial distribution. It is left to the reader to verify that (3.46) also provides the solution for the corresponding continuous-time model with $W(G) = sG$.

*4. RECOMBINATION AND SELECTION

We shall now study the effects of recombination and linkage disequilibrium. After settling the notation and presenting the recursion equations, we shall investigate the relations between various distinct measures of linkage disequilibrium. Finally, we explore the evolution to and in quasi-linkage equilibrium.

Our approach is based on that of Turelli and Barton (1990, 1994) and Barton and Turelli (1991), who extended Lande's (1976) analysis to higher moments (cf. also Bulmer 1980). We assume that recombination acts on the distribution of allelic effects after selection. In the present context, the effect of recombination is best described in terms of central (or non-central) moments. Since the selection response, both on the level of genotypes and phenotypes, is best described in terms of multivariate cumulants, and since no simple recursions for the cumulants under recombination are available, cumulants after selection have to be converted to moments upon which the recombination recursions are applied. Then the moments after recombination are converted back to cumulants. This somewhat complicated procedure is easily and efficiently automated using a formula manipulation program such as *Mathematica* (Wolfram 1996). It is also justified by the significant advantages the cumulants bring about in describing the dynamics of the distribution of breeding values or deviations from normality.

We shall continue in using the multi-index notation. However, throughout this section $\boldsymbol{j}$, $\boldsymbol{k}$, $\boldsymbol{n}$, etc., denote multi-indices of length ℓ (ℓ the number of loci), and thus refer to gametes and not to diploid genotypes. Hence, $c_{\boldsymbol{n}}$ denotes the multivariate cumulant of order $\boldsymbol{n}$ of the frequency distribution of gametes (paternal or maternal, since we assume no sex differences). Cross-gamete cumulants (cumulants of the distribution of diploid genotypes) are always denoted by $c_{\boldsymbol{j},\boldsymbol{k}}$, etc., and $c_{\boldsymbol{j},\boldsymbol{0}} = c_{\boldsymbol{0},\boldsymbol{j}} = c_{\boldsymbol{j}}$. Since we assume random mating and since cumulants satisfy (D.15), all true cross-gamete cumulants among offspring are zero, i.e., $c_{\boldsymbol{j},\boldsymbol{k}} = 0$, unless $\boldsymbol{k} = \boldsymbol{0}$ or $\boldsymbol{j} = \boldsymbol{0}$.

Linkage disequilibria are quantified by cumulants $c_{\boldsymbol{j}}$, where $\boldsymbol{j}$ is a multi-index of length ℓ with two or more positive entries. By property (D.15) and as already mentioned above, the loci in a subset S of L are in linkage equilibrium if all cumulants $c_{\boldsymbol{n}}$ vanish that satisfy $n_i = 0$ if $i \notin S$ and $n_i > 0$ for at least two $i \in S$.

Similarly, $m_{\boldsymbol{n}}$ denotes the central moment of order $\boldsymbol{n}$ of the gametic frequency distribution, and $m_{\boldsymbol{j},\boldsymbol{k}}$ is a cross-gamete central moment. Moments and cumulants after selection, but before recombination, are designated by the superscript $^{(s)}$, e.g., $m_{\boldsymbol{n}}^{(s)}$, and after selection and recombination by the superscript $^{(r)}$, e.g., $m_{\boldsymbol{n}}^{(r)}$. If there is only selection and recombination, but no mutation, then $m_{\boldsymbol{n}}^{(r)}$ is already the value in the newly produced generation.

A recombination event splits the genome into two disjoint sets. For two multi-indices $\boldsymbol{j}$ and $\boldsymbol{k}$, let $r_{\boldsymbol{j},\boldsymbol{k}}$ denote the frequency of recombination events that separate the loci referred to in $\boldsymbol{j}$ from those in $\boldsymbol{k}$.[3] By symmetry, $r_{\boldsymbol{j},\boldsymbol{k}} = r_{\boldsymbol{k},\boldsymbol{j}}$ so we have to consider only one of $r_{\boldsymbol{j},\boldsymbol{k}}$ and $r_{\boldsymbol{k},\boldsymbol{j}}$. If, for one or more loci, $\boldsymbol{j}$ and $\boldsymbol{k}$ have positive entries in common (i.e., $\boldsymbol{j} \bullet \boldsymbol{k} = \sum_{i=1}^{\ell} j_i k_i > 0$), then $r_{\boldsymbol{j},\boldsymbol{k}} = 0$. Furthermore, if $\boldsymbol{j}$ and $\boldsymbol{j}'$ have exactly the same zero entries, but possibly different positive entries, and the same holds for $\boldsymbol{k}$ and $\boldsymbol{k}'$, then $r_{\boldsymbol{j},\boldsymbol{k}} = r_{\boldsymbol{j}',\boldsymbol{k}'}$. For instance, if there are three loci, then $r_{(2,0,3),(0,2,0)} = r_{(1,0,1),(0,1,0)}$. For any multi-index $\boldsymbol{n}$ with at least two positive entries, we define the total recombination frequency by

$$r_{\boldsymbol{n}} = \tfrac{1}{2} \sum_{\boldsymbol{j}:\boldsymbol{0}<\boldsymbol{j}<\boldsymbol{n}} r_{\boldsymbol{j},\boldsymbol{n}-\boldsymbol{j}} \,. \tag{4.1}$$

Thus, $r_{\boldsymbol{n}}$ is the probability that recombination separates any of the loci referred to in $\boldsymbol{n}$, and the frequency of gametes that are 'nonrecombinant' with respect to all loci is $r_{\boldsymbol{n},\boldsymbol{0}} = 1 - r_{\boldsymbol{n}}$. If $\boldsymbol{n}$ has only one positive entry, e.g., $\boldsymbol{n} = n\boldsymbol{e}_i$, we define $r_{\boldsymbol{n}} = 0$.

As for the cumulants, we shall use the short-hand notation r_{ij}, r_{ijk}, $r_{i,jk}$, etc. Thus, if $\boldsymbol{j} = \boldsymbol{e}_i + \boldsymbol{e}_j$ and $\boldsymbol{k} = \boldsymbol{e}_k$, we write $r_{ij,k} = r_{\boldsymbol{j},\boldsymbol{k}}$ for the frequency of all recombination events in which loci i and j are derived from one haploid genome and k is derived from the other. From (4.1) we have, for instance,

$$r_{ij} = r_{i,j} \,, \tag{4.2a}$$

$$r_{ijk} = r_{i,jk} + r_{j,ki} + r_{k,ij} \,, \tag{4.2b}$$

where r_{ij} is the recombination frequency between loci i and j. Furthermore, as a direct consequence of the definition of recombination frequencies, the identities

$$r_{ij} = r_{i,jk} + r_{j,ki} \,, \tag{4.3a}$$

$$r_{k,ij} = r_{ijk} - r_{ij} \,, \tag{4.3b}$$

$$2r_{ijk} = r_{ij} + r_{jk} + r_{ki} \tag{4.3c}$$

hold.

In the terminology of Chapter II.2, we have for any subset $I \subseteq L$ and all pairs of multi-indices $\boldsymbol{j}$ and $\boldsymbol{k}$ such that $\boldsymbol{j}$ refers to I and $\boldsymbol{k}$ to $L \setminus I$:

$$r_I = r_{\boldsymbol{j},\boldsymbol{k}} \,.$$

Similarly, for any nonempty proper subset Q of S, $S \subseteq L$, and all $\boldsymbol{j}, \boldsymbol{k}$ such that $\boldsymbol{j}$ refers to Q and $\boldsymbol{k}$ refers to $S \setminus Q$, we have

$$r_Q^{(S)} = r_{\boldsymbol{j},\boldsymbol{k}} \,;$$

[3] For a subset $I \subseteq L = \{1, \ldots, \ell\}$, we say that a multi-index $\boldsymbol{j}$ refers to the loci in I if $j_i \neq 0$ for every $i \in I$, and $j_i = 0$, otherwise.

cf. II(4.8). Clearly, r_{ij} has the same meaning in both terminologies. Finally, $r_{\boldsymbol{n}} = r_{\text{tot}}^{(N)}$, where N is the set of loci i with $n_i > 0$.

Newly produced gametes carrying genes at a set of loci specified by the positive entries of $\boldsymbol{n}$ are nonrecombinant with probability $1 - r_{\boldsymbol{n}}$; they occur through recombination between the loci referred to in the 'parental multi-indices' $\boldsymbol{j}$ and $\boldsymbol{n} - \boldsymbol{j}$ with probability $r_{\boldsymbol{j},\boldsymbol{n}-\boldsymbol{j}}$. Taking the expectation of $(\boldsymbol{X} - \overline{\boldsymbol{X}})^{\boldsymbol{n}}$, one obtains the change of the central moments [cf. (D.5)] caused by recombination:

$$m_{\boldsymbol{n}}^{(\mathrm{r})} = \tfrac{1}{2} \sum_{\boldsymbol{j}:0\leq \boldsymbol{j}\leq \boldsymbol{n}} r_{\boldsymbol{j},\boldsymbol{n}-\boldsymbol{j}}\, m_{\boldsymbol{j},\boldsymbol{n}-\boldsymbol{j}}^{(\mathrm{s})} \tag{4.4a}$$

$$= (1 - r_{\boldsymbol{n}}) m_{\boldsymbol{n}}^{(\mathrm{s})} + \tfrac{1}{2} \sum_{\boldsymbol{j}:0< \boldsymbol{j}< \boldsymbol{n}} r_{\boldsymbol{j},\boldsymbol{n}-\boldsymbol{j}}\, m_{\boldsymbol{j},\boldsymbol{n}-\boldsymbol{j}}^{(\mathrm{s})} \,. \tag{4.4b}$$

Because these are central moments and $r_{\boldsymbol{j},\boldsymbol{n}-\boldsymbol{j}} = 0$ if $\boldsymbol{j} \bullet (\boldsymbol{n} - \boldsymbol{j}) > 0$, the summation actually extends only over those indices $\boldsymbol{j}$ satisfying $2 \leq |\boldsymbol{j}| \leq |\boldsymbol{n}| - 2$ and $\boldsymbol{j} \bullet (\boldsymbol{n} - \boldsymbol{j}) = 0$ (cf. Turelli and Barton 1990).

Since recombination does not alter the means κ_i, the same recursion holds for moments about an arbitrary point. For within-locus moments, i.e. $\boldsymbol{n} = n\boldsymbol{e}_i$, (4.4) reduces to $m_{n\boldsymbol{e}_i}^{(\mathrm{r})} = m_{n\boldsymbol{e}_i}^{(\mathrm{s})}$, reflecting the well known fact that recombination does not alter allele frequencies. It follows that

$$\Delta c_{n\boldsymbol{e}_i} = \Delta_{\mathrm{s}} c_{n\boldsymbol{e}_i} \,, \tag{4.5}$$

where $\Delta = \Delta_{\mathrm{s,r}}$ denotes the change across generations under the combined action of selection and recombination. The relation (4.5) holds in particular for κ_i and κ_{ii}.

From (4.4), the recursion relations for the change of the cumulants between generations are obtained using (D.10) and (D.12) (or, rather, the corresponding version for central moments). This produces complex equations for the change of multivariate cumulants across generations which, for cumulants up to order four, are given by

$$\Delta\kappa_{ij} = -r_{ij}\kappa_{ij} + (1 - r_{ij})\Delta_{\mathrm{s}}\kappa_{ij} + r_{ij}\Delta_{\mathrm{s}}\kappa_{i,j} \,, \tag{4.6}$$

$$\begin{aligned}\Delta\kappa_{ijk} &= -r_{ijk}\kappa_{ijk} + (1 - r_{ijk})\Delta_{\mathrm{s}}\kappa_{ijk} \\ &\quad + r_{i,jk}\Delta_{\mathrm{s}}\kappa_{i,jk} + r_{j,ki}\Delta_{\mathrm{s}}\kappa_{j,ki} + r_{k,ij}\Delta_{\mathrm{s}}\kappa_{k,ij} \,,\end{aligned} \tag{4.7}$$

$$\begin{aligned}\Delta\kappa_{ijkl} &= -r_{ijkl}\kappa_{ijkl} + (1 - r_{ijkl})\Delta_{\mathrm{s}}\kappa_{ijkl} \\ &\quad + (r_{i,jkl}\Delta_{\mathrm{s}}\kappa_{i,jkl} + {*}{*}{*}) + (r_{ij,kl}\Delta_{\mathrm{s}}\kappa_{ij,kl} + {*}{*}) \\ &\quad + (\kappa_{ij} + \Delta_{\mathrm{s}}\kappa_{ij})(\kappa_{kl} + \Delta_{\mathrm{s}}\kappa_{kl})R(ij;kl) + {*}{*} \\ &\quad + \Delta_{\mathrm{s}}\kappa_{i,j}(\kappa_{kl} + \Delta_{\mathrm{s}}\kappa_{kl})R(i,j;kl) + {*}{*}{*}{*}{*} \\ &\quad + \Delta_{\mathrm{s}}\kappa_{i,j}\Delta_{\mathrm{s}}\kappa_{k,l}R(i,j;k,l) + {*}{*} \,,\end{aligned} \tag{4.8}$$

where $**$ ($***$, $*****$) denotes two (three, five) similar terms with permuted subscripts and $R(ij;kl) = (1-r_{ijkl})+r_{ij,kl}-(1-r_{ij})(1-r_{kl})$, $R(i,j;kl) = r_{i,jkl}+r_{j,ikl}-r_{ij}(1-r_{kl})$, $R(i,j;k,l) = r_{ik,jl} + r_{il,jk} - r_{ij}r_{kl}$. The factor $R(ij;kl)$ is a measure of association between the pair of loci $\{ij\}$ and the pair $\{kl\}$; it is zero if the event that $\{ij\}$ stay together is independent of the event that $\{kl\}$ stay together. This will be true if the pairs do not overlap on the same chromosome (e.g., if they are in the order i, j, k, l, but not i,k,j,l). The same holds for $R(i,j;kl)$ and $R(i,j;k,l)$. Thus, in the last and third but last line of (4.8), one of the three terms is zero, whereas in the second but last line, two of the six terms are zero. Only for free recombination are all these factors zero and a simple recursion for the cumulants can be derived (see below).

With the symbolic language *Mathematica* (Wolfram 1996), a procedure is easily implemented to calculate efficiently the recursion relations for higher-order cumulants (*Mathematica* notebooks may be obtained from the author on request).

The change of the cumulants under weak selection and recombination, $\tilde{\Delta} = \tilde{\Delta}_{\mathrm{s,r}}$, is given by equations formally identical to (4.5)–(4.8), with Δ_{s} replaced by $\tilde{\Delta}_{\mathrm{s}}$ and terms like $-r_{ij}\kappa_{ij}$ replaced by $-\overline{W}r_{ij}\kappa_{ij}$. Then, by (3.8), Δ is related to $\tilde{\Delta}$ according to

$$\overline{W}\Delta\kappa_i = \tilde{\Delta}\kappa_i \,, \tag{4.9}$$

$$\overline{W}\Delta\kappa_{ij} = \tilde{\Delta}\kappa_{ij} - \overline{W}\Delta\kappa_i\Delta\kappa_j \,, \tag{4.10}$$

$$\begin{aligned}\overline{W}\Delta\kappa_{ijk} = {} & \tilde{\Delta}\kappa_{ijk} - \Delta\kappa_i(\tilde{\Delta}\kappa_{jk} + \overline{W}r_{jk}\kappa_{jk}) \\ & - \Delta\kappa_j(\tilde{\Delta}\kappa_{ki} + \overline{W}r_{ki}\kappa_{ki}) - \Delta\kappa_k(\tilde{\Delta}\kappa_{ij} + \overline{W}r_{ij}\kappa_{ij}) \\ & + 2\overline{W}\Delta\kappa_i\Delta\kappa_j\Delta\kappa_k \,. \end{aligned} \tag{4.11}$$

Although the above equations provide explicit recursions for the change of the cumulants across generations, no closed system of equations is obtained, because this change depends on cumulants of higher order in the previous generation. Therefore, additional assumptions are needed to justify neglecting higher-order cumulants. Such assumptions are, for example, that at each locus there is only a finite number of alleles, because then only finitely many cumulants are needed to specify the distribution, or assumptions about the form of the distribution. This will be discussed in more detail in connection with specific applications.

The recursion relations for *pure recombination* are not as simple as (4.6) and (4.7) might suggest. Denoting cumulants after recombination by $\kappa^{(\mathrm{r})}$, we find

$$\kappa_{ij}^{(\mathrm{r})} = (1 - r_{ij})\kappa_{ij} \quad \text{and} \quad \kappa_{ijk}^{(\mathrm{r})} = (1 - r_{ijk})\kappa_{ijk} \tag{4.12}$$

if Hardy–Weinberg proportions are assumed before recombination. With four loci, we obtain from (4.8), by setting all selection responses $\Delta_{\mathrm{s}} = 0$ and assuming that the order of loci along the chromosome is i, j, k, l:

$$\kappa_{ijkl}^{(\mathrm{r})} = (1 - r_{ijkl})\kappa_{ijkl} + R(ik;jl)\kappa_{ik}\kappa_{jl} + R(il;jk)\kappa_{il}\kappa_{jk} \,. \tag{4.13}$$

If each of the pairs of loci i,j and k,l is completely linked, then (4.13) reduces to $\kappa_{iikk}^{(\mathrm{r})} = (1 - r_{ik})(\kappa_{iikk} + 2r_{ik}\kappa_{ik}^2)$, indicating that even for two loci the recursion relations for the higher cumulants become nonlinear. (In a similar way, $c_{\boldsymbol{n}}^{(\mathrm{r})}$ can be computed for any $\boldsymbol{n}$ from $c_{\tilde{\boldsymbol{n}}}^{(\mathrm{r})}$, where $\tilde{\boldsymbol{n}}$ is the multi-index of length $|\boldsymbol{n}|$ all whose entries are 1.) Thus, cumulants do not appear to be the ideal set of measures of linkage disequilibrium. It can be shown, however, that all cumulants referring to more than one locus decay to zero at a geometric rate (see Section 4.3). This is not the case for the moments.

For *free recombination* among loci, the following explicit recursion relations hold for cumulants $c_{\boldsymbol{s}}$, where all entries of $\boldsymbol{s}$ are 0 or 1:

$$c_{\boldsymbol{s}}^{(\mathrm{r})} = 2^{-|\boldsymbol{s}|} \sum_{\boldsymbol{j}:\boldsymbol{0}\leq\boldsymbol{j}\leq\boldsymbol{s}} c_{\boldsymbol{j},\boldsymbol{s}-\boldsymbol{j}}^{(\mathrm{s})} \,. \tag{4.14a}$$

Here, $c_{\boldsymbol{j},\boldsymbol{s}-\boldsymbol{j}}^{(\mathrm{s})}$ denotes a cumulant before recombination that is not assumed to be zero (for example because it is measured after selection or after non-random mating, i.e., Hardy–Weinberg proportions are not assumed before recombination). We note that in the summation, the relation $\boldsymbol{j} \bullet (\boldsymbol{s}-\boldsymbol{j}) = 0$ is automatically satisfied because of the assumption on $\boldsymbol{s}$. The proof of (4.14a) is given in Section 4.1. If random mating is assumed, then (4.14a) shows that the change between generations is

$$\Delta c_{\boldsymbol{s}} = -(1-2^{-|\boldsymbol{s}|+1}) c_{\boldsymbol{s}} + 2^{-|\boldsymbol{s}|} \sum_{\boldsymbol{j}:\boldsymbol{0}<\boldsymbol{j}<\boldsymbol{s}} \Delta_{\mathrm{s}} c_{\boldsymbol{j},\boldsymbol{s}-\boldsymbol{j}} \,, \tag{4.14b}$$

because $c_{\boldsymbol{j},\boldsymbol{s}-\boldsymbol{j}} = 0$. For the pure recombination model, in which Hardy–Weinberg proportions obtain before recombination, (4.14) yields

$$c_{\boldsymbol{s}}^{(\mathrm{r})} = 2^{-|\boldsymbol{s}|+1} c_{\boldsymbol{s}} \,. \tag{4.15}$$

*4.1 PROOF OF (4.14A)

The proof is by induction on $|\boldsymbol{s}|$ and uses

$$r_{\boldsymbol{j},\boldsymbol{s}-\boldsymbol{j}} = \begin{cases} 2^{-|\boldsymbol{s}|+1} \,, & \text{if } \boldsymbol{0} \leq \boldsymbol{j} \leq \boldsymbol{s} \text{ and } \boldsymbol{j} \bullet (\boldsymbol{s}-\boldsymbol{j}) = 0 \,, \\ 0 \,, & \text{otherwise} \,, \end{cases} \tag{4.16}$$

which holds because we assume free recombination. Employing in successive lince (D.10); (4.4a) with moments about zero, (4.16), and induction (i.e., (4.14a) applied to $c_{\boldsymbol{k}_i}^{(\mathrm{r})}$); and again (D.10), we obtain

$$\begin{aligned} c_{\boldsymbol{s}}^{(\mathrm{r})} &= m_{\boldsymbol{s}}^{0\,(\mathrm{r})} - \sum_{\nu=2}^{|\boldsymbol{s}|} \sum_{\boldsymbol{k}_1+\cdots+\boldsymbol{k}_\nu=\boldsymbol{s}} \prod_{i=1}^{\nu} c_{\boldsymbol{k}_i}^{(\mathrm{r})} \\ &= 2^{-|\boldsymbol{s}|} \sum_{\boldsymbol{j}:\boldsymbol{0}\leq\boldsymbol{j}\leq\boldsymbol{s}} m_{\boldsymbol{j},\boldsymbol{s}-\boldsymbol{j}}^{0\,(\mathrm{s})} - \sum_{\nu=2}^{|\boldsymbol{s}|} \sum_{\boldsymbol{k}_1+\cdots+\boldsymbol{k}_\nu=\boldsymbol{s}} \prod_{i=1}^{\nu} 2^{-|\boldsymbol{k}_i|} \sum_{\boldsymbol{l}_i:\boldsymbol{0}\leq\boldsymbol{l}_i\leq\boldsymbol{k}_i} c_{\boldsymbol{l}_i,\boldsymbol{k}_i-\boldsymbol{l}_i}^{(\mathrm{s})} \\ &= 2^{-|\boldsymbol{s}|} \sum_{\boldsymbol{j}:\boldsymbol{0}\leq\boldsymbol{j}\leq\boldsymbol{s}} c_{\boldsymbol{j},\boldsymbol{s}-\boldsymbol{j}}^{(\mathrm{s})} + 2^{-|\boldsymbol{s}|} \sum_{\boldsymbol{j}:\boldsymbol{0}\leq\boldsymbol{j}\leq\boldsymbol{s}} \sum_{\nu=2}^{|\boldsymbol{s}|} \sum_{\boldsymbol{h}_1+\cdots+\boldsymbol{h}_\nu=(\boldsymbol{j},\boldsymbol{s}-\boldsymbol{j})} \prod_{i=1}^{\nu} c_{\boldsymbol{h}_i}^{(\mathrm{s})} \\ &\quad - \sum_{\nu=2}^{|\boldsymbol{s}|} \sum_{\boldsymbol{k}_1+\cdots+\boldsymbol{k}_\nu=\boldsymbol{s}} \sum_{\boldsymbol{r}_1:\boldsymbol{0}\leq\boldsymbol{r}_1\leq\boldsymbol{k}_1} \cdots \sum_{\boldsymbol{r}_\nu:\boldsymbol{0}\leq\boldsymbol{r}_\nu\leq\boldsymbol{k}_\nu} 2^{-\sum_i|\boldsymbol{k}_i|} \prod_{i=1}^{\nu} c_{\boldsymbol{r}_i,\boldsymbol{k}_i-\boldsymbol{r}_i}^{(\mathrm{s})} \,, \end{aligned} \tag{4.17}$$

where the $\boldsymbol{h}_i$ are multi-indices of length 2ℓ. Thus, since $|\boldsymbol{s}| = \sum_{i=1}^{\nu} |\boldsymbol{k}_i|$, it suffices to prove that for every ν with $2 \leq \nu \leq |\boldsymbol{s}|$,

$$\sum_{\boldsymbol{j}:\boldsymbol{0}\leq\boldsymbol{j}\leq\boldsymbol{s}} \; \sum_{\boldsymbol{h}_1+\cdots+\boldsymbol{h}_\nu=(\boldsymbol{j},\boldsymbol{s}-\boldsymbol{j})} \; \prod_{i=1}^{\nu} c_{\boldsymbol{h}_i}^{(\mathrm{s})} = \sum_{\boldsymbol{k}_1+\cdots+\boldsymbol{k}_\nu=\boldsymbol{s}} \; \sum_{\boldsymbol{r}_1:\boldsymbol{0}\leq\boldsymbol{r}_1\leq\boldsymbol{k}_1} \cdots \sum_{\boldsymbol{r}_\nu:\boldsymbol{0}\leq\boldsymbol{r}_\nu\leq\boldsymbol{k}_\nu} \; \prod_{i=1}^{\nu} c_{\boldsymbol{r}_i,\boldsymbol{k}_i-\boldsymbol{r}_i}^{(\mathrm{s})} . \tag{4.18}$$

The left-hand side of (4.18) can be written as

$$\sum_{\boldsymbol{j}:\boldsymbol{0}\leq\boldsymbol{j}\leq\boldsymbol{s}} \; \sum_{(\boldsymbol{r}_1,\boldsymbol{t}_1)+\cdots+(\boldsymbol{r}_\nu,\boldsymbol{t}_\nu)=(\boldsymbol{j},\boldsymbol{s}-\boldsymbol{j})} \; \prod_{i=1}^{\nu} c_{\boldsymbol{r}_i,\boldsymbol{t}_i}^{(\mathrm{s})} , \tag{4.19}$$

where $(\boldsymbol{r}_1, \boldsymbol{t}_1) + \cdots + (\boldsymbol{r}_\nu, \boldsymbol{t}_\nu)$ is a partition of $(\boldsymbol{j}, \boldsymbol{s} - \boldsymbol{j})$, and $\boldsymbol{r}_i \bullet \boldsymbol{t}_j = 0$ for every i and j (because $\boldsymbol{j} \bullet (\boldsymbol{s} - \boldsymbol{j}) = 0$). The right-hand side of (4.18) equals

$$\sum_{(\boldsymbol{r}_1+\boldsymbol{t}_1)+\cdots+(\boldsymbol{r}_\nu+\boldsymbol{t}_\nu)=\boldsymbol{s}} \; \prod_{i=1}^{\nu} c_{\boldsymbol{r}_i,\boldsymbol{t}_i}^{(\mathrm{s})} , \tag{4.20}$$

where partitions differing only by the order of terms $(\boldsymbol{r}_i + \boldsymbol{t}_i)$ are counted once and $\boldsymbol{r}_i \bullet \boldsymbol{t}_i = 0$ for every i. Since all the components of $\boldsymbol{s}$ are 0 or 1, we have $(\boldsymbol{r}_i + \boldsymbol{t}_i) \bullet (\boldsymbol{r}_j + \boldsymbol{t}_j) = 0$ in (4.19) and (4.20) for every pair $i \neq j$ with $1 \leq i, j \leq \nu$. Therefore, each term $\prod_{i=1}^{\nu} c_{\boldsymbol{r}_i,\boldsymbol{t}_i}^{(\mathrm{s})}$ occurs through permutations of $\boldsymbol{r}_i$ with $\boldsymbol{t}_i$ precisely 2^ν times in (4.19) and in (4.20), whence equality of (4.19) and (4.20) follows.

*4.2 RELATIONS BETWEEN MEASURES OF LINKAGE DISEQUILIBRIUM

Several distinct sets of measures of multilocus linkage disequilibrium may be found in the literature (e.g., Bennett 1954; Slatkin 1972; Turelli and Barton 1990; Lyubich 1992; Nagylaki 1993; Christiansen 1987, 1999; Baake 2000). Here, we shall provide a brief account of some of their relations and assume a finite number of alleles per locus.

Among the best known sets of such measures is that of Slatkin (1972), which was defined for diallelic loci. Several of their properties have been explored by Christiansen (1987, 1999), who also introduced another, closely related set.

Nagylaki (1993) generalized Slatkin's (1972) linkage disequilibria to multiple alleles by defining, for a given gamete $\boldsymbol{x} = (x_1, \ldots, x_\ell)$, the following allelic indicator random variables:

$$\chi_{x_i}^{(i)} = \begin{cases} 1 & \text{if } \mathcal{A}_{x_i}^{(i)} \text{ is the allele at locus } i\,, \\ 0 & \text{otherwise}\,. \end{cases} \tag{4.21}$$

Then, for a given subset of loci, $S \subseteq L$, linkage disequilibria are defined by

$$\Delta_{\boldsymbol{x}_S}^{(S)} = \mathrm{E}\left[\prod_{i\in S}\left(\chi_{x_i}^{(i)} - p_{x_i}^{(i)}\right)\right] , \tag{4.22}$$

where, as in Chapter II.4, $\boldsymbol{x}_S$ denotes the vector with components x_i for every i in S. In contrast to the linkage disequilibria $\Theta^{(S)}_{\boldsymbol{x}_S}$ defined in II(4.6), these satisfy $\sum_{x_i} \Delta^{(S)}_{\boldsymbol{x}_S} = 0$ for every $i \in S$. It can further be shown that the linkage disequilibria $\Delta^{(S)}_{\boldsymbol{x}_S}$ are related to the linkage disequilibria $\theta^{(S)}_{\boldsymbol{x}_S}$, defined in II(6.6), by

$$\Delta^{(S)}_{\boldsymbol{x}_S} = \sum_{R \subseteq S} (-1)^{|T|} \theta^{(R)}_{\boldsymbol{x}_R} \prod_{i \in T} p^{(i)}_{\boldsymbol{x}_i} , \tag{4.23}$$

where R is an arbitrary subset of S, $T = S \setminus R$, and $|T|$ is the number of elements (loci) in T (Nagylaki 1993). For two and three loci this gives

$$\begin{aligned} \Delta^{(ij)}_{x_i x_j} &= \theta^{(ij)}_{x_i x_j} = D^{(ij)}_{x_i x_j} , \\ \Delta^{(ijk)}_{x_i x_j x_k} &= \theta^{(ijk)}_{x_i x_j x_k} - p^{(i)}_{x_i} \theta^{(jk)}_{x_j x_k} - p^{(j)}_{x_j} \theta^{(ik)}_{x_i x_k} - p^{(k)}_{x_k} \theta^{(ij)}_{x_i x_j} , \end{aligned} \tag{4.24}$$

where $D^{(ij)}_{x_i x_j}$ is the classical (marginal) linkage disequilibrium of gametes containing alleles x_i and x_j at loci i and j, respectively; cf. II(3.46).

Lyubich's (1992, p. 257) linkage disequilibria $E_K(p)$ agree with the $\theta^{(S)}_{\boldsymbol{x}_S}$ for $K = S$. The relation between the $\Theta_{\boldsymbol{x}}$ and the $\theta_{\boldsymbol{x}}$ is complicated; see II(6.8a).

For a given subset S of L, let $\boldsymbol{s}$ denote the multi-index with $s_i = 1$ if $i \in S$, and $s_i = 0$ otherwise. Further, let $\boldsymbol{\chi}_{\boldsymbol{x}}$ be the random vector $\boldsymbol{\chi}_{\boldsymbol{x}} = (\chi^{(1)}_{x_1}, \dots , \chi^{(\ell)}_{x_\ell})$. Then, using the multi-index notation $\boldsymbol{n}^{\boldsymbol{k}} = \prod_i n_i^{k_i}$, we have $\mathrm{E}[\boldsymbol{\chi}_{\boldsymbol{x}}] = (p^{(1)}_{x_1}, \dots , p^{(\ell)}_{x_\ell})$, $\mathrm{E}[\boldsymbol{\chi}^{\boldsymbol{s}}_{\boldsymbol{x}}] = p^{(S)}_{\boldsymbol{x}_S}$, and

$$\theta^{(S)}_{\boldsymbol{x}_S} = \mathrm{E}[\boldsymbol{\chi}^{\boldsymbol{s}}_{\boldsymbol{x}}] - \left(\mathrm{E}[\boldsymbol{\chi}_{\boldsymbol{x}}]\right)^{\boldsymbol{s}} ; \tag{4.25}$$

cf. II(6.6a). Similarly, the linkage disequilibrium measure $\Delta^{(S)}_{x_S}$ is obtained by taking the central moment $m_{\boldsymbol{s}}$ of $\boldsymbol{\chi}_{\boldsymbol{x}}$. Upon observing that

$$\boldsymbol{X} = \sum_{\boldsymbol{x}} \boldsymbol{x} \prod_{i=1}^{\ell} \chi^{(i)}_{x_i} = \sum_{\boldsymbol{x}} \boldsymbol{x} \boldsymbol{\chi}^{\mathbf{1}}_{\boldsymbol{x}} , \tag{4.26}$$

where $\mathbf{1} = (1, \dots , 1)$, and $\mathrm{E}[\boldsymbol{\chi}^{\mathbf{1}}_{\boldsymbol{x}}] = p_{\boldsymbol{x}} = p(\boldsymbol{x})$, we obtain for the moments of $\boldsymbol{X}$ about zero:

$$m^0_{\boldsymbol{s}} = \mathrm{E}[\boldsymbol{X}^{\boldsymbol{s}}] = \sum_{\boldsymbol{x}} \boldsymbol{x}^{\boldsymbol{s}} \theta^{(S)}_{\boldsymbol{x}_S} + \overline{\boldsymbol{X}}^{\boldsymbol{s}} . \tag{4.27}$$

Similarly,

$$m_{\boldsymbol{s}} = \mathrm{E}[(\boldsymbol{X} - \overline{\boldsymbol{X}})^{\boldsymbol{s}}] = \sum_{\boldsymbol{x}} \boldsymbol{x}^{\boldsymbol{s}} \Delta^{(S)}_{\boldsymbol{x}_S} \tag{4.28}$$

can be proved. Thus, we have derived simple relations between the different sets of moment-based measures of linkage disequilibrium.

Now let us turn to the cumulant measures of linkage disequilibrium. From (D.10) and (4.27) it follows immediately that

$$c_{\boldsymbol{s}} = \sum_{\boldsymbol{x}} \boldsymbol{x} \theta^{(S)}_{\boldsymbol{x}_S} - \sum_{\nu=2}^{|\boldsymbol{s}|-1} \sum_{\boldsymbol{k}_1 + \cdots + \boldsymbol{k}_\nu = \boldsymbol{s}} c_{\boldsymbol{k}_1} \cdot \ldots \cdot c_{\boldsymbol{k}_\nu} . \tag{4.29}$$

(It may be observed that $|\boldsymbol{s}|$ is the number of loci in the S.) As explained in Appendix D below (D.11), a formula for a cumulant $c_{\boldsymbol{n}}$ with an arbitrary multi-index $\boldsymbol{n}$ can be obtained formally from the above, by taking $\boldsymbol{s}$ to be the multi-index consisting of $|\boldsymbol{n}|$ 1's and appropriately identifying cumulants on the right-hand side that correspond to the loci referred to in $\boldsymbol{n}$. For instance, the following relations are obtained:

$$\kappa_{ij} = \sum_{x_i,x_j} x_i x_j \theta^{(ij)}_{x_i x_j} , \tag{4.30a}$$

$$\kappa_{iij} = \sum_{x_i,x_j} x_i^2 x_j \theta^{(ij)}_{x_i x_j} - 2\kappa_i \kappa_{ij} , \tag{4.30b}$$

$$\kappa_{iijj} = \sum_{x_i,x_j} x_i^2 x_j^2 \theta^{(ij)}_{x_i x_j} - 2(\kappa_i \kappa_{ijj} + \kappa_j \kappa_{iij}) - 4\kappa_i \kappa_j \kappa_{ij} - 2\kappa_{ij}^2 , \tag{4.30c}$$

$$\kappa_{ijk} = \sum_{x_i,x_j,x_k} x_i x_j x_k \theta^{(ijk)}_{x_i x_j x_k} - (\kappa_i \kappa_{jk} + \kappa_j \kappa_{ik} + \kappa_k \kappa_{ij}) , \tag{4.30d}$$

$$\begin{aligned}\kappa_{ijkl} = &\sum_{x_i,x_j,x_k,x_l} x_i x_j x_k x_l \theta^{(ijkl)}_{x_i x_j x_k x_l} - (\kappa_{ij}\kappa_{kl} + \kappa_{ik}\kappa_{jl} + \kappa_{il}\kappa_{jk}) \\ &- (\kappa_i \kappa_{jkl} + ***) - (\kappa_i \kappa_j \kappa_{kl} + *****) .\end{aligned} \tag{4.30e}$$

The stars indicate the number of additional terms obtained by permutations.

Whereas, the relation (4.29) between the cumulants and the linkage disequilibria $\theta^{(S)}_{\boldsymbol{x}_S}$ is obtained from the relation (D.10) between cumulants and moments about zero, the relation between the cumulants and the linkage disequilibria $\Delta^{(S)}_{\boldsymbol{x}_S}$ is obtained from the relation between cumulants and central moments about zero [see Appendix D, following (D.11)]. Thus, if on the right-hand side of (4.30) all terms involving means (κ_i, etc.) are omitted and each θ is replaced by a Δ, then the correct relations are obtained.

Despite their formal similarity, there is a conceptual difference between the linkage disequilibria $\Delta^{(S)}_{x_S}$ and the multivariate moments $m_{\boldsymbol{n}}$ and cumulants $c_{\boldsymbol{n}}$. The former are defined for every gamete, whereas the latter represent an overall measure of linkage disequilibrium. In particular, $\kappa_{ij} = 0$ does not imply that all gametic linkage disequilibria vanish. For this it is necessary that $m_{\boldsymbol{n}}$ factorize for multi-indices $\boldsymbol{n}$ of the form $\boldsymbol{n} = n_i \boldsymbol{e}_i + n_j \boldsymbol{e}_j$ or, equivalently, that the corresponding $c_{\boldsymbol{n}}$ vanish.

Bennett (1954) introduced principal components, $L(x_S)$, that were defined such as to satisfy $L(\boldsymbol{x}_S)^{(\prime)} = (1 - r_{\boldsymbol{s}})L(\boldsymbol{x}_S)$ for every S-locus gamete $\boldsymbol{x}_S$. However, he did not provide general formulas for these principal components but worked them out for up to six loci. Recently, Dawson (2000) provided a recursive method for obtaining explicit expressions for Bennett's principal components in terms of the moments of the indicator variables $\chi_{\boldsymbol{x}}$. He also derived the inverse transformation. Despite Slatkin's (1972) remark, for more than three loci the measures $\Delta^{(S)}_{\boldsymbol{x}_S}$ are different from Bennett's principal components, because then the latter depend on the recombination rates.

Baake (2000) found a set of measures of linkage disequilibrium that simultaneously linearize and diagonalize a system of differential equations describing evolution at an arbitrary number of loci under weak mutation and tight linkage in the sense that multiple crossovers are ignored. These linkage disequilibria, which are simpler than those of Bennett, and their decay rates are given in closed form.

As noted by Turelli and Barton (1994), and implied by (4.13), the cumulant measures of disequilibrium can in general not be principal components (in the sense of linearizing and diagonalizing the recursion relations for cumulants under recombination); however, the principal components can be expressed in terms of cumulants. For instance, with four loci

$$\begin{aligned} L(ijkl) = \kappa_{ijkl} &+ \frac{R(ik;jl)}{(1-r_{ijkl})-(1-r_{ik})(1-r_{jl})}\kappa_{ik}\kappa_{jl} \\ &+ \frac{R(il;jk)}{(1-r_{ijkl})-(1-r_{il})(1-r_{jk})}\kappa_{il}\kappa_{jk} \end{aligned} \tag{4.31}$$

is obtained if the order of loci along the chromosome is i, j, k, l, so that $R(ij;kl) = 0$. For free recombination, the cumulants of the form $c_{\boldsymbol{s}}$, $\boldsymbol{s}$ a multi-index consisting only of 0's and 1's, are principal components, as mentioned by Turelli and Barton (1994) and shown by (4.15).

*4.3 EVOLUTION TO AND IN QUASI-LINKAGE EQUILIBRIUM

First, we investigate the decay of the cumulant measures of linkage disequilibrium under recombination and selection. Then we derive approximate expressions for the dynamics of multivariate cumulants in quasi-linkage equilibrium. Apart from being of independent interest, these expressions will enable us in the next chapter to calculate equilibrium distributions of multilocus systems if selection is weak relative to recombination. Basically, we use the results from Chapter II.6 to put the approach of Turelli and Barton (1990) on a firmer mathematical basis. The key for the weak-selection, QLE approximation is to recognize that the effects of alleles at distinct loci are probabilistically dependent only because selection generates linkage disequilibrium, whereas recombination drives cross-locus cumulants quickly toward QLE values. The latter can be easily approximated.

In the following, it is assumed that at each locus only a finite number of alleles occur, which implies that the range of genotypic values is bounded. In this subsection, let $c_{\boldsymbol{s}}$ and $c_{\boldsymbol{m}}$ denote cross-locus cumulants (referring to at least two loci), so that all entries of $\boldsymbol{s}$ are zero or one, whereas $\boldsymbol{m}$ may contain higher-order entries. Arbitrary cumulants are designated by $c_{\boldsymbol{n}}$. It is assumed that $r_{\boldsymbol{m}} > 0$ for every such $\boldsymbol{m}$. Let s be a measure of the strength of selection that is small relative to all pairwise recombination rates and hence, by (4.1), relative to all $r_{\boldsymbol{m}}$. In addition, it is assumed that the selection coefficients s_k in (1.8) satisfy $\sum_k |s_k| = O(s)$ as $s \to 0$, so that the weak-selection hypothesis of Chapter II.6 is fullfilled.

First we show that the cross-locus cumulants decay to order $O(s)$ at a geometric rate. Indeed, for any two loci, i and j, one obtains from (4.30a) and II(6.9a)

$$|\kappa_{ij}(t)| \le \sum_{x_i, x_j} |x_i x_j| \, (\lambda_L^t a_{x_i x_j}^{(ij)} + s b_{x_i x_j}^{(ij)}) \, ,$$

where $a_{x_i x_j}^{(ij)}$ and $b_{x_i x_j}^{(ij)}$ are constants independent of s and t, and $\lambda_L = 1 - r_{\min}$; cf. II(2.17). Since each of the product terms in (4.29) contains at least one cumulant referring to more than one locus, induction on the number of loci and repeated application of II(6.9a) proves that cross-locus cumulants $c_{\boldsymbol{s}}$ satisfy

$$|c_{\boldsymbol{s}}(t)| \leq \sum_{\boldsymbol{x}_S} |\boldsymbol{x}^{\boldsymbol{s}}| \, (\lambda_L^t a_{\boldsymbol{x}_S}^{(S)} + s b_{\boldsymbol{x}_S}^{(S)}) \,, \tag{4.32}$$

where $a_{\boldsymbol{x}_S}^{(S)}$ and $b_{\boldsymbol{x}_S}^{(S)}$ are constants independent of s and t. Extension to arbitrary cross-locus cumulants $c_{\boldsymbol{m}}$ is straightforward. Since the number of loci and number of alleles are finite, it follows that all such cumulants decay to $O(s)$ within the short time period t_1 given by II(6.10), i.e.,

$$c_{\boldsymbol{m}} = O(s) \,, \quad t \geq t_1 \,. \tag{4.33}$$

If there is no selection, then the decay to zero occurs at the geometric rate λ_L^t; cf. Chapter II.2.2.

Once linkage disequilibria have been reduced to $O(s)$, induction on the number of loci and resorting to (4.29) and II(6.13a) implies that for any finite subset S of loci and corresponding multi-index $\boldsymbol{s}$,

$$|\Delta c_{\boldsymbol{s}}(t)| \leq s \sum_{\boldsymbol{x}_S} |\boldsymbol{x}^{\boldsymbol{s}}| \, (\lambda_L^t \tilde{a}_{\boldsymbol{x}_S}^{(S)} + s \tilde{b}_{\boldsymbol{x}_S}^{(S)}) \,, \quad t \geq t_1 \,, \tag{4.34}$$

where $\tilde{a}_{\boldsymbol{x}_S}^{(S)}$ and $\tilde{b}_{\boldsymbol{x}_S}^{(S)}$ are constants independent of s and t. By lumping together appropriate terms, (4.34) extends to arbitrary cross-locus cumulants $c_{\boldsymbol{m}}$.

As cross-locus cumulants $c_{\boldsymbol{m}}$ satisfy (4.32) and (4.33), so that products of such terms are $O(s^2)$ if $t \geq t_1$, and because cumulants refering to a single locus satisfy (3.12) and (4.5), we obtain

$$\Delta c_{\boldsymbol{n}} = \tilde{\Delta} c_{\boldsymbol{n}} + O(s^2) \,, \quad t \geq t_1 \,; \tag{4.35}$$

cf. (4.9)–(4.11). Next, we show that after $t_2 \approx 2t_1$ generations [cf. II(6.13)], the cross-locus cumulants change very slowly, i.e.,

$$\Delta c_{\boldsymbol{m}} = \tilde{\Delta} c_{\boldsymbol{m}} + O(s^2) = O(s^2) \,, \quad t \geq t_2 \,. \tag{4.36}$$

(Note, however, from (3.34)–(3.39) that even at linkage equilibrium the change caused by selection alone, $\tilde{\Delta}_S c_{\boldsymbol{m}}$, is typically of order s). Indeed, (4.29) shows that

$$\Delta c_{\boldsymbol{s}} = \sum_{\boldsymbol{x}} \boldsymbol{x} \Delta \theta_{\boldsymbol{x}_S}^{(S)} - \sum_{\nu=2}^{|\boldsymbol{s}|-1} \sum_{\boldsymbol{k}_1 + \cdots + \boldsymbol{k}_\nu = \boldsymbol{s}} [(c_{\boldsymbol{k}_1} + \Delta c_{\boldsymbol{k}_1}) \cdot \ldots \cdot (c_{\boldsymbol{k}_\nu} + \Delta c_{\boldsymbol{k}_\nu}) - c_{\boldsymbol{k}_1} \cdot \ldots \cdot c_{\boldsymbol{k}_\nu}] \,.$$

The first sum is $O(s^2)$ because of II(6.13b). The terms in brackets are $O(s^2)$ because the two products $c_{\boldsymbol{k}_1} \cdot \ldots \cdot c_{\boldsymbol{k}_\nu}$ cancel, and each other product contains either a term $c_{\boldsymbol{k}_j}$ or $\Delta c_{\boldsymbol{k}_j}$ refering to at least two loci (because $\nu < |\boldsymbol{s}|$), which are $O(s)$ and $O(s^2)$, respectively. In the second case, we are finished. In the first case, however, there is at least one other term of the form $\Delta c_{\boldsymbol{k}_i}$ in this product, which is of order O(s) (if $\boldsymbol{k}_i$ refers only to a single locus) or smaller. This proves the assertion. The extension to general cross-locus cumulants $c_{\boldsymbol{m}}$ is obvious.

From these results, we derive a general, simple, and useful approximation for the cross-locus cumulants at quasi-linkage equilibrium. On account of (4.4) and (D.12), the change of the cross-locus cumulants can be expressed as

$$\Delta c_{\boldsymbol{m}} = -r_{\boldsymbol{m}} c_{\boldsymbol{m}} + h_{\boldsymbol{m}}(\mathcal{C}, t) \ , \tag{4.37}$$

where $h_{\boldsymbol{m}}(\mathcal{C}, t)$ includes all other terms in the recursion, and $\mathcal{C}$ denotes the entire collection of cumulants. If, for $t \geq t_1$, every term $\Delta c_{\boldsymbol{k}}$ occuring in $h_{\boldsymbol{m}}$ is substituted by its weak-selection approximation $\tilde{\Delta}_{\mathrm{s}} c_{\boldsymbol{k}}$, then (4.35) shows that the error made in (4.37) by replacing $h_{\boldsymbol{m}}$ by the resulting $\tilde{h}_{\boldsymbol{m}}$ is of order $O(s^2)$. Moreover, since all cross-locus cumulants are of order s if $t \geq t_1$, they contribute only terms of order s^2 to $\tilde{h}_{\boldsymbol{m}}$, because they are always multiplied with some factor s_k. Therefore, up to an error of $O(s^2)$, $\tilde{h}_{\boldsymbol{m}}$ can be replaced by its much simpler linkage-equilibrium approximation $\tilde{h}_{\boldsymbol{m},\mathrm{LE}}$, which is the expression obtained from $\tilde{h}_{\boldsymbol{m}}$ by omitting all cross-locus cumulants. Hence, following (4.36) and setting the left-hand side of (4.37) equal to $O(s^2)$, we obtain the QLE approximation

$$c_{\boldsymbol{m}}(t) = \frac{\tilde{h}_{\boldsymbol{m},\mathrm{LE}}(\mathcal{C}, t)}{r_{\boldsymbol{m}}} + O(s^2) \ , \quad t \geq t_2 \tag{4.38}$$

(cf. Turelli and Barton 1990). Let us illustrate this procedure for cumulants of order two, i.e., for covariances between loci. At linkage equilibrium, (3.35) shows that $\tilde{\Delta}_{\mathrm{s,LE}} \kappa_{ij} = \tilde{\Delta}_{\mathrm{s,LE}} \kappa_{i,j}$. Therefore, (4.6) implies $\tilde{h}_{ij,\mathrm{LE}} = \tilde{\Delta}_{\mathrm{s,LE}} \kappa_{ij}$, where the latter expression is given by (3.35). Hence, (4.38) yields the QLE approximation

$$\kappa_{ij} = \frac{\tilde{\Delta}_{\mathrm{s,LE}} \kappa_{ij}}{r_{ij}} + O(s^2) \ , \quad t \geq t_2 \ . \tag{4.39}$$

Now we assume a quadratic fitness function, i.e., $s_k = 0$ for every $k > 2$, and $s = |s_1| + |s_2|$. Then (4.39) becomes

$$\kappa_{ij} = \frac{2 s_2 \kappa_{ii} \kappa_{jj}}{r_{ij}} + O(s^2) \ , \quad t \geq t_2. \tag{4.40}$$

For the fourth-order cumulants and a quadratic fitness function, the weak-selection version of (4.36) is obtained from (4.8), (4.2), and (4.3):

$$\begin{aligned} \tilde{\Delta} \kappa_{iijk} = & - \overline{W} r_{ijk} \kappa_{iijk} + (1 - r_{ijk}) \tilde{\Delta}_{\mathrm{s}} \kappa_{iijk} \\ & + r_{i,jk} \tilde{\Delta}_{\mathrm{s}} \kappa_{ii,jk} + r_{j,ik} \tilde{\Delta}_{\mathrm{s}} \kappa_{j,iik} + r_{k,ij} \tilde{\Delta}_{\mathrm{s}} \kappa_{k,iij} \ , \end{aligned} \tag{4.41a}$$

$$\tilde{\Delta} \kappa_{iiij} = - \overline{W} r_{ij} \kappa_{iiij} + (1 - r_{ij}) \tilde{\Delta}_{\mathrm{s}} \kappa_{iiij} + r_{ij} \tilde{\Delta}_{\mathrm{s}} \kappa_{j,iii} \ , \tag{4.41b}$$

$$\tilde{\Delta} \kappa_{iijj} = - \overline{W} r_{ij} \kappa_{iijj} + (1 - r_{ij}) \tilde{\Delta}_{\mathrm{s}} \kappa_{iijj} + r_{ij} \tilde{\Delta}_{\mathrm{s}} \kappa_{ii,jj} \ . \tag{4.41c}$$

From (3.16d), (3.17c), and (3.17d), it follows that at linkage equilibrium $\tilde{\Delta}_{\mathrm{s}} \kappa_{iijk} = \tilde{\Delta}_{\mathrm{s}} \kappa_{ii,jk} = \tilde{\Delta}_{\mathrm{s}} \kappa_{j,iik} = \tilde{\Delta}_{\mathrm{s}} \kappa_{k,iij} = 0$ holds if i, j, and k are pairwise distinct. Similarly, $\tilde{\Delta}_{\mathrm{s}} \kappa_{iiij} = \tilde{\Delta}_{\mathrm{s}} \kappa_{j,iii} = 2 s_2 \kappa_{iiii} \kappa_{jj}$ if $i \neq j$, and $\tilde{\Delta}_{\mathrm{s}} \kappa_{iijj} = \tilde{\Delta}_{\mathrm{s}} \kappa_{ii,jj} = 2 s_2 \kappa_{iii} \kappa_{jjj}$ if $i \neq j$.

Since the left-hand sides of (4.41) are $O(s^2)$ if $t \geq t_2$, and in view of (4.38), we obtain the QLE approximations

$$\kappa_{iijk} = O(s^2) , \quad t \geq t_2 , \tag{4.42a}$$

$$\kappa_{iijj} = \frac{2s_2\kappa_{iii}\,\kappa_{jjj}}{r_{ij}} + O(s^2) , \quad t \geq t_2 , \tag{4.42b}$$

$$\kappa_{iiij} = \frac{2s_2\kappa_{iiii}\,\kappa_{jj}}{r_{ij}} + O(s^2) , \quad t \geq t_2 . \tag{4.42c}$$

In a similar way, (3.36) and (3.39) combined with (4.7) and (4.8), show that at QLE and for pairwise different indices, we also have $\kappa_{ijk} = \kappa_{ijkl} = O(s^2)$. All these expressions for the cumulants depend only indirectly on the directional selection coefficient s_1, namely through the (evolution of) single-locus cumulants. The approximations (4.40) and (4.42) were found by formal arguments by Turelli and Barton (1990). Their validity requires that $s \ll r_{ij}$.

For the continuum-of-alleles model no rigorous proof of the above theory is available. Additional technical assumptions would be required. For instance, the fitness function $W(G)$ had to be a polynomial of even degree and negative leading coefficient. (If the leading coefficient is positive or of odd degree, solutions exist whose cumulants tend to infinity.) Furthermore, it had to be supposed that all solutions converged to an equilibrium state and that uniform bounds for $|S_k(\boldsymbol{n},t)|$, defined in (3.11), exist. Finally, we note that the above theory relies on the assumption of a fixed number of loci. Thus, the infinitesimal model of Fisher (1918) and Bulmer (1971a, 1980) cannot be derived rigorously from the above.

5. MUTATION

The mutation model for the multilocus setting is based on that developed in Section 2.1, Chapter IV. We suppose that mutation acts directly on gamete frequencies; therefore $\boldsymbol{x}$, $\boldsymbol{y}$, and $\boldsymbol{\xi}$ denote vectors of length ℓ, which represents the number of loci, and $\boldsymbol{n}$ is a multi-index of length ℓ. We first consider the pure-mutation model, and denote by $\mu(\boldsymbol{x})$ the mutation rate of gamete $\boldsymbol{x}$, and by $u(\boldsymbol{y},\boldsymbol{x})$ the probability (distribution) that a $\boldsymbol{y}$-gamete mutates to an $\boldsymbol{x}$-gamete, conditional on a mutation event. Then the change in the gamete frequencies caused by mutation is

$$\Delta_\mu p(\boldsymbol{x}) = \int p(\boldsymbol{y})\mu(\boldsymbol{y})u(\boldsymbol{y},\boldsymbol{x})\,\mathrm{d}\boldsymbol{y} - \mu(\boldsymbol{x})p(\boldsymbol{x}) . \tag{5.1}$$

We designate the mutation rate at locus i by μ_i (assuming that all alleles have the same mutation rate), and the (conditional) probability distribution at locus i for mutations from type y_i to type x_i by $u_i(y_i,x_i)$, so that

$$\int u_i(y_i,x_i)\,\mathrm{d}x_i = 1 \qquad \text{for all } y_i , \tag{5.2}$$

cf. IV(2.1). Furthermore, let $\delta^{(i)}(\boldsymbol{y},\boldsymbol{x})$ be the Dirac delta function (or the Kronecker delta, in the case of discrete alleles) for the vectors $(y_1,\dots,y_{i-1},y_{i+1},\dots,y_\ell)$ and $(x_1,\dots,x_{i-1},x_{i+1},\dots,x_\ell)$.

We assume that mutation acts independently at each locus and ignore the probability that two or more mutations at different loci occur simultaneously. Then we have

$$\mu(\boldsymbol{y})u(\boldsymbol{y},\boldsymbol{x}) = \sum_{i=1}^{\ell} \mu_i u_i(y_i, x_i)\delta^{(i)}(\boldsymbol{y},\boldsymbol{x}) \,. \tag{5.3}$$

Passing over to the moment generating function and assuming the random-walk mutation model, $u_i(y_i, x_i) = u_i(x_i - y_i)$, IV (2.3), we obtain from (5.1) by straightforward integration

$$\Delta_\mu \psi(\boldsymbol{\xi}) = \psi(\boldsymbol{\xi}) \sum_{i=1}^{\ell} \mu_i[\varphi_i(\xi_i) - 1] \,, \tag{5.4}$$

where φ_i denotes the moment generating function of u_i. Since mutation is (almost) always a weak evolutionary force, we can approximate the change of the cumulant generating function under mutation, $\Delta_\mu \Psi = \ln[1 + (\Delta_\mu \psi/\psi)]$, by

$$\tilde{\Delta}_\mu \Psi(\boldsymbol{\xi}) = \frac{\Delta_\mu \psi(\boldsymbol{\xi})}{\psi(\boldsymbol{\xi})} = \sum_{i=1}^{\ell} \mu_i[\varphi_i(\xi_i) - 1] \,. \tag{5.5}$$

Denoting the nth moment about zero of u_i by $v_n^{(i)}$, we obtain the mutational change of the multivariate cumulants:

$$\tilde{\Delta}_\mu c_{\boldsymbol{n}} = \begin{cases} \mu_i v_n^{(i)} & \text{if } \boldsymbol{n} = n\boldsymbol{e}_i \text{ for some } i \,, \\ 0 & \text{otherwise} \,. \end{cases} \tag{5.6}$$

Hence, all cross-locus cumulants remain unchanged by mutation under the random-walk model. This is not the case for the central moments where, for instance, $\tilde{\Delta}_\mu m_{iiij} = 3\mu_i v_2^{(i)} m_{ij}$ if $i \neq j$.

For the HC-mutation model, the change of the cumulant generating function is

$$\Delta_\mu \Psi \approx \tilde{\Delta}_\mu \Psi = \sum_{i=1}^{\ell} \mu_i \Big[\varphi_i(\xi_i) \frac{\psi(\xi_1, \dots, \xi_{i-1}, 0, \xi_{i+1}, \dots, \xi_\ell)}{\psi(\boldsymbol{\xi})} - 1\Big] \,. \tag{5.7}$$

This gives more complicated formulas for $\tilde{\Delta}_\mu c_{\boldsymbol{n}}$ than in the random-walk mutation model. For example, the mean changes according to

$$\tilde{\Delta}_\mu \kappa_i = \mu_i v_1^{(i)} - \kappa_i \,, \tag{5.8}$$

and the variance according to

$$\tilde{\Delta}_\mu \kappa_{ii} = \mu_i(v_2^{(i)} - \kappa_{ii} + 2\kappa_i^2 - 2v_1^{(i)}\kappa_i) \,. \tag{5.9}$$

To combine selection, recombination, and mutation, we introduce a further simplification by ignoring the terms arising from the interaction of mutation with selection and recombination. Thus, if $\tilde{\Delta}_{\text{s,r}}$ is the change attributable to weak selection and recombination, we approximate $\tilde{\Delta}$, the weak-selection weak-mutation change between generations, by

$$\tilde{\Delta} c_{\boldsymbol{n}} = \tilde{\Delta}_{\text{s,r}} c_{\boldsymbol{n}} + \tilde{\Delta}_\mu c_{\boldsymbol{n}} \,, \tag{5.10}$$

where $\tilde{\Delta}_{\text{s,r}} c_{\boldsymbol{n}}$ was computed in Section 4.

6. THE DYNAMICS OF PHENOTYPES

For the additive model we are now able to extend the results of Chapter II.5.1 by deriving recursion relations not only for mean, but also for the variance, and the first few cumulants of the distribution of phenotypic values that take into account selection, recombination, and mutation. If the fitness function can be approximated by a polynomial of low degree, then fairly simple recursions are obtained. Because of the additivity property of the cumulants for sums of independent random variables (Appendix D), and because of the assumed properties of the distribution of environmental effects, the change of the cumulants of the distribution of breeding values is identical to the change of the cumulants of the phenotypic distribution [cf. (2.19)]. Henceforth, we shall formulate (almost) all results for the distribution of the breeding values. As a preparatory step we calculate the selection response of the breeding values.

6.1 THE SELECTION RESPONSE OF THE BREEDING VALUES

Selection changes the distribution $f(G)$ of breeding values according to

$$\Delta_s f(G) = f(G)\,\frac{W(G) - \overline{W}}{\overline{W}} \tag{6.1a}$$

which, under weak selection, can be approximated by

$$\tilde{\Delta}_s f(G) = f(G)[W(G) - \overline{W}]\,. \tag{6.1b}$$

Further, let ψ_G and Ψ_G denote the moment and cumulant generating function, respectively, of $f(G)$. As before, C_n denotes the nth cumulant of $f(G)$. For the mean C_1 and the genetic variance C_2, we write $\overline{G}$ and σ_G^2, respectively. Beause the equations (6.1) are formally identical to those for a haploid asexual population, all the results in Sections 2 and 3 remain valid if applied to $f(G)$ instead of $p_i(x_i)$, the distribution of a lone haploid locus. Thus, by (3.2), we have

$$\Delta_s\psi_G(\xi) = \frac{1}{\overline{W}}\int e^{\xi G} W(G) f(G)\,\mathrm{d}G - \psi(G)\,, \tag{6.2}$$

and, by (3.3),

$$\Delta_s\Psi_G = \ln\!\left(1 + \frac{\Delta_s\psi}{\psi_G}\right)\,. \tag{6.3}$$

In particular, if the weak-selection response is defined as in (3.5), i.e., by

$$\tilde{\Delta}_s C_n = \overline{W}\,D^n\!\left(\frac{\Delta_s\psi_G}{\psi_G}\right)(0)\,,$$

then we have by (3.7),

$$\Delta_s C_n = \sum_{\nu=1}^{n}(-1)^{\nu-1}\,\frac{1}{\nu}\,\overline{W}^{-\nu}\sum_{k_1+\ldots+k_\nu=n}\frac{n!}{k_1!\cdot\ldots\cdot k_\nu!}\,\tilde{\Delta}_s C_{k_1}\cdot\ldots\cdot\tilde{\Delta}_s C_{k_\nu}\,, \tag{6.4}$$

where the sum is over all integers $k_i > 0$ satisfying the indicated constraint. For low-order cumulants, these relations are simple; in particular, we have $\Delta_s\overline{G} = \tilde{\Delta}_s\overline{G}/\overline{W}$ [cf. (3.8)]. If one is interested only in the weak-selection response, thus willing to ignore terms of order $O(s^2)$, then such terms should be omitted on the right-hand side of (6.1b).

For arbitrary fitnesses, a simple, exact formula can be derived for the selection response of the mean. Indeed, multiplying the equations (6.1a) and (6.1b) by G and integrating, we obtain

$$\Delta_s\overline{G} = \tilde{\Delta}_s\overline{G}/\overline{W} = \mathrm{Cov}_G(G,W)/\overline{W} \tag{6.5a}$$

$$= \frac{1}{\overline{W}}\sum_{k=0}^{K} s_k M_{k+1}^0 - \overline{G}\ , \tag{6.5b}$$

where the second equality follows by series expansion of $W(G)$ according to (1.8), and M_k^0 denotes the kth moment about zero of $f(G)$.

In a similar way, the weak-selection response of the genetic variance (which in the additive model coincides with the additive genetic variance) is calculated to be

$$\tilde{\Delta}_s\sigma_G^2 = \tilde{\Delta}_s C_2 = \mathrm{Cov}_G[(G-\overline{G})^2, W] \tag{6.6a}$$

$$= \sum_{k=1}^{K} s_k\left[M_{k+2}^0 - 2\overline{G}M_{k+1}^0 + (\overline{G}^2 - \sigma_G^2)M_k^0\right]\ . \tag{6.6b}$$

For higher-order cumulants, more complicated expressions than suggested by (6.5a) and (6.6a) are obtained. Equation (6.6a) cannot be obtained directly from II(5.10) because $(G-\overline{G})^2$ is not additive. Of course, the right-hand sides of (6.5) and (6.6) can be expressed in terms of cumulants by transforming the moments (Appendix D).

The weak-selection response of arbitary cumulants C_n is obtained immediately from the results about asexually reproducing populations in Section 2.2 and 2.3, because we have $\tilde{\Delta}_s C_n = \frac{\mathrm{d}}{\mathrm{d}t}C_n$, as noted below (3.13). Thus, (2.24) yields

$$\tilde{\Delta}_s C_n = \sum_{k=0}^{K} s_k S_{nk}\ , \tag{6.7}$$

where S_{nk} is defined in (2.14) and computed in (2.15). Comparision of (6.5b) with (6.7) shows that

$$S_{1k} = M_{k+1}^0 - M_k^0\overline{G}\ , \quad k \geq 1\ . \tag{6.8}$$

Similarly, from (6.6b) we obtain

$$S_{2k} = M_{k+2}^0 - 2M_{k+1}^0\overline{G} + M_k^0(2\overline{G}^2 - M_2^0)\ , \quad k \geq 1\ . \tag{6.9}$$

In general, the selection responses of the higher cumulants are rather complicated expressions, except when the fitness function can be approximated by a polynomial of low degree. If the S_{nk} are expressed in terms of cumulants, as in (2.15), then the weak-selection response of the first four cumulants becomes

$$\begin{aligned}\tilde{\Delta}_s C_1 = {} & s_1C_2 + s_2(C_3 + 2C_1C_2) + s_3(C_4 + 3C_1C_3 + 3C_2^2 + 3C_1^2C_2) \\ & + s_4(C_5 + 4C_1C_4 + 10C_2C_3 \\ & \quad + 6C_1^2C_3 + 4C_1^3C_2 + 12C_1C_2^2) + \dots\ ,\end{aligned} \tag{6.10}$$

$$\begin{aligned}\tilde{\Delta}_s C_2 &= s_1 C_3 + s_2(C_4 + 2C_1C_3 + 2C_2^2)\\ &\quad + s_3(C_5 + 3C_1C_4 + 9C_2C_3 + 3C_1^2C_3 + 6C_1C_2^2)\\ &\quad + s_4(C_6 + 4C_1C_5 + 14C_2C_4 + 6C_1^2C_4 + 10C_3^2\\ &\qquad + 36C_1C_2C_3 + 4C_1^3C_3 + 12C_2^3 + 12C_1^2C_2^2) + \dots , \end{aligned} \tag{6.11}$$

$$\begin{aligned}\tilde{\Delta}_s C_3 &= s_1 C_4 + s_2(C_5 + 2C_1C_4 + 6C_2C_3)\\ &\quad - s_3(C_6 + 3C_1C_5 + 12C_2C_4 + 3C_1^2C_4 + 9C_3^2\\ &\qquad + 6C_2^3 + 18C_1C_2C_3) + \dots , \end{aligned} \tag{6.12}$$

$$\tilde{\Delta}_s C_4 = s_1 C_5 + s_2(C_6 + 2C_1C_5 + 8C_2C_4 + 6C_3^2) + \dots . \tag{6.13}$$

Hence, the responses to selection can be expressed solely in terms of the distribution of the breeding values (or the phenotypic values) before selection, regardless of the distribution of allelic effects at particular loci. For sexually reproducing populations, this is not true for the change between generations (see below).

Finally, we observe that in view of (3.9), the selection response of the cumulants of breeding values is related to that of the multivariate cumulants by

$$\tilde{\Delta}_s C_n = \sum_{\boldsymbol{n}:|\boldsymbol{n}|=n} \frac{n!}{\boldsymbol{n}!}\, \tilde{\Delta}_s c_{\boldsymbol{n}} , \tag{6.14}$$

where $\boldsymbol{n}$ denotes multi-indices of length 2ℓ. A structurally identical formula, with the $\tilde{}$ omitted, holds for the strong-selection response Δ_s. In particular, it follows that $\tilde{\Delta}_s \overline{G} = 2\sum_{i=1}^{\ell} \tilde{\Delta}_s \kappa_i$, the latter being given by (3.15a), or (3.21a).

6.2 THE RECURSION RELATIONS ACROSS GENERATIONS

With selection, recombination, and mutation, the recursion relations for the mean, the genetic variance, and the third and fourth cumulants of the distribution of breeding values (and phenotypic values) are obtained by appropriate summations over equations (4.5) to (4.8), as well as (5.6). All mutation terms in this section are based on the random-walk mutation model and calculated from (5.6). For simplicity, we assume subsequently that the mutation distributions $u_i(x_i)$ are symmetric, hence $v_1^{(i)} = v_3^{(i)} = \dots = 0$. (Of course, there are traits where there is a bias in the direction of mutation.) As previously, we write $\gamma_i^2 (= v_2^{(i)})$ for the variance of the mutation distribution.

Since, by (4.5), recombination and, by our assumption, mutation do not affect the mean, it evolves according to

$$\Delta\overline{G} = \Delta_s \overline{G} = \mathrm{Cov}_G(G, W)/\overline{W} . \tag{6.15}$$

This is Robertson's (1966, 1968) Secondary Theorem of Natural Selection, II(5.10). Alternative representations of $\Delta\overline{G}$ are provided by (6.5b) and (6.10). Of course, by (4.5) and (6.14), the change of the mean breeding value between generations is simply the sum of all single-locus changes,

$$\Delta\overline{G} = 2\sum_{i=1}^{\ell} \Delta\kappa_i = 2\sum_{i=1}^{\ell} \Delta_s \kappa_i . \tag{6.16}$$

A further alternative can be obtained in terms of the selection gradients $\mathcal{L}_k$, by summing (3.31) over all loci i and recalling that $\sigma_G^2 = 2\sum_i \kappa_{i.}$, $C_3 = 2\sum_i \kappa_{i..}$. Then the response of the mean can be expressed as

$$\Delta\overline{G} = \sigma_G^2\mathcal{L}_1 + C_3\mathcal{L}_2 + C_4\mathcal{L}_3 + \dots \; . \tag{6.17}$$

Next, we derive the exact change of the genetic variance across generations under selection, recombination, and mutation. Since we assume random mating, zygotes are in Hardy–Weinberg proportions. Therefore, $\kappa_{i,j} = 0$ for every i and j, and by (3.9) the genetic variance can be written as

$$\sigma_G^2 = C_2 = 2\sum_{i,j=1}^{\ell} \kappa_{ij} \; . \tag{6.18}$$

Summation of (4.10) over all pairs of loci i, j yields the exact change of the variance across generations for strong selection:

$$\Delta\sigma_G^2 = \frac{1}{\overline{W}}\tilde{\Delta}\sigma_G^2 - \tfrac{1}{2}(\Delta\overline{G})^2 \; . \tag{6.19}$$

To calculate the change $\tilde{\Delta}\sigma_G^2$ under weak selection, recombination, and mutation, we first note that $\tilde{\Delta}_{\mathrm{s}}\sigma_G^2 = 2\sum_{i,j}(\tilde{\Delta}_{\mathrm{s}}\kappa_{ij} + \tilde{\Delta}_{\mathrm{s}}\kappa_{i,j})$, because selection induces deviations from Hardy–Weinberg proportions. Then we obtain $\tilde{\Delta}\sigma_G^2$ from the weak-selection selection version (4.6) and from (5.6) by summation:

$$\tilde{\Delta}\sigma_G^2 = \tilde{\Delta}C_2 = 2\sum_{i,j}\tilde{\Delta}\kappa_{ij} \tag{6.20a}$$

$$= \tfrac{1}{2}\tilde{\Delta}_{\mathrm{s}}\sigma_G^2 - 2\overline{W}\sum_{i,j} r_{ij}\kappa_{ij} + \sum_{i,j}(1-2r_{ij})(\tilde{\Delta}_{\mathrm{s}}\kappa_{ij} - \tilde{\Delta}_{\mathrm{s}}\kappa_{i,j}) + 2\sum_i \mu_i\gamma_i^2 \; , \tag{6.20b}$$

where all sums are from 1 to ℓ. More explicit expressions can be obtained by substituting (6.6) or (6.11) for $\tilde{\Delta}_{\mathrm{s}}\sigma_G^2$, and (3.16b) and (3.17a) for $\tilde{\Delta}_{\mathrm{s}}\kappa_{ij}$ and $\tilde{\Delta}_{\mathrm{s}}\kappa_{i,j}$, respectively. A substantial simplification occurs for free recombination between all loci, i.e., for $r_{ij} = \frac{1}{2}$ if $i \neq j$, and $r_{ii} = 0$. Then (6.20b) reduces to

$$\tilde{\Delta}\sigma_G^2 = \tfrac{1}{2}\tilde{\Delta}_{\mathrm{s}}\sigma_G^2 + \tfrac{1}{2}\overline{W}(\sigma_{\mathrm{G,LE}}^2 - \sigma_G^2) + \sum_i \tilde{\Delta}_{\mathrm{s}}(\kappa_{ii} - \kappa_{i,i}) + 2\sum_i \mu_i\gamma_i^2 \; , \tag{6.21}$$

where

$$\sigma_{\mathrm{G,LE}}^2 = \sigma_{\mathrm{A,LE}}^2 = 2\sum_{i=1}^{\ell}\kappa_{ii} \tag{6.22}$$

is the linkage-equilibrium variance defined in II(4.26).

Equations (6.20) and (6.21) show that, in contrast to the mean, the change of the variance across generations cannot be solely expressed in terms of the distribution of the breeding values (or the phenotypic values), even under the simplifying assumptions of weak selection and free recombination. The change of the variance inevitably depends on the distribution of allelic effects at each locus, and can thus be predicted precisely only if these distributions are known.

At quasi-linkage equilibrium, (6.20) can be simplified substantially. Indeed, after at most t_2 generations, when the change of linkage disequilibria has been reduced to order $O(s^2)$ [cf. II(6.13)], we obtain by applying in successive lines (4.35); (6.20a) and (4.36); (4.6); and the fact that at QLE, $\tilde{\Delta}_{\mathrm{s}}\kappa_{ii} = \tilde{\Delta}_{\mathrm{s,LE}}\kappa_{ii} + O(s^2) = \dot{c}_2^{(i)}$ (because by (4.33) cross-locus cumulants are of order $O(s)$),

$$\begin{aligned}\Delta\sigma_G^2 &= \tilde{\Delta}\sigma_G^2 + O(s^2)\\ &= 2\sum\nolimits_i \tilde{\Delta}\kappa_{ii} + O(s^2)\\ &= 2\sum\nolimits_i \tilde{\Delta}_{\mathrm{s}}\kappa_{ii} + O(s^2)\\ &= 2\sum_{i=1}^{\ell} \dot{c}_2^{(i)} + O(s^2)\ , \quad t \geq t_2\ ,\end{aligned} \tag{6.23}$$

where $\dot{c}_2^{(i)}$ is given by (2.16b). Therefore, at QLE the change of the genetic variance across generations is, to order $O(s^2)$, given by

$$\Delta\sigma_G^2 = s_1 C_3 + s_2\Big(C_4 + 2C_1C_3 + 4\sum_{i=1}^{\ell}\kappa_{ii}^2\Big) + \cdots + 2\sum_{i=1}^{\ell}\mu_i\gamma_i^2\ , \quad t \geq t_2\ , \tag{6.24}$$

and additional selection terms can be obtained from (2.16b). The same expression is obtained by applying the QLE approximations (4.40) and (4.42) to (6.20b). Comparison with (6.11) shows that the terms beginning with s_2 differ from those under selection alone. For exchangeable loci, (6.24) reduces further to

$$\Delta\sigma_G^2 = s_1 C_3 + s_2\big(C_4 + 2C_1C_3 + \tfrac{1}{\ell}\sigma_G^4\big) + \cdots + 2\ell\mu\gamma^2\ , \quad t \geq t_2\ . \tag{6.25}$$

However, (6.25) shows that even under weak selection in quasi-linkage equilibrium, and under simplifying assumptions about the genetic system, at least the number of loci must be known to predict the change of the variance. It may be noted that, in general, (6.24) and (6.25) do not hold at linkage equilibrium because, at linkage equilibrium, the change in cross-locus cumulants may be of order s; see Section 3.5. The reader may also observe from (4.40) that at QLE, the difference $\sigma_{\mathrm{G,LE}}^2 - \sigma_G^2$ may be larger than $O(s^2)$.

The result II• **6.2** implies that for the evolutionarily long period $t_2 \leq t \leq K_0/s$, the genetic variance satisfies

$$\sigma_G^2(t) = \sigma_{\mathrm{G,LE}}^2(t) + O(s) = 2\sum_{i=1}^{\ell} c_2^{(i)}(t) + O(s)\ , \tag{6.26}$$

where the single-locus variances $c_2^{(i)}(t)$ evolve according to (2.16b), or (2.13), subject to the initial condition $c_2^{(i)}(t_2)$. As discussed below II(6.17), the restriction $t \leq K_0/s$ will rarely be necessary. The problem, however, that the change of the variance depends on higher cumulants cannot be resolved without additional information about the allele distribution at individual loci.

The recursion relation for the third cumulant is obtained by summing (4.11) over all triplets of loci,

$$\Delta C_3 = \frac{1}{\overline{\overline{W}}}\tilde{\Delta} C_3 - 3\Delta\overline{G}\Big(\frac{1}{2\overline{\overline{W}}}\tilde{\Delta}\sigma_G^2 + \sum_{i,j} r_{ij}\kappa_{ij}\Big) + \tfrac{1}{2}(\Delta\overline{G})^3 \,, \tag{6.27}$$

where $\tilde{\Delta} C_3$ is found from (the weak-selection version) of (4.7). For free recombination and distinct i, j, and k, we have $r_{ij} = \frac{1}{2}$, $r_{ijk} = \frac{3}{4}$, and $r_{i,jk} = \frac{1}{4}$. Then $\tilde{\Delta} C_3$ simplifies to

$$\begin{aligned}\tilde{\Delta} C_3 = \tfrac{1}{4}\tilde{\Delta}_s C_3 + \tfrac{3}{4}\overline{W}\Big(2\sum_{i,j}\kappa_{iij} - C_3\Big) \\ + \tfrac{3}{2}\sum_{i,j}\big(\tilde{\Delta}_s\kappa_{iij} + \tilde{\Delta}_s\kappa_{i,jj} - 2\tilde{\Delta}_s\kappa_{i,ij}\big) \,.\end{aligned} \tag{6.28}$$

At quasi-linkage equilibrium and for exchangeable loci under weak selection, the following formula for the evolution of higher-order cumulants is easily derived from (2.13) and (2.14) by following the arguments leading to (6.25):

$$\begin{aligned}\Delta C_n = s_1 C_{n+1} + s_2\Big(2C_1 C_{n+1} + C_{n+2} + \frac{1}{2\ell}\sum_{j=1}^{n-1}\binom{n}{j} C_{j+1}C_{n+1-j}\Big) \\ + \ldots + 2\ell\mu v_n + O(s^2) \,.\end{aligned} \tag{6.29}$$

For $\ell = \frac{1}{2}$, this reduces to the haploid dynamics (2.24).

6.3 GAUSSIAN DISTRIBUTION OF BREEDING VALUES

Much of quantitative genetics theory has assumed a normal distribution of breeding values. The reasons are, first, that often an appropriate scale transformation can be found that renders breeding values nearly Gaussian (Falconer and Mackay 1996) and, secondly, that this can be expected on the basis of the Central Limit Theorem (Feller 1971), because quantitative traits are polygenic. A major theoretical problem, however, is that (in discrete time) exponential and Gaussian fitness are the only known forms of selection that preserve a Gaussian distribution (cf. Section 3.7, and Chapters VII.4.2 and VII.7.3). Recombination, however, induces departures from normality, unless the population is in linkage equilibrium after selection. Thus, in general, the breeding values can at best be approximately normally distributed.

We shall investigate the consequences of the normality assumption on their evolution, and show that for a normal distribution of breeding values, simple and easily applicable equations for the response $\Delta\overline{G}$ across generations can be derived. In this case, S_{1k}, given by (6.8), is computed to be

$$S_{1,2k-1} = (2k-1)\sigma_G^2\Bigg[\sum_{j=0}^{k-1}\frac{(2k-2)!}{(2j)!(k-j-1)!}\left(\tfrac{1}{2}\sigma_G^2\right)^{k-1-j}\overline{G}^{2j}\Bigg] \,, \tag{6.30a}$$

$$S_{1,2k} = 2k\,\overline{G}\sigma_G^2\Bigg[\sum_{j=0}^{k-1}\frac{(2k-1)!}{(2j+1)!(k-j-1)!}\left(\tfrac{1}{2}\sigma_G^2\right)^{k-1-j}\overline{G}^{2j}\Bigg] \,. \tag{6.30b}$$

Therefore, (6.15) and (6.7) yield

$$\tilde{\Delta}\overline{G} = \sigma_G^2[s_1 + 2s_2\overline{G} + 3s_3(\overline{G}^2 + \sigma_G^2) + 4s_4\overline{G}(\overline{G}^2 + 3\sigma_G^2) + \dots] \tag{6.31}$$

for the weak-selection response of the mean breeding value, with higher-order terms given by (6.30).

Equivalently, for a Gaussian distribution of breeding values, (6.17) reduces to

$$\Delta\overline{G} = \sigma_G^2\mathcal{L}_1 \tag{6.32}$$

which, on account of (3.29), agrees with (6.31). In practice, it is useful to express the selection gradient $\mathcal{L}_1$ through properties of the distribution of phenotypic values. If it is Gaussian, as it will be if the distribution of breeding values is, then (3.28a) implies

$$\mathcal{L}_1 = \frac{\Delta_s\overline{P}}{\sigma_P^2} . \tag{6.33}$$

Hence, the classical quantitative-genetic equation for the response of the mean to selection follows:

$$\Delta\overline{G} = \frac{\sigma_G^2}{\sigma_P^2}\,\Delta_s\overline{P} . \tag{6.34a}$$

The fraction $h^2 = \sigma_G^2/\sigma_P^2 = \sigma_A^2/\sigma_P^2$ of the phenotypic variance attributable to additive genetic effects is called the (narrow sense) heritability. In the quantitative genetics literature, $\Delta_s\overline{P}$ is frequently called the selection differential. Because of (1.3), $\overline{G}$ can be replaced by $\overline{P}$, so that (6.34a) reads

$$\Delta\overline{P} = h^2\Delta_s\overline{P} . \tag{6.34b}$$

This is often called the *breeder's equation* (cf. Bulmer 1980, Chapter 9; Falconer and Mackay 1996, Chapter 11). The importance of (6.34) arises from the fact that it allows predictions for the response of the mean across generations from the readily measurable phenotypic quantities h^2 and $\Delta_s\overline{P}$.

This chapter ends with a warning note. If genes interact epistatically, then $\Delta\overline{G}$ no longer depends only on changes in gene frequencies [see II(5.7)], but also on the change of gamete frequencies (see Chapter II.6.3). Thus, because of changes in linkage disequilibria, $\overline{G}$ may change even if $\sigma_G^2 = 0$ and $\Delta_s P = 0$ (cf. Griffing 1960, Bulmer 1980, Gimelfarb 1989).

7. SUMMARY

For an additively determined quantitative trait subject to selection, recombination, and mutation, the exact dynamic equations for the change of the distributions of genotypes, of genotypic and of phenotypic values have been derived from genetic principles. This has been accomplished by describing these distributions in terms of (multivariate) cumulants. Under the assumptions of global linkage equilibrium and weak selection, the dynamics of the distribution of genotypic values is readily obtained [(2.18)–(2.23)].

As a special case, the evolution of asexual populations is given by (2.24). In the presence of linkage disequilibria, which are generated by selection, the situation is much more complicated. The strong-selection response, Δ_s, for the multivariate cumulants and the cumulants of breeding values can be calculated from certain weak-selection responses, $\tilde{\Delta}_s$, according to (3.7) and (6.4), respectively. The general equations for the weak-selection response of the multivariate cumulants are given by (3.10), (3.11) and (3.15). For the change of the low-order cumulants, (3.16), (3.17), (3.22) and (3.23) provide explicit formulas. In Section 4, the combined effects of recombination and selection are treated. The general recursions are given by (4.4)–(4.8). The relations between various measures of linkage disequilibrium are highlighted in (4.24)–(4.31). Under the assumption of weak selection, quasi-linkage approximations for the cross-locus cumulants (the linkage disequilibria) are derived in (4.38)–(4.42). The change in the cumulants caused by mutation is given by (5.6) for the random-walk mutation model and by (5.7)–(5.9) for the HC-mutation model. Section 6 is devoted to the dynamics of breeding and phenotypic values under selection, recombination, and mutation. Equation (6.7) together with (2.14) provides general formulas for their response to selection, (6.10)–(6.13) the leading terms. The change of the mean across generations is given by (6.15), with (6.5b), (6.10), and (6.17) providing alternative representations. It is equivalent to Robertson's Secondary Theorem of Natural Selection and can be expressed in purely phenotypic terms. For a Gaussian distribution of breeding values, the classical breeder's equation, (6.34), is recovered. The change of the (additive) genetic variance across generations is much more complicated and, in general, given by (6.19) and (6.20). For freely recombining loci or in quasi-linkage equilibrium, much simpler expressions are obtained, namely (6.21) and (6.24)–(6.26), respectively. Nevertheless, the prediction of the change of the genetic variance requires knowledge about details of the distribution at individual loci and is not possible from purely phenotypic observations.

VI

Stabilizing Selection and Genetic Variation in Large Populations

Many quantitative characters in natural populations are apparently subject to stabilizing selection toward an intermediate optimum (e.g., Endler 1986). This means that extreme phenotypes have a lower fitness than those near the population mean. Such selection has also been called optimizing, centripetal, or nor-optimal, selection and is considered to exhaust genetic variation. This view has been suggested by analyses that were based on the assumption of a large number of independent loci which, individually, contribute to the genetic variance only a very small amount (e.g., Fisher 1930, Robertson 1956, Bulmer 1971b). Further support came from Wright's (1935b) study of the so-called quadratic optimum model, in which two diallelic loci control the character, whose fitness deviates in a quadratic way from its maximum value (see also Lewontin 1964). By contrast, most quantitative traits exhibit relatively high levels of genetic variability in nature.

This apparent contradiction has been a fundamental problem in evolutionary biology, and the central question that has to be answered is: "What are the mechanisms that maintain genetic variability?". Several mechanisms that can potentially contribute genetic variability under stabilizing selection have been proposed and investigated. In principle, variability can be maintained by either forces acting directly on the considered trait or as a side effect of genetic variation that is independent of the observed character. Among the direct mechanisms are overdominance, migration, mutation, frequency-dependent selection, fluctuating environments, genotype-environment interaction, or epistasis. The ultimate source of genetic variability is mutation; but can a balance between mutation and stabilizing selection account for a significant fraction of the observed levels of variability? This simple and appealing hypothesis was promoted by Lande (1975) on the basis of a mathematical analysis and a review of empirical data. Since then it has been the object of intense investigation and debate. This, and part of the subsequent, chapter is devoted to the exploration of models that have been constructed to quantify the amount of genetic variability that can be maintained when stabilizing selection is acting on a polygenic trait.

In this chapter, populations are infinitely large, mate at random, and are monoecious or have equivalent sexes. Unless posited otherwise, the trait under consideration is determined additively, without dominance or epistasis. It is assumed to be under direct stabilizing selection. Alternatively, and discussed in Chapter VII, a trait may be under apparent stabilizing selection caused by selection on pleiotropic effects. Due to the complexity and analytical intractability of true multilocus models, further simplifying assumptions have to be imposed to obtain quantitative results about the amount of genetic variance that is maintained under stabilizing selection alone or under a balance between selection and mutation.

The first three sections investigate the ability of stabilizing selection *per se* for maintaining polymorphisms and genetic variability in quantitative traits controlled additively by two or more loci. Conflicting results have been obtained, depending on the model assumptions about the number of involved loci, magnitude of allelic effects, and linkage equilibrium. Our analysis will shed more light on this topic. In Section 1, a model is treated based on the assumption that each of a large number of loci in linkage equilibrium contributes only an infinitesimally small amount to the total genetic variance. Under this model, which reduces the analysis to a single locus, selection depletes all genetic variability. In Section 2, the trait is supposed to be determined by two linked loci. In this model, which extends Wright's (1935b) quadratic optimum model, large amounts of genetic variance can be maintained by stabilizing selection unless the loci have equal (or similar) effects and are completely additive, as assumed by Wright. Section 3 investigates stabilizing selection on diallelic multilocus systems. First, a model with exchangeable unlinked loci is treated analytically. Then statistical results are presented about the stable equilibria and the genetic variability expected in two-, three-, four-, and five-locus models with arbitrary recombination and allelic effects varying between loci.

The rest of the chapter is devoted to investigating how much genetic variability can be maintained under different mutation-stabilizing-selection models, a topic that has given rise to sparkling debate and has initiated new research programs in evolutionary biology. In Section 4, recurrent mutation is introduced into the general two-locus model of Section 2, and its effects on the maintenance of genetic variation are explored. Section 5 studies mutation-selection balance for an arbitrary number of diallelic loci under the assumption of global linkage equilibrium. Section 6 is dedicated to the analysis of mutation-selection balance under the haploid continuum-of-alleles model, but also reviews related results derived for models with discrete alleles. This provides the basis for Section 7, in which the maintenance of genetic variation is studied if many stochastically independent or weakly linked loci contribute to the trait under stabilizing selection. Various conflicting approximations are compared and placed in perspective. The present chapter is primarily intended to lay the mathematical foundations, whereas a comprehensive discussion of the biological consequences and conclusions is deferred to the next chapter.

1. ANALYSES FOR A SINGLE LOCUS

The model reviewed in this section and various extensions were investigated by Fisher (1930), Haldane (1932), Robertson (1956), Bulmer (1971b, 1980), Kimura and Crow (1978), Kimura (1981), Nagylaki (1984, 1992), Hastings (1990b), and Walsh (1990), among others. The idea underlying this approach is to establish the relation between selection on a quantitative trait and gene-frequency change at the loci contributing to the trait. This requires approximating the conditional distribution of phenotypes given that a particular genotype or allele is present at a given locus. We outline the model and the results, and refer to Hastings (1990b) and Nagylaki (1992, Chapter 10.5) for the complete proofs.

Let us consider a randomly mating population with discrete nonoverlapping generations that is large enough for random genetic drift to be ignored. It may be monoecious or dioecious without sex differences. Viability selection acts on a quantitative trait that is determined additively by ℓ loci and an environmental component as in V(1.1). We shall admit dominance, but assume that the loci contributing to the trait are in linkage equilibrium. The model rests on the assumption that each of a large number of loci contributes only an infinitesimally small amount to the character. We consider a fixed but arbitrary locus i, and denote by $G^{(i)}$ its contribution to the trait Y, and by $Q = Y - G^{(i)}$ the residual consisting of the contributions of all other loci and the environment. By our assumptions, $G^{(i)}$ and Q are independent random variables; both have mean zero. We denote the densities of Y and Q (before selection) by f_P and q, respectively. According to the basic assumption of the model, we assume that the variance at locus i, $\sigma^2 = \mathrm{Var}[G^{(i)}]$, is small. If f_P has a continuous second derivative, designated f_P'', it can be shown that (Hastings 1990b, Nagylaki 1992)

$$q(y) = f_P(y) - \tfrac{1}{2}\sigma^2 f_P''(y) + O(\sigma^3) \qquad \text{as } \sigma \to 0 \, . \tag{1.1}$$

This equation enables us to evaluate the changes in gene frequencies to second order.

For a given genotype $\mathcal{A}_j\mathcal{A}_k$ at locus i, let g_{jk} denote the genotypic deviation from the mean at locus i (cf. Chapter I.3.1). Then the genetic variance contributed by locus i is $\sigma^2 = \sum_{jk} g_{jk}^2 p_j p_k$, where p_j is the frequency of allele $\mathcal{A}_j$. From (1.1), we obtain the second-order approximation of the density $h_{jk}(y)$ of Y in individuals with genotype $\mathcal{A}_j\mathcal{A}_k$ at locus i:

$$\begin{aligned} h_{jk}(y) &= f_P(y - g_{jk}) - \tfrac{1}{2}\sigma^2 f_P''(y - g_{jk}) + O(\sigma^3) \\ &= f_P(y) - g_{jk} f_P'(y) + \tfrac{1}{2}(g_{jk}^2 - \sigma^2) f_P''(y) + O(\sigma^3) \, . \end{aligned} \tag{1.2}$$

Let $W_P(y)$ and W_{jk} be the average fitness of an individual with phenotypic value y and with genotype $\mathcal{A}_j\mathcal{A}_k$, respectively. denote by $\overline{W} = \int W_P(y) f_P(y)\, dy$ the mean fitness, and set

$$I_1 = -\int W_P(y) f_P'(y)\, dy \, , \tag{1.3a}$$

$$I_2 = -\int W_P(y) f_P''(y)\, dy \, . \tag{1.3b}$$

Then

$$W_{jk} = \int W_P(y) h_{jk}(y)\, dy \, , \tag{1.4}$$

and the marginal fitness $W_k = \sum_j W_{jk} p_j$ of allele A_k is found to be (Hastings 1990b)

$$W_k = \overline{W} + g_k I_1 - \tfrac{1}{2}(V_k - \sigma^2) I_2 + O(\sigma^3) \, , \tag{1.5}$$

where g_k is the average excess of $\mathcal{A}_k$ and $V_k = \sum_j g_{jk}^2 p_j$. Inserting (1.5) into I(9.7) shows that the change in the frequency of $\mathcal{A}_k$, as a result of one generation of selection, is

$$\Delta p_k = \frac{p_k}{\overline{W}}[g_k I_1 - \tfrac{1}{2}(V_k - \sigma^2) I_2] + O(\sigma^3) \, . \tag{1.6}$$

If we assume that dominance is absent, and scale the genotypic contributions according to $G_{11} = 0$, $G_{12} = c$, and $G_{22} = 2c$, we obtain by a straightforward calculation (cf. Chapter I.3.1, Example 2)

$$\Delta p = \frac{p(1-p)}{\overline{W}}[-I_1 c - \tfrac{1}{2} I_2 c^2 (1-2p)] , \tag{1.7}$$

which is correct to second order in σ, This implies that the equilibrium frequency $\hat{p}$ of an unfixed locus is

$$\hat{p} = \tfrac{1}{2} + \frac{I_1}{I_2 c} , \tag{1.8}$$

provided $\hat{p}$ is between 0 and 1.

Now let us assume in addition that Gaussian stabilizing selection acts on the trait, i.e., $W_P(y) = \exp[-s(y - P_O)^2]$, and denote the variance and the third moment of the distribution of phenotypes by σ_P^2 and M_3, respectively. Then, if s is small (selection weak), or if $(y - P_O)$ is small for all but a small fraction of the population, then integration by parts yields (Hastings 1990b)

$$\overline{W} \approx 1 - s(\sigma_P^2 + P_O^2) , \tag{1.9a}$$

$$I_1 \approx 2sP_O + 2s^2(M_3 - P_O^3 - 6\sigma_P^2) , \tag{1.9b}$$

$$I_2 \approx 2s - 6s^2(\sigma_P^2 + P_O^2) . \tag{1.9c}$$

If $s(\sigma_P^2 + P_O^2)$ is sufficiently small, then (1.9c) implies that $I_2 > 0$. (This will also be the case for any other fitness function with the same curvature near the optimum.) Simple algebra shows that the equilibrium (1.8) is unstable whenever it exists, and the boundary equilibria $\hat{p} = 0$ and $\hat{p} = 1$ are asymptotically stable. Therefore, any genetic variability will be lost in this model of stabilizing selection (Bulmer 1971b). Elementary but lengthy algebra shows that this conclusion remains valid with dominance, i.e., if $G_{12} = hc$ and $0 \le h \le 2$.

Equation (1.7) was obtained by Bulmer (1971b) on the basis of the first-order approximation

$$h_{jk}(y) \approx f_P(y - g_{jk}) . \tag{1.10}$$

As is easy to verify, this first-order approximation produces the correct second-order approximation (1.6) for the gene-frequency change, but it does not yield the correct approximation (1.5) for the marginal fitness W_k: the term with σ^2 would be missing. This leads to various inconsistency problems discussed by Hastings (1990b), Walsh (1990), and Nagylaki (1992, Chapter 10.5). For truncation selection and a normal distribution of phenotypes, Latter (1965) obtained (1.6) by using the approximation (1.2), but with the term proportional to σ^2 missing. Finally, we remark that the validity of the first-order approximation (1.10) has not yet been demonstrated with epistasis. It generally fails in the presence of linkage disequilibrium (Nagylaki 1984, 1992).

2. TWO-LOCUS MODELS OF STABILIZING SELECTION

In this section, we investigate two-locus models that generalize Wright's (1935b) quadratic optimum model and we study conditions under which stabilizing selection *per se* can maintain genetic variability. In contrast to the classical view, as mainly based on the conclusions from the above model (or its ancestors), we shall see that large amounts of genetic variation can be maintained under a variety of two-locus models and parameter combinations. The basic model is a special case of the symmetric viability model. Because in its most general form only little more can be said about its properties than what has been summarized in Chapter II.1.3, further assumptions or restrictions on the parameters are needed to obtain a deeper insight.

One possibility is to specify the fitness function. For example, Wright (1935a,b, 1952) studied a model in which fitness decreases quadratically as the phenotype deviates from its optimum value, and Gale and Kearsey (1968) and Kearsey and Gale (1968) assumed a triangular fitness function. Detailed analyses of the quadratic optimum model with the optimum at the value of the double heterozygote were performed by Gavrilets and Hastings (1993, 1994b) and Bürger and Gimelfarb (1999). This model is investigated in Section 2.2. Another possibility is to assume that selection is much weaker than recombination such that linkage disequilibrium can be neglected. This was done by Nagylaki (1989a), who obtained a complete classification of possible convergence patterns under general forms of stabilizing selection. Section 2.3 summarizes the main results. In Section 2.4, very strong selection is briefly treated. Finally, in Section 2.5, taking up and complementing the approaches of Hastings and Hom (1990) and Gavrilets and Hastings (1993), we investigate a model in which the position of the fitness optimum may be arbitrary, i.e., the phenotypic optimum may deviate from the genotypic value of the double heterozygote.

2.1 THE MODEL

Let us consider a quantitative trait that is controlled additively by two diallelic loci. The alleles at the first locus are labeled $\mathcal{A}_1$ and $\mathcal{A}_2$, and those at the second locus $\mathcal{B}_1$ and $\mathcal{B}_2$. The basic model, the recursion relations for the gamete frequencies, and the notations are as in Chapter II.1. Let the contributions of these alleles to the genotypic value G of the trait be $\beta - \frac{1}{2}\gamma_1$, $\beta + \frac{1}{2}\gamma_1$, $-\beta - \frac{1}{2}\gamma_2$, and $-\beta + \frac{1}{2}\gamma_2$, respectively, where β is an arbitrary constant. We assume that the alleles determine the genotypic value G purely additively. Then the effects of the gametes $\mathcal{A}_1\mathcal{B}_1$, $\mathcal{A}_1\mathcal{B}_2$, $\mathcal{A}_2\mathcal{B}_1$, and $\mathcal{A}_2\mathcal{B}_2$ are $-\frac{1}{2}(\gamma_1 + \gamma_2)$, $-\frac{1}{2}(\gamma_1 - \gamma_2)$, $\frac{1}{2}(\gamma_1 - \gamma_2)$, and $\frac{1}{2}(\gamma_1 + \gamma_2)$. The resulting genotypic values are shown in Table 2.1. For definiteness, we assume $\gamma_1 \geq \gamma_2$ and refer to these loci as major and minor, respectively. The parameters γ_1 and γ_2 are the effects

Table 2.1 The genotypic values in the additive model.

	$\mathcal{B}_1\mathcal{B}_1$	$\mathcal{B}_1\mathcal{B}_2$	$\mathcal{B}_2\mathcal{B}_2$
$\mathcal{A}_1\mathcal{A}_1$	$-\gamma_1 - \gamma_2$	$-\gamma_1$	$-\gamma_1 + \gamma_2$
$\mathcal{A}_1\mathcal{A}_2$	$-\gamma_2$	0	γ_2
$\mathcal{A}_2\mathcal{A}_2$	$\gamma_1 - \gamma_2$	γ_1	$\gamma_1 + \gamma_2$

of allelic substitution at locus one and two, respectively. For brevity, we call them the (substitutional) effects of the loci. The trait is assumed to be under stabilizing selection toward an intermediate phenotypic optimum. The fitness of individuals with genotypic value G is designated as $W(G) \geq 0$.

In Sections 2.2–2.4, we assume that the fitness function has its optimum at zero, the genotypic value of the double heterozygote. We scale the fitness function such that $W(0) = 1$, and suppose that it is symmetric with respect to the optimum, i.e., $W(-G) = W(G)$, and decreases monotonically from the optimum. This yields the general symmetric viability model with the genotypic fitness values as in II(1.22) and $a = 1 - W(\gamma_1 - \gamma_2)$, $b = 1 - W(\gamma_1)$, $c = 1 - W(\gamma_2)$, and $d = 1 - W(\gamma_1 + \gamma_2)$. Here, these parameters satisfy $0 \leq a, c \leq b < d \leq 1$, $c > 0$, and $b > a$.

2.2 QUADRATIC STABILIZING SELECTION

A particularly simple model of stabilizing selection is obtained if the fitness of individuals with genotypic value G is assumed to deviate quadratically as G deviates from the optimum, i.e.,

$$W(G) = 1 - sG^2\,, \qquad |G| \leq s^{-1/2}\,. \tag{2.1}$$

This is called quadratic stabilizing selection, or the *quadratic optimum model*, and yields the genotypic fitness values of II(1.22), where

$$a = s(\gamma_1 - \gamma_2)^2\,,\ b = s\gamma_1^2\,,\ c = s\gamma_2^2\,,\ d = s(\gamma_1 + \gamma_2)^2\,. \tag{2.2}$$

These parameters satisfy the relation

$$a + d = 2(b + c)\,. \tag{2.3}$$

Therefore, in the terminology of Chapter II.1.3, we have $l = 0$. The requirement $W(G) \geq 0$ implies $d = s(\gamma_1 + \gamma_2)^2 \leq 1$. This model was first investigated by Wright (1935b, 1952), and more recently by Hastings (1987) and Gavrilets and Hastings (1993, 1994b), who found all equilibria and studied the transient dynamics of the mean genotypic value and of the genetic variance.

Our aim is to explore how much genetic variability can be maintained by selection in this model. Therefore, we first determine the possibly stable equilibria, then present their stability conditions, and finally investigate how the level of genetic variability depends on the recombination rate and on the effects of the alleles. Proofs are mainly based on Karlin and Feldman's (1970) analysis of the symmetric viability model, and may be found in Bürger and Gimelfarb (1999).

Equilibria

We assume $r > 0$. Then there are up to nine equilibria, seven of which may be stable (but not simultaneously). We denote the possibly stable ones by E_1 – E_7. There are four types of equilibria: (i) a fully polymorphic equilibrium that is symmetric every allele frequency equal to $\frac{1}{2}$ (it will be denoted by E_1); (ii) a pair of fully polymorphic equilibria, which, in the language of the symmetric viability model, are called unsymmetric because their coordinates do not satisfy simple symmetry relations (they are denoted by E_2 and E_3); (iii) a pair of equilibria (E_4 and E_5) with the major locus polymorphic and the minor locus fixed; (iv) two monomorphic equilibria corresponding to fixation of A_1B_2 or A_2B_1 (E_6 and E_7). Two further monomorphic equilibria, corresponding to fixation of A_1B_1 and A_2B_2, always exist but are unstable.

The symmetric equilibrium E_1 always exists. It is easily calculated from II(1.25) and II(1.27), and given by

$$E_1: \quad \hat{x}_1 = \hat{x}_4 = \tfrac{1}{4} + \hat{D}_1 \,, \ \hat{x}_2 = \hat{x}_3 = \tfrac{1}{4} - \hat{D}_1 \,, \tag{2.4a}$$

where

$$\hat{D}_1 = \frac{1}{4s\gamma_1\gamma_2}\left[r - \sqrt{s^2\gamma_1^2\gamma_2^2 + r^2}\right] \tag{2.4b}$$

is the linkage disequilibrium.

Two further interior equilibria, E_2 and E_3, which are unsymmetric, may exist. Introducing the coordinates

$$x = x_1 - x_4 \,, \ y = x_2 - x_3 \,, \ z = x_1 + x_4 - x_2 - x_3 \,, \tag{2.5}$$

one of them is given by

$$E_2: \quad \hat{x} = \frac{(\gamma_1 + \gamma_2)[r - s(\gamma_1 - \gamma_2)^2]\sqrt{R}}{8s^{1/2}r^{3/2}\gamma_1\gamma_2} \tag{2.6a}$$

$$\hat{y} = \frac{(\gamma_1 - \gamma_2)[-r + s(\gamma_1 + \gamma_2)^2]\sqrt{R}}{8s^{1/2}r^{3/2}\gamma_1\gamma_2} \tag{2.6b}$$

$$\hat{z} = -\left(\frac{\gamma_1}{\gamma_2} + \frac{\gamma_2}{\gamma_1}\right) + \frac{3r}{4s\gamma_1\gamma_2} + \frac{s\gamma_1\gamma_2}{4r}\left(\frac{\gamma_1}{\gamma_2} - \frac{\gamma_2}{\gamma_1}\right)^2 , \tag{2.6c}$$

where

$$R = 3r^2 + 2rs(\gamma_1^2 + \gamma_2^2) - s^2(\gamma_1^2 - \gamma_2^2)^2 \,, \tag{2.7}$$

and the other, E_3, is obtained from E_2 by the interchanges $\hat{x}_1 \leftrightarrow \hat{x}_4$ and $\hat{x}_2 \leftrightarrow \hat{x}_3$ (or, equivalently, taking $-\hat{x}$ and $-\hat{y}$). It is not difficult to show that E_2 and E_3 exist if and only if

$$r_1 < r < r_2 \,, \tag{2.8}$$

where

$$r_1 = -\tfrac{1}{3}s(\gamma_1^2 + \gamma_2^2) + \tfrac{2}{3}s\sqrt{\gamma_1^4 - \gamma_1^2\gamma_2^2 + \gamma_2^4} \tag{2.9}$$

is the positive root of the equation $R = 0$ and

$$r_2 = \min\{s(\gamma_1 - \gamma_2)^2, \tfrac{1}{3}s(\gamma_1^2 - \gamma_2^2)\} \,. \tag{2.10}$$

Next, there may be two edge equilibria, E_4 and E_5, with the major locus polymorphic:

$$E_4: \quad \hat{x}_1 = \hat{x}_3 = 0 \,, \ \hat{x}_2 = \tfrac{1}{2} + \frac{\gamma_2}{\gamma_1} \,, \ \hat{x}_4 = \tfrac{1}{2} - \frac{\gamma_2}{\gamma_1} \,, \tag{2.11}$$

$$E_5: \quad \hat{x}_2 = \hat{x}_4 = 0 \,, \ \hat{x}_1 = \tfrac{1}{2} - \frac{\gamma_2}{\gamma_1} \,, \ \hat{x}_3 = \tfrac{1}{2} + \frac{\gamma_2}{\gamma_1} \,. \tag{2.12}$$

They exist (i.e., are in the simplex S_4) if and only if $\gamma_1 > 2\gamma_2$.

Finally, there are four corner (vertex) equilibria, corresponding to fixation of one of the gametes. They always exist, but only the vertices corresponding to fixation of $\mathcal{A}_1\mathcal{B}_2$ or $\mathcal{A}_2\mathcal{B}_1$ can be stable. These are denoted by E_6 and E_7, respectively, and their coordinates are:

$$E_6: \quad \hat{x}_2 = 1\,,\ \hat{x}_1 = \hat{x}_3 = \hat{x}_4 = 0\,, \tag{2.13}$$

$$E_7: \quad \hat{x}_3 = 1\,,\ \hat{x}_1 = \hat{x}_2 = \hat{x}_4 = 0\,. \tag{2.14}$$

If $r = r_1$, then $E_1 = E_2 = E_3$, and if $r = r_2$, then E_2 and E_3 coincide either with E_4 and E_5 (if $\gamma_1 > 2\gamma_2$) or with E_6 and E_7 (otherwise). All polymorphic equilibria exhibit negative (repulsion) linkage disequilibrium and all orbits eventually enter the region where $D \le 0$.

▷ To prove the last statement (which was not proved in Bürger and Gimelfarb 1999), we denote $Z = x_2x_3/x_1x_4$ and observe that $D = x_1x_4(1-Z)$. Thus, $D \le 0$ if and only if $Z \ge 1$. The proof consists of two steps. First, we show that every trajectory enters the region $D \le 0$, then that every trajectory remains there. The first assertion is easily proved because a straightforward calculation shows that $W_2W_3 \ge W_1W_4 + s\gamma_1\gamma_2$. Then we obtain from the recursion relations II(1.4) that $Z' \ge ZW_2W_3/(W_1W_4) > Z$ if $D > 0$, i.e., Z is a Lyapunov function on the region $D > 0$.

The proof of positive invariance of $D \le 0$ is more difficult. First, we observe that

$$(x_2W_2 + rD)(x_3W_3 + rD) = (x_1W_1 - rD)(x_4W_4 - rD) + A\,,$$

where $A = x_2x_3W_2W_3 - x_1x_4W_1W_4 + rD\overline{W}$. The assertion is proved if we can show that $A \ge 0$ if $D \le 0$, because then we have

$$Z' = 1 + A/[(x_1W_1 - rD)(x_4W_4 - rD)] \ge 1\,.$$

The expression A can be written as

$$A = 2\gamma_1\gamma_2x_1x_4A_1 - D(A_2 + sA_3) - 2Ds^2A_4 + 2D^2s^2(\gamma_1-\gamma_2)^2\,,$$

where A_1, $A_2 + sA_3$, and A_4 are given and estimated below. It is sufficient to show that they are nonnegative. We have

$$\begin{aligned} A_1 &= 2 + s[\gamma_1\gamma_2(x_1^2 - x_2^2 - x_3^2 + x_4^2) \\ &\qquad - 4\gamma_1^2(x_2x_4 + x_1x_3 + 2x_1x_4) - 4\gamma_2^2(x_3x_4 + x_1x_2 + 2x_1x_4)] \\ &\ge s[2(\gamma_1-\gamma_2)^2(x_1-x_4)^2 + \text{positive terms}]\,, \end{aligned}$$

where $2 \ge 2s(\gamma_1+\gamma_2)^2$ has been used. Next we obtain, by assuming $1 \ge 2s(\gamma_1+\gamma_2)^2$ (which is somewhat stronger than the assumption in the model),

$$\begin{aligned} A_2 + sA_3 &= 1 + (\gamma_1-\gamma_2)^2(x_2^2 + x_3^2) \\ &\quad + s[(\gamma_1+\gamma_2)^2(x_1^2+x_4^2) + 2\gamma_1^2(x_1x_2 + x_3x_4) + 2\gamma_2^2(x_1x_3 + x_2x_4) \\ &\qquad - 2(\gamma_1-\gamma_2)^2 - 4\gamma_1\gamma_2(x_1+x_4)] \\ &\ge (\gamma_1-\gamma_2)^2(x_2^2+x_3^2) + s(\gamma_1+\gamma_2)^2(x_1^2+x_4^2) \\ &\quad + 2s[\gamma_1^2(x_1x_2 + x_3x_4) + \gamma_2^2(x_1x_3 + x_2x_4) + 2\gamma_1\gamma_2(1 + x_2 + x_3)]\,. \end{aligned}$$

Finally, we have

$$A_4 = \gamma_1^2\gamma_2^2(x_1^2 + x_4^2) + (\gamma_1 - \gamma_2)^2[\gamma_1^2(x_1x_3 + x_2x_4) + \gamma_2^2(x_1x_2 + x_3x_4)] + 2(\gamma_1 - \gamma_2)^4 x_1x_4 \; .$$

Since all these expression are nonnegative, the proof is finished. It is not difficult to *check* the validity of the above equations and estimates with *Mathematica* (Wolfram 1996). To this end, it is advisable to make all polynomials homogeneous. ◁

Stability

The stability properties of these equilibria depend only upon the relation between the parameters γ_1, γ_2, and r/s. They are summarized below. A graphical representation is displayed in Figure 2.1.

The polymorphic symmetric equilibrium E_1 is asymptotically stable if and only if

$$r \le r_1 \; . \tag{2.15}$$

Numerical iterations of the recursion relations suggest that it is globally stable.

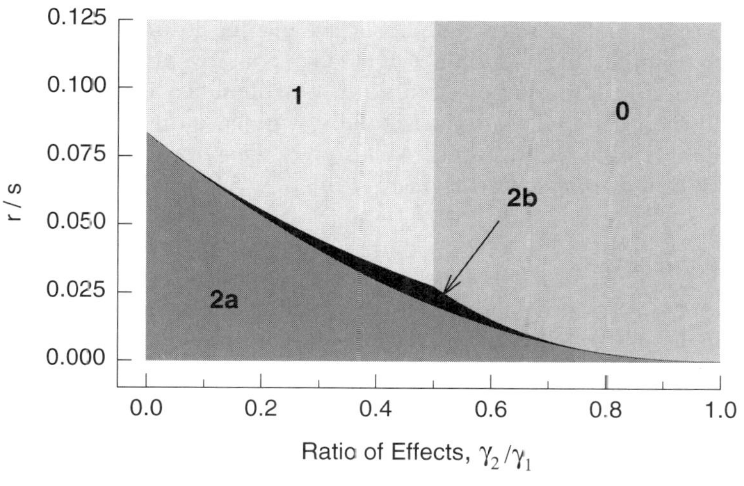

Figure 2.1 Regions of stability of the four types of equilibria as a function of the ratio of allelic effects, γ_2/γ_1. The effects are normalized so that $\gamma_1 + \gamma_2 = \frac{1}{2}$. Thus, for any choice of γ_2/γ_1, the most extreme phenotypes have fitness $1 - \frac{1}{4}s$. The following abbreviations are used for the regions of stability. 0: two monomorphic equilibria; 1: two equilibria with one (the major) locus polymorphic; 2a: the symmetric equilibrium with both loci polymorphic; 2b: two unsymmetric equilibria with both loci polymorphic. The upper boundary of region 2a is given by r_1/s, (2.9), the upper boundary of 2b by r_2/s, (2.10). The boundary between the regions 0 and 1 is at $\gamma_2/\gamma_1 = \frac{1}{2}$ and intersects with region 2b at $r/s = r_2/s = \frac{1}{36}$. It may be noted that $0 \le r/s \le \frac{1}{8}$ covers the whole range only if $s = 4$, which is the strongest possible selection as it renders the most extreme genotypes lethal. For smaller s (weaker selection), the maximum value of r/s is larger than $\frac{1}{8}$, hence the proportion of parameter values yielding stable two-locus polymorphisms is correspondingly smaller.

The polymorphic unsymmetric equilibria, E_2 and E_3, apparently are stable whenever they exist, i.e., if and only if (2.8) holds. The stability analysis is complicated because not all eigenvalues can be determined explicitly. Local stability can be proved if r is slightly larger than r_1, and if r is slightly smaller than r_2. Numerical iterations of II(1.4) suggest that the unsymmetric equilibria are stable for all r between r_1 and r_2.

The one-locus polymorphic equilibria, E_4 and E_5, are asymptotically stable if and only if they exist and $r \geq \frac{1}{3}s(\gamma_1^2 - \gamma_2^2)$, i.e., if

$$\gamma_1 > 2\gamma_2 \qquad \text{and} \quad r \geq r_2 \, . \tag{2.16}$$

The monomorphic equilibria, E_6 and E_7, are stable if

$$\gamma_1 \leq 2\gamma_2 \qquad \text{and} \quad r \geq r_2 \, . \tag{2.17}$$

Thus, if $0 < r \leq r_1$, then E_1 is asymptotically stable (and apparently globally stable). When $r = r_1$, E_1 loses its stability and the unsymmetric equilibria E_2 and E_3 arise by a pitchfork bifurcation (i.e., when the bifurcation parameter r assumes the value r_1, the stable equilibrium E_1 splits into three equilibria, two new ones, E_2 and E_3, which are stable for $r > r_1$, and the old one, E_1, which becomes unstable). For $r > r_1$, E_2 and E_3 are asymptotically stable and converge to E_6 and E_7 as $r \to s(\gamma_1 - \gamma_2)^2$ if $\gamma_1 \leq 2\gamma_2$, whereas they converge to E_4 and E_5 as $r \to \frac{1}{3}s(\gamma_1^2 - \gamma_2^2)$ if $\gamma_1 > 2\gamma_2$. In the first case, the equilibria E_4 and E_5 do not exist. When the pair of unsymmetric equilibria hits a pair of boundary equilibria (either E_4 and E_5, or E_6 and E_7), a transcritical bifurcation occurs, i.e., the boundary equilibria become stable and the equilibria E_2 and E_3 loose their stability and leave the simplex. The boundary equilibria remain stable for all larger values of r. Therefore, in this model no more than two equilibria can be simultaneously stable.

The genetic variance

The most important quantity for measuring genetic variability is the additive genetic variance, because it determines the response to natural and artificial selection. In the present additive model, total genetic and additive genetic variance coincide. The mean genotypic value and the genetic variance are calculated before selection, so that the population is in Hardy–Weinberg equilibrium. Because dominance is absent, it is sufficient to calculate the genetic variance among (haploid) gametes and multiply it by two. The mean haploid genotypic value (among gametes), $\overline{G}_h$, is simply

$$\overline{G}_h = \tfrac{1}{2}(\gamma_1 + \gamma_2)(x_4 - x_1) + \tfrac{1}{2}(\gamma_1 - \gamma_2)(x_3 - x_2) \, . \tag{2.18}$$

From this, the deviation of the mean genotypic value at equilibrium from the optimum is calculated to be 0, $\sqrt{R}/(2\sqrt{rs})$, γ_2, and $\gamma_1 - \gamma_2$ for E_1, E_2 and E_3, E_4 and E_5, and E_6 and E_7, respectively.

The genetic variance of the character is computed to be

$$\begin{aligned}\sigma_G^2 = &\, \mathrm{E}[\gamma^2] + \gamma_1\gamma_2(x_1 + x_4 - x_2 - x_3) \\ &- \tfrac{1}{2}[(\gamma_1 + \gamma_2)(x_4 - x_1) + (\gamma_1 - \gamma_2)(x_3 - x_2)]^2 \, ,\end{aligned} \tag{2.19}$$

where

$$\mathrm{E}[\gamma^2] = \tfrac{1}{2}(\gamma_1^2 + \gamma_2^2) \tag{2.20}$$

is the average squared substitutional effect. This yields the following equilibrium genetic variances:

$$\hat{\sigma}_G^2(\mathrm{E}_1) = \mathrm{E}[\gamma^2] + 4\gamma_1\gamma_2\hat{D}_1 \,, \tag{2.21}$$

$$\hat{\sigma}_G^2(\mathrm{E}_2, \mathrm{E}_3) = \frac{3}{8rs}[r^2 - 2rs(\gamma_1^2 + \gamma_2^2) + s^2(\gamma_1^2 - \gamma_2^2)^2] \,, \tag{2.22}$$

$$\hat{\sigma}_G^2(\mathrm{E}_4, \mathrm{E}_5) = \tfrac{1}{2}(\gamma_1^2 - 4\gamma_2^2) \,, \tag{2.23}$$

and

$$\hat{\sigma}_G^2(\mathrm{E}_6, \mathrm{E}_7) = 0 \,. \tag{2.24}$$

Although two equilibria may be simultaneously stable, it is important to note that alternative stable equilibria have the same (absolute value of the) mean phenotype and the same genetic variance. The reason is that the equilibria are symmetric with respect to the interchanges $\hat{x}_1 \leftrightarrow \hat{x}_4$ and $\hat{x}_2 \leftrightarrow \hat{x}_3$.

Figure 4.1 displays the deviation of the equilibrium mean phenotype from the optimum, the genetic variance, and the amount of linkage disequilibrium as a function of the ratio of effects, γ_2/γ_1, and compares it with the corresponding quantities in a model with mutation.

Using the formulas of Chapter II.1.3, the mean fitness is easily calculated at the equilibria. It is independent of r at E_4, E_5, E_6, and E_7, and strictly monotone decreasing as a function of r at E_1, E_2, and E_3. This provides a simple example for the frequently observed fact that recombination may reduce mean fitness at equilibrium and induce a so-called recombination load.

For equal effects and arbitrary symmetric fitness functions, Hastings (1987) proved that the mean phenotype converges to its optimum monotonically. With unequal effects this is no longer the case, because at the equilibria $\mathrm{E}_2 - \mathrm{E}_7$ the mean phenotype is generally not zero and, as shown by numerical iteration, the approach to equilibrium is not always monotone.

The transient dynamics was studied by Gavrilets and Hastings (1994b, 1995). They identified different time scales for the convergence to equilibrium. For equal effects, the mean phenotype approaches the optimum quickly, faster than other phenotypic quantities or the gamete frequencies, so that after a short time the mean is very near to the optimum, but quantities such as the genetic variance still have changed little (cf. Bulmer 1980). Gale and Kearsey (1968) and Kearsey and Gale (1968) were the first to observe that stabilizing selection can maintain two-locus polymorphisms if the effects are sufficiently unequal. They performed numerical calculations for a triangular fitness function, i.e., one that decreases linearly from its optimum. Apparently, they failed to discover the unsymmetric equilibria, but concluded that convergence to equilibrium does not always occur. Numerical iteration of the difference equations II(1.4) with quadratic selection shows that all orbits converge to equilibrium, but convergence may be very slow, in particular, for tightly linked loci.

Here is a summary of the main findings:

• **2.1** *If the effects of the two loci are sufficiently different ($\gamma_1 > 2\gamma_2$), then substantial genetic variance is maintained for any recombination rate. In this case, one or both loci, depending on the recombination rate, are stably polymorphic.*

The reason is that no doubly homozygous genotype is close to the optimum and, from double heterozygotes, a high recombination rate would generate many double homozygotes with extreme phenotype and low fitness. Therefore, if recombination is strong relative to selection, the minor locus is fixed and the major locus is kept polymorphic by overdominance. With tight linkage, both gametes $\mathcal{A}_1\mathcal{B}_2$ and $\mathcal{A}_2\mathcal{B}_1$ are maintained at high frequency.

• **2.2** *If the effects of the two loci are not very different ($\gamma_1 < 2\gamma_2$), then tight linkage, i.e., $r < s(\gamma_1 - \gamma_2)^2$, is necessary for maintaining genetic variation. In this case, one of the doubly homozygous genotypes $\mathcal{A}_1\mathcal{A}_1\mathcal{B}_2\mathcal{B}_2$ or $\mathcal{A}_2\mathcal{A}_2\mathcal{B}_1\mathcal{B}_1$ is close to the optimum and, therefore, fixed unless linkage is extremely tight.*

• **2.3** 1. *If $r < r_2$, an asymptotically stable two-locus polymorphism always exists, whereas if $r > \frac{1}{3}$, stable two-locus polymorphisms never exist.*
2. *For any value of r, the equilibrium genetic variance decreases from its maximum value $\mathrm{E}[\gamma^2]$ to 0 as γ_2/γ_1 increases from 0 to 1.*
3. *For fixed γ_2/γ_1, the genetic variance has a maximum for intermediate recombination rates. If the effects are not very different ($\gamma_1 < 2\gamma_2$), then the variance is lower under loose linkage than under very tight linkage; otherwise the reverse is true. Thus, tighter linkage may increase the degree of polymorphism at equilibrium, but at the same time reduce the genetic variance.*

• **2.4** *The mean phenotype approaches its equilibrium value rapidly. However, for unequal effects the equilibrium does not coincide with the optimum unless linkage is extremely tight. Convergence of the genetic variance and of the gamete frequencies may be very slow. Therefore, if there is substantial initial genetic variability and the selection regime is such that it eventually depletes genetic variability, this process may be very slow even for relatively strong selection.*

2.3 INDEPENDENT LOCI UNDER STABILIZING SELECTION

The assumption of a quadratic fitness function is mathematically convenient and is also a good approximation to smooth concave fitness functions as long as selection is weak. However, it is important to know to what extent the detailed shape of the fitness function influences the equilibrium structure. Nagylaki (1989a) investigated the two-locus model introduced above for general fitness functions that are symmetric with respect to the optimum and decrease monotonically from it. To make the analysis tractable, he assumed that linkage disequilibrium can be neglected, which will be the case if selection is weak compared with recombination. With this assumption, the dynamics of the model simplifies considerably, because it can be described in terms of allele frequencies, instead of gamete frequencies.

Let the frequencies of alleles $\mathcal{A}_1$ and $\mathcal{B}_1$ (the 'minus' alleles at the major and minor locus, respectively) be denoted by

$$p_1 = p_{\mathcal{A}_1} = x_1 + x_2 \quad \text{and} \quad p_2 = p_{\mathcal{B}_1} = x_1 + x_3 , \tag{2.25}$$

respectively, and those of $\mathcal{A}_2$ and $\mathcal{B}_2$ by $q_1 = 1-p_1$ and $q_2 = 1-p_2$. Then an elementary calculation shows that the dynamics of p_1 and p_2 can be expressed as

$$p_i' = p_i \frac{\partial \overline{W}}{\partial p_i} \Big/ \sum_{i=1}^{2} \left(p_i \frac{\partial \overline{W}}{\partial p_i} + q_i \frac{\partial \overline{W}}{\partial q_i} \right) , \tag{2.26}$$

where all allele frequencies are treated as independent in the partial differentiations and

$$\overline{W} = 1 - a(p_1^2 q_2^2 + q_1^2 p_2^2) - 2b(p_1^2 + q_1^2)p_2 q_2 - 2c p_1 q_1 (p_2^2 + q_2^2) - d(p_1^2 p_2^2 + q_1^2 q_2^2) \tag{2.27}$$

is the mean fitness. The state space is the unit square in the (p_1, p_2)-plane. Formally, the recursion (2.26) is similar to the single-locus recursion I(9.14). In particular, it has the property that mean fitness is nondecreasing. Therefore, locating all the equilibria and determining their stability properties provides a global analysis of the evolution of the population.

Nagylaki (1989a) showed that there are six possible equilibrium structures, or convergence patterns, and provided examples of each of them. The following general conclusions can be drawn:

• **2.5** *If a fitness function has the property that $W(G) \to 0$ as $|G| \to \infty$, and if for fixed γ_2 the ratio γ_1/γ_2 of effects of the major to the minor locus exceeds a critical value, then the symmetric equilibrium $(p_1 = p_2 = \frac{1}{2})$ is globally asymptotically stable; hence, a two-locus polymorphism is maintained.*

This does not contradict the analysis of the quadratic optimum model (2.1), where for $r > \frac{1}{3}$ no stable interior equilibrium exists, because the nonnegativity of $W(G)$ imposes the restriction $s(\gamma_1^2 + \gamma_2^2) = d \le 1$ on the effect γ_1 of the major locus.

Next, let us assume that the fitness function can be written in the form $W(G) = 1 - s\,|G|^\kappa + o(|G|^\kappa)$, $\kappa > 0$.

• **2.6** *Let selection be weak at the minor locus (γ_2 small).*
1. *If $\kappa < 2$, i.e., if the fitness function decreases rapidly near the optimum, then the symmetric equilibrium is globally asymptotically stable for every fixed γ_1 and s, provided γ_1/γ_2 exceeds a critical value*
2. *If $\kappa \ge 2$, in particular if $W(G)$ is smooth near the optimum, then a stable two-locus polymorphism exists only if selection at the major locus is sufficiently strong (γ_1 sufficiently large).*

For a *quadratic fitness function*, the results obtained from an analysis of (2.26) are in accordance with the more general results for $r > \frac{1}{3}$ in Section 2.2: if $\gamma_1 \le 2\gamma_2$, then the only stable equilibria are those where both loci are monomorphic, i.e., $(p_1, p_2) = (1, 0)$, $(0, 1)$ (E_6, E_7); if $\gamma_1 > 2\gamma_2$, then two stable equilibria (corresponding to E_4 and E_5) exist with the major locus kept polymorphic by marginal overdominance, and the minor locus monomorphic.

For the widely used *Gaussian fitness function*

$$W(G) = e^{-sG^2} , \tag{2.28}$$

different results are obtained. If the effects of both loci are sufficiently small (basically such that fitness is concave on the range of all genotypic values), then the same two cases as for quadratic fitness occur. However, for strong selection at the major or at both loci, the symmetric equilibrium may be simultaneously stable with either E_4 and E_5 or with E_6 and E_7, and a pair of interior unstable unsymmetric equilibria exists. If selection is very strong and the effects are sufficiently unequal (approximately, $\gamma_1 > 2\gamma_2$), then the symmetric equilibrium is globally asymptotically stable. The fitness function $W(G) = 1/(1+G^2)$ has a qualitatively identical equilibrium structure.

For a *triangular fitness function*, there are four possible convergence patterns. With increasing γ_1/γ_2, first the two patterns occuring in the quadratic optimum model arise successively. Then, if $\gamma_1/\gamma_2 > \frac{5}{2}$, the symmetric equilibrium is simultaneously stable with the one-locus polymorphisms E_4 and E_5. Finally, for very different effects, $\gamma_1/\gamma_2 > 2 + \sqrt{2}$, the symmetric equilibrium is globally asymptotically stable. Comparison with the numerical results of Gale and Kearsey (1968) suggests that the inclusion of linkage disequilibrium would relax the conditions for the existence of stable two-locus polymorphisms.

An important general conclusion can be drawn from a result of Karlin and Feldman (1970) for the symmetric viability model, which states that for sufficiently tight linkage and any set of selection coefficients there is always an asymptotically stable symmetric equilibrium (cf. Chapter II.1.3). This implies that, for *any* symmetric fitness function and r small enough, both loci are stably polymorphic (Gavrilets and Hastings, 1993).

2.4 STRONG SELECTION ON ALLELES WITH LARGE EFFECTS

In accordance with intuition, in the models considered so far, all equilibria exhibit either linkage equilibrium or negative linkage disequilibrium. This, however, is not necessarily the case in the full two-locus model II(1.4) if the fitness function is such that the parameters b and d in the symmetric viability model satisfy $2b > d$. This inequality holds, for instance, if both loci have equal but large effects, and the fitness function is Gaussian. With equal effects, the condition $2b > d$ means that the fitness function is not concave on the range of genotypic values. Actually, Gavrilets and Hastings (1994b) proved that if $2b > d$ and if selection is sufficiently strong relative to recombination, a asymptotically stable symmetric equilibrium with positive (coupling) linkage disequilibrium exists. Gimelfarb (1996a) extended this result to alleles with unequal effects and showed that under sufficiently strong Gaussian selection three symmetric equilibria coexist, two of them being stable. Linkage disequilibrium at one of these is negative, whereas it is positive at the other. He also showed numerically that in a three-locus model weaker selection suffices for maintaining a stable equilibrium with positive linkages disequilibrium.

As an illustration, let us consider double truncation selection, which, effectively, is equivalent to extremely strong Gaussian stabilizing selection on both loci. For equal effects at both loci, the symmetric viability model with $a = 0$ and $b = c = d = 1$ is obtained in the most extreme case. For unequal effects, one obtains $a = b = c = d = 0$.

Both models are easily analyzed. In the first case, after one generation of selection, $x_1 = x_4$ holds. In this plane there exist three interior symmetric equilibria with $x_2 = x_3$ if r is below a critical value $\tilde{r}$, where $\tilde{r} \approx 0.107$ (otherwise there is only one), and the two chromosome fixation equilibria $x_2 = 1$ and $x_3 = 1$. The latter two are asymptotically stable, whereas from the three symmetric equilibria only the one with positive linkage disequilibrium (the one nearest to $x_1 = x_4 = \frac{1}{2}$) is asymptotically stable.

For unequal effects, after one generation of selection $x_1 = x_4$ and $x_2 = x_3$ holds and, provided $r < \frac{1}{4}$, three symmetric equilibria exist with $\hat{D} = 0$ and $\hat{D} = \pm\frac{1}{4}\sqrt{1-4r}$. The latter two are asymptotically stable.

This kind of selection is unlikely to be of much relevance in the context of quantitative traits, because environmental effects tend to smooth and weaken selection on the genotypic values relative to selection on the phenotypic values (cf. Chapter V.1.2).

2.5 *ARBITRARY POSITION OF THE OPTIMUM*

The models investigated so far assumed that the optimum coincides with the value of the double heterozygote. If this assumption is not made, the genotypic fitnesses no longer conform with the symmetric viability model. Let us consider the quadratic optimum model with arbitrary position P_O of the optimum. Then fitness is given by

$$W(G) = 1 - s(G - P_O)^2 \ . \tag{2.29}$$

From this, the genotypic fitnesses can be calculated using Table 2.1, because we still assume additive gene action.

The resulting model is complex and no complete analysis is available. Hastings and Hom (1990) performed a complete analysis of this generalized quadratic optimum model under the assumptions of weak selection and no linkage disequilibrium. Gavrilets and Hastings (1993) analyzed the local stability properties of the boundary equilibria for the general model with linkage and performed a bifurcation analysis of certain interior equilibria. We report their basic results and explore how the equilibrium structure and the amount of genetic variation maintained depends upon the position of the optimum.

Let us first assume that recombination is strong relative to selection. Then no interior equilibria are stable. We consider the equilibrium structure as a function of P_O. For reasons of symmetry, it is sufficient to let P_O increase from zero. The special case $P_O = 0$ was treated above: depending on γ_2/γ_1, either two monomorphic equilibria (E_6 and E_7) or two equilibria with the major locus polymorphic (E_4 and E_5) are stable. They can also be stable for a range of positive values of P_O. As in (2.25), let us denote by p_1 and p_2 the frequencies of $\mathcal{A}_1$ and $\mathcal{B}_1$, respectively. Then the monomorphic equilibrium E_6, with allele frequencies $(\hat{p}_1, \hat{p}_2) = (1, 0)$, is stable if and only if

$$-\gamma_1 + \tfrac{1}{2}\gamma_2 \le P_O \le \min\left\{-\tfrac{1}{2}\gamma_1 + \gamma_2, -\tfrac{1}{2}\gamma_1 + \tfrac{1}{2}\gamma_2 + \frac{r}{2s(\gamma_1 - \gamma_2)}\right\}, \tag{2.30}$$

and E_7, with $(\hat{p}_1, \hat{p}_2) = (0, 1)$, is stable if and only if

$$\max\left\{\tfrac{1}{2}\gamma_1 - \gamma_2, \tfrac{1}{2}\gamma_1 - \tfrac{1}{2}\gamma_2 - \frac{r}{2s(\gamma_1 - \gamma_2)}\right\} \le P_O \le \gamma_1 - \tfrac{1}{2}\gamma_2 \ . \tag{2.31}$$

The coordinates of the two equilibria with the major locus polymorphic and the minor monomorphic are given by

$$\hat{p}_1 = \tfrac{1}{2} + \frac{\gamma_2}{\gamma_1} - \frac{P_O}{\gamma_1}\,, \quad \hat{p}_2 = 0 \tag{2.32}$$

(this corresponds to E_4) and

$$\hat{p}_1 = \tfrac{1}{2} - \frac{\gamma_2}{\gamma_1} - \frac{P_O}{\gamma_1}\,, \quad \hat{p}_2 = 1 \tag{2.33}$$

(which corresponds to E_5). The first one is stable if and only if

$$-\tfrac{1}{2}\gamma_1 + \gamma_2 < P_O \leq \tfrac{1}{2}\gamma_2\Big[3 - \frac{s}{r}(\gamma_1^2 - \gamma_2^2)\Big]\,, \tag{2.34}$$

and the other is stable if and only if

$$-\tfrac{1}{2}\gamma_2\Big[3 - \frac{s}{r}(\gamma_1^2 - \gamma_2^2)\Big] \leq P_O < \tfrac{1}{2}\gamma_1 - \gamma_2\,. \tag{2.35}$$

Clearly, this requires that $r \geq s\gamma_2(\gamma_1 - \gamma_2)$. The reader may note that for $P_O > 0$, there are parameter regions, one where only one of these equilibria is stable and one where two different types (monomorphic and one-locus polymorphic) coexist.

Beyond $P_O = \gamma_1 - \frac{1}{2}\gamma_2$, the monomorphic equilibrium E_7 loses its stability, and an equilibrium with the minor locus polymorphic and the major one monomorphic moves into the (boundary of the) simplex and exchanges stability with E_7. It has coordinates

$$\hat{p}_1 = 0\,, \quad \hat{p}_2 = \tfrac{1}{2} + \frac{\gamma_1}{\gamma_2} - \frac{P_O}{\gamma_2}\,, \tag{2.36}$$

and exists and is stable if and only if

$$\gamma_1 - \tfrac{1}{2}\gamma_2 < P_O < \gamma_1 + \tfrac{1}{2}\gamma_2\,. \tag{2.37}$$

For certain values of γ_2/γ_1 this is the only stable equilibrium, whereas for other values it stably coexists with (2.33).

If $P_O \geq \gamma_1 + \frac{1}{2}\gamma_2$, then both loci are under directional selection and the gamete $\mathcal{A}_2\mathcal{B}_2$ is eventually fixed. No other equilibria can then be stable, because p_1 decreases from generation to generation.

In general, the interior equilibria cannot be calculated explicitly, and equilibria with different mean phenotype and genetic variance may be simultaneously stable. Tighter linkage reduces the range of stability of the boundary equilibria. Actually, as in the model with $P_O = 0$, two unsymmetric interior equilibria may bifurcate either from $(1, 0)$ and $(0, 1)$ or from the equilibria (2.32) and (2.33) as r decreases below the critical value $s\gamma_2(\gamma_1 - \gamma_2)$ given by the respective stability conditions. The unsymmetric equilibria cannot be calculated explicitly. For very tight linkage, a stable interior equilibrium (corresponding to E_1) exists that appears to be the unique stable equilibrium (cf. Gavrilets and Hastings 1993). Under the assumption of independent loci, a complete classification of the stability properties was obtained by Hastings and Hom (1990) and represented as a case map. This applies in the present model if $r/s \gg 1$.

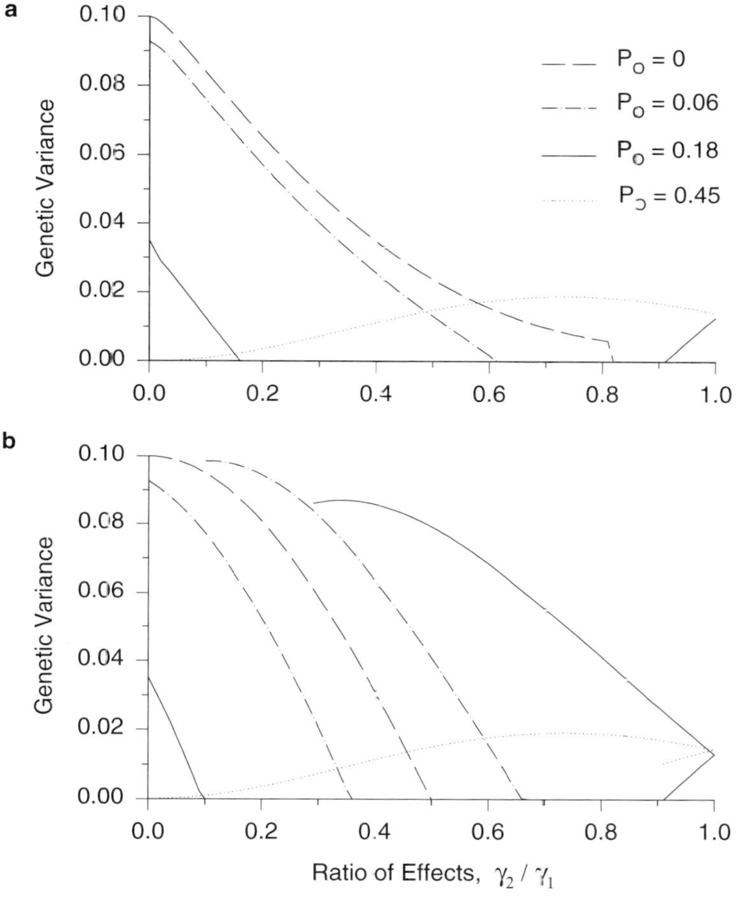

Figure 2.2 Equilibrium genetic variance for the following positions of the optimum: $P_O = 0$ (dashed line), $P_O = 0.06$ (dash-dotted line), $P_O = 0.18$ (solid line), $P_O = 0.45$ (dotted line). The selection coefficient is $s = 0.25$, the average squared effect is fixed to $\mathrm{E}[\gamma^2] = 0.1$. **a:** $r = 0.001$; **b:** $r = 0.5$.

There is no simple way to describe the position of the mean phenotype at equilibrium or the equilibrium genetic variance, because for most parameter combinations simultaneously stable equilibria exist that exhibit different levels of genetic variability. Explicit formulas for the mean phenotype and the genetic variance at the boundary equilibria are easily derived.

For the general model with linkage, numerical iteration of the recurrence relations II(1.4) with proper choice of the fitnesses is required to determine these characteristics. Figure 2.2a displays the equilibrium genetic variance for tight linkage ($r = 0.001$) as a function of γ_2/γ_1 for several values of the position P_O of the optimum. In this case, a globally asymptotically stable equilibrium always appears to exist. For $r = 0.5$, numerical calculations show that no interior stable equilibrium exists. This is in accordance with the analysis of Hastings and Hom (1990) for independent loci under weak selection. However, two equilibria with different levels of genetic variance may be simultaneously stable. Figure 2.2b illustrates the possible values of the equilibrium genetic variance for four positions of the fitness optimum.

The above results can be summarized as follows:

- **2.7** 1. *The genetic variance maintained at equilibrium may increase or decrease as the position of the fitness optimum deviates from the value of the double heterozygote.*
2. *For small or moderate deviations, substantial genetic variance is maintained for a large range of parameters, and simultaneously stable equilibria with different mean phenotype and different values of genetic variance coexist.*
3. *For sufficiently large deviations ($|P_O| \geq \gamma_1 + \frac{1}{2}\gamma_2$), both loci are under directional selection and the gamete with the most extreme genotypic value is eventually fixed.*

The general conclusion that may be drawn from all these two-locus analyses is the following:

- **2.8** *In contrast to the classical opinion based on the study of one-locus models, there is a wide variety of two-locus models and parameter combinations that maintain significant levels of genetic variation under weak as well as strong stabilizing selection.*

3. MULTILOCUS MODELS OF STABILIZING SELECTION

The final and main conclusion of the above section is in sharp contrast to the widespread view that stabilizing selection alone cannot maintain genetic variation in additive traits. It is also in contrast to analyses of mutation-selection-balance models that assume many loci in linkage equilibrium with mutational effects drawn from a continuous probability distribution (see Section 7 below). Here we study multilocus models, and attempt to answer the following questions: How many loci have to contribute to an additive quantitative trait so that no (or almost no) genetic variance is maintained at equilibrium by stabilizing selection? How does this depend on the distribution of allelic effects and the strength of stabilizing selection? What is the role of linkage? More generally, what can be said about the equilibrium structure of multilocus systems under stabilizing selection on an additive quantitative trait? Finally, we shall briefly discuss the role of dominance, epistasis, and pleiotropy.

3.1 MULTIPLE UNLINKED EQUIVALENT DIALLELIC LOCI

Wright's (1935a,b) analysis of the quadratic optimum model was extended by Latter (1960), Bulmer (1972), and Barton (1986) to multiple loci. Barton's treatment is considerably more general than those of the previous authors, because he did not assume that at equilibrium the mean phenotype coincides with the optimum. Our presentation largely follows his approach.

Consider a single character, G, which is determined additively (no dominance or epistasis) by ℓ loci, at which there are two alleles with contributions $-\frac{1}{2}\gamma$ and $\frac{1}{2}\gamma$. The population is assumed to be in global linkage equilibrium. If the frequency of the '+' allele at locus i is denoted by p_i, then the mean phenotype, $\overline{G}$, and the genetic variance, σ_G^2, which is the sum of the single-locus variances, are readily calculated to be

$$\bar{G} = 2\gamma\Big(\sum_i p_i - \tfrac{1}{2}\ell\Big) \tag{3.1}$$

and

$$\sigma_G^2 = 2\gamma^2 \sum_i p_i(1-p_i) \,. \tag{3.2}$$

The quadratic optimum model with arbitrary position of the optimum P_O is assumed, so that the fitness of individuals with phenotypic (=genotypic) value G is given by (2.29). The selection coefficient s is sufficiently small that $W(G) > 0$ for all possible phenotypes. Then the mean fitness, $\overline{W}$, of the population is

$$\overline{W} = 1 - s(\bar{G} - P_O)^2 - s\sigma_G^2 \,. \tag{3.3}$$

If selection is weak, the evolution of allele frequencies p_i at each locus can be approximated by the continuous-time dynamics I(10.17b) with $\overline{m} = \overline{W}$. Since this is a Svirezhev–Shahshahani gradient system, all trajectories converge to equilibria (Appendix A.3). For our purpose, it is useful to write the dynamics in the form

$$\dot{p}_i = s\gamma^2 p_i(1-p_i)(2p_i - 1 - 2\delta) \,, \tag{3.4}$$

where

$$\delta = \frac{\bar{G} - P_O}{\gamma} = 2\sum_i (p_i - \tfrac{1}{2}) - \frac{P_O}{\gamma} \tag{3.5}$$

is the deviation of the mean from the optimum relative to the effect of a single locus.

▷ From (3.4) we see that there may exist three equilibria at each locus. Their allele frequencies are given by $\hat{p} = 0$, $\frac{1}{2} + \delta$, or 1. The number of loci at these frequencies will be denoted by m, ν, and M, respectively. Then the equilibrium state of the whole system is described by these three numbers. For each triple (m, ν, M) with $m+\nu+M = \ell$, a large number of equilibria exists, corresponding to all different permutations of loci. However, since the effects of all loci have been assumed to be identical, these permutations have identical properties. We will therefore be concerned only with the differences between different classes of equilibria as characterized by different (m, ν, M). For given (m, ν, M), the equilibrium deviation of the mean phenotype from the optimum is

$$\delta = \left(M - m - \frac{P_O}{\gamma}\right) / (1 - 2\nu) \,. \tag{3.6}$$

The stability of the corresponding class of equilibria is determined by the eigenvalues of the Jacobian matrix $(\partial^2 \overline{W}/\partial p_i \partial p_j)$. These are:

$$\lambda = \begin{cases} -2(\frac{1}{2} + \delta) & m \text{ times} \,, & (3.7a) \\ -2(\frac{1}{2} - \delta) & M \text{ times} \,, & (3.7b) \\ -(4\nu - 2)(\frac{1}{4} - \delta^2) & \text{once if } \nu \ge 1 \,, & (3.7c) \\ 2(\frac{1}{4} - \delta^2) & \nu - 1 \text{ times if } \nu \ge 2 \,. & (3.7d) \end{cases}$$

The first point to note is that, by (3.7a,b), the only stable equilibrium for $\delta > \frac{1}{2}$ is $(\ell, 0, 0)$, whereas for $\delta < -\frac{1}{2}$, $(0, 0, \ell)$ is the single stable equilibrium. As may be seen from (3.6), $|\delta| > \frac{1}{2}$ holds if and only if $|P_O| > (\ell + \frac{1}{2})\gamma$, hence all loci are under directional selection (cf. Section 2.5).

For equilibria with $|\delta| < \frac{1}{2}$, (3.7d) implies that stability is possible only if $\nu = 0$ or 1. In both cases, (3.6) can be rearranged to show that the optimum must lie inside the range $\gamma(M - m - \frac{1}{2}) < P_O < \gamma(M - m + \frac{1}{2})$. Hence, for any particular value of P_O, stable equilibria are only possible for one combination of (m, ν, M).

From (3.6) it is obvious that if $-\frac{1}{2} \leq P_O/\gamma \leq \frac{1}{2}$, then the only stable class of equilibria is given by $(\frac{1}{2}\ell, 0, \frac{1}{2}\ell)$, provided ℓ is even, as we shall henceforth assume. If $\frac{1}{2} \leq P_O/\gamma \leq \frac{3}{2}$, then, again by (3.6), the only stable class is $(\frac{1}{2}\ell - 1, 1, \frac{1}{2}\ell)$. In general, if $2k - \frac{1}{2} \leq P_O/\gamma \leq 2k + \frac{1}{2}$, the only stable class of equilibria is given by $(\frac{1}{2}\ell - k, 0, \frac{1}{2}\ell + k)$, whereas for the other (positive) values of P_O/γ, the class $(\frac{1}{2}\ell - k - 1, 1, \frac{1}{2}\ell + k)$ is stable. An analogous argument applies to negative values of P_O/γ. ◁

Thus, we have proved:

• **3.1** *If a character under quadratic stabilizing selection is controlled by an arbitrary number of independent, exchangeable, diallelic loci, then for half of the possible positions of the optimum (within the range $|P_O| < (\ell + \frac{1}{2})\gamma$), selection maintains no genetic variation at all, whereas for the other half, variation is maintained at exactly one locus.*

Clearly, this is a highly symmetric model and it is not obvious that the results are structurally stable. In view of the results in Section 2, it would also be of interest to study the present model for a Gaussian fitness function. In Section 5, the above model will be investigated with recurrent mutation.

3.2 STATISTICAL ANALYSIS OF LINKED DIALLELIC LOCI

The mathematical analysis of multilocus systems under stabilizing selection is intractable without numerous simplifying assumptions. In addition, because the dimensionality of the parameter space and of the space of gamete frequencies increases rapidly as the number of loci increases, an explicit and analytical characterization of the equilibrium properties of multilocus models in terms of all parameters and initial conditions would be of limited value, even if the necessary analytical methods were available. Finally, parameters of genetic systems controlling quantitative traits are usually unknown or can be inferred only indirectly. Therefore, Bürger and Gimelfarb (1999) used a different approach and obtained statistical results by evaluating numerically the quantities of interest for a large number of randomly chosen parameter sets and initial conditions. This produces empirical distributions of these quantities. Given our ignorance of genetic parameters in multilocus systems, such a model provides a kind of null hypothesis. Here, we concentrate on various measures of the amount of genetic variability that is maintained under stabilizing selection, and how they depend on the number of loci if linkage and nonequivalent loci are admitted. We shall mainly, but not exclusively, report the averages of these quantities; more detailed information about their distribution may be found in Bürger and Gimelfarb, from where all reported data are taken.

As before, a quantitative character is controlled additively by ℓ diallelic loci. The contribution of one allele at each locus i is zero, whereas the contribution β_i of the other allele is a random number between zero and one. To facilitate comparison between models with different numbers of loci, it is assumed that the minimum and maximum genotypic values are always zero and one. Therefore, the actual contribution by the second allele at locus i is scaled to be $\gamma_i = \frac{1}{2}\beta_i / \sum_{j=1}^{\ell} \beta_j$. We call γ_i the allelic contribution, or the effect, of the locus. This implies that the genotypic value of the total heterozygote is always $\frac{1}{2}$, and the average allelic contribution among the ℓ loci controlling the trait is $\bar{\gamma} = 1/(2\ell)$. The quadratic optimum model is assumed with the position of the optimum coinciding with the total heterozygote, i.e.,

$$W(G) = 1 - s(G - \tfrac{1}{2})^2 . \tag{3.8}$$

The strongest possible selection occurs for $s = 4$, when the minimum and maximum phenotypes are rendered lethal. This normalization has the advantage that the strength of selection on genotypes can be compared for different numbers of contributing loci.

The statistical approach is as follows. For a genetic system with a given number of loci ($\ell = 2, 3, 4, 5$), 4000 parameter sets were constructed (effects of loci, recombination frequencies between adjacent loci, the strength of stabilizing selection). For each parameter set, allelic contributions were obtained by generating the β_i ($i = 1, 2, \ldots, \ell$) as independent random variables, uniformly distributed between 0 and 1, and transforming them into the actual allelic contribution γ_i. The strength of stabilizing selection, s, was obtained as a random variable uniformly distributed between 1 and 4, with $s = 4$ corresponding to the strongest possible selection, and $s = 1$ represents 'weak' selection (the fitness of extreme phenotypes is 75% of the fitness of the best fit phenotype). On the basis of such obtained values of allelic contributions and the strength of selection, genotypic fitnesses were calculated and substituted into the recursion relations II(2.8). Recombination frequencies between adjacent loci, $r_{i,i+1}$ ($i = 1, \ldots, \ell - 1$), were obtained as independent random variables, uniformly distributed between 0 and 0.5 (no interference). In addition, 4000 parameter sets were constructed for two-, three-, four-, and five-locus systems with random recombination between adjacent loci and with genotypic fitness values chosen as independent random variables uniformly distributed between 0 and 1, as described by Gimelfarb (1998).

For each of the 4000 parameter sets, the recursion relations II(2.8) were numerically iterated starting from 20 random initial distributions of gametes until equilibrium was reached. For each parameter set, the number of different equilibria, the gametic frequencies at each equilibrium, and the number of trajectories (initial distributions) converging to each equilibrium were recorded. Using this database, the equilibrium properties of multilocus genetic systems were analyzed.

A general feature of the equilibrium structure of genetic systems under stabilizing selection is that the probability to maintain more than one polymorphic locus drops rapidly as the number of loci increases. In fact, whereas the probability to maintain both loci polymorphic in a two-locus system is 17% (if parameters are randomly chosen as described above), in a 3-, 4-, and 5-locus system, the probability of maintaining a polymorphism in two or more loci is 3%, 1%, and less than 0.5%, respectively. The same is not true under selection with randomly assigned fitnesses, when the probability to maintain more than one locus polymorphic is higher for genetic systems with more loci.

Table 3.1 Genetic variability at stable equilibria. Parameters of stable equilibria expected for genetic systems with a given number of loci under stabilizing selection of random strength, $1 \leq s \leq 4$, and under selection with random genotypic fitnesses. Recombination is random. Entries are averages over all trajectories unless indicated otherwise.

	Stabilizing Selection				Random Fitnesses			
	Number of loci ℓ				Number of loci ℓ			
	2	3	4	5	2	3	4	5
Number of stable equilibria [1]	1.85	2.35	3.87	5.52	1.90	2.72	3.53	3.88
Polymorphic fraction of genome	0.34	0.17	0.11	0.08	0.30	0.26	0.22	0.20
Mean fitness	0.89	0.98	0.99	0.99	0.87	0.87	0.87	0.87
Deviation from optimum	0.07	0.05	0.03	0.02				
Genetic variance	0.031	0.007	0.002	0.001				
Genetic variance/$V_{\max}$	0.27	0.11	0.05	0.03				
Effect of polymorphic locus [2]	1.48	1.06	0.99	0.85				

[1] average over all parameter sets
[2] ratio of the average allelic contribution among polymorphic loci to the average allelic contribution among all loci

In the present model of stabilizing selection, the probability of maintaining a single locus polymorphic is 35%, 45%, 41%, and 41% in a 2-, 3-, 4-, and 5-locus system, respectively (for more detailed data, see Table 1 of Bürger and Gimelfarb 1999).

Table 3.1 presents some of the recorded quantities that describe the properties of stable equilibria expected for genetic systems with a given number of loci: the number of different stable equilibria per parameter set; the polymorphic fraction of the genome (the probability for a locus to be polymorphic); and the mean fitness. For stabilizing selection, the table also presents the expected deviation of the mean from the optimum; the genetic variance at equilibrium; the ratio of the genetic variance at equilibrium to the maximum genetic variance that can be maintained by the given genetic system; and the ratio of the allelic contribution at a polymorphic locus to the average allelic contribution among all loci in the system. The maximum genetic variance that can be maintained in linkage equilibrium by an additive trait controlled by ℓ loci with effects $\{\gamma_i\}$ is $V_{\max} = \frac{1}{2}\sum_i \gamma_i^2$, whereas the average allelic contribution among loci is $\bar{\gamma} = 1/(2\ell)$. If the selection intensity is fixed at either $s = 4$ or $s = 1$, and only allelic effects and recombination rates are randomly chosen, then for all quantities shown, the results deviate by less then 3% from those in the table (see Table 3 in Bürger and Gimelfarb).

Table 3.1 demonstrates that with increasing number of loci, the polymorphic fraction of the genome (i.e., the expected proportion of polymorphic loci) decreases at a much higher rate under stabilizing selection than under selection with random fitnesses. Indeed, as a function of ℓ, the polymorphic fraction of the genome under stabilizing selection, is very closely approximated by $0.99\ell^{-1.58}$, whereas for random fitnesses it decreases proportionally to $\ell^{-0.45}$. The table also shows that with increasing number of loci in the genetic system, the expected number of simultaneously stable equilibria increases faster under stabilizing selection than with randomly assigned fitnesses, and for four and five loci, it is actually higher under stabilizing selection. The maximum number of stably coexisting equilibria detected by numerical iteration starting with 20 initial distributions was 6, 12, and 15, for $n = 3$, 4, and 5, respectively, and there was exactly one polymorphic locus at all these equilibria. In addition, the probability of several simultaneously stable equilibria with different degrees of polymorphism, while zero for two loci and very low for three loci, becomes substantial if the number of loci increases (Bürger and Gimelfarb 1999).

All these results indicate that historic effects may be of paramount importance in the evolution of populations under stabilizing selection. Therefore, it cannot be expected that closely related (but isolated) populations that have experienced similar selective regimes are genetically identical at all or most loci contributing to traits under stabilizing selection, even if the populations are large so that effects of random drift are negligible.

Since recombination rates between adjacent loci were generated as independently and uniformly distributed random variables, the probability of obtaining an ℓ-locus system with all loci tightly linked decreases exponentially as ℓ increases. Iterations for 4-locus genetic systems with completely linked loci show that all trajectories converge to fully polymorphic equilibria. However, only two complementary types of gametes, e.g., 0110 and 1001, are present at these equilibria and have equal frequency. Such equilibria correspond to the symmetric equilibrium E_1 in the two-locus case. Sometimes, convergence to such equilibria is extremely slow. By perturbation arguments (Karlin and McGregor 1972), these findings should extend to sufficiently small recombination rates.

The average amount of linkage disequilibrium was calculated for equilibria with at least two polymorphic loci and was found to be very small (data not shown). Given that the probability of a stable equilibrium with two or more polymorphic loci is very small if more than two loci control the trait, linkage disequilibrium can be neglected in multilocus systems under stabilizing selection.

Although for two loci, the equilibrium structure may strongly depend on the relative strength of recombination and selection (see Figure 2.1), several of the expected equilibrium properties are nearly independent. Notably, the average number of stable equilibria, the average degree of polymorphism, the average deviation of the mean phenotype from the optimum, and the average genetic variance at equilibrium are almost independent of the strength of selection and the recombination frequency. This is shown by statistical results obtained for four loci (see Table 3 of Bürger and Gimelfarb 1999) and by the analytical theory for the two-locus model of Section 2.2. For the latter, the equilibrium genetic variance, $\hat{\sigma}_G^2$, averaged over all trajectories and all parameter sets $\{r, \gamma_1, \gamma_2\}$ decreases from $\hat{\sigma}_G^2 = 0.0311$ for $s = 4$ to $\hat{\sigma}_G^2 = \ln \frac{3}{2} - \frac{3}{8} \approx 0.0305$ as $s \to 0$. Presumably, the reason is that most genetic variance comes from single-locus polymorphisms which are stable because of overdominance. In this case, allele frequencies are independent of s, but depend only on the relative disadvantage of the two homozygotes.

The explored values of selection intensities ($1 \leq s \leq 4$) cover the range from weak to the strongest possible selection, but not extremely weak or no selection. Inclusion of values $s < 1$ would change the results very little because it would not increase the parameter range much, and because all the results indicate that the equilibrium structure is nearly uneffected by changes in s, unless selection is very strong relative to recombination. Actually, Table 3.1 shows that selection imposes only a small genetic load on the population. Thus, purely on the basis of observation, one might conclude that these populations are under weak selection, even if in fact selection is strong.

The equilibrium genetic variance and the deviation of the mean from the optimum are strongly affected by the diversity of locus effects γ_i, although in a complicated manner, because of the influence of recombination and because of the frequent coexistence of stable equilibria with different levels of genetic variation; consult the figures and their discussion in Bürger and Gimelfarb (1999). The only simple conclusion that can be drawn is that the highest levels of genetic variation are maintained in systems, where one locus has a very large effect, whereas all other loci have very small effects. In this case, there is overdominance at the major locus.

A rough impression of the dependence of the equilibrium variance and the deviation of the mean from the optimum on the number of loci can be obtained from their averages over all parameter sets. Table 3.1 shows that the average deviation from the optimum decreases as the number of loci increases. It declines slightly faster than the average allelic contribution $\bar{\gamma}$. This is due to the fact that the optimum can be matched more closely in systems with more loci of varying effects.

Since the scaling is such that the average allelic contribution, $\bar{\gamma} = 1/(2\ell)$, is smaller for systems with more loci controlling a trait, it is expected that the genetic variance at equilibrium, $\hat{\sigma}^2(\ell)$, averaged over all trajectories and parameter sets $(s, \{r_i\}, \{\gamma_i\})$, must decline with increasing number of loci, ℓ. Table 3.1 shows that this is indeed the case. However, the decline occurs at a rate much faster than expected. To a close approximation, we have $\hat{\sigma}^2(\ell) = 0.52\ell^{-4.0}$. If the genetic variance is scaled relative to the maximum genetic variance that can be maintained for a given genetic system in linkage equilibrium, $V_{\max}$, the resulting overall average value of $\sigma^2(\ell)/V_{\max}$ (with $\sigma^2(\ell)$ being the variance for a given parameter set, averaged over all trajectories) still decreases in proportion to $\ell^{-2.4}$.

What is the reason for such a fast decline? First, we have already seen that under stabilizing selection, the polymorphic fraction of the genome decreases with increasing number of loci at a rate of approximately $\ell^{-1.58}$, which is much higher than under selection with random fitnesses. Secondly, if a locus maintains a polymorphism, i.e., contributes to the genetic variance, the effect of such a locus (as compared to $\bar{\gamma}$) is smaller for systems with more loci. If a trait is controlled by two loci then, as we have seen in Section 2.2, it is either the major locus or both loci that are segregating at a polymorphic equilibrium. If $\ell = 3$ or 4, Table 3.1 shows that the effect of a polymorphic locus is expected to be approximately equal to $\bar{\gamma}$, whereas if $\ell = 5$, it is expected to be smaller than the average ($0.85\bar{\gamma}$). It is also interesting to note that, in contrast to the two-locus case, for $\ell \geq 3$ the expected genetic variance at an equilibrium is higher if this equilibrium maintains several loci polymorphic. For example, with $\ell = 4$, the expected genetic variance at equilibria with two polymorphic loci is almost four times as large as at equilibria with one polymorphic locus (data not shown).

What are the most important conclusions from this statistical investigation? First, it has been demonstrated that the expected genetic variance maintained under stabilizing selection declines very rapidly from a high value in two-locus systems to an extremely low value in five-locus systems, thus providing quantitative support for the view that stabilizing selection *per se* rarely maintains genetic variation. In particular, already four- and five-locus systems exhibit equilibrium properties that are similar to those expected under the infinitesimal model. In addition, with more than three loci, the probability of equilibria involving at least two polymorphic loci is almost negligible. This implies that in such systems no linkage disequilibrium is to be expected. A further interesting finding is that many quantities, e.g., the average polymorphic fraction of the genome and the average genetic variance are virtually independent of the strength of stabilizing selection and the level of recombination.

These results suggest:

• **3.2** *Unless genetic systems underlying quantitative characters are very different from those assumed in this section, it is very unlikely, although not impossible, that appreciable levels of genetic variation are maintained at equilibrium by the selective forces resulting from direct stabilizing selection on the trait.*

3.3 NONADDITIVE GENE INTERACTION

The models investigated so far in this chapter assume that the genotypic values are determined purely additively by the genes. It is important to note that the above stabilizing-selection models are additive only on the level of genotypic values, but that there is dominance and epistasis in fitness because of the nonlinearity of the fitness function.

Wright (1935b) investigated the quadratic optimum model with exchangeable loci and completely additive effects, as well as with complete dominance and additivity between loci. In both cases he found that no genetic variance can be maintained at equilibrium. Wright's approach was extended by Kojima (1959), Lewontin (1964), and Singh and Lewontin (1966), who found through numerical simulation that, with partial dominance, genetic variation can be maintained if the optimum is shifted towards the extreme phenotypes. Reducing recombination between loci increases the range of parameters allowing for a stable polymorphic equilibrium. However, in accordance with the results of Section 3.2, an increase in the number of loci leads to a decrease in the genetic variance that is maintained.

Recent experimental results indicate that typical quantitative traits may be under control of loci with substantial epistatic effects (Chapter VII.1.2). Indeed, epistasis appears to be a potent mechanism for maintaining genetic variability. A simple additive-multiplicative model, containing the additive model as a special case, was suggested by Gimelfarb (1989). He assumed exchangeable loci, absence of dominance, and that the genotypic values can be represented as

$$G = (1-\eta)(G_{\mathcal{A}} + G_{\mathcal{B}}) + \eta(G_{\mathcal{A}}G_{\mathcal{B}}) \,, \tag{3.9}$$

where $G_{\mathcal{A}}$ and $G_{\mathcal{B}}$ denote the genotypic effects of the (diploid) genotypes at the $\mathcal{A}$ and $\mathcal{B}$-locus, respectively. The parameter η, $0 \le \eta \le 1$, describes the 'strength' of epistasis, $\eta = 0$ corresponding to complete additivity. This model is readily extended to multiple loci. Gimelfarb assumed quadratic stabilizing selection with the optimum in the middle of the range of genotypic values, and performed numerical iterations of the recursion relations with two, three, and four loci. He found that relatively large amounts of genotypic variation are maintained at equilibrium unless epistasis is very weak. In sharp contrast to the additive model (with and without dominance), with epistatic interaction the genetic variance increases almost geometrically when the number of loci increases.

3.4 PLEIOTROPY

Pleiotropy refers to the phenomenon of genes affecting different traits. It is believed that pleiotropic effects are ubiquitous because of the complexity and inter-relation of biochemical processes underlying the development of quantitative traits. In fact, genetic correlations between different traits are widespread, and pleiotropic effects are among their major causes (cf. Wright 1968, Lande 1980a, Falconer and Mackay 1996, this Chapter VII.1.2). Numerous, quite different models of pleiotropy have been suggested and investigated. Many of these are formulated in terms of quantitative genetics (see Chapters VII.4 and VII.7.5), but some are directly formulated in the framework of classical two- or multilocus models (Rose 1982; Gimelfarb 1986, 1992, 1996b; Hastings and Hom 1990).

As an example of how the two-locus symmetric viability model emerges in this context, we briefly describe a model suggested by Gimelfarb (1996b). The rational for this model is provided by experimental findings, suggesting that there are quantitative traits which appear to be under the control of a few major genes supported by numerous genes with smaller effects (Chapter VII.1.2). Assume that ℓ diallelic loci contribute additively to ℓ quantitative characters. Each locus has a major effect on one of the characters and a minor effect on the rest of them, such that each character is controlled by one major locus and by $\ell - 1$ minor loci. At each locus, one of the alleles contributes zero to the trait, whereas the contribution of the other allele depends upon whether the locus is major or minor for the trait. If i is a minor locus for trait G_j $(j \neq i)$, its contribution is assumed to be a fraction γ_{ij} of the contribution to the trait by the major locus j. If $\gamma_{ij} = 0$ for every $j \neq i$, then locus i has no pleiotropic effects; if $\gamma_{ij} = 1$ for every j, then locus i contributes equally to all traits. Each trait is assumed to be under quadratic stabilizing selection with the optimum coinciding with the genotypic value of the total heterozygous genotype. The fitness of an individual with character values $(G_1, \ldots, G_\ell)$ is $\prod_{j=1}^{\ell} W(G_j)$.

Gimelfarb showed that in this model of pleiotropy an asymptotically stable polymorphic equilibrium is maintained for a wide range of parameters. For example, if the relative contributions of all minor loci are equal and less than 25% of that of a major locus, $\gamma_{ij} = \gamma < \frac{1}{4}$ $(i \neq j)$, an ℓ-locus polymorphism with allele frequencies equal to $\frac{1}{2}$ in all ℓ loci is maintained for practically any recombination rate and any strength of selection. If the loci contribute to fewer than ℓ traits (some $\gamma_{ij} = 0$), such a polymorphism is maintained under an even wider range of parameters (an extreme and trivial case is that of every locus contributing to a different trait).

Hastings and Hom (1989, 1990) investigated certain additive multilocus-multi-character models under the assumption of linkage equilibrium, and concluded that if stabilizing selection is weak relative to recombination, then the number of polymorphic loci is at most the number of characters. For sufficiently strong selection, however, multiple asymptotically stable polymorphisms with high degrees of linkage disequilibrium can be maintained in many more loci than the number of traits they control if some of the loci have antagonistic effects on the traits (Gimelfarb 1992, Gavrilets and Hastings 1994a). As a consequence, pleiotropy might be an important factor in maintaining genetic variation.

Zhivotovsky and Gavrilets (1992) and Gavrilets (1993) derived general conditions for existence and stability of multilocus polymorphisms for a class of fitness functions that includes dominance and pairwise additive-by-additive epistasis in fitness. As special cases, this class contains the quadratic optimum model with additive effects, certain pleiotropic models, as well as some other models from quantitative genetics. In some of these cases, high levels of polymorphism are maintained (see Chapter VII.6.3).

4. MUTATION-SELECTION BALANCE: TWO LINKED DIALLELIC LOCI

Here, we introduce mutation into the two-locus model of quadratic stabilizing selection studied in Section 2.2. The primary aim is to explore the amount of genetic variation that can be maintained by a balance between mutation and stabilizing selection, and how it depends on linkage and on the magnitude of the effects of the loci.

4.1 FORMULATION OF THE MODEL

We assume quadratic stabilizing selection according to (2.1), and the basic model and notations of Sections 2.1 and 2.2. In addition, mutation occurs with equal forward and backward mutation rate at each locus, with rate μ_1 at the first locus and μ_2 at the second. Therefore, we may call γ_1 and γ_2 the mutational effects of the loci. We denote by $U = 2(\mu_1 + \mu_2)$ the mutation rate per zygote. Under the assumption that generations are discrete, and that in the life cycle selection is followed by recombination and then mutation, the genotype frequencies are easily shown to evolve according to

$$\overline{W}x_1' = x_1W_1 - r_UD + \mu_1(x_3W_3 - x_1W_1) + \mu_2(x_2W_2 - x_1W_1)\ , \tag{4.1a}$$
$$\overline{W}x_2' = x_2W_2 + r_UD + \mu_1(x_4W_4 - x_2W_2) + \mu_2(x_1W_1 - x_2W_2)\ , \tag{4.1b}$$
$$\overline{W}x_3' = x_3W_3 + r_UD + \mu_1(x_1W_1 - x_3W_3) + \mu_2(x_4W_4 - x_3W_3)\ , \tag{4.1c}$$
$$\overline{W}x_4' = x_4W_4 - r_UD + \mu_1(x_2W_2 - x_4W_4) + \mu_2(x_3W_3 - x_4W_4)\ , \tag{4.1d}$$

where $r_U = r(1 - U)$, and the probability that two mutations occur in one gamete is neglected, as is appropriate if $U \ll 1$. These recursion relations generalize II(1.4).

4.2 UNEQUAL EFFECTS AT THE TWO LOCI

With mutation, it is no longer possible to calculate all the equilibria and their stability properties explicitly, unless the loci have identical effects ($\gamma_1 = \gamma_2$) and mutation rates ($\mu_1 = \mu_2$); this special case is briefly outlined in Section 4.3. Therefore, numerical computations will be employed to illustrate the effect of mutation. However, parts of the analysis of the model without mutation can be extended to the present one and provide a good guide for understanding the equilibrium properties. If any of the equilibria determined in Section 2.1 is stable, then the corresponding equilibrium will be in the simplex S_4, and thus biologically meaningful for $\mu_1 . \mu_2 > 0$. The perturbed equilibrium will be stable for sufficiently small μ_1 and μ_2. However, only the symmetric equilibrium E_1 and its stability properties can be determined analytically.

The equilibrium E_1 is efficiently calculated by using the symmetric coordinates (2.5). Substituting the new coordinates into (4.1) and applying the equilibrium conditions, we obtain in generalization of (2.4) that E_1 is given by

$$\hat{x}_1 = \hat{x}_4 = \tfrac{1}{4} + \hat{D}_1 \,, \; \hat{x}_2 = \hat{x}_3 = \tfrac{1}{4} - \hat{D}_1 \,, \tag{4.2a}$$

where

$$\hat{D}_1 = \frac{1}{4s\gamma_1\gamma_2(1+U)}\left[r_U + U_s - \sqrt{s^2\gamma_1^2\gamma_2^2(1-U^2) + (r_U + U_s)^2}\right] , \tag{4.2b}$$

$U_s = U(1 - s\,\mathrm{E}[\gamma^2])$, and $\mathrm{E}[\gamma^2] = \frac{1}{2}(\gamma_1^2 + \gamma_2^2)$. In the limit of weak mutation and recombination, $U \to 0$ and $r \to 0$, which is the case of most interest because then E_1 will be stable (see below), we obtain the simple approximation

$$\hat{D}_1 = -\frac{1}{4} + \frac{r+U}{4s\gamma_1\gamma_2} - \frac{U(\gamma_1 - \gamma_2)^2}{8\gamma_1\gamma_2} + O((U+r)^2) \,. \tag{4.3}$$

With a little help from *Mathematica* (Wolfram 1996), the Jacobian matrix can be calculated explicitly by using the symmetric coordinates. It has block form, with one eigenvalue always between zero and one if $U < s\,\mathrm{E}[\gamma^2]$. By very tedious calculations and with substantial help from *Mathematica*, arguments analogous to those in Karlin and Feldman (1970, p. 57) show that the remaining two eigenvalues are less in modulus than unity if and only if a complicated quadratic polynomial in r is positive (unpublished; actually, the eigenvalues can be shown to be real). In the absence of mutation, this polynomial reduces to the polynomial R given by (2.7). It has one positive root r_c, whose first-order approximation in U and s is

$$\begin{aligned} r_c = r_1 &+ \frac{2}{9\sqrt{\gamma_1^4 - \gamma_1^2\gamma_2^2 + \gamma_2^4}}\left(\frac{\gamma_1^2}{\gamma_2^2} + 5 + \frac{\gamma_2^2}{\gamma_1^2}\right)(\gamma_1^2\mu_2 + \gamma_2^2\mu_1) \\ &+ \frac{2}{9}\left[\mu_2\frac{\gamma_1^2}{\gamma_2^2} + \mu_1\frac{\gamma_2^2}{\gamma_1^2} - 2U\right] + O(Us + U^2) \,, \end{aligned} \tag{4.4}$$

where r_1 is given by (2.9). Hence, the symmetric equilibrium is asymptotically stable if and only if

$$0 \le r \le r_c \,. \tag{4.5}$$

Its range of stability may increase or decrease with an increasing mutation rate. For a given total mutation rate U, the range of stability always increases if the mutation rate μ_1 at the major locus decreases.

Numerical computations show that if r is somewhat larger than r_c and the mutation rates at both loci are equal, then two interior unsymmetric equilibria (corresponding to E_2 and E_3) are stable. As r increases further, they converge to equilibria that are perturbations of the boundary equilibria E_4, E_5, or E_6, E_7. For larger recombination frequencies, one pair of these equilibria is asymptotically stable; which one it is, depends on the disparity of effects between loci, as in the case without mutation.

The genetic variance maintained at E_1 can be calculated by substituting (4.2) into (2.19); this yields $\hat{\sigma}_G^2 = 4\hat{D}_1\gamma_1\gamma_2 + \frac{1}{4}(\gamma_1 + \gamma_2)^2$. In the limit of weak mutation and recombination, $U \to 0$ and $r \to 0$, we obtain

$$\hat{\sigma}_G^2 = \tfrac{1}{2}(\gamma_1 - \gamma_2)^2(1-U) + \frac{r+U}{s} + O((U+r)^2) \,. \tag{4.6}$$

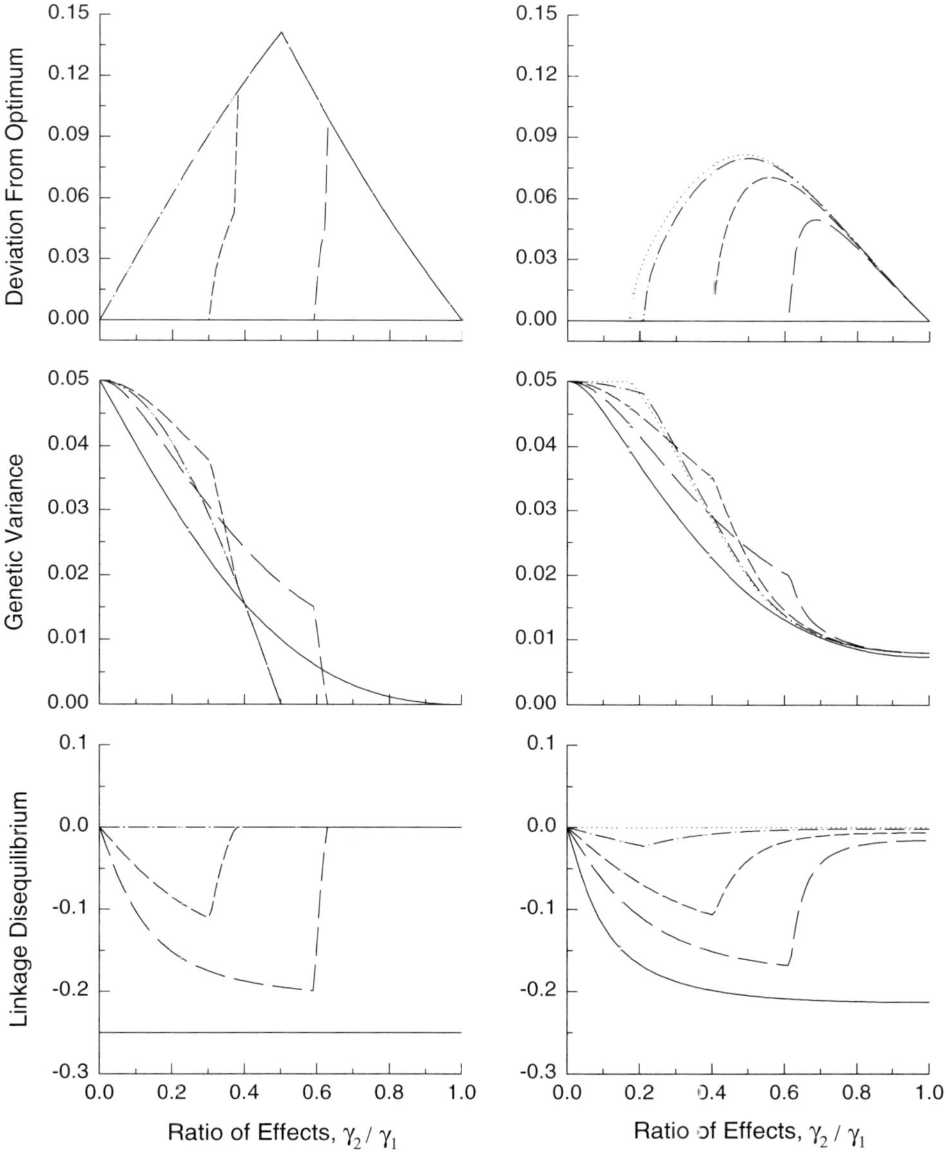

Figure 4.1 Comparison of the mean phenotype, the genetic variance, and the linkage disequilibrium maintained by quadratic stabilizing selection with mutation (right column) and without mutation (left column). The first row of figures shows the deviation of the equilibrium mean phenotype from the optimum as a function of the ratio γ_2/γ_1 of effects. The second row displays the genetic variance, and the third the linkage disequilibrium. In all figures the selection coefficient is $s = 0.1$, the mutation rate per (haploid) locus is $\mu_1 = \mu_2 = 0.0002$, and the average variance of mutational effects is $E[\gamma^2] = 0.05$. Each figure displays data for five levels of recombination. The solid line is for $r = 0$, the long dashes are for $r = 0.001$, the short dashes for $r = 0.0025$, the dash-dots for $r = 0.01$, and the dots for $r = 0.5$. In the left panel, the latter two values of r give practically the same results.

Numerical iterations of the recursion relations (4.1) were performed for several different initial values to calculate the genetic variance at other stable equilibria and to investigate its dependence on the parameters. In the case of equal mutation rates, no other equilibria than those corresponding to E_1 to E_7 were found. Further, the deviation of the equilibrium mean phenotype from the optimum, and the amount of linkage disequilibrium were calculated. Figure 4.1 compares these quantities with the corresponding ones of the quadratic optimum model without mutation. The three panels in the left column are for the case without mutation, those in the right column are with mutation. Additional computations were performed for loci with different mutation rates, but equal forward and backward rates, as well as unequal forward and backward rates. With equal forward and backward rates, results are similar to those for exchangeable loci if the total mutation rate U is the same. Compared to the right panel of Figure 4.1, only slight distortions occur. If forward and backward mutation rates are also different, then the overall picture still remains similar, but equilibria with different levels of deviation from the optimum, genetic variance, and linkage disequilibrium may be simultaneously stable, as in the model in Section 2.5, where the optimum deviates from the double heterozygous genotype (results not shown). A general tendency is that if the major locus has the higher mutation rate, then the deviation of the mean from the optimum is smaller than in the case of equal mutation rates, and it is larger otherwise. The consequences of unequal mutation rates (between loci and alleles) for the genetic variance maintained at equilibrium are, on average, very moderate if the total mutation rate U is unaltered.

4.3 EQUAL EFFECTS

If the effects and mutation rates of both loci are equal, i.e., if $\gamma_1 = \gamma_2 = \gamma$ and $\mu_1 = \mu_2 = \mu$, then all equilibria of (4.1) can be determined explicitly and a complete global stability analysis is available (for the continuous-time model it may found in Bürger 1989). The coordinates of the symmetric equilibrium E_1 and its range of stability are immediately obtained from (4.2), (4.4), and (4.5). However, note that r_c simplifies to $r_c = 8\mu^2\frac{1-b}{b} + O(\mu^3/b^2)$ as $\mu/b \to 0$, where $b = s\gamma^2$ (and $b \to 0$ is admitted). If $r > r_c$, then perturbations of the chromosome-fixation equilibria E_6 and E_7 are asymptotically stable, and E_6 attracts all trajectories from the region $x_2 > x_3$, and E_7 attracts the others. The coordinates of E_6 are

$$\hat{x}_1 = \hat{x}_4 = \frac{\mu}{b(1+2\mu)}\ , \tag{4.7a}$$

$$\hat{x}_2 = 1 - \hat{x}_1 - \hat{x}_3 - \hat{x}_4\ , \tag{4.7b}$$

$$\hat{x}_3 = \frac{\mu^2}{b^2} - \hat{D}_2 + O\left(\frac{\mu^3}{b^3}\right) \tag{4.7c}$$

as $\mu/b \to 0$, where

$$\hat{D}_2 = -\frac{2\mu^2}{rb}(1-b) + O\left(\frac{\mu^3}{b^2}\right)\ . \tag{4.7d}$$

The equilibrium E_7 is identical to E_6 with the coordinates $\hat{x}_2$ and $\hat{x}_3$ interchanged. As r converges to r_c from above, E_6 and E_7 converge to E_1. At the equilibria E_6 and E_7, the genetic variance is

$$\hat{\sigma}_G^2 = \frac{4\mu}{s(1+2\mu)} \,. \tag{4.8}$$

For loci with equal effects, the above results can be generalized to fitness functions satisfying $d \geq 2b$, which means that $W(G)$ is concave on the range of genotypic values. In particular, the equilibria and their genetic variance agree with those in the case of quadratic selection ($d = 4b$) to first order in μ. This implies that the genetic variance at equilibrium is approximately independent of the fitness of the extreme genotypes (the two double homozygotes), and quadratic stabilizing selection can be used as a close approximation to any fitness function with the same curvature near the optimum.

For arbitrary symmetric fitness functions, the present model with identical loci has the property that the mean genotypic value approaches the optimum monotonically from all initial conditions. This follows from the easily proved inequality

$$\left|\frac{x_1' - x_4'}{x_1 - x_4}\right| = (1-2\mu)\frac{1 - b(x_2+x_3) - d(x_1+x_4)}{\overline{W}} < 1 \,, \tag{4.9}$$

which holds whenever $0 < b \leq d \leq 1$. Note that (4.9) implies that all orbits converge to the plane $x_1 = x_4$, which therefore contains all equilibria. It is this fact that makes a global analysis tractable in the case $\gamma_1 = \gamma_2$. If $\mu > \frac{1}{4}b$, then E_1 is globally asymptotically stable. Proofs can be obtained from the author.

Assuming equal mutation rates, the main results can be summarized as follows:

- **4.1** *The deviation of the equilibrium mean phenotype from the fitness optimum is maximized if* $\gamma_1 \approx 2\gamma_2$, *unless linkage is very tight. If* $\mu = 0$ *and* $\gamma_1 = 2\gamma_2$, *then the deviation reaches* $\sqrt{\frac{2}{5}\,\mathrm{E}[\gamma^2]}$. *An increasing mutation rate decreases this deviation.*

- **4.2** 1. *If* $\mathrm{E}[\gamma^2]$ *is fixed, then the equilibrium genetic variance is a strictly decreasing function of the ratio of effects,* γ_2/γ_1. *If* γ_2/γ_1 *is close to zero, the genetic variance is approximately* $\mathrm{E}[\gamma^2]$. *If* γ_2/γ_1 *is close to one, it is approximately* $4\mu/s$, *unless linkage is extremely tight. This value coincides with the house-of-cards approximation (see Chapter IV.1.2 and Section 7).*

2. *Comparison with the model without mutation shows that if the effects at the two loci are very unequal* ($\gamma_1 > 2\gamma_2$), *mutation contributes only little to the genetic variance, whereas for similar effects* ($\gamma_1 < 2\gamma_2$), *most or all variance is due to mutation.*

3. *If the effects at both loci are very similar* ($\gamma_2/\gamma_1 > \frac{3}{4}$), *linkage has little influence on the equilibrium genetic variance, whereas for values of* γ_2/γ_1 *near* $\frac{1}{2}$, *the influence of linkage is not negligible. Then tighter linkage may increase or decrease the variance.*

5. MUTATION-SELECTION BALANCE: MULTIPLE UNLINKED DIALLELIC LOCI

Following Barton (1986), we now introduce mutation into the model of Section 3.1 by assuming that it occurs at the same rate μ at all loci and in both directions. Surprisingly, this leads to a radically different equilibrium structure if there are sufficiently many loci and mutation rates are at an intermediate level.

If we ignore terms of order μs, then the mutation-selection dynamics can be approximated by [cf. (3.4) and III(2.2)]

$$\dot{p}_i = s\gamma^2 p_i(1-p_i)(2p_i - 1 - 2\delta) + \mu(1-2p_i)\ , \quad i = 1, \dots, \ell\ , \tag{5.1}$$

where δ is as in (3.5) and selection is according to the quadratic optimum model (2.29) with arbitrary optimum. We shall now examine the equilibrium structure of this model.

▷ The system of differential equations (5.1) is a Svirezhev–Shahshahani gradient system in which $\overline{W} + 2\mu\sum_i[\ln p_i + \ln(1-p_i)]$ plays the role of a potential and $\overline{W}$ is as in (3.3); cf. III(2.15). Consequently, all asymptotic states are equilibria (Appendix A.3). The equilibrium allele frequencies of (5.1) are the solutions of the equations

$$p_i(1-p_i)(2p_i - 1 - 2\delta) + \beta(1-2p_i) = 0\ , \quad i = 1, \dots, \ell\ , \tag{5.2}$$

where $\beta = \mu/(s\gamma^2)$. This equation has three real solutions (for each i) if and only if

$$\delta^2 < \tfrac{1}{4}\left(1 + 10\beta - 2\beta^2 - 2\sqrt{\beta(2+\beta)^3}\right) \equiv \delta_c^2 \tag{5.3}$$

(cf. Abramowitz and Stegun 1965, 3.8.2). For $\beta \geq \frac{1}{4}$ the critical value is $\delta_c = 0$. As β decreases from $\frac{1}{4}$, the critical value δ_c increases. For $\beta \ll 1$, it is approximately given by

$$\delta_c = \tfrac{1}{2} - \sqrt{2\beta} + O(\beta)\ . \tag{5.4}$$

If $|\delta| > \delta_c$, then (5.2) has exactly one solution and, therefore, there exists a unique equilibrium of (5.1). Thus, if $\beta > \frac{1}{4}$, and hence the dynamics is dominated by mutation, a unique mutation-selection equilibrium exists. It is asymptotically stable, as demonstrated by a linear stability analysis (Barton 1986 and Hastings 1990a). Since (5.1) is a gradient system, the equilibrium is actually globally stable. If $P_O = 0$, this is the symmetric equilibrium with $p_i = \frac{1}{2}$ for every i. If $\beta < \frac{1}{4}$, then this equilibrium is unstable.

Throughout the rest of this section, it is assumed that the equilibrium deviation satisfies $|\delta| < \delta_c$. We will concentrate on equilibria with all loci near fixation and, in analogy with the pure selection model of Section 3.1, denote by m and M the number of equilibria where the '−' and the '+' allele, respectively, is near fixation. The equilibrium frequencies at these loci are denoted by $\hat{p}_-$ and $\hat{p}_+$. Let $\delta = \frac{1}{2} - \varepsilon$, where $\sqrt{2\beta} < \varepsilon \leq \frac{1}{2}$ [see (5.4)], and assume $\beta \ll 1$. Then the equilibrium frequencies can be approximated by

$$\hat{p}_- \approx \frac{\beta}{2(1-\varepsilon)} \quad \text{and} \quad \hat{p}_+ \approx 1 - \tfrac{1}{2}\left(\varepsilon - \sqrt{\varepsilon^2 - 2\beta}\right)\ , \tag{5.5a}$$

as is easily checked by substitution into (5.2). For $\delta = \frac{1}{2} + \varepsilon$ analogous approximations are obtained:

$$\hat{p}_- \approx \tfrac{1}{2}\left(\varepsilon - \sqrt{\varepsilon^2 - 2\beta}\right) \quad \text{and} \quad \hat{p}_+ \approx 1 - \frac{\beta}{2(1-\varepsilon)}\ . \tag{5.5b}$$

Now we use these approximations to find the range of equilibria that exist for given P_O and γ. The largest possible deviations from the optimum are obtained if $\varepsilon = \pm\sqrt{2\beta}$. Substituting (5.5a) and (5.5b) with these choices of ε into the definition of δ [see (3.5)], we derive from the condition $|\delta| < \delta_c$:

$$M-m-\tfrac{1}{2}-(M-1)\sqrt{2\beta}+O(m\beta) < \frac{P_O}{\gamma} < M-m+\tfrac{1}{2}+(m-1)\sqrt{2\beta}+O(M\beta)\,. \quad (5.6)$$

◁

Hence, we have shown that for each value of the optimum P_O, there are approximately $1+(\ell-2)\sqrt{2\beta}$ equilibria of type $(m,0,M)$ possible. If $\ell\sqrt{\beta}$ is large, the number of equilibria is large. Barton (1986) proved that these equilibria are stable whenever they exist. For the case $P_O = 0$, Hastings' (1990a) bifurcation analysis proves that no other equilibria can be stable than those where m loci are at one allele frequency and $M = \ell - m$ loci are at another allele frequency with $\frac{1}{3}\ell < m, M < \frac{2}{3}\ell$ (and $\nu = 0$). This implies that if $P_O = 0$, at least $\ell \geq 7$ loci are necessary in order that these equilibria exist. Barton also proved that for $\ell \geq 4$ there are values of P_O such that multiple equilibria exist.

General perturbation arguments imply that for sufficiently low mutation rates, the stable equilibria must be perturbations of the stable equilibria of the pure-selection model. Thus, for very low mutation rates there always exists a unique class of stable equilibria, and for roughly half of the possible values of P_O it is of type $(\frac{1}{2}\ell, 0, \frac{1}{2}\ell)$, whereas for the other values there is one polymorphic locus, i.e., $\nu = 1$. Actually, Hastings (1990a) proved that if $P_O = 0$, then there is a critical value β_c such that for $\beta < \beta_c$ these are the only equilibria, whereas for $\beta_c < \beta < \frac{1}{4}$ multiple stable equilibria exist, as described above. Equation (5.6) implies that the critical β_c is approximately the solution of $(\ell+1)\sqrt{2\beta} = 1$.

The Genetic Variance

If the mean is at the optimum ($\delta = 0$), then the equilibrium allele frequencies can be computed exactly from (5.2). They are $\frac{1}{2}$ and $\frac{1}{2}(1 \pm \sqrt{1-4\beta})$. Equation (3.6) shows that in this situation, only equilibria with $m = M$ are possible. Therefore, if $\delta = 0$, the equilibrium variance is

$$\hat{\sigma}_0^2 = 2\ell\gamma^2\beta = 2\ell\mu/s\,, \quad (5.7)$$

a result previously derived by Wright (1935a), Latter (1960), and Bulmer (1972). For two loci, (5.7) agrees to leading order with (4.8), which was derived for the discrete-time two-locus model under the assumption that the loci are not too closely linked ($r > r_c$), and the equilibrium allele frequencies agree to leading order with those in (4.7). The value (5.7) coincides with the multilocus house-of-cards approximation derived by Turelli (1984) for the continuum-of-alleles model (cf. Section 7).

However, stable equilibria may exist in which the variance is much greater. Assume, for instance, that $M - m$ is much larger than P_O/γ [cf. (3.6)], so that the deviation δ from the optimum is near its critical value $\delta_c = \frac{1}{2} - \sqrt{2\beta}$. Then, from (5.5a), the equilibrium allele frequencies are approximately $\hat{p}_- = \frac{1}{2}\beta$ and $\hat{p}_+ = 1 - \sqrt{\beta/2}$. Substitution into (3.2) yields the genetic variance at equilibria of this kind:

$$\hat{\sigma}_G^2 = \gamma^2 M\sqrt{2\beta} = \gamma M\sqrt{\frac{2\mu}{s}}\,. \quad (5.8)$$

For the class of equilibria with $m = M = \ell/2$, the maximum possible variance is

$$\hat{\sigma}^2_{\max} = \gamma\ell\sqrt{\frac{\mu}{2s}}\,. \tag{5.9}$$

Obviously, $\hat{\sigma}^2_{\max}$ can be much greater than $\hat{\sigma}^2_0$, particularly if μ/s is small. It is equal to one half of the variance predicted by the Gaussian allelic approximation of Kimura (1965a) and Lande (1975) for the continuum-of-alleles model (cf. Section 7).

Here are some numerical examples taken from Barton (1986) that illustrate the above findings. Let $\ell = 100$, $s = \frac{1}{2}$, $\gamma = 0.1$, and $\mu = 10^{-4}$, so that $\beta = 0.02$. If the optimum is at $P_O = 0$, there are 11 classes of stable equilibria, from $(45, 0, 55)$ to $(55, 0, 45)$. At the equilibria $(50, 0, 50)$, we have $\hat{p}_- = 0.2042$, $\hat{p}_+ = 0.9796$, $\delta = 0$, and the genetic variance is $\hat{\sigma}^2_G = 0.04$. At the equilibrium $(45, 0, 55)$, the gene frequencies are $\hat{p}_- = 0.0124$, $\hat{p}_+ = 0.9018$, the deviation from the optimum is $\delta = 0.311$, and the genetic variance is $\hat{\sigma}^2_G = 0.1084$. If the optimum is at $P_O = 5$, there are 10 possible classes of equilibria, namely from $(17, 0, 83)$ to $(26, 0, 74)$. The highest genetic variance is maintained at $(17, 0, 83)$ with $\hat{\sigma}^2_G = 0.1496$ ($\hat{p}_- = 0.0124$, $\hat{p}_+ = 0.9030$, $\delta = 0.311$) and the lowest at $(25, 0, 75)$ with $\hat{\sigma}^2_G = 0.0393$ ($\hat{p}_- = 0.0382$, $\hat{p}_+ = 0.9859$, $\delta = -0.211$).

Since this is a highly symmetric model, it is not obvious if, and to what extent, the results are structurally stable. Indeed, the bifurcation analysis of Hastings (1990a) indicates that for relevant parameter values the basin of attraction of equilibria with $M \neq m$ may be rather small. This is in agreement with simulations of a stochastic version of this (and of a more general) model showing that the system is much more likely to be found at one of the equilibria with $M = m$ (Hastings 1988) and, on average, the genetic variance is near the value $\hat{\sigma}^2_0$ (Barton 1989, and Chapter VII.2).

Let us summarize the main results of this model:

- **5.1** 1. *If there are four or more diallelic loci, then for certain values of the optimum P_O, multiple stable equilibria coexist with different levels of genetic variability. If there are seven or more loci involved, alternative equilibria with different levels of variability coexist for a wide range of values of the optimum P_O, including $P_O = 0$. Therefore, the equilibrium actually attained depends upon the history of the evolutionary process.*

2. *These equilibria exist for intermediate mutation rates, most likely including those found in nature. The genetic variance maintained at mutation-selection balance ranges from about $2\ell\mu/s$ to $\gamma\ell\sqrt{\mu/(2s)}$. However, the basin of attraction of equilibria yielding a high genetic variance appears to be small.*

6. MUTATION-SELECTION BALANCE: THE HAPLOID CONTINUUM-OF-ALLELES MODEL

The models in the previous sections were based on the assumption that only two alleles per locus can segregate. Here, resuming the approach of Chapter IV, we consider the other extreme of a continuum of possible allelic effects, and derive approximations for the variance and the higher moments of the equilibrium distribution at a haploid locus subject to a balance between mutation and stabilizing selection.

We prove that if mutation is much weaker than selection, then the variance at mutation-selection balance with Gaussian stabilizing selection and very general mutation distributions $u(x, y)$ is closely approximated by the HC-approximation of Chapter IV.1.2. Error terms of this approximation are derived for certain mutation distributions. The range of accuracy of this and other approximations, such as the Gaussian, is investigated. An approximation is suggested that is a lower bound and is very accurate for the whole range of parameters that has been explored in the literature. We shall also compute approximate formulas for the higher moments of the equilibrium distribution. First we deal with the technically more complicated discrete-time model. Then we briefly discuss the continuous-time case, in which some slightly stronger and simpler results can be derived. The technical parts of the proofs are relegated to the two final subsections. In the first two subsections we assume that alleles (allelic effects) are labeled by real numbers, i.e., $\mathcal{X} = \mathbf{R}$, as in the continuum-of-alleles model, whereas in the third we consider models with discrete alleles, for which parallel results can be proved.

6.1 DISCRETE TIME

We use the general continuum-of-alleles model as introduced in Chapter IV.2.1 with $\mathcal{X} = \mathbf{R}$ as state space. The (conditional) mutation distribution $u(x, y)$ may be general, satisfying IV(2.1), but is assumed to be strictly positive, i.e., $u(x, y) > 0$ for all x and y. The mutation rate μ is also positive. Stabilizing selection is modeled by the Gaussian fitness function

$$W(x) = \exp\left\{-\frac{(x - x_O)^2}{2V_s}\right\}, \tag{6.1}$$

where x_O is the position of the optimum and V_s is an inverse measure for the strength of selection [large V_s corresponds to weak selection; cf. V(1.13)]. This was first employed by Haldane (1954) and has become the most widely used fitness function for modeling stabilizing selection. These assumptions imply that there is a uniquely determined, globally asymptotically stable equilibrium distribution, which we denote by $\hat{p}$ (see Chapter IV.3). We recall that $\hat{p}$ is the solution of an eigenvalue problem; cf. (6.32). Its mean and variance are denoted by $\bar{x}$ and $\hat{\sigma}^2$, respectively. Although μ and V_s will be considered as parameters, and $u(x, y)$ is assumed to be fixed, we suppress this dependence in the notation.

We want to prove that $\hat{\sigma}^2$ approaches $2\mu V_s$ in the weak-mutation weak-selection limit. To this end, we need to assume that

$$V_s \to \infty \text{ and } 2\mu V_s \to 0 \quad \text{as } \mu \to 0\,, \tag{6.2}$$

i.e., selection is weak but mutation is much weaker. The anchor point for our analysis is the asymptotic estimate IV(5.5) for the equilibrium mean fitness, $\overline{W} \sim 1 - \mu$. Thus, most of the results in this section are a consequence of this generalized version of Haldane's principle about the mutation load.

▷ Substituting the expansion $W(x) = 1 - (x - x_O)^2/(2V_s) + (x - x_O)^4 h(x)/(8V_s)$ into IV(5.4), where the remainder function $h(x)$ satisfies $0 \le h(x) \le 1$, we obtain

$$\lim_{\mu \to 0} \frac{1}{2\mu V_s} \left\{ [\hat{\sigma}^2 + (\bar{x} - x_O)^2] - 2\mu V_s - \frac{1}{4V_s} \int (x - x_O)^4 h(x) \hat{p}(x)\, \mathrm{d}x \right\} = 0\,. \tag{6.3}$$

It is easily seen that the same holds in the limit $2\mu V_s \to 0$.

The subsequent analysis is based on the technical assumption that for each $n \le 4$ there is a positive constant $e_n(u) < \infty$ such that

$$W(y) \int_{-\infty}^{\infty} |x - x_O|^n \, u(y,x) \, dx \le e_n(u) \quad \text{for all } y \,. \tag{6.4}$$

This ensures that the distribution of mutants, Up [U is the mutation operator defined in IV(5.3)], has finite moments up to order four, i.e.,

$$\tilde{v}_n = \int (x - x_O)^n U\hat{p}(x) \, dx \quad \text{satisfies} \quad |\tilde{v}_n| \le \int |x - x_O|^n \, U\hat{p}(x) \, dx \le e_n(u) \,. \tag{6.5}$$

If the inequality (6.4) holds all $n \le n_0$, then (6.32) implies that $\hat{p}$ has finite moments up to order n_0. It is easily shown that (6.4) holds for any mutation distribution of house-of-cards or random-walk form that possesses moments up to order n, i.e., for all distributions $u(x,y) = u_{\mathrm{HC}}(y)$ or $u(x,y) = u_{\mathrm{RW}}(y - x)$, such that $\int |x|^n \, u_{\mathrm{HC}}(x) \, dx < \infty$ or $\int |x|^n \, u_{\mathrm{HC}}(x) \, dx < \infty$, respectively. For the HC-mutation model, $\tilde{v}_n$ is readily computed to be

$$\tilde{v}_n = \overline{W} \, v_{n,x_O} \sim v_{n,x_O} \quad \text{as } \mu \to 0 \,, \tag{6.6}$$

where v_{n,x_O} is the nth moment of u_{HC} about x_O. For the random-walk mutation model, one obtains [see (6.45) and (6.46)]

$$\tilde{v}_n = \overline{W} \sum_{j=0}^{k} \binom{n}{j} v_j \, \tilde{m}_{n-j,x_O} \sim v_n \quad \text{as } \mu \to 0 \,, \tag{6.7}$$

where v_n is the nth moment of u_{RW} about zero and $\tilde{m}_{n,x_O}$ denotes the nth moment of $W\hat{p}/\overline{W}$ about x_O.

If (6.2) and (6.4) hold, so that selection is sufficiently weak and the first four moments of $\hat{p}$ exist, then

$$\lim_{\mu \to 0} \frac{1}{8\mu V_s^2} \int (x - x_O)^4 \hat{p}(x) \, dx = 0 \,. \tag{6.8}$$

The proof is given in Section 6.6.

If the mutation distribution is symmetric around the optimum, i.e., if u satisfies $u(x_O + x, x_O + y) = u(x_O - x, x_O - y)$, then at equilibrium the mean $\bar{x}$ coincides with the optimum x_O, as follows easily from the uniqueness of the equilibrium density (cf. Chapter IV.2). The proof that for general, nonsymmetric mutation distributions, $\bar{x} \to x_O$ 'rapidly' as $\mu \to 0$ requires the assumption of a bounded mutation distribution, i.e.,

$$\|u\|_\infty = \sup_{x,y} u(x,y) < \infty \,. \tag{6.9}$$

This is the case for Gaussian and (reflected) exponential distributions in the HC- and in the random-walk mutation model, but not for all Γ-distributions. ◁

Under the assumptions (6.4) with $n = 1$ and (6.9), it is proved in Section 6.5 that the deviation of the equilibrium mean value $\bar{x}$ from the optimum x_O decreases rapidly as $\mu V_s \to 0$, i.e.,

$$\lim_{\mu V_s \to 0} \frac{(\bar{x} - x_O)^2}{2\mu V_s} = 0 \,. \tag{6.10}$$

The validity of (6.10) does not require weak selection. For a HC-mutation distribution, (6.18) generalizes and improves (6.10).

Substitution of (6.10) and (6.8) into (6.3) yields our first central result:

• **6.1** *Let the scaling assumption* (6.2) *be satisfied, so that stabilizing selection is sufficiently weak, but stronger than mutation. Then*

$$\lim_{\mu \to 0} \frac{\hat{\sigma}^2 - 2\mu V_s}{2\mu V_s} = 0 \tag{6.11}$$

holds for arbitrary symmetric and for bounded mutation distributions with finite fourth moments. Equivalently, (6.11) *can be reformulated as*

$$\hat{\sigma}^2 \sim \hat{\sigma}^2(\mathrm{HC}) = 2\mu V_s \quad \text{as } \mu \to 0 \,. \tag{6.12}$$

More generally, these approximations hold for any fitness function with Taylor approximation $W(x) \sim 1 - (x - x_O)^2/(2V_s)$ *near the optimum and vanishing sufficiently rapidly as* $|x| \to \infty$*; cf.* (6.47) *and* (6.48).

This result is a generalization of Turelli's (1984) *house-of-cards approximation* for the equilibrium variance. It shows that the HC-approximation is robust with respect to specific assumptions about the mutation distribution. Only higher-order terms depend on such details (see below). In particular, as the proof shows, the validity of the house-of-cards approximation is a consequence of the generalized version of Haldane's principle about the mutation load. The approximation $\hat{\sigma}^2(\mathrm{HC})$ for the genetic variance in the discrete-time model is identical to that for the continuous-time model [cf. IV (1.16)] if $2V_s$ is replaced by $1/s$, so that the quadratic fitness function from IV (1.2) approximates the Gaussian from (6.1).

Furthermore, by the same series expansion of $W(x)$ as leading to (6.3), IV (5.2) implies

$$\hat{\sigma}^2 + (\bar{x} - x_O)^2 \leq 2\mu V_s + (4V_s)^{-1} \int (x - x_O)^4 \hat{p}(x)\, \mathrm{d}x \,, \tag{6.13}$$

whence we obtain from (6.44)

$$\hat{\sigma}^2 + (\bar{x} - x_O)^2 \leq 2\mu V_s (1 + \mathrm{const.}/V_s) \,. \tag{6.14}$$

This holds for all $\mu \leq \frac{1}{2}$ and $V_s \geq \frac{1}{2}$. For typical parameters ($\mu V_s \ll 1$ and the variance of the mutation distribution $\ll 1$) the constant on the right-hand side of (6.14) is $\ll 1$ [see (6.44)]. Thus, we have also derived an upper bound for the equilibrium variance that is only slightly larger than the HC-approximation.

If (6.4) is satisfied for some $n \geq 3$, the following asymptotic equality for the central moments of order n of the equilibrium distribution is proved in Section 6.6:

$$\int (x - \bar{x})^n \hat{p}(x)\, \mathrm{d}x \sim 2\mu V_s\, \tilde{v}_{n-2} \quad \text{as } \mu \to 0 \,. \tag{6.15}$$

The asymptotic estimate (6.15) generalizes a corresponding result of Turelli (1984) for a Gaussian mutation distribution in the HC-model. Equation (6.15) shows that the central moments of order $n \geq 3$ are all approximately proportional to the variance, the proportionality factor being the moment $\tilde{v}_{n-2}$, (6.5), of the distribution after mutation. This implies that the kurtosis, $\hat{c}_4/\hat{\sigma}^4$ (Appendix D.3), of the equilibrium distribution is $\tilde{v}_2/(2\mu V_s)$, which is large for small μV_s.

Hence, in the HC- and in the random-walk cases, the kurtosis of the equilibrium distribution is approximately $\gamma^2/(2\mu V_s)$, where γ^2 is the variance of mutational effects. Equivalently, the fourth cumulant satisfies $\hat{c}_4 \approx \gamma^2\hat{\sigma}^2$. For a highly leptokurtic allele distribution a large fraction of genetic variance is maintained by rare alleles of large effect (cf. Appendix D.3). Barton and Turelli (1987) called an approximation on the basis of this assumption (high kurtosis) *rare-alleles approximation*. The rare-alleles approximation is a consequence of, but not identical to, the HC-approximation. Intuitively, the rare-alleles approximation will be valid whenever a locus is near fixation, but mutation produces alleles of large effects at a low rate.

The asymptotic results of Chapter IV.5.4 enable us to derive error terms for the HC-approximation (6.12). Assume first that $u(x,y)$ is bounded. Then the same procedure that led to (6.3), combined with IV(5.33) for $q = 2$, (6.38), and (6.39), yields

$$\frac{2\mu V_s - \hat{\sigma}^2}{2\mu V_s} = O\left(\mu V_s \left(\ln \frac{1}{\mu V_s}\right)^2\right) \quad \text{as } \mu \to 0 \,. \tag{6.16}$$

Thus, the error made by using the HC-approximation instead of the true equilibrium variance is approximately of order $(\mu V_s)^2$ for small μV_s. If $\bar{x} = x_O$, the logarithmic term in (6.16) can be replaced by 1. For a Gaussian mutation distribution, a slightly stronger result can be derived (see (6.26) for the random-walk mutation model and IV(1.15) for the HC-mutation model).

For the HC-mutation model and distributions of the form

$$u_{\mathrm{HC}}(x) = |x|^{-1/n} g(x) \,, \tag{6.17}$$

with $n > 1$ and $g(x)$ bounded (cf. IV(5.25) and Appendix D.3), it can be proved that

$$|\bar{x} - x_O| \leq (\text{const.})(\mu V_s)^{n/(n+1)} \quad \text{as } \mu \to 0 \,, \tag{6.18}$$

by proceeding similarly as in this Section 6.5, but using IV(5.31) instead of IV(5.34). From this, IV(5.31) for $q = 2$, and (6.39), we readily obtain

$$\frac{2\mu V_s - \hat{\sigma}^2}{2\mu V_s} = O\left((2\mu V_s)^{\frac{n-1}{n+1}}\right) \quad \text{as } \mu \to 0 \,. \tag{6.19}$$

One expects that a similar result holds for the random-walk mutation model.

For the HC-model, the above estimates can be proved under the scaling assumption

$$k_1 \mu^{d_1} \leq \frac{\gamma^2}{V_s} \leq k_2 \mu^{d_2} \quad \text{as } \mu \to 0 \,, \tag{6.20}$$

where γ^2 is the variance of the mutation distribution and k_1, k_2, d_1, d_2 are are positive constants satisfying $0 < d_2 \leq d_1 < 1$ (cf. Turelli 1984). Thus, at least in the HC-case, it is not necessary to take the variance of the mutation distribution as fixed in the above analysis. It may be noted that (6.20) implies (6.2).

The merit of the estimates (6.16) and (6.19) is that they show how the shape of the mutation distribution affects the accuracy of the HC-approximation: an increasing kurtosis of the mutation distribution (decreasing n) reduces the accuracy of the HC-approximation. The asymptotic results (6.16) and (6.19) remain valid for fitness functions satisfying (6.47) and (6.48).

6.2 CONTINUOUS TIME

For the continuous-time model of Chapter IV.2.4 slightly simpler results can be derived. Also, the proofs are simpler because no scaling assumption, such as (6.2) for the strength of stabilizing selection, is needed. They are based on the equilibrium relation

$$\hat{p}(x) = \mu \frac{\int \hat{p}(y) u(y,x)\, dy}{c_\mu - m(x)} , \tag{6.21}$$

where $m(x)$ denotes Malthusian fitness and $a_\mu = \overline{m}_\mu + \mu$. Equation (6.21) is the continuous-time analogue of (6.32) below. The reason why the continuous-time case is technically easier to handle is rather obvious: the derivation of the continuous-time model implicitly assumes weak selection and, thus, neglects terms of order $O(s^2)$ and $O(\mu s)$. As shown in Chapter IV.2.4, it approximates the discrete model in the limit of weak selection. Instead of the Gaussian fitness function (6.1), we now assume that (Malthusian) fitness is given by

$$m(x) = -s(x - x_O)^2 , \tag{6.22}$$

where s corresponds to $(2V_s)^{-1}$. By the continuous-time analogue of IV(5.5), we have $a_\mu \to 0$ as $\mu \to 0$. Hence, instead of (6.3), we obtain

$$\lim_{\mu/s \to 0} \frac{\hat{\sigma}^2 + (\bar{x} - x_O)^2 - \mu/s}{\mu/s} = 0 . \tag{6.23}$$

Therefore, it is sufficient to prove the continuous-time version of (6.10) to arrive at the desired result

$$\lim_{\mu/s \to 0} \frac{\hat{\sigma}^2 - \mu/s}{\mu/s} = 0 . \tag{6.24}$$

The proof of (6.10) for continuous time is similar to that for discrete time, but simpler (Bürger and Hofbauer 1994). In addition, instead of (6.14),

$$\hat{\sigma}^2 + (\bar{x} - x_O)^2 < \mu/s \tag{6.25}$$

is obtained, because the denominator in (6.21) must be positive. The estimate (6.25) holds for all μ and s.

The error estimates (6.16) and (6.19), with $2\mu V_s$ replaced by μ/s, remain valid in the continuous-time case. For the random-walk model and a Gaussian mutation distribution with mean zero and variance γ^2, so that $\bar{x} = x_O$, the following bounds, valid for all μ/s, can be derived (Bürger and Hofbauer 1994):

$$\frac{\mu}{s} - \frac{\pi}{2\gamma^2}\left(\frac{\mu}{s}\right)^2 \leq \sigma^2 \leq \frac{\mu}{s} - \frac{\exp(-4b^2/\gamma^2)}{2\pi\gamma^2}\left(\frac{\mu}{s}\right)^2, \tag{6.26}$$

where

$$b^2 = \min\left(\frac{\mu}{s}, \frac{\pi}{2\gamma^2}\left(\frac{\mu}{s}\right)^2\right).$$

The constants associated with the second-order term differ approximately by a factor π^2 if μ/s is small. The reader may compare (6.26) with IV(1.15), the latter showing that in the HC-model the equilibrium variance approaches the lower bound in (6.26) asymptotically. For an exponential mutation distribution, a lower bound is obtained by replacing π on the left-hand side of (6.26) by π^2.

The above results, as well as comparison of the numerical results of Turelli (1984) with those of Bürger (1986c), show that variance and kurtosis of the equilibrium distribution in discrete time are very closely approximated by the corresponding quantities in the continuous-time model, even for moderately strong selection. Since the continuous-time model is generally easier to handle than the discrete-time model and corresponding results agree to order $O(s)$, for most purposes it will be more convenient to work directly with the continuous-time model.

6.3 DISCRETE ALLELES

For models with discrete alleles, finitely or infinitely many, again the HC-approximation (6.12) is valid. The reason is, as above, the validity of Haldane's principle, i.e., $\hat{\overline{W}} \sim 1 - \mu$ (see IV(5.21) and Chapter III.3.1). Then, by a reasoning similar to that for a continuum of alleles, the equilibrium variance is given by (6.12), if the fitness function is (6.1) and if there is a (unique) allele with the maximum fitness of one. Owing to IV(5.21), the error term for discrete alleles is always of order μ^2. Since Haldane's principle does not hold if there are two or more alleles with maximal fitness (see Chapter III.3.1), the HC-approximation is not valid in such a case.

Actually, for finitely many discrete allelic types, a result analogous to (6.12) holds for any fitness function that decays away from the fitness optimum, provided there is a unique optimal allele so that all other alleles have lower fitness. Suppose that the (one or two) nearest neighbours of the optimum type have effects $(\pm)a$ and fitness $1 - s'$. Then, by a reasoning similar to that which led to III(3.3), the genetic variance is approximately $a^2\mu/s'$. For Gaussian stabilizing selection or the quadratic optimum model this is equivalent to (6.12).

The case of three alleles was treated by Turelli (1984) in some detail. Slatkin (1987) derived an approximation for the stepwise-mutation model based on the consideration of five alleles that is accurate for small and intermediate values of μV_s and converges to the HC-approximation in the limit $\mu V_s \to 0$. Moran (1976) proved for the stepwise-mutation model of Chapter III.4 with fitnesses

$$W_0 = 1 > W_1 = W_{-1} \geq W_2 = W_{-2} \geq \ldots$$

and mutation to the next neighbours, each at rate $\mu/2$, that the equilibrium variance is bounded above by $\mu(1-\mu)/(W_1 - \mu)$, in good agreement with (6.12).

Therefore, we conclude:

• **6.2** *The haploid HC-approximation* (6.12) *is robust with respect to numerous model assumptions if mutation is weak relative to selection. It holds for the continuum-of-alleles model as well as for discrete-alleles models, and in continuous as well as in discrete time.*

6.4 DOMAINS OF VALIDITY OF THE APPROXIMATIONS

Asymptotic approximations, such as (6.12), provide no hints about their actual range of validity. The value of upper or lower bounds, such as (6.14), (6.25), and (6.26), is that these hold for all parameter values. The accuracy and domain of validity of approximations for the equilibrium variance can be quantified by numerical computation of equilibrium distributions. Extensive numerical work was performed by Turelli (1984) for the discrete-time model and a Gaussian mutation distribution. This was complemented by numerical calculations for the continuous-time model (Bürger 1986c, Bürger and Hofbauer 1994). For detailed descriptions of the numerical methods refer to these articles. For typical parameters (cf. Chapter IV.1) the differences between the continuous-time and the discrete-time results are indeed negligible, as suggested by the above analysis. In particular, for a wide range of parameters, does the equilibrium distribution in discrete time depend on the compound parameter μV_s (as in continuous time), but not on μ and V_s separately (cf. Table 7.1 in Chapter VII).

For definitness, we consider the discrete-time model with Gaussian stabilizing selection, (6.1), optimum $x_O = 0$, and the random-walk mutation model with a mutation distribution with mean zero and variance γ^2. With the identification $s = (2V_s)^{-1}$, Kimura's (1965a) Gaussian allelic approximation IV(1.8) becomes

$$\hat{\sigma}^2(\mathrm{G}) = \sqrt{\mu V_s \gamma^2}\ . \tag{6.27}$$

As $\hat{\sigma}^2(\mathrm{HC}) = 2\mu V_s$ is an (approximate) upper bound for the true equilibrium variance, the Gaussian allelic approximation can never be valid if $\gamma^2 > 4\mu V_s$.

Fleming (1979) extended Kimura's analysis and derived a second-order approximation. For a haploid locus, his approximation of the equilibrium variance reduces to

$$\hat{\sigma}^2(\mathrm{F}) = \hat{\sigma}^2(\mathrm{G})\left[1 - \frac{\hat{\sigma}^2(\mathrm{G})}{V_s}\left(\frac{3+\eta}{16\mu} - \frac{19}{16}\right)\right], \tag{6.28}$$

where η is the kurtosis of the mutation distribution (see Nagylaki (1984) for a concise exposition and some extensions of Fleming's analysis). The approximation (6.28) was derived under the assumptions of a fixed mutation rate and of

$$\gamma^2/(\mu V_s) \ll 1\ . \tag{6.29}$$

For a Gaussian mutation distribution, Figure 6.1 displays data points from numerical computation of the actual genetic variance, the HC-approximation (6.12), the Gaussian allelic approximation (6.27), Fleming's approximation (6.28), and the lower bound

$$\hat{\sigma}^2(\mathrm{LB}) = 2\mu V_s\left(1 - \pi\,\frac{\mu V_s}{\gamma^2}\right) \tag{6.30}$$

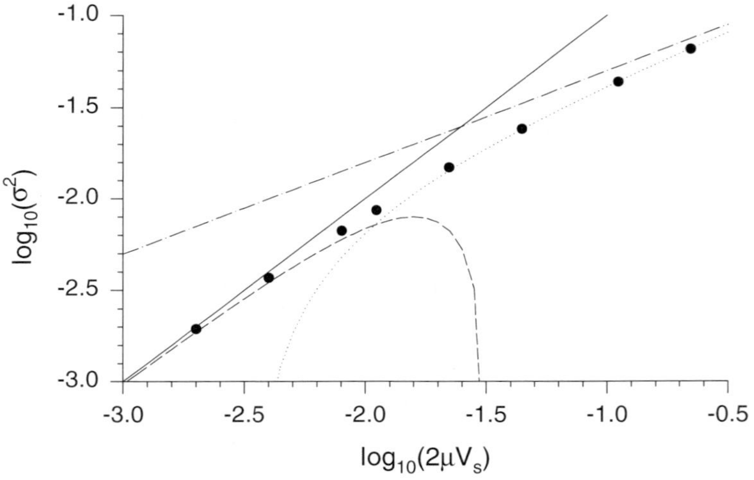

Figure 6.1 The equilibrium variance and some of its approximations for mutation-stabilizing-selection balance under the random-walk mutation model. The plot is double logarithmic (to base 10), i.e., it shows the logarithm of the equilibrium variance, $\log_{10}(\hat{\sigma}^2)$, as a function of $\log_{10}(2\mu V_s)$. The strength of selection is $V_s = 20$, and the variance of the Gaussian mutation distribution is $\gamma^2 = 0.05$. The solid line represents the HC-approximation (6.12), the dash-dotted line the Gaussian allelic approximation (6.27), the dotted line Fleming's approximation (6.28), and the dashed line the lower bound (6.30). The dots are data points obtained from numerical computation.

from (6.26). The figure, as well as the numerical results of Turelli (1984) and Bürger (1986c), suggest that the Gaussian approximation is also an upper bound for the true equilibrium variance, and they demonstrate the astonishing accuracy of Fleming's approximation. Figure 6.1 also shows that the lower bound $\hat{\sigma}^2(\mathrm{LB})$ is the best among these approximations as long as it is larger than $\hat{\sigma}^2(\mathrm{F})$; otherwise $\hat{\sigma}^2(\mathrm{F})$ is the best. A straightforward calculation shows that the point of intersection of these two approximations is given approximately by

$$2\mu V_s = \frac{\gamma^2}{4.6} \,. \tag{6.31}$$

The factor 4.6 is very accurate if $\gamma^2 \leq V_s/5$. This observation is further substantiated by Turelli's simulations and is in good agreement with his conclusions. For the published numerical data, the maximum of the lower bound, (6.30), and of Fleming's approximation, (6.28), never differs by more than 15% from the true equilibrium variance. The maximal deviation occurs for parameter values approximately satisfying (6.31). Consideration of all available numerical data shows that in the continuous-time model, the maximum of the lower bound and of Fleming's approximation is always more accurate than the HC- and the Gaussian approximation. For the discrete-time model, it may happen [cf. (6.14)] that the actual equilibrium variance is very slightly higher than the HC-approximation. The HC-approximation becomes very accurate if $20\mu V_s < \gamma^2$, which quantifies the assumption of weak mutation. The Gaussian allelic approximation becomes accurate if $\mu V_s > \gamma^2$. Summarizing, we may conclude:

- **6.3** 1. *The lower bound $\hat{\sigma}^2(\mathrm{LB})$ is a highly accurate approximation for the equilibrium variance if $9\mu V_s < \gamma^2$; otherwise Fleming's approximation $\hat{\sigma}^2(\mathrm{F})$ is better. The maximum of $\hat{\sigma}^2(\mathrm{LB})$ and $\hat{\sigma}^2(\mathrm{F})$ provides an excellent approximation for the entire parameter range explored.*
2. *The HC-approximation $\hat{\sigma}^2(\mathrm{HC})$ is very accurate if $20\mu V_s < \gamma^2$, whereas the validity of the Gaussian allelic approximation $\hat{\sigma}^2(\mathrm{G})$ requires $\mu V_s > \gamma^2$.*

Additional numerical results show that, in accordance with the second-order approximation (6.19), the equilibrium variance decreases with increasing kurtosis of the mutation distribution, and that more stringent inequalities about the relative magnitude of μV_s and γ^2 than above are required for Fleming's approximation and for the HC-approximation to be accurate in that case (cf. Chapter VII.2).

*6.5 PROOF OF (6.10)

Equations IV(5.11) to IV(5.14) show that the equilibrium density $\hat{p}$ satisfies

$$\hat{p}(x) = \frac{\mu}{1-\mu}\,\frac{U\hat{p}(x)}{\varepsilon + 1 - W(x)}\,, \tag{6.32}$$

where $\varepsilon = \varepsilon_\mu = [\overline{W}_\mu - (1-\mu)]/(1-\mu)$ and $0 < \varepsilon < \mu/(1-\mu)$. As a consequence of (6.9), we have

$$\|U\hat{p}\|_\infty \le \|uW\|_\infty < \infty\,. \tag{6.33}$$

Next, we define a function $h(x)$ by

$$\exp\Big\{-\frac{x^2}{2V_s}\Big\} = 1 - (2V_s)^{-1}x^2h(x)\,. \tag{6.34}$$

Thus, $h(x) = 1 - (4V_s)^{-1}x^2 + (24V_s^2)^{-1}x^4 - + \ldots$, $h(0) = 1$, and $0 < h(x) \le 2V_s/x^2$. Then we obtain, using (6.32) and (6.34),

$$\begin{aligned}
|\bar{x} - x_O| &\le \int_{-\infty}^{\infty} |x - x_O|\,\hat{p}(x)\,\mathrm{d}x \\
&= \frac{\mu}{1-\mu}\int_{-\infty}^{\infty} \frac{|x-x_O|}{\varepsilon + (2V_s)^{-1}(x-x_O)^2h(x-x_O)}\,U\hat{p}(x)\,\mathrm{d}x \\
&= \frac{2\mu V_s}{1-\mu}\int_{-\infty}^{\infty} \frac{|x|}{2\varepsilon V_s + x^2h(x)}U\hat{p}(x+x_O)\,\mathrm{d}x \\
&= \frac{2\mu V_s}{1-\mu}\left(\int_{|x|\le 1} \ldots\,\mathrm{d}x + \int_{|x|>1} \ldots\,\mathrm{d}x\right).
\end{aligned} \tag{6.35}$$

To derive an estimate for the first of these integrals, we assume $V_s \ge \frac{1}{2}$ and use (6.33) and $h(x) \ge 1 - (4V_s)^{-1}$ if $|x| \le 1$. Then

$$\begin{aligned}
\int_{-1}^{1} \frac{|x|}{2\varepsilon V_s + x^2h(x)}U\hat{p}(x+x_O)\,\mathrm{d}x &\le \|U\hat{p}\|_\infty \int_{-1}^{1} \frac{|x|}{2\varepsilon V_s + [1-(4V_s)^{-1}]x^2}\,\mathrm{d}x \\
&\le \|uW\|_\infty \left[1 - (4V_s)^{-1}\right]^{-1} \ln\left[1 + \frac{1}{2\varepsilon V_s}\left(1 - \frac{1}{4V_s}\right)\right] \\
&\le (\mathrm{const.}) \ln\left[(\mu V_s)^{-1}\right]\,,
\end{aligned} \tag{6.36}$$

where for the last estimate we have employed that

$$\varepsilon V_{\mathrm{s}} \geq (\text{const.})(2\mu V_{\mathrm{s}})^2$$

for all μ and V_{s} such that $u(x,y) > 0$ if $|x_O - \max(x,y)| \leq 2\mu V_{\mathrm{s}}/(1-\mu)$; cf. IV (5.34) with $q = 2$ and $s = 1/(2V_{\mathrm{s}})$. (Actually, we have assumed $u(x,y) > 0$ for all (x,y), but the above shows that $u(x,y) > 0$ in a neighborhood of (x_O, x_O) is sufficient.)

To obtain an upper bound on the second integral in (6.35), we observe that $[2\varepsilon V_{\mathrm{s}} + x^2 h(x)]^{-1} \leq 2$ if $|x| \geq 1$ and $V_{\mathrm{s}} \geq \frac{1}{2}$, and we use (6.4). This yields

$$\begin{aligned}\int_{|x|>1} \frac{|x|}{2\varepsilon V_{\mathrm{s}} + x^2 h(x)} U\hat{p}(x+x_O)\,\mathrm{d}x &\leq 2\int_{|x|>1} |x|\, U\hat{p}(x+x_O)\,\mathrm{d}x \\ &\leq 2\int_{-\infty}^{\infty}\Big[W(y)\int_{-\infty}^{\infty}|x-x_O|\,u(x,y)\,\mathrm{d}x\Big]\hat{p}(y)\,\mathrm{d}y \\ &\leq 2e_1(u)\ . \end{aligned} \tag{6.37}$$

Thus, from (6.35), together with (6.36) and (6.37), we obtain

$$|\bar{x} - x_O| \leq (\text{const.})\, 2\mu V_{\mathrm{s}} \left[\ln\Big(\frac{1}{\mu V_{\mathrm{s}}}\Big) + 1\right] \leq (\text{const.})\, 2\mu V_{\mathrm{s}} \ln\Big(\frac{1}{\mu V_{\mathrm{s}}}\Big) \tag{6.38}$$

for all μ and V_{s} such that $u(x,y) > 0$ if $|x_O - \max(x,y)| \leq 2\mu V_{\mathrm{s}}/(1-\mu)$. For sufficiently small μV_{s}, the latter inequality is automatically satisfied because of our general assumptions. The two constants in (6.38) are, of course, different and the second estimate requires $\mu V_{\mathrm{s}} < 1$. This proves (6.10).

*6.6 *PROOF OF (6.8), (6.15), AND (6.7)*

For given $n \geq 3$, we first prove

$$\int (x - x_O)^n \hat{p}(x)\,\mathrm{d}x \sim 2\mu V_{\mathrm{s}}\, \tilde{v}_{n-2} \quad \text{as } \mu \to 0\ , \tag{6.39}$$

by assuming (6.2) and that (6.4) holds for this and all smaller n. Then (6.8) is a trivial consequence for $n = 4$. Also, (6.15) follows directly from (6.39), because $(x-\bar{x})^n = \sum_{k=0}^{n}\binom{n}{k}(x-x_O)^k(x_O-\bar{x})^{n-k}$ and, by (6.38), all terms except the one with $k = n$ decay faster than $2\mu V_{\mathrm{s}}$ as $\mu \to 0$.

Let $\varepsilon = \varepsilon_\mu$ [see (6.32)] and define

$$A = (2\varepsilon V_{\mathrm{s}})^{1/4} \quad \text{and} \quad B = V_{\mathrm{s}}^{1/4}\ .$$

Using (6.32) and proceeding as in (6.35), we obtain

$$\begin{aligned}&\int_{-\infty}^{\infty}(x-x_O)^n\hat{p}(x)\,\mathrm{d}x \\ &\quad = \frac{2\mu V_{\mathrm{s}}}{1-\mu}\int_{-\infty}^{\infty}\frac{x^n}{2\varepsilon V_{\mathrm{s}} + x^2 h(x)}U\hat{p}(x+x_O)\,\mathrm{d}x \\ &\quad = \frac{2\mu V_{\mathrm{s}}}{1-\mu}\Bigg(\int_{|x|\leq A}\ldots\,\mathrm{d}x + \int_{A<|x|<B}\ldots\,\mathrm{d}x + \int_{|x|\geq B}\ldots\,\mathrm{d}x\Bigg), \end{aligned} \tag{6.40}$$

where $h(x)$ is as in (6.34). We show that the first and third integral tend to zero as $\mu \to 0$, and the second converges to $\tilde{v}_{n-2}$.

Indeed, for the first integral we obtain (by requiring $A \leq 1$, so that $h(x) \geq \frac{1}{2}$ if $|x| \leq A$)

$$\begin{aligned}\left| \int\limits_{|x|\leq A} \frac{x^n}{2\varepsilon V_s + x^2 h(x)} U\hat{p}(x+x_O)\,\mathrm{d}x \right| &\leq \frac{2A^n}{A^4+A^2} \int_{-\infty}^{\infty} U\hat{p}(x+x_O)\,\mathrm{d}x \\ &= \frac{2A^{n-2}}{A^2+1}\overline{W} \\ &\leq 2^{\frac{n+2}{4}} (2\mu V_s)^{\frac{n-2}{4}} ,\end{aligned} \tag{6.41}$$

where the last inequality holds because $\varepsilon \leq \mu/(1-\mu) \leq 2\mu$ (if $\mu \leq \frac{1}{2}$) and $\overline{W} \leq 1$.

For the second integral we observe that

$$1 \geq h(x) > h(B) \geq 1 - \frac{B^2}{4V_s} = 1 - \tfrac{1}{4}V_s^{-1/2} \geq \tfrac{1}{2}$$

if $|x| < B$ (and $V_s \geq \frac{1}{4}$), and therefore

$$x^2 h(x) \geq \tfrac{1}{2}A^2 \gg 2\varepsilon V_s \quad \text{as } \mu \to 0$$

if $A < |x| < B$. These estimates, together with $A \to 0$ and $B \to \infty$ as $\mu \to 0$, imply

$$\begin{aligned}\int\limits_{A<|x|<B} \frac{x^n}{2\varepsilon V_s + x^2 h(x)} U\hat{p}(x+x_O)\,\mathrm{d}x &\sim \int\limits_{A<|x|<B} \frac{x^{n-2}}{h(x)} U\hat{p}(x+x_O)\,\mathrm{d}x \\ &\sim \int\limits_{A<|x|<B} x^{n-2}\, U\hat{p}(x+x_O)\,\mathrm{d}x \\ &\sim \int_{-\infty}^{\infty} x^{n-2}\, U\hat{p}(x+x_O)\,\mathrm{d}x \\ &= \tilde{v}_{n-2} .\end{aligned} \tag{6.42}$$

For the third integral in (6.40), we obtain

$$\begin{aligned}\left| \int\limits_{|x|\geq B} \frac{x^n}{2\varepsilon V_s + x^2 h(x)} U\hat{p}(x+x_O)\,\mathrm{d}x \right| &\leq \frac{1}{B^2 h(B)} \int\limits_{|x|\geq B} |x|^n\, U\hat{p}(x+x_O)\,\mathrm{d}x \\ &\leq \frac{B^{-2}}{1-B^2/(4V_s)} e_n(u) \\ &= O(V_s^{-1/2}) \quad \text{as } \mu \to 0 .\end{aligned} \tag{6.43}$$

Substitution of (6.41), (6.42), and (6.43) into (6.40) completes the proof of (6.39), and thus of (6.15) and (6.8).

By replacing in the above estimates B by 1 and using $[2\varepsilon V_s + x^2 h(x)]^{-1} \leq 2$ if $|x| \geq 1$ and $V_s \geq \frac{1}{2}$ in (6.43), we obtain the estimate

$$\frac{1}{2\mu V_s} \int_{-\infty}^{\infty} (x - x_O)^n \hat{p}(x)\, dx \leq 4(2\mu V_s)^{\frac{n-2}{4}} + e_{n-2}(u) + 2e_n(u) \,, \tag{6.44}$$

and this holds for all $\mu \leq \frac{1}{2}$ and $V_s \geq \frac{1}{2}$.

The proof of (6.7) runs as follows. By the binomial theorem and by changing the sequence of integration, we get

$$\begin{aligned} \tilde{v}_n &= \sum_{j=0}^{n} \binom{n}{j} \int \left(\int (x-y)^j u_{\mathrm{RW}}(x-y)\, dx \right) (y - x_O)^{n-j} W(y) \hat{p}(y)\, dy \\ &= \overline{W} \sum_{j=0}^{n} \binom{n}{j} v_j\, \tilde{m}_{n-j,x_O} \,. \end{aligned} \tag{6.45}$$

Now, we have $\tilde{m}_{0,x_O} = 1$, $\tilde{m}_{1,x_O} = O(\mu V_s)$ by (6.38), $\tilde{m}_{2,x_O} \sim 2\mu V_s$ by (6.11) and (6.38), and

$$\tilde{m}_{n,x_O} \sim \int (x - x_O)^n \hat{p}(x)\, dx \sim 2\mu V_s\, \tilde{v}_n = O(\mu V_s) \quad \text{as } \mu \to 0 \tag{6.46}$$

for $n \geq 3$ by (6.39). This proves $\tilde{v}_n \sim \overline{W} v_n \sim v_n$, i.e., (6.7).

From the proofs of (6.10), (6.8), and (6.15), we see that these equations hold for fitness functions that are more general than the Gaussian (6.1). Indeed, they hold if

$$W(x) = 1 - \frac{(x - x_O)^2}{2V_s}\, h(x - x_O) \,, \tag{6.47}$$

where

$$h(0) = 1 \,, \tag{6.48a}$$

$$\min\left\{1, \frac{2V_s}{x^2}\right\} \geq h(x) \geq 1 - c\frac{x^{\tau_1}}{V_s^{\tau_2}} \,, \tag{6.48b}$$

for all x and $\tau_1, \tau_2, c > 0$,

$$h(x) \geq \frac{1}{2x^2} \quad \text{if } |x| \geq 1 \,, \tag{6.48c}$$

$$h(x) \sim \frac{2V_s}{x^2} \quad \text{as } |x| \to \infty \,. \tag{6.48d}$$

Conditions (6.48a)–(6.48c) are needed in the proofs. The quantities A and B have to be redefined as $A = (2\varepsilon V_s)^{1/k}$ and $B = V_s^{1/k}$, with $k > \max\{2, \tau_1/\tau_2\}$. Condition (6.48d) guarantees that $W(x) \to 0$ as $|x| \to \infty$, as required for the existence of an equilibrium distribution (cf. condition (C) in Chapter IV.2.2).

7. POLYGENIC TRAITS UNDER MUTATION-SELECTION BALANCE

We now employ the results of the previous section to derive approximations for the genetic variance of a quantitative character that is determined by many recombining loci with an arbitrary number of possible alleles generated by mutation at each locus. This is a rather complex task, because we have to deal with true multilocus systems. It may be recalled from Sections 2–5 that in simple genetic systems with two or more diallelic loci, quite complicated equilibrium behavior can occur, and that the level of the equilibrium genetic variance may depend on many details of the particular genetics. Here, we shall be concerned primarily with the continuum-of-alleles model, which will be shown to have surprisingly simple properties in this context. Nevertheless, we have to rely on simplifying assumptions to achieve mathematical progress. First, the equilibrium genetic variance is derived in the limit of small mutation rates under the assumption of global linkage equilibrium. It is shown to agree with the multilocus extension of the HC-approximation. This holds for any equilibrium and without imposing symmetry assumptions on the mutation distribution or assuming that the mean phenotype coincides with the optimum. Then, with such assumptions, we investigate the role of linkage for equilibria at which the mean coincides with the optimum, and we extend the Gaussian and the HC-approximation to incorporate the effects of linkage disequilibria.

The analysis below is based on the general additive model outlined in Chapter V.1. We posit that phenotypes experience Gaussian stabilizing selection according to V(1.12). Then the genotypic fitness function is again Gaussian,

$$W(G) = \exp\left\{-\frac{(G-P_O)^2}{2V_s}\right\}, \tag{7.1}$$

where $V_s = \omega^2 + \sigma_E^2$ (cf. Chapter V.1). Throughout this section, we use the notations $\bar{x}_i = \kappa_i$ for the allelic mean at locus i, and if there is linkage equilibrium, σ_i^2 for the halpoid variance at locus i.

7.1 ARBITRARY EQUILIBRIA IN LINKAGE EQUILIBRIUM

The results of Section 6 can be readily extended to the full multilocus setting by assuming linkage equilibrium and invoking the marginal properties of multilocus systems summarized in Chapter V.2.1. Let $\hat{\sigma}_G^2$ denote the (additive) genetic variance at *some* equilibrium, and $\hat{\bar{G}}$ the corresponding mean genotypic value. We scale the mutation rates per (haploid) locus, μ_i, such that $\mu_i = \mu\nu_i$, where the ν_i are strictly positive constants, and we set $s = 1/(2V_s)$. In view of the fact that the relevant results from discrete- and continuous-time models agree to order $O(s)$, we deal directly with the continuous-time model. This helps to avoid excessive technical complications with terms of order $O(s^2)$.

Therefore, our main aim is to prove the following multilocus HC-approximation:

$$\lim_{\mu/s\to 0} \frac{\hat{\sigma}_G^2 - 2\sum_i \mu_i/s}{2\sum_i \mu_i/s} = 0, \tag{7.2}$$

where the summation is over all ℓ loci. The general assumptions about the single-locus mutation distributions $u_i(x_i, y_i)$ are as in Section 6.

▷ The proof follows Bürger and Hofbauer (1994). We approximate (7.1) by a quadratic fitness function, so that by V(1.11)

$$W(G) = 1 - s(P_O^2 + \sigma_E^2) + 2sP_O G - sG^2 \; . \tag{7.3}$$

From V(2.8) the marginal (Malthusian) fitnesses are

$$\tilde{W}_i(x_i) = 1 - s\sigma_E^2 - s(\sigma_G^2 - \sigma_i^2) - s\left(x_i - [P_O - (\overline{G} - \bar{x}_i)]\right)^2 \; , \tag{7.4}$$

and the mean fitness is

$$\overline{W} = 1 - s\sigma_E^2 - s\sigma_G^2 - s(\overline{G} - P_O)^2 \; . \tag{7.5}$$

Therefore, we can apply the results of Section 6.2 by setting

$$x_O = P_O - (\overline{G} - \bar{x}_i) \tag{7.6}$$

and $m(x) = W_i(x_i) = -s(x_i - x_O)^2$ in (6.22). Thus, the fitness optimum 'experienced' by the maternal alleles at locus i is the difference between the phenotypic fitness optimum and the mean contribution of all other loci and the paternal alleles at locus i. Then we have $\bar{x}_i - x_O = \overline{G} - P_O$, and (6.10) implies

$$\lim_{\mu/s \to 0} \frac{(\hat{\overline{G}} - P_O)^2}{\mu_i/s} = 0 \quad \text{for every } i \; . \tag{7.7}$$

Hence, (6.24) yields

$$\lim_{\mu/s \to 0} \frac{\hat{\sigma}_i^2 - \mu_i/s}{\mu_i/s} = 0 \qquad \text{for every } i \; . \tag{7.8}$$

This proves (7.2) because

$$\left| \frac{\hat{\sigma}_G^2 - 2\sum_i \mu_i/s}{2\sum_i \mu_i/s} \right| \leq (\text{const.}) \sum_i \left| \frac{\hat{\sigma}_i^2 - \mu_i/s}{\mu_i/s} \right| \; .$$

If, in addition, the equilibrium mean phenotype coincides with the fitness optimum, $\hat{\overline{G}} = P_O$, and the mutation distribution is bounded, then the above reasoning, together with (6.16) and the remark thereafter, leads to

$$\frac{\hat{\sigma}_G^2 - 2\sum_i \mu_i/s}{2\sum_i \mu_i/s} = O\left(\frac{\mu}{s}\right) \; . \tag{7.9}$$

◁

This analysis shows that, under the assumption of linkage equilibrium, the *multilocus HC-approximation*, henceforth denoted by $\hat{\sigma}_G^2(\mathrm{HC})$, is valid to first order in μ/s:

$$\hat{\sigma}_G^2 \approx \hat{\sigma}_G^2(\mathrm{HC}) = 2\sum_{i=1}^{\ell} \frac{\mu_i}{s} = 4V_s \sum_{i=1}^{\ell} \mu_i \tag{7.10a}$$

$$= 2UV_s \; . \tag{7.10b}$$

Here, U is the total (genomic) mutation rate for the character. This approximation is valid even if the mean deviates from the optimum. As $\bar{x}_i - x_O = \overline{G} - P_O$, we infer from (6.25) that any equilibrium solution has to satisfy

$$\hat{\sigma}_i^2 + (\hat{\overline{G}} - P_O)^2 < \mu_i/s \quad \text{for every } i\,. \tag{7.11}$$

In particular, this shows that

$$\hat{\sigma}_G^2 < 2\sum_{i=1}^{\ell} \frac{\mu_i}{s} \tag{7.12}$$

and

$$(\hat{\overline{G}} - P_O)^2 < \min_i \frac{\mu_i}{s} \tag{7.13}$$

hold for any equilibrium distribution and for arbitrary parameters. The estimates (7.11), (7.12), and (7.13) do not require boundedness of the mutation distributions u_i.

The inequalities (7.12) and (7.13) are in contrast to the equilibrium properties of the diallelic two- and multilocus models analyzed in Sections 4 and 5. There, equilibria existed, where the deviation of the mean from the optimum reached the value of an allele's effect, so that the variance was higher than the HC-approximation. The reason for the discrepancy between those models and the present one, in which the deviation of the mean from the optimum is determined by the minimal mutation rate per locus, (7.13), is the following: without mutation, in the continuum-of-alleles model any given optimum (that lies in the range of variation of the trait) can be perfectly matched through the build-up of spikes at each locus such that $2\sum_i \bar{x}_i = P_O$, where the $\bar{x}_i$'s denote the positions of the spikes. Thus, no polymorphism will be maintained without mutation. Deviations from the optimum are introduced only through asymmetries of the mutation distributions. With mutation, these deviations from the optimum decrease with decreasing mutation rate. This explanation is consistent with Barton's (1986) discussion, in which he mentions that introduction of asymmetries drives populations much closer to the 'optimal' equilibrium, and to the HC-approximation.

Another interesting point is that in the diallelic model stabilizing selection leads to underdominance, i.e., to disruptive selection, at each individual locus (cf. Sections 1 and 5; Barton 1986, Keightley and Hill 1988). This is not the case in the continuum-of-alleles model, where each locus experiences stabilizing selection, as may be seen from (7.4). The phenomenon of underdominance appears to be an artifact of the assumption of global linkage equilibrium, together with the symmetries of that model. If one considers, for example, a full two-locus model, then each locus experiences either directional or overdominant selection if the other is near fixation; cf. Sections 2 and 4.

It should be emphasized that we have proved neither existence nor uniqueness of an equilibrium solution in the multilocus model. The above results apply to *any existing* equilibrium. Only in the special case that the mutation distributions at individual loci are such that at each haploid locus the mean $\hat{\bar{x}}_i$ coincides with the experienced optimum, and hence $2\sum_i \hat{\bar{x}}_i = \hat{\overline{G}} = P_O$, does existence of a multilocus equilibrium follow from the haploid existence result in a straightforward way. This includes central equilibria in models in which all single-locus mutation distributions are symmetric around $P_O = 0$, because then $\hat{\bar{x}}_i = \hat{\overline{G}} = P_O = 0$.

In fact, it can be shown that the uniqueness of equilibrium solutions may depend upon the mutation model. It is easy to see that in the random-walk mutation model, infinitely many equilibria exist. To show this, suppose that the equilibrium allelic densities $\hat{p}_1, \ldots, \hat{p}_\ell$ define an equilibrium solution, choose constants a_i such that $\sum_i a_i = 0$, and translate each $\hat{p}_i$ by a_i. This leaves all cumulants except the first unaltered and, because the right-hand side of V(2.13) is independent of $\bar{x}_i$ (c_1 in the notation there), gives a new equilibrium solution with identical phenotypic distribution. The reader may also note that this property of the continuum-of-alleles model corresponds to the first-order term of the approximation on which the analysis of the model in Section 1 was based [cf. VI(1.2), VI(1.10)].

For a Gaussian distribution of allelic effects, this translation invariance was discovered and discussed by Lande (1975, 1980b). Kimura (1981) observed and discussed a similar phenomenon. It provides the possibility of extensive neutral evolution at individual loci. Other mutation models, such as the HC-mutation model, do not have this translation-invariance property. For simplicity, consider the HC-model. Then the equations V(2.13) for the change of the cumulants due to selection and mutation, together with the translation invariance of the cumulants (Appendix D), show that after a shift of the equilibrium distribution by $(a_1, \ldots, a_\ell)$, the single-locus means and variances change according to $\tilde{\Delta}\bar{x}_i = \mu_i a_i$ and $\tilde{\Delta}\sigma_i^2 = 2\mu_i a_i(a_i + 2\bar{x}_i - v_1^{(i)})$, respectively. (Here, $v_1^{(i)}$ denotes the mean of the mutation distribution u_i.) Therefore, the response of the phenotypic mean is $\tilde{\Delta}\overline{G} = 2\sum_i \mu_i a_i$. Thus, the changes at individual loci are very small, so that some neutral evolution in finite populations may still be possible. For the two-locus HC-mutation model, it is easy to show directly from (6.21) that only a single equilibrium can exist.

7.2 CENTRAL EQUILIBRIA FOR LINKED LOCI

In exploring the role of recombination, we focus our attention on 'central' equilibria, at which the population mean coincides with the optimum, i.e., $\hat{\overline{G}} = P_O$. Our aim is to derive weak-selection, Quasi-Linkage Equilibrium (QLE) approximations for the cumulants at such equilibria if selection is weak relative to recombination. In a slightly different way, these approximations were first obtained by Turelli and Barton (1990). The present derivation is an application of the cumulant equations derived in Chapter V. Since we assume weak selection, we deal directly with the quadratic optimum model, (7.3), and we scale the optimum such that $P_O = 0$. Therefore, all selection coefficients s_k in V(1.8) are zero, except s_0 (which does not enter the recursion equations) and $s_2 = -s = -1/(2V_\mathrm{s})$. We assume the random-walk mutation model and a symmetric mutation distribution. With the notation of Chapter V.5, but denoting the variance $v_2^{(i)}$ of the mutation distribution by γ_i^2, we require for the moments about zero of the mutation distribution

$$v_{2k+1}^{(i)} = 0 \ (k \geq 0) \ \text{ and } \ v_4^{(i)} \ll \gamma_i^2 \,. \tag{7.14}$$

With a finite number of alleles, (7.14) is only an approximation, because then mutation may alter the mean at equilibrium. Furthermore, we assume that the population mean coincides with the optimum ($\overline{G} = C_1 = 0$), and we ignore small terms arising from the interaction of mutation with selection and recombination, thus assuming that the multivariate cumulants change according to V(5.10).

With these assumptions, the recursion equations for the haploid single-locus means are computed to be

$$\tilde{\Delta}\kappa_i = -s\kappa_{i..} \,, \tag{7.15}$$

where V(5.10), the weak-selection analogue of V(4.5), and V(3.16a) have been used. Similarly, for the haploid single-locus variances, we obtain from V(5.10), from the weak-selection analogue of V(4.6), from V(3.16b) and V(3.17a):

$$\tilde{\Delta}\kappa_{ij} = -\overline{W}r_{ij}\kappa_{ij} - s[(1-r_{ij})\kappa_{ij..} + 2\kappa_{i.}\kappa_{j.}] + \delta_{ij}\mu_i\gamma_i^2 \,. \tag{7.16}$$

A simple consequence of (7.15) and (D.14) is that, at equilibrium, the skewness of the phenotypic distribution vanishes. By setting $\tilde{\Delta}\kappa_{ii} = 0$ in (7.16), we obtain the equilibrium conditions

$$\hat{\kappa}_{ii..} + 2\hat{\kappa}_{i.}^2 = \frac{\mu_i\gamma_i^2}{s} \,. \tag{7.17}$$

First, it is instructive to reconsider the case of global linkage equilibrium, in which all κ_{ij}, κ_{ijk}, etc. vanish, except those with identical subscripts. Then (7.15) implies that at equilibrium every single-locus third-order cumulant vanishes, $\hat{\kappa}_{iii} = 0$ (actually, the equilibrium distributions are symmetric), and (7.17) yields the relation

$$\hat{\kappa}_{iiii,\mathrm{LE}} + 2\hat{\sigma}_i^4 = \frac{\mu_i\gamma_i^2}{s} \,. \,^{1} \tag{7.18}$$

Assuming a Gaussian distribution of allelic effects at locus i, so that $\hat{\kappa}_{iiii} = 0$, we recover the Gaussian allelic approximation IV(1.8). Assuming $\hat{\kappa}_{iiii,\mathrm{LE}} \approx \gamma_i^2\hat{\sigma}_i^2$ (6.15), as is the case if most variability is maintained by rare alleles of large effects, and $\hat{\sigma}_i^2 \ll \gamma_i^2$, the haploid HC-approximation IV(1.16) is recovered.

We shall now derive QLE approximations for the genetic variances $\hat{\kappa}_{ii}$. This will be done by substituting in (7.17) the QLE approximations of Chapter V.4.3 for the covariances $\hat{\kappa}_{ij}$ and for the fourth-order cumulants $\hat{\kappa}_{iijk}$, $\hat{\kappa}_{iijj}$, and $\hat{\kappa}_{iiij}$. From V(4.40) we obtain the QLE-approximation for the equilibrium covariance between loci i and j to leading order in s:

$$\hat{\kappa}_{ij} \approx -\frac{2s\hat{\sigma}_i^2\hat{\sigma}_j^2}{r_{ij}} \,. \tag{7.19}$$

By V(4.42), the QLE values for the fourth-order cross cumulants are, to leading order in s and for pairwise distinct i, j, and k,

$$\hat{\kappa}_{iijk} \approx 0 \,, \tag{7.20a}$$

$$\hat{\kappa}_{iijj} \approx -\frac{2s\hat{\kappa}_{iii,\mathrm{LE}}\,\hat{\kappa}_{jjj,\mathrm{LE}}}{r_{ij}} \,, \tag{7.20b}$$

$$\hat{\kappa}_{iiij} \approx -\frac{2s\hat{\kappa}_{iiii,\mathrm{LE}}\,\hat{\sigma}_j^2}{r_{ij}} \,. \tag{7.20c}$$

---

[1] We use the notation $\hat{\kappa}_{ii}$ for the marginal haploid equilibrium variance at locus i to signify that linkage equilibrium is not assumed. For the linkage-equilibrium approximation, we write $\hat{\sigma}_i^2$ and, for higher-order cumulants, we use subscripts LE.

Substitution of (7.19) and (7.20) into (7.17), and omission of terms of order s^2, yields the equilibrium relation

$$\hat{\kappa}_{iiii} + 2\hat{\kappa}_{ii}^2 \approx \frac{\mu_i \gamma_i^2}{s}\left(1 + 4s \sum_{j:j\neq i} \frac{\hat{\sigma}_j^2}{r_{ij}}\right) + 2s\hat{\kappa}_{iii,\mathrm{LE}} \sum_{j:j\neq i} \frac{\hat{\kappa}_{jjj,\mathrm{LE}}}{r_{ij}} \,. \tag{7.21}$$

For a continuum of alleles, the sum with the third-order cumulants vanishes, because in linkage equilibrium we have $\hat{\kappa}_{ijk} = 0$ for $j, k \neq i$ and, as noted above, $\hat{\kappa}_{i\cdot\cdot} = 0$. Therefore, $\hat{\kappa}_{iii,\mathrm{LE}} = 0$. The above expressions can be explicitly evaluated by applying, for example, the Gaussian or the rare-alleles approximation.

The Gaussian QLE approximation

If the allelic distribution at each locus is assumed to be Gaussian, so that $\hat{\kappa}_{iiii} = 0$ and $\hat{\sigma}_i^2 = \hat{\sigma}_i^2(\mathrm{G}) = \sqrt{\mu_i V_\mathrm{s} \gamma_i^2}$, we obtain from (7.19) the Gaussian QLE approximation for the covariances:

$$\hat{\kappa}_{ij} \approx \hat{\kappa}_{ij}(\mathrm{G}) = -\frac{1}{r_{ij}}\sqrt{\mu_i \mu_j \gamma_i^2 \gamma_j^2} \quad \text{if } i \neq j \,. \tag{7.22}$$

In the same way, we obtain from (7.21) the Gaussian QLE approximation for the variance per haploid locus:

$$\hat{\kappa}_{ii} \approx \hat{\kappa}_{ii}(\mathrm{G}) = \sqrt{\mu_i V_\mathrm{s} \gamma_i^2}\left(1 + \frac{1}{\sqrt{V_\mathrm{s}}} \sum_{j:j\neq i} \frac{\sqrt{\mu_j \gamma_j^2}}{r_{ij}}\right) . \tag{7.23}$$

Since the genetic variance is $\sigma_G^2 = 2\sum_{i,j} \kappa_{ij}$, its QLE approximation is

$$\hat{\sigma}_G^2 \approx \hat{\sigma}_G^2(\mathrm{G}) = 2\sum_{i=1}^{\ell} \hat{\sigma}_i^2(\mathrm{G}) = 2\sum_{i=1}^{\ell} \sqrt{\mu_i V_\mathrm{s} \gamma_i^2} \,, \tag{7.24}$$

which is independent of the linkage map. We call $\hat{\sigma}_G^2(\mathrm{G})$ the (multilocus) Gaussian allelic approximation. Introducing the so-called mutational variance

$$\sigma_\mathrm{m}^2 = 2\sum_{i=1}^{\ell} \mu_i \gamma_i^2 \,, \tag{7.25}$$

which measures the variance produced by a cohort of new mutations that exist entirely as heterozygotes, and assuming exchangeable loci ($\mu_i = \mu$ and $\gamma_i^2 = \gamma^2$), the Gaussian allelic approximation can be written as

$$\hat{\sigma}_G^2(\mathrm{G}) = \sqrt{2\ell V_\mathrm{s} \sigma_\mathrm{m}^2} \,. \tag{7.26}$$

For $\ell = \frac{1}{2}$, this reduces to the haploid result (6.27).

These approximations agree with the expressions obtained by Lande (1975) to the order considered here. In this seminal paper, Lande used Kimura's (1965a) analysis (cf.

Chapter IV.1.1) and assumed that the equilibrium distribution of allelic effects in the gametes is approximately multivariate normal. This is often called the Gaussian allelic model and was applied by Lande to derive the equilibrium genetic variance and the correlations between the effects of alleles at different, but linked, loci. Among others, his analysis is based on the assumption that mutational effects are small compared with the single-locus equilibrium variances. Combining his analysis with a review of available data, Lande (1975) concluded that high levels of genetic variation can be maintained through a balance between stabilizing selection and mutation (cf. Chapter VII).

Fleming's QLE approximation

Fleming (1979) derived the following more accurate approximation for the equilibrium variance at locus i which generalizes (6.28):

$$\hat{\kappa}_{ii}(\mathrm{F}) = \sqrt{\mu_i V_\mathrm{s} \gamma_i^2}\left[1 - \sqrt{\frac{\mu_i \gamma_i^2}{V_\mathrm{s}}}\left(\frac{\eta_i + 3}{16\mu_i} - \frac{19}{16}\right) + \sum_{j:j\neq i}\sqrt{\frac{\mu_j \gamma_j^2}{V_\mathrm{s}}}\left(1 + \frac{1}{r_{ij}}\right)\right], \tag{7.27}$$

where $\eta_i = v_4^{(i)}/\gamma_i^4$ denotes the kurtosis of the mutation distribution (see Nagylaki 1992, Chapter 10.6, for a concise exposition). Fleming's approximation is based on the assumption of constant mutation rates and on $\gamma_i^2 \ll \mu_i V_\mathrm{s}$. For the equilibrium covariance, Fleming also obtained (7.22). These results imply that the genetic variance is

$$\hat{\sigma}_G^2 \approx \hat{\sigma}_G^2(\mathrm{F}) = \hat{\sigma}_G^2(\mathrm{G})\left(1 + \frac{\hat{\sigma}_G^2(\mathrm{G})}{2V_\mathrm{s}}\right) - \frac{1}{8V_\mathrm{s}}\sum_{i=1}^{\ell}\hat{\sigma}_i^4(\mathrm{G})\left(\frac{\eta_i + 3}{u_i} - 3\right), \tag{7.28}$$

which again is independent of the recombination distribution. The numerical results of Chapter VI.6.4 and of VII.2.3 suggest that for a Gaussian mutation distribution, Fleming's approximation (7.28) becomes accurate as $\gamma_i^2 < 9\mu V_\mathrm{s}$. Sometimes, it will be convenient to refer to the approximations (7.22)–(7.28) collectively as the KLF-approximations.

The house-of-cards QLE approximation

Under the rare-alleles approximation, we obtain the following QLE approximations, by substituting in the right-hand sides of (7.19) and (7.21) the HC-approximation $\hat{\sigma}_i^2(\mathrm{HC}) = 2\mu_i V_\mathrm{s}$ for $\hat{\sigma}_i^2$, and using $\hat{\kappa}_{iiii} \approx \gamma_i^2 \hat{\kappa}_{ii}$ for the left-hand side of (7.21):

$$\hat{\kappa}_{ij} \approx \hat{\kappa}_{ij}(\mathrm{HC}) = -\frac{4V_\mathrm{s}\mu_i\mu_j}{r_{ij}} \quad \text{if } i \neq j, \tag{7.29}$$

and

$$\hat{\kappa}_{ii} \approx \hat{\kappa}_{ii}(\mathrm{HC}) = 2\mu_i V_\mathrm{s}\left(1 + 4\sum_{j:j\neq i}\frac{\mu_j}{r_{ij}}\right). \tag{7.30}$$

By summation, it follows that the equilibrium genetic variance is

$$\hat{\sigma}_G^2 \approx 4V_s \sum_{i=1}^{\ell} \mu_i \left[1 + 2 \sum_{j:j \neq i} \frac{\mu_j}{r_{ij}} \right] . \tag{7.31}$$

For exchangeable loci, the house-of-cards QLE approximation (7.31) simplifies to

$$\hat{\sigma}_G^2 \approx 2UV_s \left(1 + \frac{2(\ell - 1)\mu}{r_H} \right) , \tag{7.32}$$

where r_H denotes harmonic mean of the $\ell(\ell-1)/2$ recombination frequencies r_{ij}, $i < j$.

With a finite number of alleles per locus, the recursion relation (7.15) for the mean is only an approximation, because mutation may alter the mean. Therefore, the last term in (7.21), which involves the third-order cumulants, is not necessarily zero. For identical diallelic loci, Turelli and Barton (1990) showed that (7.32) has to be replaced by

$$\hat{\sigma}_G^2 \approx 2UV_s \left(1 + \frac{2(\ell - 2)\mu}{r_H} \right) . \tag{7.33}$$

If $\ell = 2$, this agrees to leading order in μ with the exact two-locus result (4.8).

If r_{ij} approaches zero for one pair of loci, i and j, the approximations (7.22) to (7.33) approach infinity, which is obviously incorrect. This indicates that the above QLE approximations break down when linkage becomes too tight, which is not surprising because the derivation had assumed selection to be weak relative to recombination.

Collectively, we shall refer to approximations yielding $\hat{\sigma}_G^2 \approx 2UV_s$, such as (5.7), (7.10), (7.32), or (7.33), as weak-mutation or rare-alleles approximations, because they apply if mutation rates per locus are relatively low and most variance is maintained by rare alleles of large effect.

Whereas the KLF-approximations (7.24) and (7.28) for the genetic variance are independent of the linkage map, the HC-approximation (7.31) predicts a slight increase in genetic variance as linkage becomes tighter. However, as r_H decreases to zero, the true equilibrium genetic variance should eventually decrease to the one-locus equilibrium variance. Numerical calculations of Turelli (1984), Turelli and Barton (1990), and Bürger (1989) confirm this. The numerical results of Turelli for up to six loci indicate that (7.32) somewhat overestimates the increase in genetic variance caused by linkage (cf. Chapter VII.2.3). Numerical and analytical results with diallelic loci show that the precise value of the equilibrium genetic variance may depend upon several details of the genetic system if linkage is tight, because several stable equilibria may coexist (Turelli and Barton 1990, and Sections 4, 5).

Bulmer (1974) showed that even for genomes such as that of *Drosophila melanogaster*, with effectively no recombination in males and only three major chromosomes, the harmonic mean recombination rate will be about 0.1 if the relevant loci are scattered randomly throughout the genome. For species with more chromosomes such values will be much closer to 0.5. This is in accordance with an analytical approximation of the harmonic mean recombination rate in Turelli (1984, Eq. 4.10). Thus for gametic mutation rates on the order of 0.01, (7.32) predicts that the actual equilibrium variance is unlikely to exceed the linkage-equilibrium prediction (7.10) by more than 20%.

The approximations (7.32) and (7.33) are also consistent with an analysis of Hastings (1989) which assumed weak mutation relative to selection, but allowed for dominance and epistasis. On the basis of induced fitness arguments, he showed that if the induced selection coefficients, s, at single loci are small relative to the harmonic mean recombination rate r_{H}, then the contribution of genetic variance caused by linkage disequilibrium is negligible if the total mutation rate is much smaller than r_{H}, i.e., if $U/r_{\mathrm{H}} \ll 1$. If the strength of selection is on the same order of magnitude as r_{H}, Hastings showed that linkage disequilibrium is unimportant if $U/s \ll 1$. Obviously, the latter condition is much more restrictive.

All these results suggest that for realistic parameter values linkage is likely to have little effect on polygenic mutation-selection balance. Summarizing, we may conclude:

• **7.1** *The (additive) genetic variance maintained by mutation-stabilizing-selection balance in a large population is only weakly dependent on linkage relations, unless linkage is very tight. Thus, for understanding polygenic mutation-selection balance, it is in general sufficient to extrapolate the single-locus predictions, e.g. the HC-approximation* (6.12), *the Gaussian allelic approximation* (6.27), *or Fleming's approximation* (6.28), *to their respective multilocus linkage-equilibrium versions* (7.10), (7.24), *and* (7.28).

This will be confirmed in Chapter VII.2, where small populations are considered.

7.3 THE DYNAMICS OF THE MEAN AND VARIANCE

If the equilibrium distribution of genotypic values has been slightly perturbed, but the population is still in QLE, simple equations for its rate of return to equilibrium can be derived. Under the assumption of a symmetric phenotypic distribution ($C_3 = 0$) and weak selection (such that the Gaussian fitness function can be approximated by a quadratic function), it follows from V(6.15), V(6.10), and V(1.14) that the mean phenotype changes according to

$$\tilde{\Delta}\bar{P} = \tilde{\Delta}\bar{G} = -\frac{1}{V_{\mathrm{s}}}(\bar{P} - P_O)\sigma_G^2 , \tag{7.34}$$

where $\tilde{\Delta}$ denotes the change across generations based on the weak-selection assumption; cf. V(2.22). Under the additional assumptions of exchangeable loci, V(6.24) implies that the change of genetic variance can be approximated by

$$\tilde{\Delta}\sigma_G^2 = -\frac{1}{2V_{\mathrm{s}}}\left(C_4 + \frac{1}{\ell}\sigma_G^4\right) + \sigma_{\mathrm{m}}^2 . \tag{7.35}$$

If the distribution of breeding values is Gaussian ($C_4 = 0$), this simplifies to

$$\tilde{\Delta}\sigma_G^2 = -\frac{1}{2\ell V_{\mathrm{s}}}\sigma_G^4 + \sigma_{\mathrm{m}}^2 , \tag{7.36}$$

supporting the assumption that the change of genetic variances caused by selection is very small in polygenic systems (cf. Lande 1976, Bulmer 1980). At equilibrium, (7.36) yields the Gaussian approximation (7.26).

If it is assumed that most genetic variance is contributed by rare alleles, then at each locus the fourth cumulant satisfies $c_4 \approx \gamma^2 c_2$ [see the paragraph after (6.15)]. The additivity property of the cumulants (Appendix D.1) implies that the same relation, $C_4 \approx \gamma^2\sigma_G^2$, holds for the fourth cumulant of the distribution of breeding values.

Substituting this relation into (7.35) and assuming, as is appropriate under the HC-approximation, that σ_G^4 can be neglected relative to ℓC_4, one obtains (Barton 1989)

$$\tilde{\Delta}\sigma_G^2 = -\frac{\gamma^2}{2V_s}\sigma_G^2 + \sigma_m^2 \,. \tag{7.37}$$

Both (7.36) and (7.37) show that in the absence of mutation the variance tends to zero.

8. SUMMARY AND CONCLUSIONS

The study of the consequences of stabilizing selection on the evolution of quantitative traits, and of the role of mutation in maintaining genetic variation, has been one of the prime research topics in population genetics since the mid 1960s. In this chapter, we investigated the (presumably) most important models that have been developed to this end. Although the analyses show that in two-locus models high levels of genetic variation can be maintained by stabilizing selection *per se* under a variety of assumptions and for large ranges of parameter combinations, this appears not to be the case in systems with four or more additive loci. For additive loci, the expected polymorphic fraction of the genome and the genetic variance seem to decrease very rapidly as the number of loci increases above two. For instance, in a five-locus system less than 0.5% of randomly chosen parameter sets yield stable polymorphisms involving two or more loci. The reason is that if many loci with different effects contribute to a trait, then the optimum phenotype will, in general, be closely matched by some homozygous genotype.

In multilocus models that are based on a continuum of possible alleles, the optimum phenotype can be matched perfectly and, in general, no genetic variance will be maintained at equilibrium. A proof, however, is only available under the assumption of linkage equilibrium. Thus, it seems likely that the continuum-of-alleles model provides a better approximation to reality than models with exchangeable diallelic loci.

The analysis of mutation-selection models produced highly model-dependent results. If stabilizing selection alone maintains a stable polymorphism, as in many of the two-locus models, then mutation adds only little to increase the genetic variance at equilibrium. In the multilocus diallelic model that was investigated, multiple simultaneously stable equilibria may coexist. These may differ greatly in their amount of genetic variation, which, basically, depends on the deviation of the equilibrium mean phenotype from the optimum. The lowest value of genetic variance maintained at any of the equilibria in this model coincides with the house-of-cards approximation (7.10); the highest possible value agrees with the Gaussian approximation (7.26). The basins of attraction of the former kind of equilibria appear to be much larger than those of the latter. In addition, the basins of attraction of equilibria of the latter kind decrease if the loci exhibit increasingly different effects. These and other results suggest that equilibria with a high level of genetic variance will rarely be attained.

For the multilocus continuum-of-alleles model, conflicting approximations have been proposed in the past. One is Kimura's Gaussian approximation, which was extended to linked loci by Lande, and improved by Fleming; the other is Turelli's HC-approximation. From the mathematical point of view both are valid approximations, though, in different regions of the parameter space. The KLF-approximations are accurate if the effects of newly produced mutants are smaller than the existing variance at a locus, whereas the HC-approximation is accurate in the opposite case. In particular, the HC-approximation becomes exact in the limit of small mutation rates per locus relative to the strength of stabilizing selection and the variance of mutational effects. Interestingly, the HC-approximation is much smaller than the Gaussian in its domain of validity, and it is much larger than the Gaussian if the latter is valid (cf. Figure 6.1). Both sorts of approximations can be extended to weakly linked loci, and both predict that linkage has only a minor effect on the equilibrium level of genetic variation.

Comparison of the results for the continuum-of-alleles model with those for the diallelic models shows that the different approximations are not a consequence of the number of alleles per locus, but rather are the consequence of different assumptions about the magnitude and effect of new mutations. Basically, all the approximations derived in this chapter for the genetic variance maintained under a balance between mutation and stabilizing selection can be grouped in weak-mutation, or rare-alleles approximations, and KLF-approximations. Roughly, the former apply if $10\mu_i V_s < \gamma_i^2$, otherwise, the latter are fairly accurate. This statement will be supported further in the next chapter.

The rather elaborate analyses that have been developed to determine the level of genetic variance at mutation-stabilizing-selection balance have led not only to a much better theoretical understanding, but also to new experimental work because, by producing conflicting results, they helped to identify open problems about the magnitude of biological parameters involved in the genetics of quantitative traits.

VII

Quantitative Variation and Selection

Genetic variation is an indispensable prerequisite for biological evolution. Its amount and the mechanisms by which it is maintained determine the course of evolution in response to selection. This is particularly valid for quantitative traits, which have a polygenic basis. Whereas the short-term response of the mean value of a quantitative trait can be predicted from purely phenotypic measurements (see Chapter V.6), it is in general impossible to predict evolutionary processes extending over more than a few generations accurately, without having detailed information about the genetics of the traits involved, and the forces by which genetic variation is maintained and, possibly, constrained. This applies not only to the medium or long-term response of quantitative traits to directional selection, as exerted by animal or plant breeders and experimentalists, but also to such diverse problems as the determination of rates of environmental change with which a population is able to cope without going extinct, problems of speciation, or models offering explanations for the advantage of sex and recombination.

The purpose of this chapter is two-fold. First, the potential of various evolutionary and genetic mechanisms to maintain heritable variation in quantitative traits is explored, especially the properties of the resulting equilibrium distributions. Secondly, some of their evolutionary consequences are investigated and discussed, such as the response of quantitative traits to directional selection, and the advantage of sex and recombination. As hypotheses about the maintenance of quantitative variation can only be evaluated on the basis of empirical knowledge about the underlying genetics and the magnitude of the involved parameters, Section 1 presents a review of the relevant empirical data and of some of the methodological problems.

Many traits are, or appear to be, under stabilizing selection. Evidence for the ubiquity of stabilizing selection on quantitative traits comes from the reduced fitness of extreme phenotypes and from the constancy of morphological form over geological time (see Charlesworth *et al.* 1982, Maynard Smith 1983, Endler 1986). However, a trait may be under stabilizing selection because of a causal relationship between the trait and fitness, or because mutations affecting the trait have pleiotropic effects on fitness (Robertson 1956, 1967). If these, as most mutational effects, are deleterious, then extreme phenotypes may have reduced fitness, because they are likely to carry many mutants. Also, the observed genetic variation of a trait may be a mere side effect of genetic variation that is maintained for reasons which are independent of the observed character.

Among the direct mechanisms, mutation-stabilizing-selection balance is the best studied. This is so because of its conceptual simplicity, and because it provides a kind of null model: the variation thus maintained can be either diminished or increased by other selective forces and genetic mechanisms. This alleged importance is reflected by the attention mutation-selection models have received in the literature, and may justify the elaborate treatment presented in this book. In Section 2, we therefore extend some of the models of Chapter VI by exploring the effects of population size and genetic drift on mutation-stabilizing-selection balance; and we discuss the biological implications.

The potential of balancing selection, such as overdominance, heterogeneous or varying environments, genotype-environment interaction, or frequency-dependent selection, for maintaining quantitative variation is investigated in Section 3. In Section 4, the important topic of pleiotropy is examined by considering multivariate characters under stabilizing selection that are affected by pleiotropic mutations. Fitness itself is a quantitative character and exhibits substantial variation, apparently, because most mutations are slightly deleterious. Section 5 reviews some simple models exploring the impact of deleterious mutations on a population and outlines some of the most important evolutionary consequences. Section 6 combines several of the ideas explicated in the previous sections and surveys models of apparent stabilizing selection in which mutations affecting a quantitative trait have pleiotropic side effects on fitness. In Section 7, the response of quantitative traits to various forms of directional selection is studied, in particular, how this depends on assumptions about the underlying genetics. The chapter ends with a brief summary and conclusions.

This chapter is conceptually different from all previous ones in that the emphasis here is on biological consequences and conclusions that can be drawn from the models rather than on their analysis. Therefore, much of the theory developed here is more heuristic and only outlines the basic models and the results. As a consequence, the treatment is not as self-contained as in the first six chapters, but much more accessible to the mathematically less sophisticated reader.

1. A REVIEW OF EXPERIMENTAL DATA AND RESULTS

The kind of data and parameters required to address the question of the maintenance of genetic variation have been defined by numerous theoretical studies. These include: measures of the distribution of a quantitative character such as the additive genetic variance; the number of loci at which mutations affecting the trait arise; dominance and pleiotropic effects of mutations; epistatic interaction of mutations; mutation rates and information about the distribution of mutational effects; the amount of genetic variation in the trait produced each generation by mutation; linkage relations between loci; the form of natural selection on the trait (direct or via pleiotropic effects, stabilizing, directional, or else); the strength of selection on new mutations. In the following, we give a brief summary of what is known and what is not known, and we provide references to the empirical literature.

1.1 VARIABILITY OF CHARACTERS

The variability of a character is measured by the phenotypic variance, its heritable component by the genetic or additive genetic variance. For many purposes, the additive genetic variance is more informative because it determines the response to selection (see Chapters II.5, II.6, V.6, VI.7.3, and Section 7). Actually, as is evident from the breeder's equation V(6.33), it is the (narrow-sense) heritability, $h^2 = \sigma_A^2/\sigma_P^2$, that relates the selection response of the mean phenotype to the selection differential. The heritability can be measured from estimating correlations between relatives (Chapter I.3.2), although, obtaining such estimates from wild populations is often difficult (Falconer and Mackay 1996, Chapter 10; Lynch and Walsh 1998, Chapter 17).

The study of Mousseau and Roff (1987) and Roff and Mousseau (1987) is based on more than 1100 heritability estimates obtained from the literature. It shows that in natural populations almost any level of heritability can be found. For instance the heritabilities of 56 skeletal metric traits from *Rhesus Macaques* of Cayo Santiago range from zero to 87% (mean $= 0.32 \pm 0.08$ s.e.m., Cheverud and Buikstra 1982). Even the heritability of homologous traits may show a comparable variation between species. An example is the tarsus length in different bird species, which has an average heritability of 0.63 ± 0.09 s.e.m., but may range from 0.15 in *Branta canadensis* (Lessels 1982) up to 0.92 in *Puffinus puffinus* (Brooke 1977).

Part of this variation among estimates will be due to measurement errors, but there are some rules of thumb concerning the magnitude of heritabilities. First, characters strongly correlated with fitness, such as life-history traits (e.g., litter size or egg production), tend to be less heritable ($h^2 < 0.2$) than characters not so strongly related to fitness, such as morphological traits (e.g., tail length or number of abdominal bristles). Physiological and behavioral traits have, on average, intermediate heritabilities. Second, heritabilities measured in random-bred laboratory populations are often similar to that in natural populations. The typical h^2 is between 0.2 and 0.6 for natural populations as well as for laboratory populations which live in a highly standardized environment. For recent reviews consult Roff (1997) and Lynch and Walsh (1998). Notable exceptions were found in Darwin finches, in which species that are subject to fluctuating selection with a strong frequency-dependent component, or to introgression from species on different islands, have very high heritabilities, whereas closely related species that are specialists, and apparently under strong stabilizing selection, have much lower heritabilities (cf. Grant and Price 1981, Boag 1983, Grant 1983, Gibbs and Grant 1987).

In summary, exceptionally high levels of heritability (> 0.6) are found in populations in which environmental factors such as hybridization or fluctuating selection may promote the maintenance of genetic variation. But still, considerable amounts of heritability are found in natural populations in which environmental fluctuation and heterogeneity play only a minor role, or in random-bred laboratory populations. These data suggest that universal genetic causes, such as mutation-selection balance, should explain at least heritabilities on the order of 0.2. This hypothesis will be examined in subsequent sections.

Little is known about total fitness, because usually only fitness components are amenable to analysis (cf. Falconer and Mackay 1996). Such fitness components are reflected by many life-history traits which, in general, have rather low heritabilities.

Houle (1992) suggested to use the additive genetic coefficients of variation, $\sigma_A/\overline{P}$, for comparing variation between traits. He made the interesting observation that the additive coefficients of variation of life-history traits are, on average, about three times as high as those of morphological traits, and the residual coefficients of variation,

$\sqrt{\sigma_P^2 - \sigma_A^2}/\overline{P}$, are an order of magnitude larger. This shows that life-history traits have lower heritabilities than morphological traits, not because they harbor less additive genetic variation, but because their residual variability is much higher. Most of this increased residual variance appears to be due to environmental causes or genotype-environment interaction, rather than to nonadditive variance (Houle *et al.* 1996). Hence, it is likely that, relative to morphological traits, a larger number of loci and a wider range of environmental factors influence fitness traits, because they are affected by many different developmental processes. A recent review of the quantitative genetics of life-history traits may be found in Charlesworth and Hughes (1999).

Finally, the ratio σ_G^2/σ_m^2 of the standing genetic variance, σ_G^2, to the variance generated by mutation each generation, the mutational variance σ_m^2 defined in VI(7.25), is a useful measure (Barton 1990). If some sort of mutation-selection balance is responsible for most genetic variance, then σ_G^2/σ_m^2 should be much smaller than if other forces, such as balancing selection, maintain most genetic variation. A recent review of data (Houle *et al.* 1996) found an overall median of σ_G^2/σ_m^2 of less than 100 (generations), and less than 50 for life-history traits. All estimates are less than 1000, and some 90% are greater than 25. This supports the hypothesis that life-history traits are large mutational targets, i.e., they are affected by a larger proportion of the genome than the average morphological trait. Among the largest values of σ_G^2/σ_m^2 that were found are those for abdominal and sternopleural bristles, with 692 and 362 generations, respectively.

1.2 THE GENETIC BASIS OF QUANTITATIVE VARIATION

Since most quantitative traits exhibit almost continuous variation, quantitative traits have been considered to be typically influenced by a large number of loci with individually small effects, so-called polygenes. This view has been expressed in the infinitesimal model, which assumes an infinite number of additive loci, each with an infinitesimally small effect on the trait (Fisher 1918, Bulmer 1971a, 1980). Notable exceptions have been known for a long time – genes with major effects on quantitative traits – one of the most remarkable examples being the Booroola gene that (as a homozygote) increases mean litter size in the Australian Merino sheep by approximately 1.2 (Piper and Bindon 1988; for additional examples, see Falconer and Mackay 1996). Hence, another popular view has been that most variation is due to a few loci of large effects, which are supported by many loci with very small effects. As the amount of equilibrium genetic variation and the medium and long-term response to directional selection of quantitative traits depend strongly on the underlying genetics (see Section 7), it is important to investigate this genetics.

Information about the location and number of loci affecting a quantitative trait can be obtained by assaying the effects of segments of chromosomes that are identified by marker genes on the character of interest. An ideal setting to locate Quantitative Trait Loci (QTL) is provided by demonstrating associations between linked markers and the trait through crosses between inbred lines, because they show complete linkage disequilibrium for genes differing between lines and affecting the trait. Whereas classical methods were restricted by the number of available markers, the advent of powerful molecular techniques brought about the availability of large numbers of molecular markers. Therefore, it is now becoming possible to map QTL to smaller and smaller chromosomal regions. In particular, these methods allow us to estimate the distribution of the effects of loci, or more precisely of genetic factors (chromosomal regions with undetected recombination), that are involved in the observed variation of a trait, as well as their dominance properties and their epistatic interaction.

Yet, these methods are still in their infancy and similar to classical methods, though to a lesser extent, the regions to which QTL are typically mapped are often so large (on the order of a few centimorgans) that they may contain hundreds of loci. A persisting difficulty with the new methods is the detection of genes of small effect and the discrimination between multiallelism, tightly linked loci, and epistatic interactions. Thus, the number of detected factors may substantially underestimate the number of contributing loci, and the magnitude of their effects may be severely overestimated. Indeed, mapped QTL of large effect may turn out to be caused by more than one variable site at a genetic locus (Long *et al.* 1998). Moreover, the statistical methods are based on a number of genetic assumptions which, if violated, can yield false results, such as locating QTL to regions where in fact there are none. For descriptions and critical discussions of these methods, consult Lander and Botstein (1989), Zeng (1994), Wright and Kong (1997), Franklin and Mayo (1998), Kearsey and Farquhar (1998), Lynch and Walsh (1998), Mayo and Franklin (1998), Long and Langley (1999), and Zeng *et al.* (1999). A concise review of data available until 1989 and of the (mostly still persisting) problems in unraveling the genetic basis of quantitative variation may be found in Barton and Turelli (1989).

One of the currently most promising approaches is to map QTL to small regions that contain candidate loci. These are loci known from the laboratory, for instance through mutations of large effects, and suspected to influence character expression. Recent experiments mapped QTL to locations consistent with candidate genes, thus suggesting that alleles of small effect at such candidate loci contribute to standing natural variation (Long *et al.* 1996, Bohuon *et al.* 1998, Kearsey and Farquhar 1998, Long *et al.* 1998, Lyman and Mackay 1998, Gurganus *et al.* 1999, Nuzhdin *et al.* 1999).

In a survey of 52 QTL mapping experiments, Lynch and Walsh (1998, Chapter 15) reported that 45% of the 222 investigated traits had a QTL accounting for at least 20% of the total phenotypic variance. In general, there is little correlation between the number of detected QTL and the total percentage of variation they explain. This indicates that loci with major effects may not be uncommon. However, some caution is necessary because such experiments tend to overestimate the effects of the detected QTL. Shrimpton and Robertson (1988a,b) detected a minimum of 17 factors on the third chromosome contributing to differences in sternopleural bristle number between two populations of *Drosophila melanogaster*. Extrapolation to the whole genome would yield approximately 60 factors. In their analysis, factors contributing less than 0.3 phenotypic standard deviations were statistically not detectable. The results of mutation-accumulation experiments, in which highly inbred lines of *D. melanogaster* were selected for up to 200 generations for high and low bristle number, indicate that most of the response (the high and low sternopleural bristle lines had diverged by an average of 17 bristles) is caused by a few factors of individually large effects (Fry *et al.* 1995; Mackay and Fry 1996). The results are consistent with previous findings (Frankham *et al.* 1980) that some of these loci, or factors, may have high mutation rates, in excess of 10^{-4} (for instance, the *bobbed* locus, where unequal crossing over generates high variability). Such high mutation rates would give rise to multiallelism in large populations. Indeed, there is some indication that isoalleles at major loci may not be uncommon, and are often associated with pleiotropic effects (Lynch and Walsh 1998). A further indication that much of the response in such experiments comes from the selection of new mutations of relatively large effect is that the response is in general episodic. This was also the case in an experiment in which body weight in mice was divergently selected, and most of the response could be attributed to two major and one minor factor (Keightley 1998a). Keeping the caveats discussed above in mind, these results are compatible with the view that quantitative traits are under control of a few loci with major effects that are supported by a (possibly large) number of loci with small effects.

However, in some experiments much higher numbers of QTL were detected. In a recent high-resolution analysis, 26 QTL affecting response to selection for abdominal and sternopleural bristle number in *D. melanogaster* were mapped onto the X and the third chromosome (Nuzhdin *et al.* 1999). Most of the mapped positions were consistent with candidate genes and with locations previously inferred to contain bristle QTL. Interestingly, the QTL contributing to response to selection for high bristle number were not the same as those contributing to response for low bristle number. Cheverud *et al.* (1996) and Vaughn *et al.* (1999) examined crosses between inbred lines of mice that had been selected for separately large and small size about 50 years ago. They detected up to 20 QTL affecting weight or age at specific ages, all of them having small effects.

These seemingly conflicting results raise another problem: mutants of large effects respond to selection much faster than those with small effects. In addition, selectively favored mutations of small effect are more easily lost by random drift and, thus, remain undetected. Hence, methods based on crosses between lines under strong divergent selection may lead to a bias by over-representing genes of large effect.

All such experiments provide only lower bounds for the number of loci. Some of the reasons have already been mentioned. Additionally, detection of variation at loci with low, but perhaps typical, mutation rates on the order of 10^{-5} – 10^{-6} (Drake *et al.* 1998) requires large numbers of individuals to be examined if the locus is in mutation-selection balance. However, there is evidence reviewed by Turelli (1984) that more than 10^4 loci may affect viability in *Drosophila*, and hundreds of loci affect certain yield characters of maize. Various theoretical arguments suggest that the number of loci affecting quantitative traits is likely to be large (see below).

Existing data show that *dominance* is common with a tendency for mutations of large effect to be partly recessive. Mutations of small effect show highly variable degrees of dominance with an average that may be close to additivity (for reviews, see Caballero and Keightley 1994, Lynch and Walsh 1998, and García-Dorado *et al.* 1999).

Strong *epistatic interactions* between QTL have been demonstrated only rarely. As interactions are difficult to detect, the rarity of epistasis may not reflect biological reality and recent analyses provide a different picture. For instance, an analysis of body weight in mice demonstrated potentially extensive epistasis between 9% of all possible pairs of 19 detected loci (Routman and Cheverud 1997). In *Drosophila*, strong epistatic interactions have been found between factors affecting bristle number (Long *et al.* 1996), metabolic traits (Clark and Wang 1997), and olfactory behavior (Fedorowicz *et al.* 1998). A particularly striking example is reported by Li *et al.* (1997) for three grain yield components in rice, in which almost 75% of the 63 markers (distributed throughout the whole genome) influenced the trait(s) primarily through interactions.

Unfortunately, very little is known about epistatic interactions between deleterious mutations, although this would be crucial to many issues of evolutionary interest (see Kondrashov 1993, Whitlock *et al.* 1995, and Section 5). The reason are substantial difficulties in measuring such interactions (West *et al.* 1998). Clear evidence for widespread and strong epistasis between the fitness effects of pairs of mutations comes from experiments with bacteria, algae, and fungi (Elena and Lenski 1997, de Visser *et al.* 1997a,b) However, no net epistasis was found because, apparently, positive and negative interactions tended to cancel the effects of each other (cf. Otto 1997).

Deviations from the additive model, V(1.1) and V(1.3), may occur because genotypes respond differently to environmental change. It may happen that in one environment a genotype has a lower phenotypic value than another genotype, whereas the reverse is true in a different environment. This phenomenon is called *genotype-environment interaction* (see Section 3). It has been frequently reported in the empirical literature on varieties of crops and domesticated animals, and is common in life-history traits (Stearns 1992, Lynch and Walsh 1998). However, it is much more difficult to demonstrate that genotype-environment interaction influences QTL expression. Nevertheless, a number of recent experiments have detected substantial genotype-environment and genotype-sex interaction for newly arising mutations affecting various fitness-related and morphological traits in *Drosophila* (e g., Fry *et al.* 1996, Fernández and López-Fanjul 1997, Gurganus *et al.* 1998, Mackay and Lyman 1998, Thompson *et al.* 1998, Wayne and Mackay 1998), and for survival and fertility-related traits in the nematode *Caenorhabditis elegans* (Shook and Johnson 1999). As mutations were found that have a noticeable effect only in certain environments, experiments performed in a constant environment may underestimate the number of QTL and the mutational variance $\sigma_{\rm m}^2$.

Many major mutants are known to have effects on several traits and many traits are correlated. Moreover, the biochemical processes underlying the development of quantitative traits are interconnected. Therefore, *pleiotropy* is considered to be a widespread phenomenon (e.g., Wright 1968, Lande 1980a, Falconer and Mackay 1996). Evidence from experiments, however, is not always as clear as one would wish (Barton and Turelli 1989). Recently, widespread pleiotropic effects of individual QTL were actually measured (Cheverud *et al.* 1997, Leamy *et al.* 1999). They were restricted to sets of functionally and developmentally related traits as suggested by some theories of the evolution of genetic systems (Riedl 1978, Wagner 1996a, and references therein). Many mutations, in particular those with large effects, have deleterious pleiotropic effects on fitness (see below).

1.3 MUTATIONAL PARAMETERS

The most extensive data set exists for the mutational variance, $\sigma_{\rm m}^2$, the amount of additive genetic variation in phenotypic traits produced by mutation per generation. This is an important parameter, because it occurs in models of mutation-selection balance (see Chapter VI.7, and Sections 2, 4, 6), and because $2\sigma_{\rm m}^2 t$ is the expected variance among originally isogenic lines after t generations of divergent neutral evolution. This latter fact is exploited for its measurement (Lynch and Walsh 1998, Chapter 12).

The ratio of the mutational variance to the environmental variance, is called the mutational heritability, $h_{\mathrm{m}}^2 = \sigma_{\mathrm{m}}^2/\sigma_E^2$. Estimates of h_{m}^2 typically range from 10^{-3} to 10^{-2} for a variety of organisms and traits. Some life-history traits in plants and viability in *D. melanogaster* have lower mutational heritabilities. However, estimates of h_{m}^2 may be biased downwards if mutants are nonadditive, nonneutral, or if there is genotype-environment interaction (Houle *et al.* 1996; Lynch and Walsh 1998, Chapter 12; Mackay and Lyman 1998; Lynch *et al.* 1999).

Whereas mutational variances can be measured directly, most other parameters of interest can be inferred only indirectly. The probably most important method is the mutation-accumulation experiment which starts with genetically identical lines and, after a number (t) of generations, measures the among-line variance, given by $2\sigma_{\mathrm{m}}^2 t$, and the change in the mean genotypic value, given by $U\bar{\gamma}t$, where $U = 2\sum_{i=1}^{\ell} \mu_i$ is the total (genomic) mutation rate for the character, μ_i is the mutation rate per (haploid) locus, and $2\bar{\gamma}$ is the difference between the mean phenotypes of mutant and nonmutant homozygotes. From these compound values, lower-bound estimates for the total mutation rate, U_{min}, and upper-bound estimates for the average mutational effect, $\bar{\gamma}_{\mathrm{max}}$, can be obtained by the Bateman–Mukai method (e.g., Lynch and Walsh 1998). Estimates of U and $\bar{\gamma}$ can be obtained by maximum likelihood and maximum distance methods; however, they require assumptions about the distribution of mutational effects and may yield different results (see García-Dorado 1997, Keightley 1998b, Lynch *et al.* 1999). Alternative approaches, applicable to certain natural populations, infer mutation parameters from inbreeding depression, i.e., from the frequently observed reduced fitness of offspring resulting from matings between inbred parents (see Charlesworth *et al.* 1990; Deng and Fu 1998; Lynch and Walsh 1998, Chapter 12; Li *et al.* 1999). These approaches require information (or assumptions) about the average degree of dominance and suppose that most variation is due to mutation-selection balance.

Total (genomic) mutation rates per character, U, are estimated to be in the range 10^{-2} to 10^{-1} (Lynch and Walsh 1998, Chapter 12), with some estimates exceeding this value (Clark *et al.* 1995). These figures are difficult to reconcile with typical mutation rates for enzyme loci on the order of 10^{-5} to 10^{-6}, a paradox discussed in detail by Turelli (1984) and Barton and Turelli (1989). If these estimates are correct, then between 10^3 and 10^5 loci have to contribute to quantitative traits, or mutation rates of QTL must be very high. Indeed, the hypothesis that mutations of small effect occur much more frequently than those of large effects, and that at least some QTL may exhibit higher mutation rates than enzyme loci (cf. Lande 1988), has received experimental support. The data of Frankham *et al.* (1980), Fry *et al.* (1995), and Mackay and Fry (1996) suggest mutation rates (or, rather, rates of variability generated by processes such as unequal crossing over) in excess of 10^{-4} for some loci/factors contributing to bristle number (e.g., *bobbed*); see also Lynch and Walsh (1998, Chapter 12).

Few data are available for the distribution of mutational effects and for γ^2, the expected squared (heterozygous) effect of a mutation. Given the above estimates of h_{m}^2 and U, it follows that γ^2/σ_E^2 can be expected to be between 10^{-2} and 1. Indeed, experiments studying the effects of transposable-element insertions suggest values for γ^2/σ_E^2 of approximately 0.07 and 0.2 for abdominal and sternopleural bristle numbers (Lyman *et al.* 1996). For various enzyme activities, estimates on the order of 0.015 were obtained (Clark *et al.* 1995). These, as well as other experiments in *Drosophila*

(Caballero *et al.* 1991, Mackay *et al.* 1992, Santiago *et al.* 1992, López and López-Fanjul 1993b) suggest that the distribution of mutants is highly leptokurtic, i.e., most variance is contributed by few mutants of large effect. There is also evidence from RNA viruses that mutations of small effect are much more common than those of large effect (Burch and Chao 1999).

Recent experiments indicate that a substantial fraction of mutations contributing to bristle number are deleterious, in particular those having a large effect on the character; some are even recessive lethals (Mackay *et al.* 1992; López and López-Fanjul 1993a,b; Nuzhdin *et al.* 1995). Such mutations, however, cannot easily be used for adaptation, unless selection is strong. The variance generated by nearly neutral mutations may be smaller by an order of magnitude than the total $h^2_{\rm m}$, the mean squared effect of such mutations may be as low as $\gamma^2 = 0.01\sigma^2_E$, and their genomic mutation rate may be on the order of $U = 0.01$ (Lande 1995).

Most mutations that affect fitness are deleterious. The spontaneous lethal mutation rate in *D. melanogaster* is between 0.01 and 0.025 per gamete and generation (Crow and Simmons 1983, Fry 2000). Their majority is recessive with heterozygous effects of only a few percent (Simmons and Crow 1977). However, controversial results about the rate and distribution of deleterious (but nonlethal) mutations have been obtained. The mutation-accumulation experiments of Mukai and colleagues (Mukai 1964, Mukai *et al.* 1972, Ohnishi 1977) on egg-to-adult viability in *Drosophila* suggest a lower bound for the total mutation rate of $U_{\rm min} \approx 0.6$, an upper bound for the average homozygous fitness effect of $2\bar{s}_{\rm max} \approx 0.05$, and a dominance coefficient of $h \approx 0.35$, yielding a heterozygous effect of approximately 0.02 (Crow and Simmons 1983, Lynch *et al.* 1995b; but see García-Dorado *et al.* 1999). Such results are in accordance with estimates from plants obtained by inferences from inbreeding depression, suggesting deleterious mutation rates of about 0.5 (Charlesworth *et al.* 1990, Johnston and Schoen 1995; a recent mutation-accumulation experiment in *Arabidopsis* by Schultz *et al.* 1999 yielded a lower estimate of $U_{\rm min} \approx 0.1$).

This, as well as other (often circumstantial) evidence, led to the view that deleterious mutation rates in higher organisms are on the order of one or larger, and that most mutations are slightly deleterious (see Charlesworth and Charlesworth 1998, Kondrashov 1998, Lande 1995, Lynch *et al.* 1999). For humans, molecular data indicate that the rate of mutations under selective constraint (those in coding regions of DNA that are effectively removed by selection, which requires that their deleterious effects are larger than $10/N_{\rm e}$) is at least 1.6 per genome per generation (Eyre-Walker and Keightley 1999). The additional contribution from mutations in noncoding regions is currently unknown. Various consequences of the high deleterious mutation rate for human evolution and human health are discussed by Kondrashov (1995) and Crow (1999).

However, the classical experiments in *Drosophila*, in particular their statistical evaluation, were criticized by Keightley (1996), García-Dorado (1997), Caballero and Keightley (1998), and García-Dorado *et al.* (1999), who argued that the lower-bound estimates for the deleterious mutation rate, $U_{\rm min}$, are rather on the order of 0.01 to 0.05, and the upper-bound estimates for deleterious (homozygous) effects, $\bar{s}_{\rm max}$, are on the order of 0.1 to 0.2. Indeed, a recent replication of a 'Mukai-like' experiment by Fry *et al.* (1999) yielded a lower bound for the deleterious mutation rate of chromosome two of only 0.02, which is much smaller than Mukai's average value of 0.16 and

Ohnishi's 0.07. Moreover, Fry *et al.* obtained $\bar{s}_{\max} \approx 0.11$, which is five times as large as the classical estimates. Since a haploid second chromosome is about one-fifth of the diploid genome, these mutation rates have to be multiplied by five to obtain genomic estimates. Such differences are important, because slightly deleterious mutations can pose a threat to the long-term persistence of populations (see Section 5). Which of these values is closer to reality is currently unclear because of several uncertainties concerning the validity of some assumptions underlying the statistical procedures (cf. Keightley 1998b, Lynch *et al.* 1999).

Fry (2000) reanalyzed the previous mutation-accumulation experiments with *Drosophila* (except Ohnishi's) by correcting various errors in the analyses, by applying the same viability measure to all of them, and by using an order method to measure viability that reduces biases caused by evolution in control lines. Moreover, he performed an additional experiment. His revised estimates show much less variation between experiments than the original ones. The new estimates of $U_{\min}$ range from 0.14 to 0.68 with a mean of 0.42, and the new $\bar{s}_{\max}$ estimates range from 0.038 to 0.117 with a mean of 0.070. In particular, in all these experiments, the revised estimates for the average reduction in fitness per generation are very similar, ranging from 0.81% to 1.48% with a mean of 1.12%. This indicates that accumulation of deleterious mutations can indeed lead to a substantial erosion of mean fitness in sexual populations. Fry further argued that a substantial part of the variation between the experiments is caused by different distributions of deleterious effects in the lines.

Estimates of average heterozygous effects on life-history traits in *Daphnia* (Lynch *et al.* 1998) and of average effects on cell division rate in *Escherichia coli* (Kibota and Lynch 1996) are between 0.01 and 0.02, similar to Mukai's values. The lower bound for the deleterious mutation rate in *E. coli*, however, is only about $U_{\min} = 0.0002$. These data come from mutation-accumulation experiments, which, at least in *E. coli*, could not detect mutations of deleterious effect below 0.01, and in which mutations of effect larger than 0.06 had a high probability of being selectively lost. This total genomic deleterious mutation rate cannot be directly compared with the *Drosophila* data, because of the number of germ line cell divisions in *Drosophila*, and because the *E. coli* genome is much smaller, but contains only some 12% noncoding DNA, whereas *Drosophila* contains about 80% noncoding DNA, and because in *Drosophila* studies mutations with large deleterious effects are maintained. Using molecular methods, Keightley and Eyre-Walker (1999) estimated the deleterious mutation rate in *E. coli* to be on the order of 0.002, which is about ten times larger than the lower-bound estimate from the mutation-accumulation experiment of Kibota and Lynch (1996). Mutation-accumulations experiments in *C. elegans* reported very low estimates of $U_{\min}$ for various life-history traits, including life-time productive output (fitness), mostly in the range between 10^{-2} and 10^{-3}, but large homozygous effects, between 0.05 and 0.24 (Keightley and Caballero 1997, Vassilieva and Lynch 1999). Davies *et al.* (1999) combined molecular methods with mutation-accumulation experiments and concluded that less than 4% of all deleterious mutations are detected in mutation-accumulation experiments with *C. elegans*.

These results show not only substantial variation in mutational parameters between species, but also between methodologically similar experiments using different base populations of the same species, and between different methods applied to the same population. A further cumbersome issue is that already the older, high estimates of deleterious mutation rates are hardly compatible with data about inbreeding depression, and estimates of U of less than 0.1 are definitely incompatible.

It seems likely that most current estimates of $U_{\min}$ are substantial underestimates of the true deleterious genomic mutation rate, because of variation of mutational effects, and because none of the mutation-accumulation experiments can detect mutations with an effect of $s < 10^{-3}$. Such mutations, however, may constitute the vast majority and could be the evolutionarily most important ones in natural populations. Therefore, we are left with a number of important open problems.

Based on our current knowledge, it may well be reasonable to distinguish mutations according to their effects into three classes: lethal or highly deleterious ones occurring at relatively low frequency, a larger class with moderate effects, leading to a fitness reduction in homozygotes between a few and some 25%, and a great majority of nearly neutral mutations which are undetectable in the laboratory because their effects are so tiny, or because they are only expressed under rare environmental conditions. For recent discussions and, in part, divergent conclusions, we refer to Charlesworth and Charlesworth (1999), García-Dorado *et al* (1999), Keightley and Eyre-Walker (1999), Lynch *et al.* (1999), and Fry (2000). A comprehensive review on mutation rates, from the molecular to the phenotypic level, and from viruses to humans, can be found in Drake *et al.* (1998).

1.4 STRENGTH AND FORM OF NATURAL SELECTION

Natural selection acts in many different ways on the phenotype of an organism. As the phenotype is described by a potentially infinite number of characters, many of them correlated, it is difficult to disentangle direct effects of selection from indirect effects (Lande and Arnold 1983). In principle, selection can be described by a fitness function that relates fitness of individuals to the (quantitative) traits under selection (cf. Chapter V.1.2). A trait, or a linear combination of traits, is under directional selection if the fitness function is an increasing or decreasing function of the trait value. Fitness itself is such a trait. By definition, its fitness function is linear. A trait is under stabilizing selection, if the fitness function has a mode, or optimum, and decreases away from this mode. Disruptive selection occurs if fitness of extreme individuals is greater than that of intermediates, i.e., the fitness function has a minimum between two peaks. In practice, fitness functions can frequently be determined only up to a quadratic approximation, and often traits experience combinations of these simple forms of selection. Even if a correlation between a trait and fitness can be demonstrated, this does not imply a causal relationship (e.g., Nuzhdin *et al.* 1995, and this Section 6). For theory and applications of the measurement of selection, see Lande and Arnold (1983), Endler (1986), Schluter (1988), Rausher (1992) Kingsolver and Smith (1995), Grant and Grant (1995), and Falconer and Mackay (1996).

As already exploited in previous chapters, a Gaussian fitness function is convenient for modeling stabilizing selection [cf. V(1.12)]. Its strength on the genotypic values is described by the parameter V_s, the width of the fitness function [cf. V(1.13)]. Stabilizing selection is weak if V_s is large, and it is strong if V_s is small. Reviews of Turelli (1984) and Endler (1986) suggest that typical values of V_s are between $10\sigma_E^2$ (strong selection) and $100\sigma_E^2$ (very weak selection).

2. A SINGLE TRAIT UNDER MUTATION AND STABILIZING SELECTION

In the previous chapter, the mathematical theory of the balance between mutation and stabilizing selection on an additive character was developed in considerable detail under the assumption that populations are sufficiently large that random genetic drift can be ignored. The purpose of this section is, first, to extend that theory by including random drift and exploring the role of population size in maintaining genetic variation. Secondly, the mathematical approximations will be compared with results from (more realistic) computer simulations. Thirdly, on the basis of these results, we shall discuss to what extent observed levels of genetic variation can be explained by a balance between mutation and direct stabilizing selection on an additive trait.

2.1 APPROXIMATING THE STATIONARY DISTRIBUTION IN A FINITE POPULATION

Random sampling in a finite population induces stochasticity in the gene frequencies and, hence, in the evolution of the distribution of phenotypic values. Therefore, a proper understanding of phenotypic evolution requires the knowledge of the sampling distribution of the average phenotype, of the genetic (and hence phenotypic) variance, and of the higher cumulants. For this, no mathematically rigorous theory exists that is based on genetic principles. The reason is that the stochastic theory of selection and mutation in a finite population is difficult, and is only developed for simple cases such as two alleles per locus (see Ewens 1979). A theory for the sampling distribution of the mean of the phenotypic distribution was derived by Lande (1976) under the simplifying assumption of a constant genetic variance and a normal distribution of phenotypic values. This and some generalizations will be reviewed below. Accurate approximations for the expected variance maintained under a balance between stabilizing selection, recombination, mutation, and random genetic drift can be deduced from the deterministic theory by simple but nonrigorous arguments. As in Chapter VI.7, we consider a randomly mating population with discrete generations, the same genotype frequencies in both sexes, and a trait that is determined additively without dominance by ℓ recombining loci. Stabilizing selection acts on phenotypes according to the Gaussian fitness function V(1.12). Then, by V(1.13), genotypic values experience the fitness function

$$W(G) = \exp\left\{ -\frac{(G - P_O)^2}{2V_s} \right\} . \tag{2.1}$$

The expected genetic variance

For the simple binomial sampling model of Chapter I.5, we showed that random genetic drift reduces the heterozygosity each generation by the factor $1 - 1/(2N)$. Because of I(3.15), the (additive) genetic variance is proportional to the heterozygosity if there are two alleles at a locus; therefore the variance is reduced by the same factor. For the multinomial Wright–Fisher model with multiple alleles at a single locus (Appendix E.3), it can also be proved that genetic drift reduces the genetic variance of an additive trait by the factor $1 - 1/(2N)$ (Latter and Novitski 1969, Nagylaki 1992). Basically, the (variance) effective population size, N_e, is defined so that this property extends to populations with a more complex breeding structure. (The concept of effective population size is briefly explained in Appendix E.1.) Throughout, we suppose that our population has constant effective size N_e and that the average reduction of genetic variance per generation caused by random genetic drift is given by the relation

$$\mathrm{E}\left[\Delta_{\mathrm{D}}\sigma_G^2|\mathcal{P}\right] = -\frac{\sigma_G^2}{2N_\mathrm{e}}, \tag{2.2}$$

where the expectation is conditional on the set $\mathcal{P}$ of genotype frequencies before sampling. It appears to be unknown how linkage disequilibrium, and thus selection, affect the validity of this equation. Nevertheless, for our theoretical treatment of the multilocus case we assume (2.2) as an approximation, because in many cases results derived by using (2.2) have been confirmed by simulations (see also below).

Equation (2.2) can be combined with the results in Chapters VI.7, to derive the change of the genetic variance under the joint action of selection, recombination, mutation, and drift for populations in quasi-linkage equilibrium. In an infinite population near linkage equilibrium and with symmetric mutation distributions at all loci, the odd cumulants can be neglected at central equilibria, for which the mean phenotype coincides with the optimum (Chapter VI.7.2). Because random drift does not induce asymmetries, we can assume that this continues to be valid near stochastic equilibrium (stationarity), at least if the basin of attraction of other equilibria is small. By VI(7.35), which assumes quasi-linkage equilibrium, and by (2.2), the expected change of the genetic variance of a quantitative trait determined by ℓ exchangeable loci can be approximated by

$$\mathrm{E}\left[\Delta\sigma_G^2|\mathcal{P}\right] = -\frac{1}{2V_\mathrm{s}}\left(C_4 + \tfrac{1}{\ell}\sigma_G^4\right) + \sigma_\mathrm{m}^2 - \frac{\sigma_G^2}{2N_\mathrm{e}}, \tag{2.3}$$

where Δ denotes the change between generations under selection, recombination, mutation, and random genetic drift, and C_4 is the fourth cumulant of the distribution of breeding values. In (2.3), terms arising from the interaction of genetic drift with selection, mutation, and recombination have been ignored. By taking expectations in (2.3) and setting the left-hand side to zero, we obtain the equilibrium relation

$$\mathrm{E}[\sigma_G^4] + \frac{\ell V_\mathrm{s}}{N_\mathrm{e}}\,\mathrm{E}[\sigma_G^2] + \ell\,\mathrm{E}[C_4] = 2\ell V_\mathrm{s}\sigma_\mathrm{m}^2, \tag{2.4}$$

which contains the unknowns $\mathrm{E}[\sigma_G^4]$, $\mathrm{E}[\sigma_G^2]$ and $\mathrm{E}[C_4]$. Therefore, the expected genetic variance at mutation-selection-drift balance, $\hat{\sigma}_G^2 = \mathrm{E}[\sigma_G^2]$, can be calculated from this equation only under additional assumptions. One assumption that is applied to all subsequent approximations is that $\mathrm{E}[\sigma_G^4] \approx (\mathrm{E}[\sigma_G^2])^2$, i.e., $\mathrm{Var}\left[\sigma_G^2\right]$ can be neglected in (2.4). This is certainly the case if (2.12) holds, because then we have $\gamma^2/[1+(N_\mathrm{e}\gamma^2/V_\mathrm{s})] \ll \ell V_\mathrm{s}/N_\mathrm{e}$.

If the distribution of breeding values is Gaussian, then $\mathrm{E}[C_4] = 0$ and the *stochastic* version of the *Gaussian allelic approximation* for the expected equilibrium variance is obtained from (2.4):

$$\hat{\sigma}_G^2 \approx \hat{\sigma}_G^2(\mathrm{SG}) = \sqrt{\left(\frac{\ell V_\mathrm{s}}{2N_\mathrm{e}}\right)^2 + 2\ell V_\mathrm{s}\sigma_\mathrm{m}^2} - \frac{\ell V_\mathrm{s}}{2N_\mathrm{e}} \tag{2.5}$$

(Latter 1970, Lynch and Lande 1993). In the limit of $N_e \to \infty$, this reduces to the deterministic Gaussian approximation VI(7.26). By VI• **6.3**, the stochastic Gaussian approximation (2.5) can be expected to be accurate if the loci are not too tightly linked, and if the mutation distribution is normal and $\mu V_s > \gamma^2$ holds for all loci.

In the absence of selection, i.e., in the limit $V_s \to \infty$, (2.4), or (2.5), yields the *neutral approximation* for the expected variance of a trait under a balance between mutation and drift:

$$\hat{\sigma}_G^2 \approx \hat{\sigma}_G^2(\mathrm{N}) = 2N_e\sigma_m^2 \,. \tag{2.6a}$$

This is nearly identical to the result of Lynch and Hill (1986), that

$$\hat{\sigma}_G^2(\mathrm{N}) = \sigma_m^2 \frac{(2N_e - 1)^2}{2N_e} \,, \tag{2.6b}$$

which was derived for a diallelic model (see also Clayton and Robertson 1955). The assumption of two alleles per locus will be appropriate if $4N_e\mu < 1$, because then the waiting time for a new mutation will be larger than the average fixation time (see Appendix E.2). Actually, Lynch and Hill (1986) developed a rather general model for phenotypic evolution by neutral mutations, allowing for dominance, linkage, and various mating systems. They showed that the approximation (2.6) is fairly robust if most mutations are nearly additive or recessive. However, complete dominance of all new mutations would lead to an average additive genetic variance that is larger than (2.6) by a factor of 2, whereas completely recessive mutations would lead to a reduction of (2.6) by a factor of $\frac{2}{3}$.

From (2.4), the *stochastic* version of *Fleming's approximation* is found to be:

$$\hat{\sigma}_G^2 \approx \hat{\sigma}_G^2(\mathrm{SF}) = \sqrt{\left(\frac{\ell V_s}{2N_e}\right)^2 + \hat{\sigma}_G^4(\mathrm{F})} - \frac{\ell V_s}{2N_e} \,, \tag{2.7}$$

where, $\hat{\sigma}_G^4(\mathrm{F})$ is the square of Fleming's deterministic approximation VI(7.28). By VI• **6.3**, the approximation (2.7) can be expected to be accurate if loci are not too tightly linked, and if the mutation distribution is normal and $10\mu V_s > \gamma^2$ holds for all loci (cf. Section 2.3).

Finally, we deduce a stochastic version of the HC-approximation. Under this approximation, the equilibrium distribution at each haploid locus is highly leptokurtic in a large population. Combining (2.2) with VI(7.37), the so-called *Stochastic House-of-Cards (SHC) approximation* for the expected equilibrium variance under mutation-stabilizing selection-drift balance is readily obtained:

$$\hat{\sigma}_G^2 \approx \hat{\sigma}_G^2(\mathrm{SHC}) = \frac{2N_e\sigma_m^2}{1 + (N_e\gamma^2/V_s)} = \frac{2UV_s}{1 + V_s/(N_e\gamma^2)} \,. \tag{2.8}$$

In the limit of infinite population size, (2.8) reduces to the deterministic HC-approximation VI(7.10), and in the limit of weak selection it reduces to the neutral approximation (2.6a). It is half of the harmonic mean of the neutral and the deterministic HC-approximation, i.e., $\hat{\sigma}_G^2(\mathrm{SHC}) = \hat{\sigma}_G^2(\mathrm{N})\hat{\sigma}_G^2(\mathrm{HC})/[\hat{\sigma}_G^2(\mathrm{N}) + \hat{\sigma}_G^2(\mathrm{HC})]$. Actually, the SHC-approximation was found independently, by different methods, and almost at the same time by four (groups of) authors (Bürger 1988b, Keightley and Hill 1988, Barton 1989, Bürger *et al.* 1989, Houle 1989).

By VI• **6.3**, the SHC-approximation (2.8) can be expected to be valid if $20\mu V_s < \gamma^2$ holds at each locus, if loci are not too tightly linked, and if the mutation distribution is not, or only weakly, leptokurtic. This inequality will be satisfied if mutation rates per locus are on the order of 10^{-4} or lower, and if stabilizing selection is not extremely weak. Thus, the SHC-approximation will apply to sexually reproducing populations if the trait is influenced by many unlinked or loosely linked genes that are not hypervariable. Computer simulations show that this is indeed the case (see Section 2.3 and Figures 2.1, 2.2).

By contrast, the stochastic versions of the Gaussian and of Fleming's approximation will apply to traits in asexually reproducing populations with a total mutation rate in excess of 10^{-3}, because then the genome can be identified with a single haploid locus subject to a mutation rate of $U = 2\ell\mu$. Formally, one has to set $\ell = \frac{1}{2}$ in (2.5) and (2.7). Figures 2.1a and 2.2b demonstrate the accuracy of the stochastic Fleming approximation (2.7) for this and other cases. Similarly, the (stochastic) KLF-approximations will apply to sexually reproducing populations with no recombination, because then the genetics is equivalent to a single diploid locus with a mutation rate per (haploid) locus of $\mu = U/2$. Then one has to set $\ell = 1$ in (2.5) and (2.7); cf. Lynch *et al.* (1991) and Charlesworth (1993).

An approximation for the equilibrium genetic variance based on the infinitesimal model was developed by Santiago (1998). It produces accurate results if the trait is determined by a large number of linked loci; for unlinked loci, it is very similar to the stochastic Gaussian approximation.

The distribution of the mean phenotype and the genetic variance

Under the assumption of a normal distribution of phenotypic values and a constant genetic variance, Lande (1976) showed that in a population of effective size N_e, the sampling distribution of the mean phenotype, conditional on the mean phenotypic value $\bar{P}$ in the preceeding generation, is normal; the expected mean evolves according to the deterministic recursion VI(7.34), and the variance is equal to σ_G^2/N_e.[1] If the distribution of $\bar{P}$ is stationary, then its expected variance is

$$\mathrm{Var}\left[\bar{P}\right] = \frac{(\hat{\sigma}_G^2 + V_s)^2}{N_e(\hat{\sigma}_G^2 + 2V_s)} \tag{2.9a}$$

$$\approx \frac{V_s}{2N_e} \qquad \text{if } \hat{\sigma}_G^2 \ll V_s\,. \tag{2.9b}$$

Lande's theory depends solely on the assumption that selection acts only on the phenotype, and that nothing other than selection and drift affects the mean phenotype. It is independent of the mechanisms that maintain variation.

Barton (1989) pursued a different approach based on the observation that under the multinomial Wright–Fisher model, the covariance between the allele-frequency

[1] In general, this formula for sampling the mean is only an approximation; see the discussion in Nagylaki 1994, pp. 362–363.

fluctuations at a locus is given by

$$\mathrm{Cov}[\Delta_{\mathrm{D}} p_i, \Delta_{\mathrm{D}} p_j | \mathcal{P}] = \frac{1}{2N_{\mathrm{e}}} p_i(\delta_{ij} - p_j) \,. \tag{2.10}$$

The reader may note that the $p_i(\delta_{ij} - p_j)$ are not only the diffusion coefficients of the diffusion operator (E.10), but also coincide with the coefficients g^{ij} of the genetic covariance matrix defined in I(10.13) that mediates the effects of selection.

A straightforward calculation shows that random fluctuations in the mean and variance of an additive trait satisfy

$$\mathrm{E}\left[(\Delta_{\mathrm{D}}\overline{G})^2|\mathcal{P}\right] = \sigma_G^2/N_{\mathrm{e}} \,, \tag{2.11a}$$

$$\mathrm{E}\left[(\Delta_{\mathrm{D}}\sigma_G^2)^2|\mathcal{P}\right] = \gamma^2\sigma_G^2/N_{\mathrm{e}} \,, \tag{2.11b}$$

$$\mathrm{E}\left[\Delta_{\mathrm{D}}\overline{G}\,\Delta_{\mathrm{D}}\sigma_G^2|\mathcal{P}\right] = 0 \,. \tag{2.11c}$$

These relations extend to multiple loci in linkage equilibrium. Combining them with the equations VI(7.34) and VI(7.37) for the deterministic change of the mean and variance near equilibrium under stabilizing selection, recombination, and mutation, where VI(7.37) requires that most genetic variance at individual loci is maintained by rare alleles, the recursion relations for the change of the variance of the mean and the variance of the variance can be derived. At stationarity, when the distribution of phenotypic values is nearly symmetric, and for exchangeable loci, the variance of the mean is found to be given by (2.9b), and the variance of the variance is

$$\mathrm{Var}\left[\sigma_G^2\right] = \frac{\hat{\sigma}_G^2(\mathrm{SHC})\gamma^2}{1 + (N_{\mathrm{e}}\gamma^2/V_{\mathrm{s}})} \tag{2.12}$$

(Barton 1989). In the limit of $V_{\mathrm{s}} \to \infty$, (2.12) reduces to $\hat{\sigma}_G^2(\mathrm{N})\gamma^2$, which agrees to leading order with the expression derived by Lynch and Hill (1986) for a neutral character, but differs from that of Zeng and Cockerham (1991) obtained under a different model.

The expected mean fitness

For a Gaussian distribution of phenotypic values with mean $\overline{P}$ and variance σ_P^2, the mean fitness in an infinite population under Gaussian stabilizing selection of the form V(1.12) is

$$\overline{W} = \frac{1}{\sqrt{1 + \sigma_P^2/\omega^2}} \exp\left\{ -\frac{(P_O - \overline{P})^2}{2(\sigma_P^2 + \omega^2)} \right\} . \tag{2.13}$$

In a finite population in which the mean phenotype is normally distributed with expected value $\mathrm{E}[\overline{P}] = P_O$ and variance given by (2.9b), and in which the genetic variance is constant, the expected mean fitness at stochastic equilibrium is

$$\mathrm{E}[\overline{W}] = 1\Big/\sqrt{1 + (\mathrm{Var}\left[\overline{P}\right] + \sigma_P^2)/\omega^2} \tag{2.14a}$$

$$\approx 1 - \frac{\sigma_P^2}{2\omega^2} - \frac{1}{4N_{\mathrm{e}}} \,, \tag{2.14b}$$

where we recall that $\omega^2 = V_s - \sigma_E^2$. The approximation (2.14b) holds if $\omega^2 \gg \mathrm{Var}[\bar{P}] + \sigma_P^2$, as will usually be the case. It shows that the load induced by drift is almost independent of the genetic variance and the strength of selection (Lande 1980b). Stochasticity in the genetic variance slightly increases this load.

2.2 THE SIMULATION MODEL

The quantitative-genetic model underlying the above approximations is very complex, and some of the approximations have not been derived rigorously, or only under rather restrictive assumptions. Therefore, computer simulations are necessary to test these approximations, and to determine their range of validity. The model described below is rather flexible, and has been applied to several problems in evolutionary biology. It is a descendent of that developed by Bürger *et al.* (1989).

Direct stochastic (Monte-Carlo) simulation is used to model (some of) the events occurring in natural populations. Each individual and each gene are represented. The genotypic value of the character is determined by ℓ additive loci without dominance and epistasis. The random-walk mutation model (Chapters IV.1, IV.2) with a continuum of possible alleles at each locus is simulated by drawing the effect of each mutation from a continuous distribution. Therefore, the number of possible segregating alleles per locus is only limited by population size. The phenotypic value of an individual is obtained from the genotypic value by adding a random number drawn from a normal distribution with mean zero and variance $\sigma_E^2 = 1$. The generations are discrete, and the life cycle consists of three stages:

(i) *Population regulation and random sampling of breeding pairs.* The surviving offspring of the preceding generation form the potential breeding pool. If at this point, the number of individuals is K (= carrying capacity) or less, then all individuals (with the possible exception of one because of monogamy) serve as parents for the next generation, and the appropriate number of breeding pairs is formed. Otherwise, K individuals are sampled without replacement to constitute the next generation of parents. The sex-ratio is always 1 : 1. Thus, the mating system is dioecious and monogamous, and the population size among adults is limited by the number of nesting places.

(ii) *Production of offspring.* Each breeding pair produces the same number, $2B$, of offspring. Therefore, neither differences nor stochasticity in fertility or fecundity occur. The genotype of each descendant is obtained from that of its parents through recombination of a given rate between 'neighboring' loci without interference, and alleles are subject to mutation with genic mutation rate μ. If a mutation occurs, then its effect is obtained by adding a random number drawn from a given distribution with mean zero and variance γ^2 to the current allelic effect.

(iii) *Viability selection.* Selection acts solely through viability selection prior to reproduction. This is imposed by assigning a fitness value to each phenotype according to a given fitness function, for instance the Gaussian. The survivors, determined by drawing uniformly distributed random numbers and comparing them with fitness, serve as the potential breeding pool for the next parental generation as described above. By this means, viability selection induces demographic stochasticity. For the simulations presented in Section 2.3, always more than 80% of the individuals survived selection. As $B \geq 2$ was chosen, the number of breeding adults was always K.

For each parameter combination, a certain number of replicate runs with stochastically independent initial populations were performed, each run being over 10^5 generations. Initial populations were obtained from a preceding initial phase of several hundreds or thousands of generations (depending on K) during which mutation-selection balance had been reached. The number of replicate runs per parameter combination ranged from 5 for the largest population sizes, to 100 for the smallest. This yielded very small standard errors among replicate runs, e.g., 1% or less for the variance. The statistics are evaluated after reproduction but before selection. Observed values are averages over all replicates and all generations.

Since each breeding pair produces exactly $2B$ offspring and, in the absence of selection, N parents are sampled without replacement from the pool of BN potential parents, the family size, i.e., the number of offspring produced per individual, follows a hypergeometric distribution with mean 2 (the population replaces itself) and variance $V_k = 2(1 - 1/B)[1 - (2B - 1)(BN - 1)]$. This yields an effective population size of approximately $N_e = 4(N - 1)/(V_k + 2)$ (Appendix E.1). If $BN \gg 1$, then $N_e \approx 2B(N - 1)/(2B - 1)$. In the simulations, the effective population size was calculated by computing the number of offspring per individual that survived to contribute to the next generation. These N_e values differ from the theoretical values by less than 1%, for large N by less than 0.1%.

An asexual version of the program has also been developed.

2.3 SIMULATION RESULTS

Comprehensive simulations using this model were performed by Bürger *et al.* (1989), Bürger (1989), Bürger and Lande (1994). Here, we summarize the most important findings and add some new numerical results. Numerical support for the assertions made, beyond that displayed in the figures, can be found in these references.

The expected genetic variance

The probably most important general conclusion is that in freely recombining populations, the SHC-approximation (2.8) is accurate under a wide range of parameters (basically, under the same assumptions as the HC-approximation), for small and large population sizes. This is documented by Figure 2.1a which displays the theoretically expected and the observed equilibrium variances in freely recombining sexual populations and in genetically equivalent (same ℓ, μ, etc.), but nonrecombining asexual, populations. Figure 2.1a also demonstrates that the stochastic version of Fleming's approximation is very accurate for asexual populations without recombination for the reasons discussed at the end of Section 2.1. For sexually reproducing populations without recombination, it is similarly accurate (results not shown). Figure 2.1a visualizes the extent to which sexual and asexual populations differ in the their genetic variance that is maintained by mutation under additive gene action. Sexual populations have a reduced mean fitness, because segregation and recombination break up favorable gene combinations and cause an increase of variance. It also shows that asexual populations approach their infinite-size expectation for the variance at population sizes of a few hundred individuals, whereas sexual populations of the same size have variances considerably lower than their infinite-size expectation. This is most pronounced with

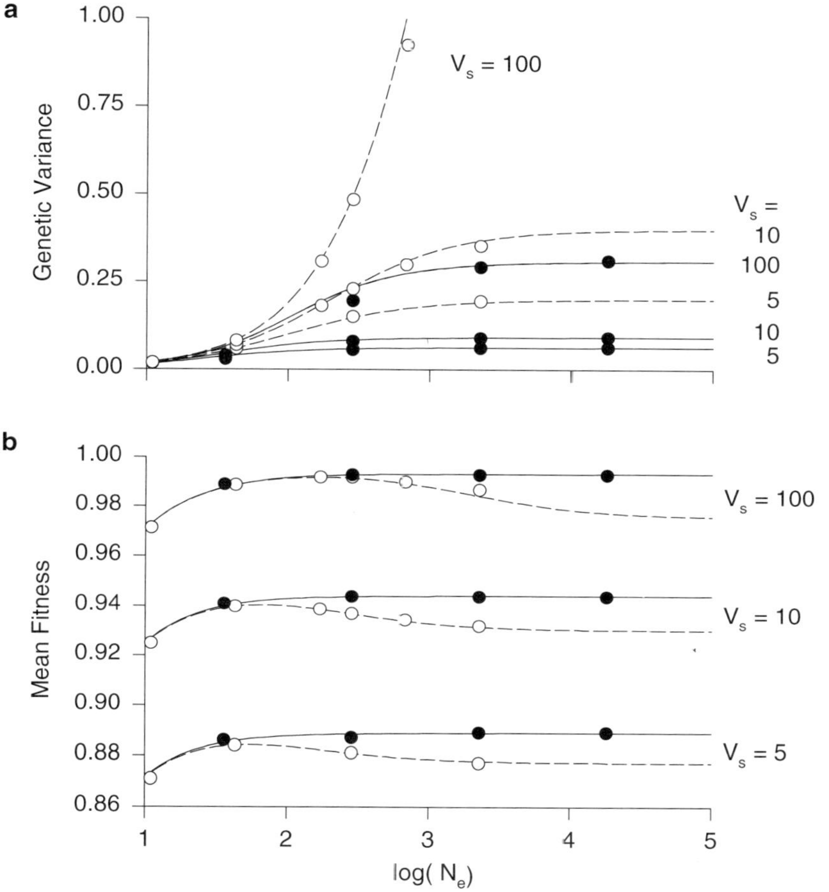

Figure 2.1 Genetic variance and mean fitness at mutation-selection-drift balance for recombining sexual and nonrecombining asexual populations. For all data shown, the genomic mutation rate is $U = 0.02$, the input of mutational variance is $\sigma_m^2 = 0.001$, the mutation distribution is normal with mean zero and variance $\gamma^2 = 0.05$, and the number of loci is $\ell = 50$. Figure **a** displays the equilibrium genetic variance in a freely recombining sexual population (dashed lines and open symbols) and in a nonrecombining asexual population (solid lines and black symbols) as a function of the decadic logarithm of effective population size for three strength of stabilizing selection: $V_s = 5$, $V_s = 10$, $V_s = 100$. The dashed lines for the variance are calculated from the SHC-approximation (2.8), the solid lines from the stochastic Fleming approximation (2.7) with $\ell = \frac{1}{2}$ and $\mu = U$. Figure **b** displays the corresponding equilibrium mean fitness. The theoretical expectations are based on (2.14a), with the variance calculated from (2.8) (dashed lines) and (2.7) (solid lines). Symbols represent data from simulations, with some data taken from Bürger and Lande (1994) and Bürger and Lynch (1995).

moderate or weak stabilizing selection.

As noted previously, the SHC-approximation tends to overestimate the observed average variance slightly, because the deterministic HC-approximation and the neutral approximation do so. In accordance with the deterministic theory of Chapter VI.6, this overestimate increases if the kurtosis of the mutation distribution increases. In a large population and for a reflected Γ-distribution [IV(2.6)] with $\theta = 0.5$ (hence, a kurtosis of $\frac{26}{3}$), the actual variance may be reduced by nearly 40% relative to the SHC-approximation (see Figure 2.2). The simulation results show that the neutral prediction is accurate if $N_e\gamma^2/V_s < 0.1$, as expected from a comparison of (2.6) and (2.8).

Simulation studies of Keightley and Hill (1988) and Barton (1989) showed less agreement with the SHC-prediction, presumably because these authors used diallelic models with identical and fixed allelic effects.

The roles of linkage and of the number of loci

All theoretical approximations for the equilibrium variance predict that $\hat{\sigma}_G^2$ is proportional to the number of loci affecting the trait if loci are unlinked. This has been strongly supported by computer simulations (Keightley and Hill 1988, Bürger *et al.* 1989, Bürger 1989, Foley 1992, Bürger and Lande 1994). For loosely linked loci, the sum of single-locus variances [also called the linkage-equilibrium variance, cf. V(6.22) and II(4.26)], is always slightly larger than the average genetic variance, but never by more than a few percent. This shows that Hardy–Weinberg and linkage disequilibria can safely be ignored in this case.

As discussed in Chapter VI.7.2, the KLF-approximations are independent of the recombination distribution as long as linkage is not tight, whereas under the rare-alleles approximation the variance increases with decreasing harmonic-mean recombination frequency r_H. Equation VI(7.33) predicts that this increase is very small if $U/r_H \ll 1$.[2] Indeed, Figure 2.2a shows that decreasing the recombination rate from 0.5 leads to a slight increase in genetic variance, but that VI(7.33) always produces an overestimate. Interestingly, extremely tight linkage between neighboring loci is required to reduce the genetic variance below that for independent loci. For 50 loci, the single-locus prediction is approached only as the recombination frequency between neighboring loci decreases below $r = 10^{-4}$ (yielding $r_H = 7 \times 10^{-4}$). As long as $r > 0.01$ ($r_H = 0.066$), only little (repulsion) linkage disequilibrium builds up. The effects of linkage are qualitatively similar for both mutation distributions. Drawing the recombination frequencies from an exponential distribution has very little effect on the equilibrium distribution relative to the case of identical recombination frequencies (results unpublished).

A difficult and empirically unresolved problem is the number of factors or loci contributing to a trait. Figure 2.2b shows how, for a given mutational input to the trait (fixed U and σ_m^2), the equilibrium genetic variance depends on the number of loci (factors) and on the recombination distribution. Again, the figure demonstrates that for a Gaussian mutation distribution, the stochastic version of Fleming's approximation is very accurate if $10\mu V_s > \gamma^2$, whereas the SHC-approximation is otherwise more appropriate. In terms of number of loci and total mutation rate, this inequality may be rewritten as $5UV_s > \ell\gamma^2$. If the total mutation rate U is lower than 0.02, the

[2] A straightforward calculation shows that for ℓ loci with recombination frequency r between neighboring loci and no interference, the harmonic mean recombination frequency of all $\frac{1}{2}\ell(\ell-1)$ pairs of loci is $\frac{1}{2}\ell(\ell-1)/\sum_{k=1}^{\ell-1}(\ell-k)/r(k)$, where $r(k) = \frac{1}{2}[1-(1-2r)^k]$ is the recombination frequency between a pair of loci separated by $k-1$ loci.

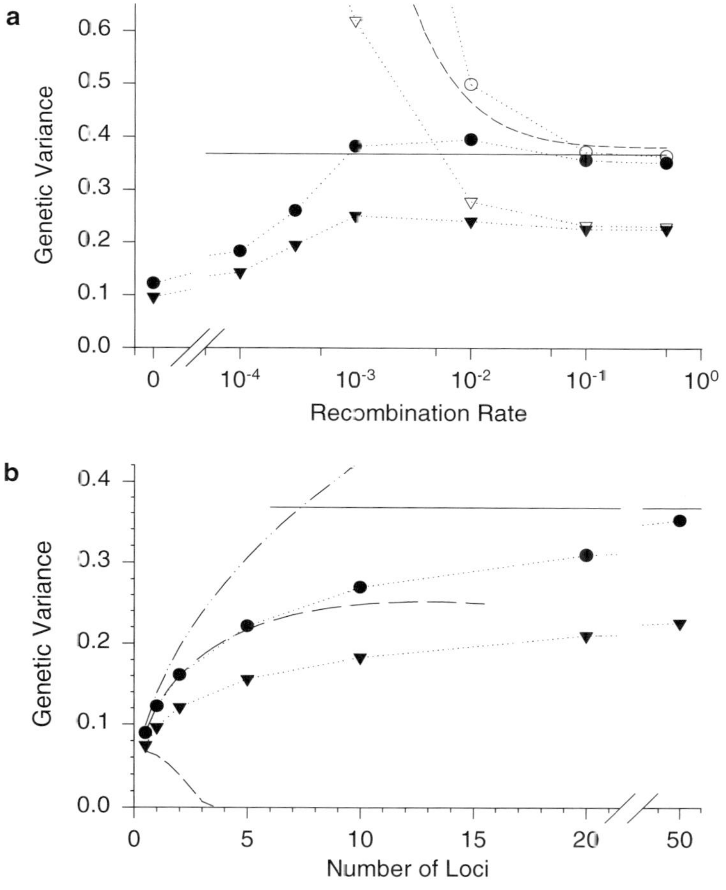

Figure 2.2 The genetic variance at mutation-selection-drift balance as a function of the recombination rate (**a**) and the number of loci (**b**). For all data, $U = 0.02$, $\sigma_{\mathrm{m}}^2 = 0.001$, $N_{\mathrm{e}} = 2276$, $V_{\mathrm{s}} = 10$, and $B = 5$. Circles (solid or open) represent data for a Gaussian mutation distribution with mean zero and variance $\gamma^2 = 0.05$, triangles represent data for a reflected Γ-distribution with the same mean and variance, and $\theta = 0.5$ (thus, its kurtosis is $\frac{26}{3}$). Filled symbols indicate the observed equilibrium genetic variance, open symbols (only in Figure **a**) represent the linkage-equilibrium variance. The solid line is the SHC-approximation (2.8). In Figure **a**, the recombination rate is between neighboring loci and there is no interference. The dashed line is the finite-population version of the approximation VI(7.33) that takes linkage into account. In Figure **b**, symbols are displayed for $\ell = \frac{1}{2}, 1, 2, 5, 10, 20, 50$, and genic mutation rates of $\mu = U/(2\ell)$, except for $\ell = \frac{1}{2}$ which represents an asexual diploid population with no recombination and a genic mutation rate of $\mu = U/2$. The long- and short-dashed lines represent the stochastic Fleming approximation (2.7) for the Gaussian and the reflected Γ-distribution, respectively. The dash-dotted line is the stochastic Gaussian approximation (2.5).

value taken in Figure 2.2, then the range of validity of the SHC-approximation extends to lower numbers of loci, whereas the opposite is the case for higher U. With increasing kurtosis of the mutation distribution, higher mutation rates per locus are required for

Fleming's approximation to be accurate, and lower such values are required for the SHC-approximation. Thus, the higher the kurtosis of the mutation distribution, the larger is the range of values of $\mu V_s/\gamma^2$ for which none of the approximations is very accurate.

A genetic factor consisting of several linked loci yields the same (low) equilibrium variance as a single locus with the same (total) mutation rate only if the loci within the factor are extremely tightly linked. For example, the variance of a trait with $U = 0.02$, a normal mutation distribution, $V_s = 10$, and $N_e = 2276$, that is determined by 50 loci, grouped into five clusters (factors) with free recombination between the clusters and recombination frequencies of $r = 10^{-4}$ between neighboring loci in the cluster, and no interference (yielding $r_H = 2.3 \times 10^{-4}$), is 0.257, whereas the variance for completely linked loci within the cluster (corresponding to the five-locus case in Figure 2.2b) is 0.221. With $r = 10^{-3}$ within the cluster ($r_H = 2.3 \times 10^{-3}$), the variance is 0.377 compared with 0.352 for unlinked loci. Therefore, a single locus with a high mutation rate contributes slightly less to the genetic variance than several linked loci with the same total mutation rate.

The expected mean fitness

Figure 2.1b shows that (2.14a) provides an accurate approximation for the observed equilibrium mean fitness at mutation-selection-drift balance. Due to the load induced by phenotypic variance, the mean fitness of freely recombining populations is always less than that of corresponding nonrecombining populations (cf. Charlesworth 1993, Lande and Shannon 1996). This reduction is negligible, however, in populations below a few hundred individuals. The mean fitness is nearly independent of the linkage map and the mutation distribution because for all data displayed in Figure 2.2, the average mean fitness is between 0.931 and 0.944. This is in contrast to models in which genetic variation is maintained by overdominance (Franklin and Lewontin 1970, Lande 1975).

The distribution of breeding values

The simulations show that the distribution of breeding values is close to Gaussian for a wide range of parameters. For many loci, this is a consequence of the central limit theorem. The average mean phenotype is close to zero and the average skewness of the distribution of G is vanishingly small. The kurtosis is less than one for every sexual population of Figure 2.1a and for every asexual population with $N_e \geq 285$. It is between 0 (the value of Gaussian distribution) and 0.2 for every sexual population with $N_e \geq 171$. For the asexual populations of size $N_e = 43$, it is between seven and eight. For all data in Figure 2.2 with five or more loci and $r \geq 10^{-3}$, the kurtosis is less than 0.2 for the normal mutation distribution and less then 0.6 for the reflected Γ-mutation distribution. In general, the kurtosis increases if the mutation rates per locus decrease, if the number of loci decreases, if population size becomes small (less than 100), if stabilizing selection gets extremely strong ($V_s < 5$), or if the distribution of mutational effects becomes very leptokurtic.

The distribution of the mean phenotype and the genetic variance

The distribution of the mean phenotype is almost perfectly Gaussian for all parameters checked, and equation (2.9) for the variance of the mean phenotype is very accurate. Only for weak selection ($N_e/V_s < 0.2$) is the distribution of the mean slightly platykurtic. This Gaussian phenotypic theory also works well for leptokurtic mutation distributions and for linked loci. The variance of the genetic variance is predicted well by equation (2.12) whenever the SHC-approximation applies, although it has a tendency to underestimates the observed $\text{Var}\left[\sigma_G^2\right]$ by approximately 10% (see Bürger and Lande 1994).

Heterozygosity

Whereas in diallelic models the heterozygosity is proportional to the genetic variance, this is not the case in the present model, in which every mutation leads to a novel allele. Unless selection on the gene level is strong ($N_e\gamma^2/V_s > 1$), the observed heterozygosity is close to the neutral prediction, $H = 4\mu N_e/(1 + 4\mu N_e)$, which is valid for a model in which every mutant is of a new allelic type (see, e.g., Nagylaki 1992, Chapter 9.7). Foley (1992) extended this to include the effects of stabilizing selection. His approximation has the same structure, but with the 'effective' mutation rate $\mu_e = \mu/\sqrt{1 + cN_e\sigma_m^2/V_s}$, where c is a constant which is approximately $\frac{1}{2}$. Slatkin (1987) derived a very accurate, but algebraically complex, approximation for the heterozygosity in an infinite population that is based on his five-allele approximation.

Autocorrelation and stochasticity

All simulations show high fluctuations of the mean phenotype, the genetic variance, and other quantities between generations, and substantial autocorrelation. The extent of this phenomenon is displayed in Bürger *et al.* (1989) and Bürger and Lande (1994). This phenomenon was also noted by Keightley and Hill (1983) for directional selection and by Keightley and Hill (1988) for stabilizing selection. Bürger and Lande (1994) derived the autocorrelation function of the mean phenotype and, for a neutral model with recurrent mutation, of the genetic variance. They showed that the autocorrelation time for the mean phenotype is $V_s/\hat{\sigma}_G^2$, and that for the genetic variance is $2N_e/(1 + 8N_e\mu)$. As shown by their simulations, both formulas provide rather good approximations. As expected, the autocorrelation time of the genetic variance decreases with increasing strength of stabilizing selection and with increasing average effect of mutations, because selection for or against a new mutant accelerates its fixation (Kimura and Ohta 1969). Furthermore, the autocorrelation time decreases with increasing linkage, with increasing kurtosis of the mutation distribution, and with increasing mutation rate. These high autocorrelations pose serious problems for conclusions based on measurements of genetic variance, because of the long-lasting influence of random excursions of the variance.

2.4 SUMMARY AND CONCLUSIONS

The development of the mathematical theory of mutation-stabilizing-selection balance for a purely additive trait in an asexual or a sexual population has succeeded in producing quantitatively accurate approximations for several key quantities characterizing the stationary distribution. The role of the basic genetic parameters in influencing this stationary distribution, in particular its genetic variance, is fairly well understood. Hence, the explanatory power of mutation-stabilizing-selection balance as a mechanism of generating and maintaining genetic variation can be evaluated quantitatively, provided appropriate and reliable data are available.

The distinction between the different kinds of approximations for the genetic variance (KLF-approximations versus rare-alleles approximations) lies only in the assumptions about the mutational properties of the genes or, more generally, of the genetic factors determining the trait. A factor contributes approximately the same amount of genetic variance to a trait as a single locus with the same total mutation rate if the recombination frequencies between neighboring loci within the factor are on the order of 10^{-4} or lower. (Otherwise, the loci in the factor contribute approximately as much as independent loci.) If, for each such factor, $10\mu V_s < \gamma^2$ holds, then the SHC-approximation (2.8) can be applied, whereas if $10\mu V_s > \gamma^2$, the stochastic version of Fleming's approximation (2.7) is accurate. Thus, the latter will be applicable if a few genetic factors with very high mutation rates determine the trait. This will often be the case in nonrecombining species because, essentially, there is only a single locus. The validity of the stochastic Gaussian approximation (2.5) requires $\mu V_s > \gamma^2$, and the neutral approximation (2.6) applies if $N_e\gamma^2/V_s < 0.1$. In populations of small or moderate size, the different approximations yield similar results, because the impact of the many details of the genetics is overwhelmed by random genetic drift. However, due to the stochastic nature of mutation, recombination, and selection, any single population may considerably deviate from the behavior of the 'average population'.

It is obvious from the review of data in Section 1.1 about the variability of characters that a universal explanation for the observed levels of variability will not exist. However, because mutation is the ultimate cause of genetic variability and because it is ubiquitous, mutation-selection balance should be responsible for at least a certain level of variation upon which other mechanisms may act (Lande 1975). Here, we shall examine what the levels of heritability, h^2, and of the ratio of standing genetic variance to the mutational variance, σ_G^2/σ_m^2, are that a balance between mutation and direct stabilizing selection can account for.

The following discussion is based on inferences derived from the SHC-approximation. By choosing a scale such that $\sigma_E^2 = 1$ and by approximating γ^2 by σ_m^2/U, as is exact for exchangeable loci, the expected heritability at mutation-selection balance is calculated to be

$$\hat{h}^2 = h^2(\text{SHC}) = \frac{2\sigma_m^2}{\sigma_m^2[2 + 1/(UV_s)] + N_e^{-1}} . \tag{2.15}$$

If a population's effective size satisfies $1/N_e < \frac{1}{9}\sigma_m^2[2 + 1/(UV_s)]$, then its heritability differs by less than 10% from that of an equivalent infinite population. The ratio of standing equilibrium genetic variance to the mutational variance is calculated to be

$$\frac{\hat{\sigma}_G^2}{\sigma_m^2} \approx \frac{\hat{\sigma}_G^2(\text{SHC})}{\sigma_m^2} = \frac{2N_e}{1 + 2N_e s} , \tag{2.16}$$

where $s = \gamma^2/(2V_s)$ is the deleterious effect of the 'average' mutant. If selection is strong relative to drift, then (2.16) reduces to

$$\hat{\sigma}_G^2/\sigma_m^2 = 1/s , \tag{2.17}$$

the value obtained from the deterministic HC-approximation VI(7.10). In the case of a few hypervariable loci, the stochastic versions of the Gaussian or of Fleming's approximation would give more accurate, but lower, estimates than the corresponding SHC-approximation. Somewhat lower values as suggested by (2.15), (2.16) and (2.17) apply if mutation distributions are highly leptokurtic.

In a sufficiently large population, the equilibrium heritability $\hat{h}^2$ of a neutral trait may approach unity. Therefore, high heritabilities of neutral characters are compatible with mutation-selection balance. However, under near neutrality ($2N_e s \ll 1$), $\hat{\sigma}_G^2/\sigma_m^2$ approaches $2N_e$, which is incompatible with most of the data compiled by Houle *et al.* (1996), because most of their estimates are much lower than likely values of N_e. Let us consider some numerical cases, in which we continue to scale parameters such that $\sigma_E^2 = 1$.

Examples. 1. If the total mutation rate for a trait is very small, say $U = 0.001$, then even in very large populations and under relatively weak selection, say $V_s = 20$, the heritability will be below 4%.

2. If $U = 0.02$ and $V_s = 10$, as in several of the above simulation results, then (2.15) yields $\hat{h}^2 = 2\sigma_m^2/(7\sigma_m^2 + N_e^{-1})$, which approaches 0.286 for very large populations. If, in addition, $\sigma_m^2 = 10^{-3}$ and $N_e = 1000$, then $\hat{h}^2 = 0.25$ is obtained. The same parameter set yields $\hat{\sigma}_G^2/\sigma_m^2 = 400$ (because $\gamma^2 = 0.05$), which is far above the median of the data reviewed by Houle *et al.* (1996).

3. Finally, if mutational effects are assumed to be very large ($\gamma^2 = 0.5$) and $U = 0.01$, so that $\sigma_m^2 = 0.005$, then for the relatively weak, but perhaps typical, strength of stabilizing selection of $V_s = 20$ (Turelli 1984) and for $N_e = 1000$, one obtains $\hat{h}^2 = 0.28$ and $\hat{\sigma}_G^2/\sigma_m^2 = 77$. Both values may be considered as typical for morphological traits. Stronger selection decreases both values. ◇

These examples show that a balance between mutation and direct stabilizing selection can explain values of h^2 between 5% and 30% for traits that are under appreciable stabilizing selection. However, the range of parameters for which typical values of *both* h^2 *and* σ_G^2/σ_m^2 are obtained is rather restricted. If the low empirical estimates of σ_G^2/σ_m^2 are correct, then the above results indicate that genes contributing to quantitative traits may be under stronger selection than explained by stabilizing selection alone.

Apart from the fact that many genes have effects on many characters, simple load arguments suggest that the number of traits in an organism that can be under direct stabilizing selection is limited. Equation (2.14) shows that the load caused by a single trait under direct stabilizing selection is approximately $\sigma_P^2/(2\omega^2)$, typical values of which are considered to be in the range between 1% and 10% (Lande 1980a, Turelli 1984). Thus, direct stabilizing selection of the magnitude often observed, cannot operate independently on a large number of characters (Barton 1990).

A somewhat unrealistic feature of the mutation model underlying the above theory (the random-walk mutation model) is that the fitness effect of any mutation depends on the genetic background in which it occurs. Almost all mutations occurring in individuals with a nearly optimal phenotype are deleterious, whereas almost half of these mutations are beneficial if they occur in individuals with an extreme phenotype. This may be in conflict with the observation that a large fraction of mutations are apparently unconditionally deleterious, although there is recent evidence that the availability of beneficial mutations is much higher in populations that are distant from an adaptive peak (Burch and Chao 1999). However, as shown in Chapter VI, the HC-approximation for the equilibrium variance is very robust with respect to assumptions about the mutation distribution and, of course, is valid under the HC-mutation model, which produces mutations of unconditional effect.

A point that seems worthwhile to reiterate is that direct stabilizing selection on a trait is compatible with extensive neutral evolution at the underlying loci (cf. Chapter VI.7.1). This may lead to genetic divergence and isolation between lines that are subject to the same selective pressure.

The influence of the mating system, of inbreeding, and of nonadditive gene action on the variance maintained under mutation-selection balance have also been studied to some extent. Lande (1977, 1984) showed that if the Gaussian allelic model applies, then assortative mating and inbreeding have little influence on the equilibrium variance. This is supported by an analysis of Frank and Slatkin (1990). However under the HC-approximation, Turelli (1986) and Frank and Slatkin (1990) found that strong inbreeding can reduce the variance by up to 50%. Frank and Slatkin also investigated a stepwise-mutation model in which suboptimal alleles are recessive, and found that this leads to substantial dominance variance, but the total genetic variance agreed with the HC-approximation. To date, apparently, no systematic studies of the influence of dominance or epistatic deviations on the maintenance of genetic variation under stabilizing selection have been performed. Analyses of some special models indicate that dominance has little influence on the amount of genetic variance maintained, whereas epistasis has the potential to maintain high levels of genetic variation (Lewontin 1964, Slatkin 1987, Gimelfarb 1989, Frank and Slatkin 1990, Zhivotovsky and Gavrilets 1992; see also Chapter VI.3.3 and this Section 6.3).

3. OTHER MODELS OF DIRECT SELECTION

Here, we shall give a brief overview of other mechanisms that have been proposed as having the potential to maintain genetic variation in traits under direct stabilizing selection. Most of them invoke a balance between selective forces (balancing selection) leading to protected polymorphisms, i.e., to the stable segregation of at least two alleles per locus. Although much theory has been concerned with the maintenance of polymorphisms at single loci, relatively few studies have addressed balancing selection on quantitative traits, mainly, because of the difficulties encountered by the analysis of multilocus systems. Due to the lack of space and of systematic investigations, we shall not present detailed mathematical analyses, but rather concentrate on some basic models that capture essential features, and summarize the most important conclusions that can be drawn.

3.1 OVERDOMINANCE

Overdominance, or heterozygote advantage, can maintain stable polymorphisms (see Chapters I.9.3, II.1, II.5.3, II.5.4). Therefore, it is a potent force in maintaining genetic variability. The best known example is sickle-cell anaemia, which provides heterozygotes with resistance to malaria. Overdominance may arise, as in this case, by pleiotropy, because the two alleles affect different fitness components (one homozygote induces anaemia, the other is susceptible to malaria). Overdominance may also arise at the molecular level if the two allozymes produced make the heterozygote more versatile than either homozygote. There are also other ways by which overdominance can occur, for instance through epistatic interactions, or through heterogeneous environments (see below).

Explicit models for exploring the amount of genetic variation in traits under stabilizing selection that are influenced by overdominant loci were analyzed by Bulmer (1973) and Gillespie (1984). Their basic model is the additive model of Chapter V.1.1 with Gaussian stabilizing selection on the phenotype. It assumes that, in addition to the contribution to the trait value, each homozygote reduces fitness by an amount s. Gillespie called this the optimum model with pleiotropic overdominance. Assuming linkage equilibrium, the change of allele frequencies is the sum of the components caused by stabilizing selection and by overdominance. For exchangeable diallelic loci, Bulmer (1973) showed that a stable polymorphism exists if $s > \gamma^2/(2V_s)$, i.e., if overdominance is stronger than stabilizing selection. He extended this model to include recurrent mutation and random drift, and derived a diffusion approximation for the heterozygosity and for the variance, by applying Wright's stationary distribution of allele frequencies (E.17) to a locus with a heterozygote advantage of $s - \gamma^2/(2V_s)$.

Gillespie (1984) extended Bulmer's deterministic model by admitting several alleles per locus. He derived the exact equilibrium variance and studied the transient properties. An interesting feature is that the selective disadvantage, s_i, of the homozygotes at locus i plays a similar role in determining the equilibrium genetic variance as the mutation rate, μ_i, in the mutation-stabilizing selection models. The variance caused by a single locus i decreases as the quantity $2V_s s_i$ decreases, and eventually vanishes as overdominance becomes too weak relative to stabilizing selection. The response to selection, however, for instance after a shift in the optimum, may be quite different from that expected under mutation-selection balance, because the variance can decrease under certain circumstances (see Gillespie 1984, and Section 7.6).

A different model yielding overdominance was investigated by Rose (1982). It is a model of antagonistic pleiotropy in which a locus affects two fitness components, W_1 and W_2, in such a way that $W_1(\mathcal{A}_1\mathcal{A}_1) > W_1(\mathcal{A}_1\mathcal{A}_2) > W_1(\mathcal{A}_2\mathcal{A}_2)$ and $W_2(\mathcal{A}_1\mathcal{A}_1) < W_2(\mathcal{A}_1\mathcal{A}_2) < W_2(\mathcal{A}_2\mathcal{A}_2)$. These fitness components may interact additively or multiplicatively to give the total fitness. If favorable effects are sufficiently dominant, then this model, and some of its extensions, maintains a stable polymorphism.

According to these results, large amounts of genetic variation can be maintained by pleiotropic overdominance at several loci and, indeed, heterozygote advantage was once thought to be among the most plausible explanations for genetic variation. However, despite considerable efforts, very few cases have been documented experimentally (Lewontin 1974, Maynard Smith 1998). Given the difficulties in detecting fitness differences of 1% and lower, and the fact that a minute advantage of the heterozygous genotype is sufficient to maintain a stable poylmorphism (Bulmer 1973, Gillespie 1984, and III• **2.1**), this is not a compelling argument against overdominance, but there are others. For instance, high levels of genetic variation are also reported from asexual species and from predominantly selfing plants (e.g., Lynch 1984, Mayo 1987, Houle *et al.* 1996). Another reason against an important role of overdominance results from the potentially high segregation load it causes, which is at least half of the fitness reduction of the homozygote with the smaller disadvantage (Chapter III.3.2). Hence, the number of overdominant loci in a genome is unlikely to be very high. Consult Lewontin (1974) and Kimura (1983) for further discussion. Finally, in a review of the quantitative genetics of life-history traits, Charlesworth and Hughes (1999) conclude that widespread overdominance is incompatible with the available data, which indicate low levels of dominance variance. Therefore, it seems likely that overdominance (however caused) can be ruled out as a general source of genetic variation, although it may maintain a high degree in specific cases.

3.2 HETEROGENEOUS AND VARYING ENVIRONMENTS

Populations inhabit environments that are not uniform, but may be structured and vary in time or space. Most individuals within a local subpopulation will experience similar environmental conditions changing on time scales below one generation and within the range of movement of individuals. This micro-variation, being common to most or all individuals, can usually be subsumed into the environmental variance σ_E^2. However, there is also temporal variation on time scales longer than one generation and between different patches of habitat. Such 'coarse grained' environmental variation, or macro-variation, may have a profound influence on genetic variability, because it can lead to differentiation between local populations and to fitnesses varying in time and space. For instance, if one allele is advantageous in one environmental patch and another in a different patch, then both alleles may be maintained in the population if there is migration between the two subpopulations. However, the situation turns out to be complex, because polymorphisms are maintained only under certain conditions. Experiments show that temporally and spatially varying enviroments can increase genetic variance for some traits relative to control lines in a constant environment (Mackay 1981), and there is evidence that fluctuating environments maintain polymorphisms in nature (Nevo 1978). We first briefly discuss the ability of spatial structure and of temporal environmental fluctuations to maintain genetic variance. Then we consider a further complication occurring in varying environments, namely genotype-environment interaction.

Spatial heterogeneity and migration

Most theoretical investigations have been concerned with the maintenance of single-locus polymorphisms (see Felsenstein 1976), few with quantitative traits. Two very basic and apparently similar models that lead to quite different conclusions were proposed by Levene (1953) and Dempster (1955). In their simplest version, both models consider a diallelic locus, a number of demes (local patches), and a single large random-mating population. In Levene's model, zygotes or juveniles disperse randomly into local demes and settle. Selection operates separately within each deme in a density-dependent manner: the fraction of adults in every deme is fixed. This form of selection is called soft selection. After selection, all adults from all demes join a single random-mating population to produce the next generation.

In Dempster's model, fixed proportions of juveniles settle in the demes, and the fraction of adults is proportional to the mean fitness in the deme. This is called hard selection and may be appropriate if total population size is controlled. The recurrence relations are straightforward to derive and the following conclusions emerge from a stability analysis of the equilibria (see Hartl and Clark 1997, Maynard Smith 1998 for elementary treatments, and Nagylaki 1992 for detailed analyses of more general migration models). In Dempster's hard selection model, a stable polymorphism exists only if the arithmetic mean fitness of heterozygotes across all demes is higher than the means across environments of both homozygotes. Thus, overdominance is required for a stable polymorphism. In Levene's soft selection model, it is sufficient that the harmonic mean of the heterozygotes exceeds that of both homozygotes. This is a weaker condition, but still quite stringent (see Hoekstra *et al.* 1985).

Nonrandom assortment of genotypes into demes, however, and certain spatial structures may easily lead to the maintenance of stable polymorphisms. A simple such model is one in which there is a sequence of environments, for instance along the shore of a river, where at the different ends different genotypes are selectively favored, and where dispersal of individuals is limited to neighboring environments. Mating occurs at random only within each deme. This sort of selection in a heterogeneous environment results in a cline, i.e., in a gradient of gene frequencies between the extreme localities.

Models of this kind were investigated by Felsenstein (1977) and Slatkin (1978). They suppose that in each environment, stabilizing selection of the same strength acts on the character, but the position of the optimum varies between environments. Assuming that allelic effects at each locus follow a Gaussian distribution (as in the models of Kimura and Lande), both authors demonstrated that under certain assumptions about the patterns of migration and selection, substantial genetic variation can be maintained by a balance between spatially varying selection and migration (cf. Bulmer 1980, pp. 180–184). For a one-dimensional geographic continuum (in ξ), with position of the optimum given by $P_O(\xi) = \beta\xi$ and dispersal distance following a normal distribution with mean zero and variance l^2, Slatkin (1978) showed that at equilibrium

$$\hat{\sigma}_G^2 = 2\zeta\sqrt{V_s + \zeta^2} + 2\zeta^2 \,, \tag{3.1}$$

where $\zeta = \sum_{i=1}^{\ell} \sqrt{\mu_i \gamma_i^2 + \beta^2 l^2}$. If $\beta l = 0$ (no cline), then ζ^2 can be neglected in (3.1), and (3.1) reduces to the Gaussian allelic approximation VI(7.24). Hence, in a steep environmental gradient, migration has the potential of maintaining much more genetic variation than simple mutation-selection balance. Recently, Barton (1999) extended this model and performed analyses without assuming the Gaussian allelic model. Among others, he found that with a sufficiently steep cline ($\beta l > 1$), the Gaussian approximation is accurate even when mutation and selection alone are better described by the HC-approximation. The reason is that gene flow inflates the genetic variance in the same way as a source of mutations of small effects.

In the absence of spatially varying selection, migration can also maintain genetic variation in small local populations if there is a constant input of additive genetic variance by mutation. Lande (1991, 1992) showed that, under various geographic models, any amount of migration in a stable population structure can maintain almost as much genetic variance in each local population as would be present in a panmictic population of the same total effective size (cf. Nagylaki 1994 for the analysis of an alternative model).

These results demonstrate that spatial variation of the environment *per se* does not automatically maintain genetic variation. Howver, under certain patterns of geographic variation, and certain modes of mating and migration, high levels of genetic variance can be maintained.

Temporal environmental variation

Temporal variation of environments occurs in many forms and on different time scales. Here, we are only interested in changes that occur on scales of one or more generations. Since a changing optimum will exhibit a stronger influence on the trait's distribution than a changing shape (width) of the fitness function, a simple class of models assumes that the phenotypic trait is under Gaussian stabilizing selection, (2.1), with the position of the optimum varying between generations, $P_O = P_O(t)$, whereas the width of the fitness function is constant. If the optimum moves in one direction, which may be a reasonable model for sustained long-term environmental change, such as a climatic trend, or if the optimum changes periodically, relatively long periods of directional selection may result. This sort of environmental variation may, indeed, lead to high levels of additive variation and, under certain conditions, is a potent mechanism of maintaining high heritabilities, sometimes in excess of 50%. It will be treated in detail in Section 7 in the context of directional selection.

If, by contrast, the optimum fluctuates randomly across generations without autocorrelation, for instance, so that in each generation the position of the optimum is drawn from a normal distribution, then virtually no increase of variance occurs relative to mutation-stabilizing-selection balance with a resting optimum. This has been shown by assuming the Gaussian allelic approximation (Lande 1977), the rare-alleles approximation (Turelli 1988b), and by computer simulations of the sort described in Section 2.2 (Bürger 1999). If the position of the optimum, however, changes with positive serial correlation, then the mean fitness of a population may be increased by an increasing genetic variance, and modifiers can evolve which increase the genetic variance, for instance by enhancing recombination (Slatkin and Lande 1976, Charlesworth 1993, Lande and Shannon 1996). These studies assumed discrete, nonoverlapping generations. For models of age-structured populations with overlapping generations, Ellner (1996) showed that fluctuating selection *per se* can maintain genetic variation if the generations overlap and the variance of the fluctuations is sufficiently large.

Summarizing, it may be concluded that temporal variation in fitnesses can lead to an increase of (additive) genetic variance, far above the constant-optimum mutation-selection value, if the environmental change occurs in a predictable way such that periods of directional selection induce adaptive response (see Sections 7.2–7.4, where the relevant references may also be found). Because the genetic variance decays very slowly after the mean phenotype has reached a new optimum (e.g., Gavrilets and Hastings 1994b, 1995), the variance can remain relatively high, before selection, possibly on a pleiotropically connected trait, acts to increase the variance again.

Genotype-environment interaction

All previous analyses in this chapter were based on the additive model of quantitative genetics which assumes that environmental effects are genotype independent. This, however, is not always the case, because genotypic values and environmental effects may be correlated (genotype-environment correlation), or because different environments may induce different phenotypic values associated with a given genotype. The latter phenomenon is often called environmental sensitivity and is a special case of genotype-environment interaction. In extreme cases, the ranking of genotypes may be altered by a shift in the environment, i.e., there may be a change in the order of genotypic values of a sequence of genotypes when measured in different environments. This occurs mainly in 'coarse-grained' environments, in which individuals grow up in a single environmental state. Different environmental states may either be caused by spatial variation, such as different patches of habitat, or by temporal environmental fluctuation, when different generations experience different environmental conditions. In any such case, the phenotypic value is not simply given by V(1.1), but has to be expressed as

$$P = G + I + E \,. \tag{3.2}$$

Here, G is the average genotypic value produced by its genotype, the average taken over all environmental states. The interaction term, I, is the genotype-dependent component of environmental effects. It depends on both the genotype and the state of the environment that an individual experiences. The micro-environmental component, E, is assumed to be independent of I and normally distributed with mean zero and variance σ_E^2. Then, in generalization of V(1.3), the phenotypic variance can be partitioned as

$$\sigma_P^2 = \sigma_G^2 + \sigma_I^2 + \sigma_E^2 + \mathrm{Cov}[G, E] \,. \tag{3.3}$$

For a detailed account of genotype-environment interaction, the reader is refered to Falconer and Mackay (1996) and Lynch and Walsh (1998).

Let us now briefly discuss theoretical studies on the role of genotype-environment interaction that have arrived at diverse conclusions. Gillespie and Turelli (1989) investigated a special case of the model (3.2) and (3.3) by assuming that $\mathrm{Cov}[G, E] = 0$, that G is determined additively by ℓ loci as in V(1.2), and that, analogously,

$$I = \sum_{i=1}^{\ell} (I_j^{(i)} + I_k^{(i)}) \,, \tag{3.4}$$

where j and k label the alleles at locus i. In this representation, $I_j^{(i)}$ is a random variable that reflects the genotype-dependent contribution of the environment. The model becomes analytically tractable after imposing a number of symmetry assumptions. These are: $\mathrm{E}[I_j^{(i)}] = 0$, $\mathrm{Var}[I_j^{(i)}] = V_I$, $\mathrm{Cov}[I_j^{(i)}, I_k^{(i)}] = V_I \rho_w$ if $j \neq k$, and $\mathrm{Cov}[I_j^{(i)}, I_k^{(h)}] = V_I \rho_b$ if $i \neq h$, where ρ_w and ρ_b refer to correlations within and between loci, respectively. The expectations are taken over the full distribution of environmental fluctuations and hold for all i, j, k, and h. In particular, we have $\sigma_I^2 = 2\ell V_I$.

A straightforward calculation shows that the variance of $G - I$ across environments for a particular genotype is

$$\mathrm{Var}[G + I | \text{genotype}] = \sigma_I^2 [1 + \rho_w + c(1 - \rho_w) + 2(\ell - 1)\rho_b] \,, \tag{3.5}$$

where c is the fraction of homozygous loci in the genotype. This shows that the variance of the average phenotype across environments produced by a genotype is a decreasing function of the number of heterozygous loci. Thus, this model implies increased 'developmental homeostasis' (i.e., low environmental variance in the phenotypic value produced by a genotype) for multilocus heterozygous genotypes, a hypothesis endorsed by Lerner (1954). The usual explanation for this phenomenon is that heterozygotes are better buffered against environmental variation, because of the multitude of allelic products they produce.

Gillespie and Turelli (1989) showed further that in the limit of small allelic effects and small V_I, and if stabilizing selection on phenotypes is independent of the environmental states, increased developmental homeostasis of heteroyzgotes produces heterozygote advantage, i.e., the mean fitness of a genotype is an increasing function of the number of heterozygous loci. This leads to a stable polymorphism with high allele frequencies. Although the authors concluded that large amounts of genetic variation can be maintained in their model, this was actually not proved, because in the limiting case of zero allelic effects they solved, the amount of genetic variance for the trait is zero (Gimelfarb 1990). Gimelfarb also raised other concerns, in particular, about some of the symmetry assumptions (but see Gillespie and Turelli 1990), and gave an example in which genotype-environment interaction does not lead to heterozygote advantage. Thus, although the model of Gillespie and Turelli (1989) can maintain high levels of allelic variation, it is questionable if it can maintain much quantitative variation.

A model similar in spirit was investigated by Zhivotovsky and Feldman (1992). They *assumed* that for each genotype the variance across environmental states decreases as the number of heterozygous loci of the genotype increases. Thus, they assumed a relation similar to (3.5), but more general. In their model, which is based on that of Zhivotovsky and Gavrilets (1992), effects of loci are not necessarily the same, nor is linkage disequilibrium ignored. Zhivotovsky and Feldman proved the existence of a stable polymorphism which is characterized by unequal allele frequencies, unequal proportions of complementary gametes, a reduction of genetic variance by linkage disequilibrium, and which has the potential of maintaining high heritabilities. Stability of this equilibrium requires that the effect of stabilizing selection is weak relative to the induced overdominance.

A quite different model of genotype-environment interaction was analysed by Via and Lande (1987). They considered a two-environment situation and used Falconer's (1952) idea that a metrical character expressed in two environments can be considered as a pair of genetically correlated characters, each of which is expressed in only one environment. They assumed a model of soft selection of Levene's type, in which stabilizing selection acts within each environment on the expressed character, that disruptive selection acts between the environments, and that mutations are pleiotropic. Applying Lande's (1979, 1980a) multivariate theory of normally distributed phenotypes, Via and Lande showed that if the genetic correlation between the characters is not ± 1, then the mean phenotype expressed in each environment will eventually attain the optimum value for that environment. This is the evolution of phenotypic plasticity (the property of a genotype to produce a phenotype that is shaped by the environment). At this joint optimum, there is no increase in the equilibrium genetic variability over that maintained by mutation-stabilizing-selection balance within each environment. Also, perturbations of the mean phenotype by a few phenotypic standard deviations lead only to small changes in the variances, unless the genetic correlation between the characters is high. Via and Lande showed further that a genetic correlation of ± 1 leads to an increased variance, because then the mean phenotype does not evolve to the joint phenotypic optimum and, therefore, remains under disruptive selection.

The main reason for the divergent conclusions reached by these groups of authors appears to stem from the fact that in the models of Gillespie and Turelli (1989) and Zhivotovsky and Feldman (1992) no single genotype will produce a phenotype that has highest fitness in all environments, whereas this is the case in the model of Via and Lande (1987). More recent and detailed genetic models of the evolution of quantitative traits in a spatially-structured (clinal) environment show that, in general, complete phenotypic plasticity does not evolve, and that some heritable variation may be maintained in the population by a balance between migration and local selection for different alleles in demes along the environmental gradient (e.g., Scheiner 1998). For discussion and further models of genotype-environment interaction and the evolution of plasticity, we refer to Gavrilets and Scheiner (1993) and de Jong (1999).

The models discussed here show that certain kinds of macro-environmental heterogeneity and genotype-environment interaction have the potential to maintain quantitative variation, although this may be diminished by the evolution of phenotypic plasticity. As noted in Section 1.1, heritabilities of random-bred laboratory populations are, on average, similar to that in natural populations, although some populations experiencing highly variable environments harbor large amounts of genetic variation. Also, populations kept in uniform environments for hundreds of generations may maintain considerable genetic diversity (López-Fanjul and Hill 1973). Therefore, variable or heterogeneous environments cannot be a general cause of genetic variation.

3.3 FREQUENCY- AND DENSITY-DEPENDENT SELECTION

Density-dependent selection occurs if the fitness of genotypes, or phenotypes, depends on population size because, for instance, under crowding conditions some types have reduced fitness. Frequency-dependent selection means that the fitness of phenotypes depends on their frequency distribution. This typically occurs if certain types have higher fitness when rare, e.g., because parasites attack common genotypes or because the ability to utilize different food resources depends on body size. A simple quantitative-genetic model of this kind was analyzed by Slatkin (1979). He considered an approximately normally distributed phenotypic character, P, in a population of size $N(t)$. The distribution of the character (before selection) is denoted by $f_P(P,t)$, its mean and variance by $\overline{P}(t)$ and $\sigma_P^2(t)$, respectively. The (absolute) fitness of individuals of type P is assumed to depend on $N(t)$ and $f_P(P,t)$ in the following way:

$$W(P,t) = 1 + \rho - \frac{\rho N(t)}{k(P)} \int \alpha(P-P') f_P(P',t)\,\mathrm{d}P' \,, \tag{3.6}$$

where $1+\rho$ is the maximum fitness in the absence of competition, $k(P)$ represents resources that can be utilized by an individual of type P, and $\alpha(P-P')$ represents the competition between individuals of types P and P' for the limiting resource. This functional form of W is related to the Lotka–Volterra competition equations (cf. Roughgarden 1979). As an example of $k(P)$, Slatkin uses a function proportional to a Gaussian density with mean P_O, which is the value of the character for which the maximum resources are available, and variance σ_k^2, which measures the range of available resources. Similarly, as an example of α he uses

$$\alpha(P - P') = \exp[-\tfrac{1}{2}(P - P')^2/\sigma_\alpha^2] ,$$

where σ_α^2 measures the extent of competition between individuals. Roughly, with these choices, the fitness function (3.6) leads to disruptive selection on the character if $\sigma_P^2 < \sigma_k^2 - \sigma_\alpha^2$, and to stabilizing selection otherwise (including the case $\sigma_k^2 < \sigma_\alpha^2$).

Assuming the additive genetic model, a continuum-of-alleles, and a Gaussian distribution of phenotypes, Slatkin (1979) showed that for these specific functions a stable equilibrium with positive variance exists and is stable if $\sigma_k^2 - \sigma_\alpha^2 > \sigma_E^2$ and $\rho < 2$, where the latter condition is necessary for demographic stability of the difference equation describing population growth (cf. Chapter I.8.4). At this equilibrium, the population mean satisfies $\overline{P} = P_O$ and the variance is $\sigma_P^2 = \sigma_k^2 - \sigma_\alpha^2$. However, he also proved that for less flexible genetic models (e.g., with only two alleles at a locus), often no polymorphism is maintained. Related models were investigated by Felsenstein (1977) and Bulmer (1980, pp. 171–172), who arrived at similar conclusions.

We may conclude from the results of this section that most of the discussed mechanisms have the potential to maintain variation under more or less restrictive assumptions, but none of them is able to provide a general explanation for the maintenance of genetic variation. They may, however, explain high heritabilities of some ecologically important traits.

4. MULTIVARIATE STABILIZING SELECTION AND PLEIOTROPY

All models investigated above consider a single quantitative trait. In general, this is not realistic, because many traits are correlated with other traits. In a randomly mating population, correlation between traits can be caused by pleiotropy or, to a lesser extent, by linkage disequilibrium (Lande 1980a). However, pleiotropic effects lead to detectable genetic covariance only if different genes have a common bias toward positive or negative pleiotropic effects. Otherwise, their effects may cancel, leading to no net covariance. In the absence of covariance, such 'hidden' pleiotropic effects may still cause an association between the amount of genetic variance of different characters and, thus, can influence the response to selection. Therefore, consideration of apparently independent characters in isolation can lead to wrong conclusions. Furthermore, if genes contributing to a trait of interest have unknown pleiotropic effects on another trait under selection, this can substantially affect the (apparent) strength of selection on the observed trait as well as its distribution. As it is generally believed that pleiotropic effects of mutations are ubiquitous (Section 1.2), and because selection acts on whole organisms, and hence on many traits, the importance of investigating the consequences of pleiotropic mutations on the genetic variances and covariances of a set of traits is hard to overestimate.

The first such study was performed by Lande (1980a), who established a general framework for analyzing a multivariate quantitative-genetic model based on multiple loci producing pleiotropic mutations. After outlining Lande's basic model in the special case of random mating, we investigate the consequences of hidden pleiotropic effects on the apparent strength of selection on a trait. Then we study the influence of pleiotropy on the equilibrium genetic variance under multivariate stabilizing selection. Finally, we briefly consider a model of constrained pleiotropic effects.

4.1 THE BASIC MODEL

We consider an effectively infinite, randomly mating diploid population with discrete generations and equivalent sexes. A set of n phenotypic traits is determined by ℓ loci. The pleiotropic effects of an allele from the paternal [maternal] gamete at locus i are denoted by the column vector $\boldsymbol{x}^{(i)} = (x_1^{(i)}, \ldots, x_n^{(i)})^\top$ $[\boldsymbol{x}^{*(i)} = (x_1^{*(i)}, \ldots, x_n^{*(i)})^\top]$, which represents a value of the random vector $\boldsymbol{X}^{(i)}$ $[\boldsymbol{X}^{*(i)}]$.[3] The vector of phenotypic values is denoted by $\boldsymbol{P} = (P_1, \ldots, P_n)^\top$, its genetic and environmental component by $\boldsymbol{G} = (G_1, \ldots, G_n)^\top$ and $\boldsymbol{E} = (E_1, \ldots, E_n)^\top$, respectively. The random vector $\boldsymbol{E}$ is assumed to be independent of $\boldsymbol{G}$ and multivariate normally distributed with mean vector $\mathbf{0}$ and covariance matrix $\Sigma_{\boldsymbol{E}}$. The covariance matrices of $\boldsymbol{P}$ and $\boldsymbol{G}$ are denoted by $\Sigma_{\boldsymbol{P}}$ and $\Sigma_{\boldsymbol{G}}$, respectively. Assuming the additive model, one obtains in analogy to Chapter V.1.1:

$$\boldsymbol{P} = \boldsymbol{G} + \boldsymbol{E} = \sum_{i=1}^{\ell} (\boldsymbol{X}^{(i)} + \boldsymbol{X}^{*(i)}) \, , \tag{4.1}$$

$$\overline{\boldsymbol{P}} = \overline{\boldsymbol{G}} \, , \qquad \text{and} \qquad \Sigma_{\boldsymbol{P}} = \Sigma_{\boldsymbol{G}} + \Sigma_{\boldsymbol{E}} \, , \tag{4.2}$$

where bars denote mean values. Let $\mathsf{C}^{(ij)} = \mathrm{E}[(\boldsymbol{X}^{(i)} - \overline{\boldsymbol{X}}^{(i)})(\boldsymbol{X}^{(j)} - \overline{\boldsymbol{X}}^{(j)})^\top]$. Then, because random mating is assumed, the covariance matrix of genotypic values $\Sigma_{\boldsymbol{G}}$ can be expressed as the sum of the covariance matrices of pleiotropic effects at each locus ($\mathsf{C}^{(ii)}$) and the covariance matrices between allelic effects due to linkage disequilibrium ($\mathsf{C}^{(ij)}$ for $i \neq j$):

$$\Sigma_{\boldsymbol{G}} = 2 \sum_{i=1}^{\ell} \sum_{j=1}^{\ell} \mathsf{C}^{(ij)} \, . \tag{4.3}$$

The genetic covariance matrix $\Sigma_{\boldsymbol{G}}$ is symmetric because $\mathsf{C}^{(ij)} = \mathsf{C}^{(ji)^\top}$.

To model mutation, and for the sake of simplicity, we assume that mutational effects are drawn from a continuous distribution (continuum-of-alleles model) and according to the random-walk mutation model of Chapter IV.2.1. Specifically, mutation at locus i occurs with probability $\mu^{(i)}$ and transforms an allele of effect $\boldsymbol{x}^{(i)}$ into one of effect $\boldsymbol{x}^{(i)} + \boldsymbol{\xi}^{(i)}$, where $\boldsymbol{\xi}^{(i)}$ is drawn from a multivariate normal distribution with mean $\mathbf{0}$ and covariance matrix $\Gamma^{(i)}$. The diagonal elements (mutational variances) are denoted by $\gamma_k^{2(i)}$, $k = 1, \ldots, n$. These assumptions are from Turelli's (1985) analysis and somewhat less general than those of Lande (1980a), but still every locus affecting both characters is capable of producing essentially any effect on each character. The only constraint is the (possible) existence of a correlation in the mutation-induced

[3] In this section, we use superscripts in parentheses to indicate loci. This notation is consistent with that in Chapter V.1.1 if the values x_i of the random variables X_i defined there are asumed to be in $\mathbf{R}^n$. For a single character, we simply have $\boldsymbol{x}^{(i)} = x_1^{(i)} = x_i$.

changes of the characters G_k. If a locus i does not effect every character, or if effects are strictly proportional (statistically dependent), then the matrix $\Gamma^{(i)}$ is singular. Except in Section 4.4 on constrained pleiotropic effects, we shall assume that every $\Gamma^{(i)}$ is nonsingular.

The phenotypic fitness function is assumed to be multivariate normal with the optimum scaled to zero, i.e.,

$$W_P(\boldsymbol{P}) = \exp\{-\tfrac{1}{2}\boldsymbol{P}W^{-1}\boldsymbol{P}^\top\}\,, \tag{4.4}$$

where W is a positive definite symmetric matrix generalizing the parameter ω^2 of the univariate case, V(1.12). It follows as in Chapter V.1.2 that fitnesses of genotypic values are given by

$$W(\boldsymbol{G}) = \exp\{-\tfrac{1}{2}\boldsymbol{G}\mathsf{V}_\mathrm{s}^{-1}\boldsymbol{G}^\top\}\,, \tag{4.5}$$

where $\mathsf{V}_\mathrm{s} = W + \Sigma_{\boldsymbol{E}}$. By applying a linear scale transformation, the matrix $\Sigma_{\boldsymbol{E}}$ can always be assumed to be the identity matrix (with 1's in the diagonal and 0's elsewhere). This shall, henceforth, be assumed.

4.2 THE APPARENT STRENGTH OF SELECTION

Following Turelli (1985), we consider two characters, P_1 and P_2, that are pleiotropically connected and under stabilizing selection, as in the above model. Let $V_{\mathrm{s},1}$ and $V_{\mathrm{s},2}$ denote the diagonal entries of V_s (measuring the strength of stabilizing selection) and let $\rho_\mathrm{s} = V_{\mathrm{s},12}/(V_{\mathrm{s},1}V_{\mathrm{s},2})^{1/2}$ (where $V_{\mathrm{s},12}$ denotes the offdiagonal entries of V_s) be the correlation that measures the extent to which selection acts to produce covariance between P_1 and P_2. If $\rho_\mathrm{s} = 0$, then selection acts on the traits independently.

The aim is to calculate the strength, $\tilde{V}_{\mathrm{s},1}$, of apparent stabilizing selection on G_1 in terms of the parameters of the bivariate model by supposing that P_1 is under stabilizing selection according to

$$\tilde{W}_P(P_1) = \exp\{-\tfrac{1}{2}P_1^2/\tilde{\omega}^2\}\,, \tag{4.6}$$

and ignoring selection on P_2. The strength of (apparent) stabilizing selection, $\tilde{V}_{\mathrm{s},1}$, is estimated from the within-generation reduction of phenotypic variance by assuming that the equilibrium distribution of phenotypes is Gaussian. Denoting the phenotypic variance of P_1 by $\sigma_{P,1}^2$, the change caused by selection by Δ_s, and assuming weak selection ($\tilde{V}_{\mathrm{s},1} = \tilde{\omega}^2 + 1 \approx \tilde{\omega}^2 + \sigma_{P,1}^2$), we obtain

$$\tilde{V}_{\mathrm{s},1} \approx \sigma_{P,1}^4/\Delta_\mathrm{s}\sigma_{P,1}^2 \tag{4.7}$$

(see Turelli 1985 for details). More generally, for a multivariate Gaussian distribution of phenotypes, the change in the phenotypic covariance matrix caused by selection is

$$\Delta_\mathrm{s}\Sigma_{\boldsymbol{P}} = \Sigma_{\boldsymbol{P}}(\Sigma_{\boldsymbol{P}} + W)^{-1}\Sigma_{\boldsymbol{P}} \approx \Sigma_{\boldsymbol{P}}\mathsf{V}_\mathrm{s}^{-1}\Sigma_{\boldsymbol{P}}\,, \tag{4.8}$$

where the approximation holds for weak selection (Lande 1980a). For two characters, (4.7) and (4.8) produce (Turelli 1985)

$$\tilde{V}_{\mathrm{s},1} \approx V_{\mathrm{s},1}\,\frac{1-\rho_\mathrm{s}^2}{1-2\rho_\mathrm{s}y+y^2}\,, \tag{4.9}$$

where $y = (V_{s,1}/V_{s,2})^{1/2}\,\mathrm{Cov}(P_1, P_2)/\sigma^2_{P,1}$. Thus, we have $\tilde{V}_{s,1} \le V_{s,1}$, i.e., the apparent strength of selection is equal or stronger than the actual strength, because of the indirect selection on the trait through hidden, or unknown, pleiotropic effects (cf. Lande and Arnold 1983). Equality holds if and only if $\rho_s = y$. In particular, in the absence of selection for correlation ($\rho_s = 0$) and of correlation of mutational effects ($\Gamma^{(i)}$ diagonal for every i), (4.9) yields $\tilde{V}_{s,1} = V_{s,1}$, because in this case no correlation between traits is generated.

4.3 EQUILIBRIUM GENETIC VARIANCES

We now determine the equilibrium properties of the pleiotropic model under a number of simplifying assumptions. In particular, we review approximations for the equilibrium genetic variance $\hat{\sigma}^2_{G,1}$ of trait P_1, and compare these multivariate predictions with the univariate prediction $\tilde{\sigma}^2_{G,1}$ that would be infered by neglecting pleiotropy and considering trait P_1 subject to direct stabilizing selection of strength equal to the apparent strength, $\tilde{V}_{s,1}$. By interchanging indices, analogous approximations are obtained for P_2.

The model and approximations rest on the assumption that (apparent) selection is sufficiently weak relative to recombination that linkage disequilibrium can be ignored. This has been shown to be valid for a wide range of parameters in the univariate case (see Sections 2 and VI.7), is supported by the analysis of Lande (1980a), and by computer simulations of Turelli (1985) and Krall (unpublished) for the bivariate case. With this assumption, all matrices $\mathsf{C}^{(ij)}$, $i \ne j$, in (4.3) vanish, and we can write the equilibrium genetic variance of trait P_1 as

$$\hat{\sigma}^2_{G,1} = 2\sum_{i=1}^{\ell} \hat{\sigma}_1^{2(i)}\,, \tag{4.10}$$

where $\hat{\sigma}_1^{2(i)}$ denotes the single-locus haploid prediction for the equilibrium variance contributed to P_1 by locus i. Similarly, the genetic covariance between P_1 and P_2 is the sum of the single-locus covariances. Therefore, as in Chapter VI.7, approximations for the equilibrium variances and the covariance can be obtained by summing over the corresponding single-locus haploid predictions.

The dynamics of the multivariate density of allelic effects at a haploid locus i is given by the recursion relations IV(2.8) and IV(2.7), because the model there was general enough to allow for vectors of allelic effects (see Example 4 in Chapter IV.2.1). Thus, we simply set the variable x there equal to $\boldsymbol{x}^{(i)}$ from above. Furthermore, the results IV• **2.1** and **2.3** prove that a uniquely determined, globally asymptotically stable equilibrium exists (as suggested by the simulations of Lande 1980a and Turelli 1985), and that the equilibrium mutation load at a (lone) haploid locus is always less or equal than $\mu^{(i)}$.

Approximations for the equilibrium genetic variances and covariances were derived by Lande (1980a) under the assumption of a multivariate Gaussian distribution of allelic effects, and by Turelli (1985) under the assumption that at each locus i and for every trait k, the variance of mutational effects is much larger than the existing equilibrium variance, i.e., $\gamma_k^{2(i)} \gg \hat{\sigma}_k^{2(i)}$ for every k and i, which is simply a generalization of the assumption that led to the HC-approximation in the univariate case. For sufficiently weak selection relative to recombination, Lande obtained the conceptually elegant, but algebraically cumbersome, multivariate Gaussian allelic approximation

$$\hat{\mathrm{C}}^{(ii)}(\mathrm{G}) = \mathrm{V}_{\mathrm{s}}^{1/2}\left(\mu^{(i)}\mathrm{V}_{\mathrm{s}}^{-1/2}\Gamma^{(i)}\mathrm{V}_{\mathrm{s}}^{-1/2}\right)^{1/2}\mathrm{V}_{\mathrm{s}}^{1/2}\,, \tag{4.11}$$

where the matrix square roots are taken to be positive semi-definite. Actually, (4.11) holds for an arbitrary number of traits. The general multivariate HC-approximation of Turelli (1985) is not presented, because it involves a complicated integral that can be evaluated explicitly only in special cases (see Turelli's Appendix, Eqs. 2.6 and 2.7).

Therefore, we deal explicitly only with selection acting independently on each trait (i.e., $\rho_{\mathrm{s}} = 0$ so that V_{s} is diagonal) and with uncorrelated mutational effects (i.e., the matrices $\Gamma^{(i)}$ are diagonal). Then it follows from (4.9) that apparent and actual strength of selection coincide, i.e.,

$$\tilde{V}_{\mathrm{s},1} = V_{\mathrm{s},1}\,. \tag{4.12}$$

Consequently, for both characters ($k = 1, 2$) and every locus i, the multivariate Gaussian allelic approximation (4.11) simplifies to the univariate Gaussian allelic approximation

$$\hat{\sigma}_k^{2(i)} \approx \sqrt{\mu^{(i)}\gamma_k^{2(i)}V_{\mathrm{s},k}} \tag{4.13}$$

[cf. VI(6.27)], and the correlations between the characters is zero. It follows from (4.10), (4.12), and (4.13) that in this special case the Gaussian predictions for the equilibrium variance coincide, whether they ignore hidden pleiotropy or not, i.e.,

$$\tilde{\sigma}^2_{G,1}(\mathrm{G}) = \hat{\sigma}^2_{G,1}(\mathrm{G})\,. \tag{4.14}$$

Turelli (1985) showed numerically that if selection or mutation induce correlation, then $\tilde{\sigma}^2_{G,1}(\mathrm{G})$ is within a factor of two of $\hat{\sigma}^2_{G,1}(\mathrm{G})$ for a wide range of parameter values under which the Gaussian allelic approximation applies. Therefore, hidden pleiotropic effects have only a moderate confounding influence on the equilibrium predictions if the multivariate Gaussian approximation applies.

Qualitatively different results are obtained in the parameter range in which the multivariate HC-approximation is adequate. Turelli (1985) showed that if neither selection nor mutation induce correlations, then the correlation between characters is again zero, but the true equilibrium variance is approximately

$$\hat{\sigma}^2_{G,1}(\mathrm{HC}) = 4V_{\mathrm{s},1}\sum_{i=1}^{\ell}\frac{\mu^{(i)}}{1+\beta^{(i)}}\,, \tag{4.15}$$

where

$$\beta^{(i)} = \sqrt{s_2^{(i)}/s_1^{(i)}} \quad \text{and} \quad s_k^{(i)} = \gamma_k^{2(i)}/V_{\mathrm{s},k}\,. \tag{4.16}$$

Thus, $(\beta^{(i)})^2$ measures the relative strength of selection on new mutants that is attributable to their effect on the hidden trait (P_2) *vs.* their effect on the observed trait (P_1). Obviously, $\beta^{(i)} > 1$ if and only if selection on mutational effects on the hidden trait is stronger than selection on the effects on the observed trait. In contrast to the univariate HC-approximation VI(7.10), the multivariate approximation depends on the variance of the mutation distribution.

Since in the absence of correlations (4.12) holds, and because the univariate HC-approximation yields the prediction $\tilde{\sigma}^2_{G,1}(\mathrm{HC}) = 4\tilde{V}_{\mathrm{s},1}\sum_i \mu^{(i)}$ for the variance under selection of apparent strength $\tilde{V}_{\mathrm{s},1}$, we have the inequality

$$\hat{\sigma}^2_{G,1}(\mathrm{HC}) < \tilde{\sigma}^2_{G,1}(\mathrm{HC}) \,. \tag{4.17}$$

Thus, the true variance is always overestimated by $\tilde{\sigma}^2_{G,1}(\mathrm{HC})$, despite the fact that the latter is based on $\tilde{V}_{\mathrm{s},1}$ which overestimates the true strength of selection. If all loci are exchangeable ($\beta^{(i)} = \beta$ and $\mu^{(i)} = \mu$), then the simple relation

$$\hat{\sigma}^2_{G,1}(\mathrm{HC}) = \frac{2UV_{\mathrm{s},1}}{1+\beta} = \frac{1}{1+\beta}\,\tilde{\sigma}^2_{G,1}(\mathrm{HC}) \tag{4.18}$$

is obtained. This shows that if selection on mutational effects on the hidden trait is stronger than on the observed trait ($\beta > 1$), then the multivariate approximation $\hat{\sigma}^2_{G,1}(\mathrm{HC})$ for the genetic variance can be much smaller than the prediction $\tilde{\sigma}^2_{G,1}(\mathrm{HC})$, which ignores pleiotropy and is based on the apparent strength of stabilizing selection.

Under the additional assumption of equal strength of selection on all traits ($V_{\mathrm{s},k} = V_{\mathrm{s}}$), the relation (4.18) can be extended to $n-1$ hidden traits (Turelli 1985):

$$\hat{\sigma}^2_{G,1}(\mathrm{HC}) = \frac{1}{n}2UV_{\mathrm{s}} = \frac{1}{n}\,\tilde{\sigma}^2_{G,1}(\mathrm{HC}) \,. \tag{4.19}$$

This is in sharp contrast to (4.14), which holds for the Gaussian approximation, and shows that in general, pleiotropic effects may strongly influence the genetic variance, even if they do not result in a genetic correlation between the observed trait and other traits that are under selection. Actually, in this special case, in which correlations are neither induced by mutation nor by selection, the methods of Chapter VI.6.1 can be extended to show that the sum of all true single-trait variances is bounded from above by the univariate HC-approximation, i.e.,[4]

$$\sum_{k=1}^{n} \hat{\sigma}^2_{G,k} \le 2UV_{\mathrm{s}} \,. \tag{4.20}$$

If selection or mutation induce correlations between traits, the relation between $\hat{\sigma}^2_{G,1}$ and $\tilde{\sigma}^2_{G,1}$ becomes much more complex. Supposing that loci are exchangeable, Turelli's (1985) analytical and simulation results can be roughly summarized as follows. If $\beta \ll 1$ (i.e., selection on pleiotropic loci occurs primarily through the observed character), then $\hat{\sigma}^2_{G,1}/\tilde{\sigma}^2_{G,1} \approx 1$, although this ratio may be larger or smaller than one, depending on the correlations. If $\beta \gg 1$, then $\tilde{\sigma}^2_{G,1}$ may over- or underestimate $\hat{\sigma}^2_{G,1}$: letting $\gamma_2^2/\gamma_1^2 \to \infty$ with $V_{\mathrm{s},1}/V_{\mathrm{s},2}$ fixed, then $\lim_{\beta\to\infty} \hat{\sigma}^2_{G,1}/\tilde{\sigma}^2_{G,1} = 0$; letting $V_{\mathrm{s},1}/V_{\mathrm{s},2} \to \infty$ yields $\lim_{\beta\to\infty} \hat{\sigma}^2_{G,1}/\tilde{\sigma}^2_{G,1} = \infty$. If $\beta \le 10$, then $\tilde{\sigma}^2_{G,1}$ typically overestimates $\hat{\sigma}^2_{G,1}$ by

[4] With more effort than in Chapters VI.6.1 and VI.6.3, it might be possible to prove that the multivariate HC-approximation (4.15) provides an upper bound for pleiotropically connected traits that are uncorrelated.

a factor between 1 and 5, with some values outside of this range. Hence, if most of the selection experienced by the observed character is indirect, almost any relation between $\hat{\sigma}^2_{G,1}$ and $\tilde{\sigma}^2_{G,1}$ is possible. Nonexchangeable loci do not change the results qualitatively.

Interestingly, the Gaussian and the HC-approximation produce identical predictions for the genetic correlation between traits, and they agree with Turelli's numerical results. These correlations (but not the covariances!) can be adequately calculated from (4.11). The dependence of the equilibrium genetic variance on the correlation parameters is rather complex. In most cases, correlation reduces the variance in both traits, in particular, if the covariances in the selection matrix $\mathrm{V_s}$ and in the mutation matrices $\Gamma^{(i)}$ are large and of opposite signs.

The domains of validity of the approximations were investigated by Turelli (1985). His numerical results indicate that the multivariate Gaussian allelic approximation is accurate if

$$\mu^{(i)} > 3\max\{\gamma_1^{2(i)}/V_{\mathrm{s},1}, \gamma_2^{2(i)}/V_{\mathrm{s},2}\}$$

for every locus i, whereas for the validity of the multivariate HC-approximation only the compound condition

$$3\mu^{(i)} < \sqrt{\gamma_1^{2(i)}\gamma_2^{2(i)}/(V_{\mathrm{s},1}V_{\mathrm{s},2})}$$

is required for every i. Actually, the approximation for, say, the first trait is accurate under this condition even if $\mu^{(i)} > \gamma_2^{2(i)}/V_{\mathrm{s},2}$. Thus, the multivariate HC-approximation can adequately approximate $\hat{\sigma}^2_{G,1}$ as long as $\mu^{(i)} < \gamma_1^{2(i)}/V_{\mathrm{s},1}$, irrespectively of the parameters describing the effects on the other trait.

For an asexually reproducing population, Slatkin and Frank (1990) developed a stepwise-mutation model on a two-dimensional grid that allows for a class of pleiotropic and a class of nonpleiotropic mutations. They considered the case of no net correlation between traits and showed that the multivariate HC-approximation also leads to valid predictions if the strength of selection on alleles is of the same order of magnitude (or larger) than the mutation rate, and if a substantial fraction of mutations are pleiotropic. However, there are parameter ranges where neither of the approximations is accurate.

4.4 A MODEL OF CONSTRAINED PLEIOTROPIC EFFECTS

A model conceptually different from Lande's (1980a) model of 'universal pleiotropy' was developed by Wagner (1989). His model of 'constrained pleiotropy' is based on the assumption that the possible pleiotropic effects of mutations at a polygenic locus are constrained by the developmental genes which control the expression of the locus. More precisely, it considers n quantitative traits and assumes that each locus, i, contributing to these traits codes for a single product that can be characterized by an underlying (physiological) variable, $Y^{(i)}$. This may, for instance, represent the net activity of some enzyme. The vector of genotypic values, $\boldsymbol{G}$, is assumed to be determined by a 'developmental function' that maps the values $\boldsymbol{Y} = (Y^{(1)}, \ldots, Y^{(\ell)})$ to the values of $\boldsymbol{G}$. Assuming additivity among loci is equivalent to choosing this function to be linear. Then it can be described by an $n \times \ell$ matrix B, whose coefficients $b_k^{(i)} = b_{ki}$ determine the phenotypic effect of alleles at locus i on character k, i.e.,

$$\boldsymbol{G} = \mathrm{B}\boldsymbol{Y} \; . \tag{4.21}$$

In this model, the possible effects of alleles at every locus i are restricted to the one-dimensional subspace of the n-dimensional phenotype space that is defined by the ith column of the matrix B. Thus, the ratio of pleiotropic effects of an allele on any pair of characters depends only on the locus, but not on the actual allelic state. The covariance matrix of mutational effects can be shown to be $\sum_i \Gamma^{(i)} = \mathrm{BB}^{\top}$ if the $\boldsymbol{Y}$ values are scaled appropriately.

Phenotypic values are obtained by the usual additivity assumption, $\boldsymbol{P} = \boldsymbol{G} + \boldsymbol{E}$, and the genetic covariance between two traits G_j and G_k can be expressed as $\mathrm{Cov}[G_j, G_k] = \sum_{i=1}^{\ell} b_j^{(i)} b_k^{(i)} \, \mathrm{Var}[Y^{(i)}]$, provided linkage equilibrium and random mating are assumed.

Under these assumptions, the analysis of the model can be reduced to the investigation of mutation-selection balance in each of the one-dimensional traits $Y^{(i)}$. For the case in which Gaussian stabilizing selection acts independently on two characters ($\rho_{\mathrm{s}} = 0$), Wagner (1989) derived the approximation

$$\hat{\sigma}^2_{G,1}(\mathrm{B}) = 4V_{\mathrm{s},1} \sum_{i=1}^{\ell} \frac{\mu^{(i)}}{1 + (\beta^{(i)})^2} \tag{4.22}$$

for the equilibrium genetic variance by employing a HC-approximation to the $Y^{(i)}$. In (4.22), $\beta^{(i)} = (b_2^{(i)}/b_1^{(i)})\sqrt{V_{\mathrm{s},1}/V_{\mathrm{s},2}}$, which is equivalent to the parameter $\beta^{(i)}$ defined in (4.16), because $\gamma_k^{2(i)} = (b_k^{(i)})^2$. Obviously, the approximation (4.22) differs qualitatively from the multivariate HC-approximation (4.15) for uncorrelated traits and may be much lower than that if $\beta^{(i)} > 1$.

Since it can be shown that this 'B-matrix model' is obtained from Lande's (1980a) model if the correlation between pleiotropic effects of mutations is equal to one at each locus, the approximations (4.15) and (4.22) are the extremes on a continuum. If mutation is weak relative to selection (such that the HC-approximation applies) and selection does not induce correlation, then the equilibrium genetic variance will be between the approximations (4.15), which applies if mutational effects are uncorrelated, and (4.22), which applies if they are completely correlated. Actually, the B-matrix approximation (4.22) is formally identical to an approximation obtained by Turelli (1985) by considering only five possible alleles per locus. The reason is that with such a low number of alleles, pleiotropic effects are automatically constrained in a manner analogous to the B-matrix model. It is evident by comparing (4.15) and (4.22) that with constrained pleiotropic effects, the discrepancy between $\hat{\sigma}^2_{G,1}$ and $\tilde{\sigma}^2_{G,1}$ is higher than in (4.18). As also shown by Wagner (1989), in the completely symmetric case the variance declines with increasing number of traits in the same way as in (4.19).

The above models and approximations have been extended to finite populations, and extensive Monte-Carlo simulations have been performed with variants of the program outlined in Section 2.2 (Wagner 1989, Krall, unpublished). The results are in qualitative agreement with the deterministic theory, but show that with decreasing population size the reduction of variance caused by pleiotropy becomes less severe.

They also show that results are almost independent of the recombination map, as long as the harmonic mean recombination rate is above 0.05.

4.5 CONCLUSIONS

The analysis of multivariate pleiotropic models identified two key issues:

1. If the genes determining a trait under stabilizing selection have pleiotropic effects on other traits under stabilizing selection, then the genetic variance at equilibrium is usually smaller and, possibly, much smaller than it would be without these pleiotropic effects. This is even the case if the observed trait is uncorrelated to the other, pleiotropically connected traits. For instance, in the completely symmetric case, in which genes affect n traits that all are under the same strength of selection and the covariance matrix of the mutation distribution is a multiple of the identity matrix, so that no correlation is induced, the variance of each trait is reduced by a factor of $1/n$. The effect may be much larger if selection on the disregarded traits is stronger than on the observed trait.

2. If there are unknown pleiotropic effects on other traits under selection, or if they are neglected, then the observed trait appears to be under stronger selection than it actually is. Nevertheless, the prediction for the equilibrium genetic variance inferred from the observation of only this trait may be much larger than the true equilibrium variance, in particular, if mutation rates per locus are small, so that the multivariate HC-approximation applies.

In combination with the results of Section 2, the present ones show that if pleiotropic effects are universal and many traits are under direct stabilizing selection, then mutation-selection balance cannot account for a substantial fraction of the commonly observed levels of heritable variation. However, if pleiotropic effects are not as universal as sometimes thought, but are restricted to sets, or modules, of functionally related characters, then mutation-selection balance may generate appreciable levels of quantitative variation, in particular, if only one or a few of the characters in a module are under direct selection.

In addition, and as discussed in Chapter VI.3.4, there are special models of pleiotropy that can explain high levels of polymorphism and a certain amount of (additive) genetic variation in the absence of mutation. These assume loci with antagonistic effects on the traits, or that every locus has a major effect on one of a set of traits and minor effects on the other traits (Gimelfarb 1992, 1996b; Gavrilets and Hastings 1994a).

5. DELETERIOUS MUTATIONS AND EVOLUTION

A large fraction of mutations with detectable effects are deleterious, and at least in some species, if not in many, deleterious mutations occur at a high rate (Section 1.3). Numerous biological phenomena may be the result of evolutionary processes that were driven by the need to reduce the detrimental impact of mutations on population persistence. In this section, we outline the basic models that have been advanced to explain some of these phenomena, in particular, with respect to the advantage of sex and recombination.

Surprisingly little is known about fitness itself, because it is impossible or, at least, extremely difficult to measure. Usually, only fitness components, such as reproductive success, viability, or various life-history traits, are amenable to experimental analysis (see Charlesworth and Hughes 1999, and the references therein). Similarly, total fitness cannot be handled adequately by theoretical models, because it involves many components, such as viability and fertility, and can be further complicated by sexual selection, male mating success, and others. Therefore, in all the models discussed below, fitness is equated with viability. As the vast majority of mutations are deleterious or neutral, it is of great interest to investigate the resulting equilibrium distribution of fitness. An influential model was devised by Kingman (1978) for an asexually reproducing population. He regarded (relative) fitness as a continuous trait, taking values between 0 and 1, and being subject to linear directional selection. By positing that the fitness of the gene after mutation is independent of that before, i.e., by assuming the HC-mutation model, he determined the properties of the equilibrium distribution. A different model, but similar in spirit, was analyzed by Eshel (1971). Generalized versions of such models were investigated in Chapters III and IV, where it was shown that under very general assumptions about the mutation distribution (including back mutations), the equilibrium mean (relative) fitness, $\overline{W}$, in an asexual population always satisfies $1-\mu \leq \overline{W} < 1$, where $1-\mu$ is the probability that no mutation occurs during reproduction. For a small mutation rate μ it was shown that $\overline{W} \approx 1-\mu$. Such results are, in general, not obtainable for sexually reproducing populations, except in one-locus models (Chapter III). Also, very little is known about the distribution of mutational effects on fitness.

First, we deal with some simple models that are amenable to mathematical analysis and allow us to explore the impact of deleterious mutations on the mean fitness of infinitely large asexual and sexual populations. Then we outline the potentially disastrous consequences of the interaction between the accumulation of deleterious mutations and random genetic drift for finite populations. Finally, we discuss the advantages these processes may confer to sexual reproduction and recombination.

5.1 EQUILIBRIUM DISTRIBUTIONS IN INFINITE POPULATIONS

We consider an infinitely large population with discrete generations and suppose that fitness depends only on the number of mutations an individual carries. There are ℓ loci, each with a genic mutation rate of μ. Then the genomic mutation rate is $U = \ell\mu$ in the haploid case, and $U = 2\ell\mu$ in the diploid case. All mutations are assumed to be unconditionally deleterious and to occur independently.

Asexual populations

Let p_k denote the (relative) frequency of individuals (genomes) carrying k mutations, and let W_k denote the corresponding fitness subject to the normalization $W_0 = 1$. As all mutations are assumed to be deleterious, we have

$$1 = W_0 > W_1 > W_2 > W_3 > \dots .$$

Let $\tilde{u}_{jk}$ denote the probability that a genome with j mutations mutates to a genome with k mutations. By our assumptions, the $\tilde{u}_{jk}$ follow the binomial distribution

$$\tilde{u}_{jk} = \binom{\ell - j}{k - j} \mu^{k-j} (1-\mu)^{\ell - j} , \tag{5.1}$$

where $j \leq k \leq \ell$, because back mutations to the wild-type allele are ignored. In the limit of large ℓ and small μ, but constant $U = \ell\mu$, each mutation occurs at a new locus (hence, there are at most two alleles per locus), and the number of new mutations, i, a genome incurs per generation follows an approximate Poisson distribution given by

$$u_i = e^{-U} \frac{U^i}{i!} , \quad i \geq 0 \tag{5.2}$$

(cf. Higgs 1994). The probability that no mutation occurs is $u_0 = e^{-U}$. The general recursion relation III(1.1), or III(4.6), simplifies to

$$p'_k = \frac{1}{\overline{W}} \sum_{j=0}^{k} u_{k-j} W_j p_j , \tag{5.3}$$

where $\overline{W}$ is the mean fitness.

An immediate and completely general consequence of (5.3) is that at equilibrium, the frequency of the optimal genotype satisfies $p_0 = u_0 p_0 / \overline{W}$, which implies that the equilibrium mean fitness is

$$\hat{\overline{W}} = e^{-U} . \tag{5.4}$$

This result is independent of how fitness is related to the number of mutations (cf. Kimura and Maruyama 1966). As we shall see, this is only true for (haploid) asexually reproducing populations.

It is shown in Chapters III.4 and IV.2.2 that mutation-selection models that admit a potentially infinite number of types, such as (5.3), do not necessarily possess an equilibrium distribution if fitnesses of extreme types do not decay to zero and mutation rates are above a critical 'error threshold'. In such a case, any initial distribution may show dissipative behavior, i.e., an indefinite increase in variance. This cannot happen with the multiplicative fitness function $W(k) = (1-s)^k$. To determine the equilibrium distribution, assume that the initial distribution of mutants per individual is Poisson with mean $\bar{k}$, possibly monomorphic ($\bar{k} = 0$). Then selection reduces each p_k by the factor $(1-s)^k$, so that after renormalization, a Poisson distribution with mean $\bar{k}(1-s)$ is obtained. Since the number of new mutations follows a Poisson distribution with mean U, the distribution after mutation, and hence in the next generation, is again Poisson and has mean $\bar{k}(1-s) + U$. It follows that the equilibrium distribution is Poisson, and the average number of mutations per individual is (cf. Haigh 1978)

$$\hat{\bar{k}} = U/s . \tag{5.5}$$

Various other fitness functions, intended as models of synergistic epistasis [defined below (5.10)] have been investigated, for instance, linear and quadratically decreasing fitness (Kimura and Maruyama 1966), truncation selection (Kondrashov 1982), and

$W(k) = \exp(-sk^{\alpha})$ (Higgs 1994). In general, the equilibrium distribution is not Poisson, but the numerical results of Charlesworth (1990) indicate that for $W(k) = \exp(-ak - bk^2)$ and $U/a > 10$, the equilibrium distribution is close to Gaussian, although some skewness may be present. The mean equilibrium fitness, however, is always e^{-U}.

Sexual populations

In sexually reproducing populations, simple approximations for the mean equilibrium fitness have been derived only in special cases, because mean fitness may depend on the dominance relations, on epistasis, and on the recombination distribution. The simplest case is that of a randomly mating population, in which every mutation occurs at a new locus and fitness effects within and between loci interact multiplicatively, i.e., the fitness of an individual with k mutations is $W_k = (1-s)^k$. Then, a population that initially is in linkage equilibrium will remain in linkage equilibrium if the loci are recombining (cf. Chapters II.1.2 and II.5.3). By III(2.7d) the equilibrium frequency of mutations at each locus is μ/s, by (5.5) their number per individual is Poisson distributed with mean $\hat{\bar{k}} = U/s$, and the equilibrium mean fitness is $\hat{\bar{W}} = (1-\mu)^{2\ell} \approx e^{-U}$. For this model, Dawson (1999) proved that if loci are recombining freely and the population initially is in linkage disequilibrium, then convergence to the Poisson distribution with mean U/s occurs. He derived an explicit solution of this dynamics in terms of the factorial cumulants.

In what follows, we require that $\mu \ll s < 1$. Otherwise, complex dynamic behavior and phenomena related to error thresholds can occur (see Higgs 1994, Baake and Wiehe 1997, Baake and Gabriel 1999). If dominance is admitted, but fitness between loci is multiplicative, then the fitness of an individual with i loci heterozygous for the mutant and j loci homozygous can be written as

$$W_{ij} = (1-2hs)^i(1-2s)^j . \tag{5.6}$$

Again, with a sufficiently high level of recombination, linkage equilibrium can be assumed. Then the equilibrium mean fitness is

$$\hat{\bar{W}} = \prod_{i=1}^{\ell} \hat{\bar{W}}_i , \quad \text{where} \quad \hat{\bar{W}}_i = 1 - \mu - 2sh(1-\mu)\hat{q}_i$$

denotes the equilibrium mean fitness at locus i, and $\hat{q}_i$ is the equilibrium frequency of the mutant at locus i (see III(2.4a,b), but note the different meaning of s). By series expansion, it follows that

$$\hat{\bar{W}}_i = 1 - 2\mu + [\mu^2/(2h^2 s)](1 - 2h + 4h^2 s) + O(\mu^3) ,$$

which may be slightly larger or smaller than $1-2\mu$. Therefore, in the absence of epistasis, and unless mutations are nearly completely recessive (say, $h > 2\sqrt{\mu/s}$), the equilibrium mean fitness is

$$\hat{\bar{W}} \approx (1-2\mu)^{\ell} \approx e^{-U} , \tag{5.7}$$

which is approximately equal to that of an asexual population with the same total deleterious mutation rate. In the completely recessive case ($h = 0$), III(3.20) shows that the mean fitness is higher:

$$\hat{\bar{W}} \approx (1-\mu)^\ell \approx e^{-U/2} \,. \tag{5.8}$$

Unless mutations are highly recessive, the equilibrium distribution of deleterious mutations per individual is approximately Poisson with mean

$$\hat{\bar{k}} = U/(2hs) \,. \tag{5.9}$$

This result is exact in the limit of $\mu \to 0$ and U constant (Higgs 1994). Therefore, in the absence of epistasis, the equilibrium distributions of asexual and of sexual populations are approximately Poisson, and identical if mutations have the same (heterozygous) effects and occur at the same genomic rate.

With epistasis, this is no longer the case. Assume that deleterious mutations arise at a low rate per locus, so that they occur only in heterozygous form, and that the number of new mutations per gamete follows a Poisson distribution with mean $\frac{1}{2}U$. Let the fitness of an individual with k mutations be

$$W(k) = \exp(-ak - bk^c) \,, \tag{5.10}$$

where a, b, c are constants and $a, c \geq 0$. If $b = 0$, fitness is multiplicative. If $b > 0$ and $c > 1$, fitness decreases more rapidly than exponentially (multiplicatively) and, thus, displays synergistic epistasis. This means that the effect of a deleterious mutation is magnified in a genome carrying other deleterious mutations. If $b < 0$ and $c > 1$, or if $b > 0$ and $c < 1$, then W displays 'diminishing returns' epistasis, i.e., the effect of a deleterious mutation is diminished in a genome carrying other deleterious mutations.

Under linkage equilibrium, and for $a = 0$ and $c = 2$, a simple analytical approximation for the equilibrium distribution was obtained by Higgs (1994). He showed that for small b and small U, but arbitrary U/b, the mean number of deleterious mutations per individual, $\hat{\bar{k}}$, is the positive root of

$$2\bar{k}^2 + \bar{k} - U/b = 0 \,, \tag{5.11}$$

mean fitness is

$$\hat{\bar{W}} \approx 1 - b\left(\hat{\bar{k}} + \hat{\bar{k}}^2\right) = 1 - \tfrac{1}{2}U - \tfrac{1}{8}b\left(\sqrt{1+8U/b} - 1\right) \,, \tag{5.12}$$

and the frequency of gametes carrying k mutations is

$$\hat{p}_k \approx (\text{const.})\, e^{-\hat{\bar{k}}} \frac{\hat{\bar{k}}^k}{k!} \left[1 - b(k+k^2)\right] . \tag{5.13}$$

The approximations (5.11) and (5.12) are very accurate as long as $b \leq 0.1$ and $U/b \leq 2$. With increasing b or increasing U, the root of (5.11) underestimates the true value of $\hat{\hat{k}}$. However, the true $\hat{\hat{k}}$ is always less than U/b, and the mean fitness is always higher than e^{-U}, as can be shown by considering higher-order terms in (5.12). It is remarkable that with small b, $\hat{\overline{W}} \approx 1 - \frac{1}{2}U$, i.e., the mutation load with epistasis is only half of that without epistasis (cf. Kimura and Maruyama 1966).

The analyses and numerical results of Kimura and Maruyama (1966), Kondrashov (1982), Charlesworth (1990), and Higgs (1994) all suggest that in sexual populations with recombination, the mean number of deleterious mutations per individual is always lower than in an asexual population with the same deleterious mutation rate and fitness effects; the mean fitness is always higher, provided there is synergistic epistasis. This difference increases with increasing total mutation rate. The mean fitness of a sexual population without recombination (only segregation) is always between the two cases just considered, but closer to the one with free recombination. With diminishing-returns epistasis, the mean fitness of a sexual population is lower than that of an asexual. Unfortunately, no rigorous proofs seem to be available. In accordance with the asexual case, Charlesworth's (1990) numerical results indicate that for $c = 2$, the equilibrium distribution is close to Gaussian, but slightly skewed.

5.2 ACCUMULATION OF DELETERIOUS MUTATIONS IN FINITE POPULATIONS

Frequent deleterious mutations may have dramatic consequences for finite populations in which random sampling cannot be ignored. Then detrimental mutations may accumulate and eventually become fixed, thus leading to a progressive fitness decline that can result in population extinction.

Asexual populations

In a finite population, the fittest class of individuals may be very small. Consider a population of size N and assume that the distribution of mutants is approximately Poisson with mean $\hat{\hat{k}} = U/s$, as in (5.5). Then the number of individuals in the class with the lowest number of mutations (the so-called least-loaded class) is approximately $Ne^{-U/s}$, which may be very small if $U/s \gg 1$. Due to random sampling, this class will be lost after a finite number of generations. If no back mutations occur, this class of individuals cannot be reconstituted in an asexual population, because in the absence of recombination, offspring cannot carry less mutations than their parents. Thus, the class with one more mutation will be the new least-loaded class and, after some time, suffer the same fate. This repetitive stochastic process is known as Muller's ratchet: the loss of each least-loaded class can be regarded as a turn of the ratchet (Muller 1964, Felsenstein 1974, Maynard Smith 1978). If every mutation reduces fitness by the multiplicative factor $(1 - s)$, then the ratchet (and, actually, the whole distribution) advances at a constant (average) rate, and leads to a gradual decline in mean fitness. Although, this phenomenon has been the object of numerous theoretical studies, no completely satisfactory analytical results have been established yet (e.g., Haigh 1978, Gabriel *et al.* 1993, Lynch *et al.* 1993, Stephan *et al.* 1993, Gessler 1995, Charlesworth and Charlesworth 1997, Gordo and Charlesworth 2000, and the references therein).

It is worth noting that, in principle, the ratchet can advance without fixation of mutant alleles at individual loci, i.e., in a large population many least-loaded classes can be lost without any appreciable fixation at individual loci (Charlesworth *et al.* 1993). However, as shown by Higgs and Woodcock (1995) and Charlesworth and Charlesworth (1997), losses of the least-loaded class are followed by fixations of deleterious alleles, so that the advance of the ratchet and the rate of fixation are parallel processes.

If fitnesses interact synergistically, then the ratchet may be halted. Consider the simple case of linearly decreasing fitness, and let k_0 denote the value such that $W_k = 0$ if $k > k_0$, and $W(k) > 0$ if $k \leq k_0$. Then the ratchet will proceed but, eventually, will reach the value k_0. At this point, the whole population consists of individuals carrying k_0 mutations, but no further advance is possible because any additional mutation causes lethality, and such mutations cannot be fixed. Of course, if in a single generation every individual incurs one or more mutations, then the population will be extinct. The ratchet can also effectively be stopped with other synergistic fitness functions (Kondrashov 1994). However, Butcher (1995) demonstrated that this hypothesized cessation of the ratchet under synergistic epistasis is a result of the assumption that all mutations are equivalent. He showed that if mutational effects are drawn from a continuous distribution that generates mutations of arbitrarily small effect, then mutations with increasingly small effects will continue to drive the ratchet as the more deleterious mutations become too damaging to accumulate. Although, the rate of mutation accumulation may decrease in such a case, the rate of fitness loss does not necessarily decrease.

The (numerical) results of Gessler and Xu (1999) show that a distribution of mutational effects can also lead to an increase of the rate of the ratchet relative to the case of mutations of identical effects, equal to the mean of the distribution. They considered multiplicative fitness, an approximate negative exponential distribution with mean $\bar{s}$, implying that most mutations have small deleterious effects and a few have large effects, and parameters such that $Ne^{-U/\bar{s}} < 1$. In this case, the ratchet is mainly driven by mutational pressure, and it proceeds (much) faster than if all mutations had the same effect $\bar{s}$.

Most studies of Muller's ratchet have been pursued under the assumption that population size is unaffected by the accumulation of mutations. This is clearly unrealistic if the reproductive rate is limited. In a population with density-dependent population regulation and carrying capacity K, in which adults are able to produce R offspring and $W(k) = (1-s)^k$, mutations accumulate gradually, as with constant population size, until the mean viability has been reduced to $1/R$. At this point, the average number of surviving offspring per adult becomes less than one, and the population size begins to decline. Due to the gradual reduction in population size, selection against deleterious mutations becomes weaker, and mutations accumulate more rapidly. This synergistic process leads to rapid extinction and has been called a mutational meltdown (Lynch and Gabriel 1990, Gabriel *et al.* 1993). It can drive even large asexual populations to rapid extinction, because the mean time to extinction is approximately proportional to the logarithm of population size. For instance, with a reproductive rate of $R = 100$, a mutation rate per zygote per generation of $U = 1$, a selection coefficient of $s = 0.025$, and no epistasis, a population of half a million, initially mutation-free, individuals will, on average, be extinct within 800 generations (see Lynch *et al.* 1993 for an extensive numerical study). The mean time to extinction is primarily determined by the length of the phase during which mutations accumulate, because the final phase of the mutational meltdown is very short.

Another interesting property is the existence of an intermediate selection coefficient, s_*, that minimizes the mean time to extinction. As s increases from zero (neutrality), the rate of the ratchet decreases, but the lower rate of mutation accumulation is more than offset by the greater damage the mutations inflict on the population per ratchet turn. Thus, the mean time to extinction decreases with increasing s. If the critical value s_* is exceeded, then the ratchet turns increasingly slowly and the mean extinction time increases rapidly until, with $s = 1$ (lethality), the ratchet cannot turn at all, and extinction occurs only in the exceedingly unlikely event (of probability $(1 - e^{-U})^N$) that all individuals mutate. The critical value s_* depends on population size, mutation rate, and other parameters, but is on the order of $s = 0.1$ or less under high mutation pressure, i.e., $U \geq 0.5$ (Gabriel *et al.* 1993, Butcher 1995). Therefore, more severely deleterious mutations cause a lower extinction risk for a population, and maximization of individual sensitivity to mutations could be a strategy to avoid the operation of Muller's ratchet (see also Gabriel and Bürger 2000).

It is unlikely that beneficial mutations can stop the ratchet, because they are very rare and, in an asexual population, can only become fixed if they occur in a genome that (with the beneficial mutation) has higher fitness than all other genomes. With a high mutation rate, however, the fraction of the population carrying few deleterious mutations is very small. For a quantitative genetic model, Wagner and Gabriel (1990) showed that under certain conditions compensatory mutations (mutations that compensate for the phenotypic effects of deleterious mutations) may have the potential to stop Muller's ratchet.

In summary, it seems to be generally accepted that Muller's ratchet represents a serious threat for asexually reproducing populations The model predictions are consistent with the rarity of obligate asexuality, and the fact that many clonal lineages are evolutionary very young. In a similar way, nonrecombining regions of the genome, such as the Y chromosome, or organelle genomes (mitochondria) may be affected by Muller's ratchet (see Lynch and Gabriel 1990, Lynch *et al.* 1993, Lynch and Blanchard 1998, Charlesworth and Charlesworth 1997, 1998).

Sexual populations

In recombining sexual populations, the process of Muller's ratchet will be ineffective, because parents can contribute the best portions of their genomes to (some of) their offspring. Although, sexual reproduction and recombination will slow down the rate of accumulation of deleterious mutations and increase the rate of fixation of beneficial ones, slightly deleterious mutations can become fixed by random genetic drift with appreciable probability if their effect s is smaller than $1/N_e$, N_e the effective population size. Let us consider the following simple quantitative argument. In a population of N individuals with genomic deleterious mutation rate U, the total number of new mutations per generation is UN. Suppose that fitness between loci is multiplicative, so that each mutation occuring as a homozygote reduces fitness by the factor $(1-2s)$. Then, in linkage equilibrium, the expected fitness reduction per generation due to fixations is

$$\overline{W}'/\overline{W} = (1 - 2s)^{UN\pi_{\text{fix}}} , \tag{5.14}$$

where π_{fix} denotes the probability that a mutant allele, which initially occurs as a single copy, eventually becomes fixed (Appendix E.2). In the absence of dominance and under the biologically realistic assumption of $2sN_e/N \ll 1$, (E.6) can be simplified to

$$\pi_{\text{fix}} \approx \frac{2sN_e/N}{e^{4sN_e} - 1} \,. \tag{5.15}$$

Similar to the asexual case, a mutational meltdown and rapid extinction will occur as soon as the absolute fitness declines below unity.

Let $\overline{W}_0$ be the (total) mean fitness at the beginning of the process of mutation accumulation, which may be initiated, for instance, if a population is suddenly reduced to small size or is newly founded. Then the length, T, of the phase of mutation accumulation until mean fitness equals unity can be caluculated from

$$(\overline{W}'/\overline{W})^T \, \overline{W}_0 = 1 \,. \tag{5.16}$$

Taking logarithms and substituting (5.14) and (5.15), we obtain for the 'mean time to genetic inviability' (Lande 1994)

$$T \approx \frac{\ln \overline{W}_0(e^{4N_e s} - 1)}{4UN_e s^2} \,. \tag{5.17}$$

[5]

Typically, the subsequent phase of the mutational meltdown is very short, such that (5.17) provides an accurate approximation for the mean time to extinction. Similar to the asexual case, there is a critical value s_* that minimizes the mean time to extinction. Thus, moderately deleterious mutations may be the most harmful for a population (Gabriel and Bürger 1994, Lande 1994, Lynch *et al.* 1995b).

These considerations show that sexual populations may indeed be vulnerable to the accumulation of deleterious mutations. However, (5.17) indicates that the mean time to extinction increases nearly exponentially with $N_e s$. Therefore, with presumably typical values of s of a few percent, only populations with effective sizes up to approximately 100 will be affected by this process. Given that estimates of N_e/N average about 0.1 (Frankham 1995), populations with up to 1000 individuals may be at risk.

The above model is based on numerous simplifying assumptions. It ignores epistasis, beneficial mutations, variable mutational effects, and demographic stochasticity. All these effects have been studied to a certain extent (Charlesworth *et al.* 1993; Gabriel and Bürger 1994; Lande 1994, 1998; Lynch *et al.* 1995a,b; Schultz and Lynch 1997). The main conclusion that can be drawn is that with variable mutational effects, populations of effective sizes of up to $N_e \approx 1000$ may be at a serious risk of extinction from fixation of deleterious mutations within 10^3–10^4 generations. The reason is that with a distribution of effects, mutations of small and moderate effects will occur, which are the most harmful for the population. Only very strong synergistic epistasis and a, presumably, unrealistically high rate of beneficial mutations can substantially reduce this extinction risk. Outcrossing populations in excess of $N_e = 1000$ are likely to be

[5] The reason why such an approximation is not available for asexuals is that the Wright–Fisher model with selection is amenable to a detailed mathematical analysis only if there are two alleles per locus. In an asexual population, however, each gametic type corresponds to an allele and, with a high mutation rate, there will be many such types.

immune to the threat caused by accumulation of deleterious mutations. For further discussion, in particular, in connection with conservation biological concerns, we refer to the above mentioned articles, and to Lande (1995) and Whitlock and Bürger (2000).

5.3 THE ADVANTAGE OF SEX AND RECOMBINATION

Most higher organisms (eukaryotes) reproduce sexually, and very few phylogenetically old, obligately asexual species exist. This suggests that sexual reproduction and recombination are selectively favored over asexuality. However, this selective advantage must be large, because it has to overcome the so-called twofold cost of sex that arises from the production of 'needless' males: a gene that suppresses meiosis is transmitted to its offspring with certainty instead of a probability of one half and, therefore, should spread rapidly. Other disadvantages are confered to sexual reproduction by the complexities and associated risks of the process of meiosis, and by the property of recombination to break up favorable gene combinations, thereby, reducing the fitness of progeny (often, this is called the recombination load; cf. Chapter III.3). Numerous theories and models have been proposed to explain the origin and the selective maintenance of sexual reproduction and recombination, but many open questions remain. For classical, comprehensive treatises, see Williams (1975), Maynard Smith (1978), Michod and Levin (1988); for more recent reviews, consult Kondrashov (1988, 1993), Feldman *et al.* (1997), and Barton and Charlesworth (1998).

The above models of mutation accumulation clearly demonstrate that sexual reproduction and recombination provide a long-term advantage at the group level, because they efficiently eliminate deleterious mutations, except in rather small populations. However, the selection of a mutation that suppresses meiosis or reduces recombination cannot be avoided that way; instead it would lead to the extinction of the whole population through Muller's ratchet. It has also been shown that in large populations, and with constant selection coefficients, a small amount of recombination can halt the ratchet (Pamilo *et al.* 1987, Charlesworth *et al.* 1993). Therefore, Muller's ratchet has been considered to be of little relevance for the evolution of the high recombination rates that often are observed. However, a recent numerical study of Gessler and Xu (1999) shows that this may be different if there is a distribution of mutational effects (with mean $\bar{s}$) if $U \gg \bar{s}$, and if the (approximate) average number of individuals in the least-loaded class, $Ne^{-U/\bar{s}}$, is less than one. Then, as briefly discussed in Section 5.2, the ratchet is primarily driven by mutational pressure. Gessler and Xu's results show that alleles enhancing recombination can spread when rare, and resist invasion when common. This appears to occur at an optimal level of recombination, which can be quite high. The reason seems to be that with a sufficiently high recombination rate, the best class can be restored and the population can attain a (stochastic) equilibrium. If $Ne^{-U/\bar{s}} \gg 1$, as in many previous studies, then the population is close to equilibrium for a long time before the ratchet turns and tiny recombination rates can maintain the population at equilibrium.

By contrast, the *Fisher–Muller hypothesis* states that sex and recombination are advantageous, because beneficial mutations that arise in different individuals can be united in one genome. Recombination can also free a beneficial mutation from the bad genes of the genome in which it occured and, thus, give it the chance to spread, become fixed, and balance the effects of the deleterious mutations (e.g., Peck 1994, Peck *et al.* 1997). Such a process can also lead to a substantial advantage for a recombining sexual population in a changing environment, because it may increase genetic variability and, thus, greatly enhance the rate of adaptation (see Charlesworth 1993, Kondrashov and Yampolsky 1996b, Bürger 1999, Waxman and Peck 1999, and Section 7.4). However, because in a constant environment, sex and recombination lead to increased levels of genetic variability by breaking up favorable gene combinations, they decrease the mean fitness (see Figure 2.1). Therefore, a changing environment confers a substantial advantage to sex and recombination only if the change is 'predictable' and sufficiently large, so that there are periods of directional selection that last for a dozen or more generations.

Several investigators examined the conditions that allow increased recombination to evolve by means of multilocus models in which a modifier locus alters the recombination rate between loci affecting viability (e.g., Feldman 1972, Kondrashov 1984, Barton 1995, Charlesworth and Barton 1996, Otto and Feldman 1997, and the references therein). Bascially, increased recombination is favored if there is synergistic epistasis, a high mutation rate, and the modifier is tightly linked to the viability loci. Under diminishing-returns epistasis, decreased recombination rates are favored. Intuitively, these conditions are evident from the deterministic models for the mutation load in Section 5.1, because the mutation load in sexuals is much lower than in asexuals only if mutation rates are high and deleterious mutations interact synergistically. Selection for modifiers increasing recombination may also occur in predictably changing environments (Charlesworth 1993).

In summary, there is no shortage of mechanisms by which sexual reproduction and genetic recombination may be favored by natural selection (Barton and Charlesworth 1998), but some operate only in the long run on the population level, whereas others require special conditions, such as environmental change involving directional selection, or high deleterious mutation rates and synergistic epistasis. Unfortunately, experimental evidence is inconclusive or ambigous, and critical tests to discriminate between hypotheses are difficult to perform.

There are many other biological phenomena that may be an evolutionary response to a high rate of deleterious mutations. Among these are the evolution of diploidy, the evolution of mate choice, the evolution of senescence, the evolution of inbreeding avoidance, and the degeneration of Y chromosomes. The articles in Woodruff and Thompson (1998) cover these and many other topics about the role of mutation in evolution.

6. APPARENT STABILIZING SELECTION AND PLEIOTROPY

The frequent observation of a negative correlation between fitness and the squared deviation of a quantitative trait from its population mean does not imply a causal relationship between fitness and trait, i.e., such a trait is not necessarily under direct stabilizing selection. Indeed, Robertson (1967) argued that much of the observed stabilizing selection could be caused by pleiotropic side effects of alleles that primarily contribute to fitness-related traits. As little is known about the pleiotropic effects of alleles influencing a quantitative character, and because the equilibrium genetic variance cannot be predicted accurately, unless it is known how selection acts on all fitness-related traits to which these alleles contribute (Sections 1 and 4), Hill and Keightley (1988) suggested a simple model that condenses all pleiotropic fitness effects of an allele into a net deleterious effect on overall fitness.

In this *Hill–Keightley (HK) model*, mutations at the loci contributing to a quantitative trait, which is not under direct selection, have an unconditionally deleterious effect on fitness. This model produces apparent stabilizing selection on the metric trait and leads to the maintenance of genetic variation by mutation-selection balance. Various versions of this type of model will be discussed below. Apparent stabilizing selection can also be induced by pleiotropic effects of selectively maintained polymorphisms or by correlation in nonheritable variation of characters. Such models will be outlined after the HK model.

6.1 PLEIOTROPIC EFFECTS OF DELETERIOUS MUTATIONS

An illuminating special case of the HK model that can be treated analytically was investigated by Barton (1990). He assumed that mutations at ℓ diploid loci can affect a neutral character, P, in a randomly mating population. Mutations occur at a rate μ per locus and each reduces the fitness of heterozygotes by the multiplicative factor $(1-s)$. Hence, at equilibrium, linkage equilibrium between segregating deleterious alleles can be assumed (Dawson 1999). It is further assumed that $\mu \ll s \ll 1$, so that any particular mutant is rare. Therefore, at mutation-selection balance, the number k of deleterious mutants per individual can be assumed to follow a Poisson distribution, $\hat{p}_k$, with mean

$$\hat{\bar{k}} = U_G/s \, , \tag{6.1}$$

where $U_G = 2\ell\mu$ is the total deleterious mutation rate from all loci affecting the trait. If U denotes the genomic deleterious mutation rate, then the equilibrium mean fitness is $\hat{\overline{W}} \approx e^{-U}$ (Section 5.1). Clearly, $U_G \leq U$. The additive effects of mutations on the character are drawn from a distribution symmetric about zero with variance γ^2 and fourth cumulant $\eta\gamma^4$, where η is the kurtosis ($\eta = -2$ if all effects are equal in absolute value, and $\eta = 0$ for a normal distribution; cf. Appendix D). Thus, the genotypic value of an individual carrying k mutations is $G = \sum_{i=1}^{k} X_i$, where the X_i are independent, identically distributed random variables according to the mutation distribution. The phenotypic value is $P = G + E$, where E is independent of P and normally distributed with mean zero and variance σ_E^2.

Using independence of effects X_i and additivity of cumulants, it follows that

$$\mathrm{E}[G] = \sum_k \mathrm{E}[G|k]p_k = 0 \, , \tag{6.2a}$$

$$\mathrm{Var}[G] = \mathrm{E}[G^2] = \sum_k \mathrm{E}[G^2|k]p_k = \gamma^2\bar{k} \, , \tag{6.2b}$$

$$\begin{aligned} \mathrm{Var}[G^2] &= \sum_k \left(C_4[G|k] + 3\,\mathrm{E}[G^2|k]^2\right) p_k - \mathrm{E}[G^2]^2 \\ &= \sum_k (k\eta\gamma^4 + 3k^2\gamma^4)p_k - \bar{k}^2\gamma^4 = \gamma^4(3\sigma_k^2 + \eta\bar{k} + 2\bar{k}^2) \, , \end{aligned} \tag{6.2c}$$

where p_k denotes the (nonequilibrium) distribution of the number of deleterious mutations, $\bar{k}$ and σ_k^2 are its mean and variance, respectively, and C_4 denotes the fourth cumulant of the distribution of G. At mutation-selection equilibrium, p_k is Poisson distributed, and because the mutational variance of the trait is $\sigma_{\mathrm{m}}^2 = U_G\gamma^2$, (6.1) and (6.2b) imply that the equilibrium genetic variance is

$$\hat{\sigma}_G^2 = \sigma_{\mathrm{m}}^2/s \, . \tag{6.3}$$

This is formally equivalent to the relation (2.17), which was obtained for a balance between mutation and direct stabilizing selection on the trait; but the context and the meaning of s is different here.

Since, under this model, extreme phenotypes will tend to carry many deleterious mutations, a negative correlation between fitness and deviation from the mean is expected, giving the false appearance of stabilizing selection on the character. In practice, the strength of stabilizing selection is usually measured as the regression of fitness on the squared deviation from the optimum, b_{W,P^2} (Lande and Arnold 1983). However, for theoretical purposes, fitness is often approximated by the Gaussian function $\exp\{-(P-P_O)^2/(2\omega^2)\}$, where $\omega^2 = -1/(2b_{W,P^2})$ [cf. V(1.12)]. Therefore, we have to regress the logarithm of fitness. Moreover, since in our model there is no a priori optimum, the regression of the logarithm of fitness on the squared phenotypic deviation from the mean will first be calculated. Owing to the symmetry of the model, we can scale the mean to zero and set $P_O = \bar{P} = \bar{G} = 0$. Thus, measuring in units of environmental standard deviations, we obtain

$$b_{W,P^2} = \sigma_E^2 \operatorname{Cov}[\ln W, P^2]/\operatorname{Var}[P^2] = \sigma_E^2 \operatorname{Cov}[\ln W, G^2]/\operatorname{Var}[P^2] \tag{6.4}$$

(cf. Keightley and Hill 1990). As $\sigma_P^2 = \sigma_G^2 + \sigma_E^2$, and because E is normally distributed and independent of G, it follows that

$$\operatorname{Var}[P^2] = \operatorname{Var}[G^2] + 4\sigma_G^2\sigma_E^2 + 2\sigma_E^4 = 2\sigma_P^4 + \hat{\sigma}_G^2\gamma^2(3+\eta) \, , \tag{6.5}$$

where the second equality assumes mutation-selection equilibrium and follows from (6.1), (6.2c), and (6.3). Another simple calculation shows that

$$\operatorname{Cov}[\ln W(G), G^2] = \sigma_k^2\gamma^2 \ln(1-s) \approx -\sigma_{\mathrm{m}}^2 \, , \tag{6.6}$$

where the approximation assumes $\ln(1-s) \approx -s$ and $\sigma_k^2 \approx \bar{k} \approx U_G/s$, as is appropriate for a Poisson distribution.

Using the scale on which $\sigma_E^2 = 1$, so that $h_{\mathrm{m}}^2 = \sigma_{\mathrm{m}}^2$, and neglecting the term $\hat{\sigma}_G^2\gamma^2(3+\eta)$, which typically is much smaller than σ_P^4, we can combine these results and obtain for the strength of apparent stabilizing selection caused by deleterious mutation-selection balance

$$\tilde{\omega}^2 \approx \frac{\sigma_P^4}{h_{\mathrm{m}}^2} = \frac{\sigma_P^2}{s\hat{h}^2} = \frac{h_{\mathrm{m}}^2}{s^2(\hat{h}^2)^2} \, . \tag{6.7}$$

Here, the heritability $\hat{h}^2$ is calculated from (6.3). It may be noted from (6.4) and (6.5) that the regression of the logarithm of fitness on the squared genotypic deviation, as used by Barton (1990), is always larger than b_{W,P^2}, because $\mathrm{Var}[P^2]/\mathrm{Var}[G^2] \geq 1$. Thus, in this type of model and in contrast to those of direct stabilizing selection (Chapter V.1.2), environmental effects lead to weaker apparent selection on phenotypes than on genotypes.

What inferences can be drawn from these predictions about the amount of genetic variance and the strength of apparent stabilizing selection this model of pleiotropic mutation-selection balance can account for? We shall draw on the data summarized in Section 1. First, this model can only predict very weak stabilizing selection because by (6.7), $\tilde{\omega}^2 \geq 1/h_\mathrm{m}^2$ and, typically, $10^{-3} \leq h_\mathrm{m}^2 \leq 10^{-2}$. Secondly, for morphological traits, the median of estimates of $\sigma_G^2/\sigma_\mathrm{m}^2$ is somewhat less than 100, suggesting via (6.3) that $s \approx 0.01$, which is smaller than the often quoted 'typical' estimate of $s \approx 0.02$. However, because empirical estimates of $\sigma_G^2/\sigma_\mathrm{m}^2$ lie in the range between 10 and 1000, the average deleterious effect of the alleles contributing to a quantitative trait may be any number between 0.001 and 0.1. The other way round, if we choose $s = 0.01$ and $10^{-3} \leq h_\mathrm{m}^2 \leq 10^{-2}$, then (6.3) yields $0.1 \leq \hat{\sigma}_G^2/\sigma_E^2 \leq 1$ and, thus, a heritability h^2 between 0.09 and 0.5. With $s = 0.05$, h^2 would only be between 0.02 and 0.17. Therefore, this model can maintain moderately high heritabilities if the deleterious effects of the mutations are small.

Keightley and Hill (1990) investigated an HK model in which effects of mutant alleles are sampled from a continuous bivariate distribution, assuming that mutant effects on the metric trait are additive and symmetrical about zero, and mutant effects on fitness are also additive, but unconditionally deleterious. More precisely, they chose a Wishart distribution with one degree of freedom, such that both the *absolute* effects of mutations on the trait and on fitness follow a Γ distribution with high kurtosis. The absolute values of mutant effects on the metric trait and on fitness may be correlated (measured by ρ) because "it is likely that many mutations have almost no effect at the level of the gene product and are therefore neutral or nearly neutral with respect to all traits, whereas mutations that drastically alter some gene products are likely to have large pleiotropic effects on many traits" (Keightley and Hill 1990).

Since the complexity of this model seems to preclude simple analytical predictions, Keightley and Hill used Monte-Carlo simulation for studying it. A distinctive feature of their model is that the equilibrium genetic variance $\hat{\sigma}_G^2$ tends to infinity as N_e tends to infinity, although the rate of this increase is very low if the correlation ρ is close to unity. The reason is presumably that if $\rho < 1$, then some alleles can have a large effect on the trait, but still be nearly neutral and, hence, frequent. In fact, if the correlation is unity, this model behaves similarly to one of direct stabilizing selection, and the variance quickly approaches a finite value with increasing N_e. For neutral genes, the equilibrium variance $\hat{\sigma}_G^2$ increases linearly with N_e, as in (2.6). In general, the model of Keightley and Hill (1990) predicts even weaker apparent stabilizing selection than the one discussed above with fixed deleterious effects. (For the parameter values in their Table 2, the prediction (6.7) produces $V_\mathrm{s} = 5.2/\varepsilon_s$, $\varepsilon_s = (\mathrm{E}[s^2])^{1/2}$, which is among the lowest values in their table. Their measure V_s for the strength of stabilizing selection is scaled differently than ours and corresponds to our $\tilde{\omega}^2/\sigma_P^2 = 1/(sh^2)$.) In accordance with (6.7), the numerical results of Keightley and Hill show that a decrease in the number of mutations appearing in the population per generation, or an increase in selection against deleterious alleles, or a decrease in the correlation ρ, all lead to weaker apparent selection.

Caballero and Keightley (1994) performed a simulation study of the pleiotropic model employed by Keightley and Hill (1990), but included dominance of effects on both fitness and character. They used parameter estimates for bristle number in *D. melanogaster* from accumulation of spontaneous and P element-induced mutations, and found that the degree of dominance has little effect on the equilibrium genetic variance and heritability, unless all mutants are completely recessive. However, the simulation results of Caballero and Keightley also reveal that for the two parameter sets, quite different types of alleles are responsible for the majority of the variation maintained. For P element data (large average mutant effects on both fitness and bristle number, high kurtosis, correlation of mutant effects of $\rho = 0.38$, genomic mutation rate of $U_G = 0.1$), most variation is contributed by nearly neutral mutations with substantial effects on the trait. For parameters derived from spontaneous mutation accumulation experiments (small average mutant effects on both fitness and bristle number, unknown levels of kurtosis and correlation, $U_G = 1$), in general, a large fraction of the variance is contributed by very deleterious mutations with small effects on the trait. This is in contrast to the quasi neutrality of bristle traits (Robertson 1967, Nuzhdin *et al.* 1995) and the observation of segregating genes with substantial effects on bristles (Shrimpton and Robertson 1988a,b). Therefore, the study of Caballero and Keightley underlines the fact that for pleiotropic models, the properties of the equilibrium distribution depend on the magnitude of unknown, or difficult to measure, genetic parameters.

The models of Barton (1990) and Keightley and Hill (1990) become unrealistic for deleterious genomic mutation rates in excess of one, because in both cases the total load will be close to e^{-U} since multiplicative and additive fitness effects are assumed, respectively. However, only with large U, and hence large h_{m}^2, can these models predict noticeable stabilizing selection.

Kondrashov and Turelli (1992) investigated a generalization of Barton's model that includes synergistic epistasis between fitness effects of mutations. They showed that, in accordance with (6.3), the variance is again $\hat{\sigma}_G^2 = \hat{\bar{k}}\gamma^2 = \sigma_{\mathrm{m}}^2/s$, and apparent stabilizing selection is weak and of comparable strength as in Barton's model. Since, with synergistic epistasis, a higher genomic mutation rate can be tolerated, the model of Kondrashov and Turelli can simultaneously explain high heritabilities and moderately strong stabilizing selection provided $h_{\mathrm{m}}^2 > 10^{-2}$. The question is to what extent such parameter values are compatible with data. The simulation results of Kondrashov and Turelli show that with equal effects of alleles on the metric trait, for various synergistic fitness functions, and for realistic parameter values, the equilibrium distribution of the number of deleterious mutations per genome, the distribution of breeding values, and the apparent stabilizing selection function are all nearly Gaussian.

6.2 A COMPOUND MODEL

Tanaka (1996) investigated a model in which an additive trait is under direct Gaussian stabilizing selection, but all mutations affecting the trait have deleterious effects that reduce fitness by the factor $(1 - s)$. Mutations are assumed to be rare, so that only the wild type alleles (which are optimal) mutate and that it suffices to consider the fate of cohorts of mutations. Mutational effects on the trait are drawn from a Gaussian distribution with variance γ^2. With these ingredients, and assuming linkage equilibrium, exchangeable loci, and neglecting random genetic drift, he showed that the equilibrium genetic variance can be approximated by

$$\hat{\sigma}_G^2 \approx \frac{\sigma_{\rm m}^2}{s + \gamma^2/(2V_{\rm s})}\,, \tag{6.8}$$

where $s \approx -\ln(1-s)$ is assumed. One may call $s_{\rm T} = s + \gamma^2/(2V_{\rm s})$ the total deleterious effect of the average mutation. If there are no deleterious side effects on fitness ($s = 0$), then (6.8) yields the HC-approximation VI(7.10b), whereas for a neutral trait ($V_{\rm s} = \infty$), (6.3) is obtained. Not surprisingly, in this model less genetic variation is maintained than in either the purely pleiotropic model or the one with pure stabilizing selection. As expected from the analysis of the purely pleiotropic model, the apparent strength of stabilizing selection in this model is only slightly stronger than the true strength, unless the latter is extremely weak.

The following numerical example shows that if two 'classes' of mutations contribute to a trait under moderately strong direct stabilizing selection ($V_{\rm s} = 20$, $\sigma_E^2 = 1$), those with large effects on the trait having pleiotropic deleterious effects, the others being nearly neutral, then substantial heritiabilities and low values of $\sigma_G^2/\sigma_{\rm m}^2$ can be explained simultaneously.

Example. Consider an infinite population, in which one class of mutations has large effects on an additive trait ($\gamma^2 = 0.5$) and pleiotropic, unconditionally deleterious (heterozygous) fitness effects of magnitude $s = 0.02$. Let their total mutation rate be $U = 0.01$, giving $\sigma_{\rm m}^2 = 0.005$. According to (6.8), these mutations contribute 0.11 to the equilibrium genetic variance. The second class consists of mutations of much smaller effects ($\gamma^2 = 0.01$ and $s = 0.001$), but is four times as large ($U = 0.04$). This class yields $\sigma_{\rm m}^2 = 0.0004$ and contributes 0.32 to the genetic variance. In total, this gives $\sigma_{\rm m}^2 = 0.0054$, $\hat{\sigma}_G^2 \approx 0.43$, a heritability of $\hat{h}^2 \approx 0.30$, and $\hat{\sigma}_G^2/\sigma_{\rm m}^2 \approx 80$. ◇

This is in accordance with typical observed values for morphological traits. The above calculation assumes independent loci, which is clearly impossible in view of the large number of loci that is required for such parameters. However, as is discussed in Section 2, much less genetic variability will be obtained only if a large fraction of these loci is very tightly linked ($r \le 10^{-4}$ between neighboring loci). It may also be noted that the heritability is uniquely determined by $\sigma_G^2/\sigma_{\rm m}^2$ and $\sigma_{\rm m}^2 (= h_{\rm m}^2$ because $\sigma_E^2 = 1)$, so that estimates of $\sigma_G^2/\sigma_{\rm m}^2 \approx 100$ and $h_{\rm m}^2 \le 10^{-2}$ constrain the heritability to $h^2 \le 0.5$.

6.3 EPISTASIS AND PLEIOTROPIC EFFECTS OF POLYMORPHISMS

Robertson (1956) was the first to demonstrate that apparent stabilizing selection on a metric trait can be a mere side effect of pleiotropy. He studied a model in which diallelic loci are held polymorphic by overdominance for fitness, but have pleiotropic effects on a metric trait. Clearly, such a model can maintain high levels of genetic variance for many traits, because

$$\hat{\sigma}_G^2 = 2\sum_{i=1}^{\ell} \hat{p}_i(1-\hat{p}_i)\gamma_i^2\,, \tag{6.9}$$

where $\hat{p}_i$ is the equilibrium frequency of one of the alleles and γ_i is the average effect of a substitution [cf. I(3.15)]. In this model, $\hat{\sigma}_G^2$ is roughly proportional to the number of loci, ℓ, that are polymorphic and contribute to the trait. If L_i denotes the segregation load at locus i [III(3.22)] and $\bar{L} = \sum_i L_i 2\hat{p}_i(1-\hat{p}_i)\gamma_i^2 / \sum_i 2\hat{p}_i(1-\hat{p}_i)\gamma_i^2$ the weighted mean segregation load, then the strength of apparent selection can be approximated by

$$\tilde{\omega}^2 \approx \frac{\sigma_P^2}{\bar{L}h^2} \tag{6.10}$$

(Robertson 1956, Barton 1990, Gavrilets and de Jong 1993). Formally, this is equivalent to (6.7). It can account for moderately strong selection only if $\bar{L}$ is relatively large, on the order of a few percent. However, because the real load, $L = \sum_i L_i$, is approximately $\ell\bar{L}$ and the genetic variance of the trait and the genetic variance in relative fitness are proportional to ℓ, this creates serious difficulties for the model: First, a high genetic variance requires many (overdominant) loci, for which there is little empirical support and which would cause a high segregation load. Second, with many contributing loci, the variance in relative fitness becomes very large, which is unlikely to be realistic, whereas only a fraction of approximately $1/(2\ell)$ of the genetic variance in apparent relative fitness is explained by the variance in apparent fitness. Thus, the real fitness values are scattered widely around the apparent fitness function (Barton 1990, Gavrilets and de Jong 1993).

Quantitative variation may also be maintained as a pleiotropic side effect of polymorphisms maintained by frequency-dependent selection. However, Barton (1990) argued that this would induce only weak apparent stabilizing selection.

By contrast, Gavrilets and de Jong (1993) introduced a pleiotropic model that has the potential of accounting simultaneously for high heritabilities and the appearence of strong stabilizing selection. It involves epistasis, and is conceptually more general than the models discussed above. Since it is technically quite involved, we outline only its basic features and properties.

In this model, a diploid monoecious randomly mating population is considered, and fitness measurements (viability) of every individual are given. There are ℓ diallelic loci, p_i and q_i denote the frequencies of the alleles $\mathcal{A}_i$ and $\mathcal{B}_i$, and $\chi_i(\chi_i')$ are indicator variables that equal 1 if the maternal (paternal) allele at the ith locus is $\mathcal{A}_i$, and 0 otherwise. Linkage equilibrium is assumed. The distribution of phenotypic values, P, of an additive quantitative trait is denoted by $f_P(P)$; its mean is $\bar{P} = 0$ and its variance σ_P^2. In this model, as in those treated above, different individuals with the same phenotype may have different fitnesses. The fitness of phenotype P is therefore defined as $W_P(P) = \mathrm{E}[W|P]$, where the expectation is over all individuals with phenotypic value P.

With these elements, Gavrilets and de Jong (1993) derived a general approximate expression for the 'apparent fitness function' $W_P(P)$, valid for arbitrary fitnesses and arbitrary phenotypic distributions. In the case of an additively determined trait ($P = \sum_i \gamma_i(\chi_i + \chi_i') + E$), this expression simplifies to

$$\begin{aligned} W_P(P) = \overline{W} &+ \sum_{i=1}^{\ell} \frac{\partial \overline{W}}{\partial p_i}\left[-\gamma_i p_i q_i \frac{f_P'(P)}{f_P(P)} + \tfrac{1}{2}\gamma_i^2 p_i q_i (q_i - p_i)\frac{f_P''(P)}{f_P(P)} + \dots\right] \\ &+ \left(\sum_{i,j=1}^{\ell} \tfrac{1}{2}\frac{\partial^2 \overline{W}}{\partial p_i \partial p_j}\gamma_i p_i q_i \gamma_j p_j q_j\right)\frac{f_P''(P)}{f_P(P)} + \dots\,, \end{aligned} \tag{6.11}$$

where higher-order terms in γ_i are neglected, because the effects γ_i of the loci are assumed to be small. If the genetic system is at a stable polymorphic equilibrium that is determined solely by selection, and if linkage disequilibrium can be neglected, then $\partial\overline{W}/\partial p_i = 0$ [cf. I(9.11a), II(4.18), II(4.18)] and the matrix of second-order derivatives, $\partial^2\overline{W}/\partial p_i\partial p_j$, is negative definite. Therefore, the simple sum in (6.11) vanishes, and the double sum (in parentheses) is negative. If the distribution of phenotypes is normal with mean zero, we have $f''_P(P)/f_P(P) = -1/\sigma_P^2 + (P/\sigma_P^2)^2$, and the apparent fitness function $W(P)$ displays stabilizing selection.

If the population is not at a polymorphic equilibrium determined by selection, but the allelic effects γ_i are drawn randomly from a probability distribution that is independent of fitness and has mean zero, then the terms in (6.11) proportional to $f''_P(P)/f_P(P)$ are likely to dominate the terms proportional to $f'_P(P)/f_P(P)$, and stabilizing or disruptive selection will be observed. If a population is at an equilibrium determined by mutation-selection balance, or migration-selection balance, such that at each locus one allele is very rare, then the double sum in (6.11) can be neglected, and again stabilizing selection will be observed. In this case, the model of Gavrilets and de Jong generalizes those of Section 6.1, and if $W(G|k) = (1-s)^k$, it reduces to Barton's model. These considerations suggest that the observation of stabilizing selection on an additive trait may be common if genes contributing to the trait are under selection. However, if in (6.11) the term proportional to $f'_P(P)/f_P(P)$ dominates and the distribution f_P is approximately normal, then directional selection will be observed.

Gavrilets and de Jong (1993) specifically studied a class of multilocus models, in which a polymorphism in ℓ loci is maintained by epistatic fitness interactions and the loci contribute to a not necessarily additive trait. More precisely, following Zhivotovsky and Gavrilets (1992), they assumed that fitness of genotypes can be represented by

$$W = W_0 + \sum_{i=1}^{\ell}[a_i(\chi_i + \chi_i') + 2b_i\chi_i\chi_i'] + \sum_{i,j:i\neq j} c_{ij}(\chi_i + \chi_i')(\chi_j + \chi_j') \,, \tag{6.12}$$

where the terms involving b_i are responsible for dominance and those with c_{ij} for (additive×additive) epistasis. The trait value P is assumed to be given by an analogous expression, but with other constants. Gavrilets and de Jong show, both by analytical example and numerical simulation, that epistatic fitness models of type (6.12) that maintain the same high level of genetic polymorphism as the overdominant model, can induce strong stabilizing selection on an additive trait in such a way that both the apparent load and the apparent variance in (relative) fitness are almost as high as the real load and the real variance in fitness. Basically, this requires synergistic epistasis between fitness effects, i.e., the mean of the epistatic parameters c_{ij} must be negative and the variance small. A further interesting point is that with epistatic fitnesses, the variance in fitness can be much lower than without epistasis, so that the real fitness values are clustered tightly around the apparent (nearly quadratic) fitness function. They also showed by numerical examples that polymorphisms maintained by fitnesses of the form (6.12) induce stabilizing selection on an additive trait (see above), and on a trait for which dominance and epistatic effects are random and independent of those on fitness. However, with directional dominance, or if the trait coefficients (a_i, b_i, c_{ij}) are correlated with the respective fitness coefficients, apparent directional selection is observed for equilibrium populations. Not surprisingly, examples can be constructed yielding complex apparent fitness functions.

6.4 NONHERITABLE CORRELATIONS

A quite different model was investigated by Wagner (1996b). He considered two metric traits that are phenotypically correlated by environmental covariance, the first being neutral, the second under direct stabilizing selection. He showed that in the extreme case in which none of the genes contributing to the neutral trait has pleiotropic effects on the selected trait, the strength of apparent selection can be up to half of the strength of direct stabilizing selection on the selected trait, whereas the breeding values of the first trait remain neutral. Therefore, the genetic variance of the first trait would be given by the neutral expectation (2.6), i.e., $\hat{\sigma}_G^2 = 2N_e\sigma_{m,1}^2$. In the more realistic case, in which only a fraction ϕ of genes have no pleiotropic effects, the variance is reduced to $2N_e\phi\sigma_{m,1}^2$. Therefore, this model can induce strong apparent stabilizing selection and high levels of genetic variance, but the ratio $\hat{\sigma}_{G,1}^2/\sigma_{m,1}^2$ will be $2N_e\phi$ and, thus, potentially much higher than the 'typical' values of 50–100.

6.5 SOME CONCLUSIONS

Many mechanisms can induce apparent stabilizing selection on a quantitative trait that is not causally related to fitness. The models discussed above can maintain potentially high levels of genetic variance for many traits simultaneously if pleiotropic effects are widespread. However, models in which the observed variation in a trait is a mere pleiotropic side effect of mutation-selection balance of deleterious mutants cannot easily account for the appearance of more than weak stabilizing selection, unless the genomic deleterious mutation rate is very high and there is synergistic epistasis between deleterious fitness effects. Models in which variation is a side effect of a selectively maintained polymorphism can explain high levels of genetic variation in the trait and strong apparent stabilizing selection, provided there is synergistic epistasis between fitness effects.

For additive traits under moderately strong direct stabilizing selection, the numerical example in Section 6.2 shows that models allowing for mutations with large effects on the trait and pleiotropic deleterious effects on fitness, as well as a large class of neutral mutations with very small effects on the trait, can quite easily explain heritabilities of about $\frac{1}{3}$ and ratios of σ_G^2/σ_m^2 of less than 100.

An experimental study by Nuzhdin *et al.* (1995) indicates that sternopleural and abdominal bristle number in *D. melanogaster* are under apparent stabilizing selection ($\tilde{\omega}^2/\sigma_P^2 = 51, 77$, respectively), part of which is caused by mutations of large net effect on bristle number with partly recessive, deleterious pleiotropic effects on fitness. The data of Nuzhdin *et al.* are neither compatible with a purely pleiotropic model of a balance between deleterious mutations and selection, nor with direct stabilizing selection on either of the traits; but direct selection against low abdominal bristle number cannot be excluded. Thus, in nature several mechanisms may interact to give the appearance of stabilizing selection.

7. DIRECTIONAL SELECTION

Understanding the response to directional selection is of crucial importance, both for evolution as well as for animal and plant breeding. Natural directional selection may result from biotic or abiotic changes in the environment that favor high (or low) values of one or a combination of traits, e.g., larger size. Fitness components and life-history traits are, almost by definition, under directional selection, at least for a wide range of trait values. Artificial selection experiments are usually designed to increase the yield or (economic) value of a specific breed. Such selection can be extremely successful as demonstrated, for example, by the amazing variation among present breeds of dogs, where the smallest and largest differ in weight by two orders of magnitude (Stockard 1941). Also, controlled selection experiments have produced continued responses for many generations and resulted in character values far beyond the range of variation in the base population. For example, ninety generations of two-way selection for oil content in maize seeds increased oil content from approximately 5% to over 20% in the high line and decreased it to almost zero in the low line (Dudley 1977, Dudley and Lambert 1992); up to 89 generations of selection on abdominal bristle number in *D. melanogaster* led to an increase from approximately nine to 39 bristles in females, corresponding to almost 15 standard deviations of the initial distribution (Yoo 1980a). However, there are selection experiments in which substantial initial response is observed, beyond the range of variation in the base population, but finally, a plateau is reached; and there are experiments in which no initial response occurred (see Falconer and Mackay 1996, Chapter 12).

There are two reasons, why selection can result in character values far beyond the extreme values in the base population. One is recombination, which together with many generations of selection can combine all the genes having effects in the selected direction in a single genome; the other is mutation. Almost all selection response in early generations will be due to existing variation in the base population, but the long-term response must eventually come from new mutations. This has been supported experimentally (e.g., Yoo 1980b, López and López-Fanjul 1993a,b, Mackay and Fry 1996).

A general theory for the response of the distribution of an additive quantitative trait to selection was developed in Chapter V. In this section, we shall use these and other methods and results to study the response to certain simple, but important, forms of directional selection. In particular, we shall investigate how the initial response depends on properties of the initial distribution of the trait, and how mutation, recombination, the reproductive mode, and other genetic properties influence the long-term response. Unless otherwise noted, the theory in this section assumes a randomly mating diploid population with discrete generations, no sex differences, and the additive model of quantitative genetics described in Chapter V.1.

7.1 CLASSICAL RESULTS ON TRUNCATION SELECTION

In Chapter V.6.3, the so-called breeder's equation,

$$\Delta\overline{P} = h^2 \Delta_s \overline{P} \;, \tag{7.1a}$$

was derived for an additive trait and a Gaussian distribution of phenotypes. This equation is basic to most of animal and plant breeding, because it allows us to predict the mean phenotypic value in the next generation from phenotypically measurable quantities, namely from the (narrow sense) heritability, h^2, and the selection differential, or selection response within a generation, $\Delta_s\overline{P}$, which measures the difference in the means of the whole population and the parents selected for reproduction. A frequently used dimensionless measure of the strength of selection is the *selection intensity*, i, which is defined as the standardized selection differential, $i = \Delta_s\overline{P}/\sigma_P$. Then the response to selection between generations, $\Delta\overline{P}$, can be expressed as

$$\Delta\overline{P} = ih^2\sigma_P \,, \tag{7.1b}$$

i.e., as the selection intensity times the heritability measured in units of phenotypic standard deviations σ_P.

In the quantitative genetics literature, the breeder's equation is sometimes written in the form $R = \beta S$, where $R = \Delta\overline{P}$, $S = \Delta_s\overline{P}$, and β denotes the coefficient of the regression of offspring values on mid-parental values (i.e., the arithmetic mean of the parents). The equation $R = \beta S$ is a purely statistical relation that goes back to a series of articles by Pearson at the end of the nineteenth century. Its validity requires linearity of the regression of offspring on the biparental values (P, P^*). Under various sets of assumptions (e.g., random mating and the additive model), it can be shown that $\beta = h^2$. A lucid and rigorous account is given by Nagylaki (1992, pp. 311–315); but see also Bulmer (1980, pp. 144–145), who includes sex differences; Falconer and Mackay (1996, Chapter 11) for a more intuitive approach; and Chapter I.3.2, where the heritability was related to the correlation between a single parent and its offspring. A recent experimental and theoretical study of the (non)linearity of the regression may be found in Gimelfarb and Willis (1994).

In breeding programs, *truncation selection* is typically performed. This means that the best q% of the population (i.e., those with the highest, or lowest, phenotypic values) are chosen as parents to produce the next generation. Denoting the trunction point by τ, we may write the corresponding fitness function as

$$W_P(P) = \begin{cases} 0\,, & P < \tau\,, \\ 1\,, & P \geq \tau\,. \end{cases} \tag{7.2}$$

Assuming a Gaussian distribution of phenotypic values, f_P, and writing the truncation point as $\tau = \overline{P} + t\sigma_P$, we obtain for the mean fitness

$$\overline{W} = q = \int_\tau^\infty f_P(y)\,\mathrm{d}y = \int_t^\infty \phi(x)\,\mathrm{d}x \,, \tag{7.3}$$

where ϕ denotes the standardized normal density. Then a simple exercise in integration shows that the selection intensity satisfies

$$i = \phi(t)/q \,. \tag{7.4}$$

Thus, the selection response can be easily calculated from the proportion of individuals selected, although in finite populations the selection intensity is slightly smaller (Falconer and Mackay 1996, Chapter 11). This makes (7.1) so useful.

Using the results of Chapter VI.1, we can relate the selection intensity i to the selection coefficient determining gene-frequency change. Consider an allele $\mathcal{A}_k$ occuring at an arbitrary locus, and denote its average excess by g_k and the average fitness of individuals carrying this allele by W_k. Assuming that the variance contributed by each locus is small, that the loci contribute additively, and that linkage disequilibrium can be ignored, the density $h_{jk}(P)$ of P in individuals with genotype $\mathcal{A}_j\mathcal{A}_k$ at a given locus can be approximated by $f_P(P - g_{jk})$; cf. VI(1.10). This approximation is valid to first order in g_{jk}, the average excess of $\mathcal{A}_j\mathcal{A}_k$. Inserting this into VI(1.4), we obtain from VI(1.5) the first-order approximation for W_k: $W_k \approx \overline{W} + g_k I_1$. From VI(1.3a), (7.3) and (7.4), we get $I_1 = f_P(\tau) = \phi(t)/\sigma_P = i\overline{W}/\sigma_P$. Therefore, the selection coefficient determining the change of allele frequency is

$$s_k = \frac{W_k - \overline{W}}{\overline{W}} \approx i\frac{g_k}{\sigma_P} , \tag{7.5}$$

and $\Delta p_k \approx p_k s_k$ (cf. Kimura and Crow 1978, and Nagylaki 1992, p. 318). A second-order approximation can be obtained from VI(1.5) and VI(1.3b), because a straightforward calculation shows that $I_2 = -t\phi(t)/\sigma_P^2 = -it\overline{W}/\sigma_P^2$. Latter (1965) obtained this second-order approximation, but by using a defective second-order approximation for the fitness W_{jk} (cf. Chapter VI.1).

We shall now investigate a simple form of directional selection for which exact results are easily obtained and for which the influence of assumptions about the distribution of allele frequencies on the selection response has been studied.

7.2 EXPONENTIAL SELECTION

From a theoretical point of view, a particularly simple form of directional selection is exponential selection, which is defined by the fitness function

$$W_P(P) = e^{sP} . \tag{7.6}$$

Substituting (7.6) into V(1.4), a simple exercise in integration shows that the average fitness of individuals with genotypic value G is

$$W(G) = e^{\frac{1}{2}s^2\sigma_E^2}\, e^{sG} . \tag{7.7}$$

If the distribution of breeding values is Gaussian with mean $\overline{G}$ and variance σ_G^2, then the phenotypic distribution is Gaussian with the same mean ($\overline{P} = \overline{G}$) and variance $\sigma_P^2 = \sigma_G^2 + \sigma_E^2$. The mean fitness of the population is readily calculated to be

$$\overline{W} = e^{\frac{1}{2}s^2\sigma_P^2}\, e^{s\overline{P}} . \tag{7.8}$$

Furthermore, the selection differential is

$$\Delta_s\overline{P} = \frac{1}{\overline{W}}\frac{1}{\sqrt{2\pi\sigma_P^2}}\int_{-\infty}^{\infty} P\, e^{sP}\, e^{-(P-\overline{P})^2/(2\sigma_P^2)}\,\mathrm{d}P \; - \overline{P} = s\sigma_P^2 \tag{7.9}$$

and, analogously, $\Delta_s\overline{G} = s\sigma_G^2$. Whereas the mean phenotype among offspring will be different from the mean phenotype among their (selected) parents, the mean genotypic value among offspring will be the same as among their parents [V(6.15)]. Therefore, we obtain

$$\Delta\overline{G} = \Delta_s\overline{G} = s\sigma_G^2 \,. \tag{7.10}$$

Since we always have $\overline{P} = \overline{G}$ among the offspring, it follows that $\Delta\overline{P} = \Delta\overline{G}$ and, hence, $\Delta\overline{P} = h^2\Delta_s\overline{P}$, which is the breeder's equation (7.1).

As a consequence of (7.9), the selection intensity is

$$i = s\sigma_P \,. \tag{7.11}$$

If W_k is the (marginal) fitness of an allele with average excess g_k, then $W_k/\overline{W} = e^{sg_k}$. From this we obtain

$$\frac{W_k - \overline{W}}{\overline{W}} = e^{sg_k} - 1 \,, \tag{7.12}$$

which agrees to leading order with (7.5). These results suggest that an exponential fitness function with $s = i/\sigma_P$ is an appropriate model for truncation selection as long as ig_k/σ_P is sufficiently small.

Dynamics of general distributions

Here, we concentrate on infinitely large populations, in particular, on their initial response to exponential selection. The long-term reponse in finite populations is treated in Section 7.4. The recursion relations for the response of the distribution of diploid genotypes to exponential selection were derived in V(3.44), those for the breeding values in V(3.45). In accordance with the well-known results on multiplicative fitnesses (Chapter II.5.3), these equations show that Hardy–Weinberg proportions and linkage equilibrium are preserved under exponential selection. (Under exponential selection and with additive allelic effects, the fitness of genotypes is multiplicative accross loci.) Moreover, they show that the shape of a multivariate Gaussian allelic distribution is also preserved.

Writing $\overline{G} = C_1$ for the mean and $\sigma_G^2 = C_2$ for the genetic variance, V(3.45) shows that the selection response of the first three cumulants of the distribution of genotypic (breeding) values is

$$\Delta_s\overline{G} = s\sigma_G^2 + \frac{s^2}{2}C_3 + \frac{s^3}{3!}C_4 + \dots \tag{7.13a}$$

$$\Delta_s\sigma_G^2 = sC_3 + \frac{s^2}{2}C_4 + \frac{s^3}{3!}C_5 + \dots \tag{7.13b}$$

$$\Delta_s C_3 = sC_4 + \frac{s^2}{2}C_5 + \frac{s^3}{3!}C_6 + \dots \tag{7.13c}$$

Since for a Gaussian distribution, all cumulants of order $n \geq 3$ are zero, such a distribution remains invariant under exponential selection (but not under truncation), and its mean evolves according to (7.10). For any other distribution, however, the response of the mean will depend on the variance and on the higher-order cumulants, so that (7.10) can only be an approximation. Even worse, the shape of the distribution will also be changed by selection. For instance, the variance may increase or decrease depending, to first order, on the sign of the skew of the distribution.

Therefore, we now investigate the robustness of predictions that are based on knowledge of only the first few cumulants of the initial distribution. We include mutation, which is assumed to occur at the same rate μ at all loci according to the random-walk mutation model with conditional density u. Mutation and selection are assumed to be sufficiently weak, so that their interaction (terms of order μs) can be ignored (see Bürger 1993, for details). We assume that the initial population is in linkage equilibrium and that all recombination frequencies are positive. Then the population will remain in linkage equilibrium. We denote the cumulant generating function of the initial distribution of genotypic values G by Ψ_0, the moment generating function of u by φ, and nth derivatives by D^n. Then the cumulants C_n evolve according to

$$C_n(t) = (D^n\Psi_0)(st) + \frac{2\ell\mu}{s}[D^{n-1}\varphi(st) - D^{n-1}\varphi(0)] , \qquad (7.14)$$

where the selection term is from V (3.46) and the mutation term was derived in Bürger (1993). For $\ell = \frac{1}{2}$, (7.14) describes the dynamics in a haploid asexual population. For several well known probability distributions the generating functions, and hence $C_n(t)$, can be calculated explicitly. Equation (7.14) shows that mutation always changes the cumulants and induces departures from normality. A Gaussian mutation distribution continuously increases all cumulants; explicit formulas may be found in Bürger (1993).

Interesting questions concern the influence of mutation and of the shape of the initial distribution on the response to selection. Since usually only the first few cumulants of a distribution can be infered with some accuracy, it is important to know how sensitive predictions are that are derived from such incomplete knowledge. For instance, if only the mean and the variance of the initial distribution are known, then it is reasonable to approximate the initial distribution by a Gaussian and calculate the change of the cumulants on that basis. If, however, the first four cumulants are known (and the distribution is symmetric), then a reflected Γ distribution can be used. We shall refer to the ensuing approximations as Gaussian and Γ-approximation, respectively. Detailed equations for these two approximations are straightforwardly derived from (7.14) and may be found in Bürger (1993). These approximations can be compared with the exact solution, which can be calculated numerically.

Here, we shall explore the dynamics of an initial population that has been in mutation-stabilizing-selection balance and, then, is exposed to directional selection. As discussed previously, rare alleles of large effect contribute significantly to the genetic variance maintained under mutation-stabilizing-selection balance if mutation rates are low and the variance of allelic effects is large. However, under directional selection, advantageous alleles will sweep through the population, thereby causing a large increase in genetic variance (Barton and Turelli 1987). Since in experiments such an increase is usually not observed, doubts have been raised about mutation-selection balance as a relevant mechanism for maintaining genetic variation (Barton and Turelli 1989).

A detailed numerical study of the initial response to exponential selection was performed by Bürger (1993) and is complemented here. The exact dynamics of the full frequency distribution is determined numerically by iterating a discrete version of the (asexual) mutation-selection equation IV (2.8) for which a very fine grid for the allelic effects was chosen. Mutation is assumed to occur at a constant rate μ and according to the random-walk mutation model with a Gaussian mutation distribution of mean zero

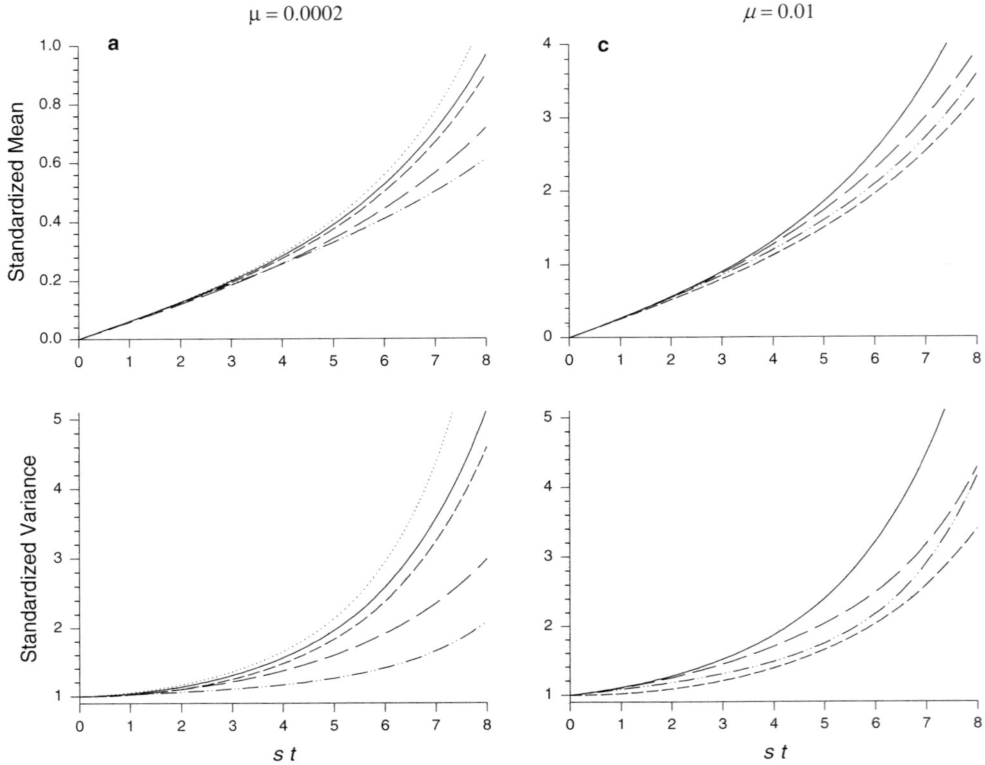

Figure 7.1 Initial response to exponential selection. Figures **a** and **c** display the evolution of the mean phenotype relative to the standard deviation of the initial population ($c_1(st)/c_2(0)^{1/2}$) as a function of st. Figures **b** and **d** display the variance in multiples of the initial variance ($c_2(st)/c_2(0)$) as a function of st. The left-hand side figures (**a** and **b**) are for a mutation rate of $\mu = 0.0002$, the right-hand side figures for $\mu = 0.01$. Initial distributions are in mutation-selection balance for these parameters with stabilizing selection on the genotypic values of strength $V_s = 10$. Solid curves indicate the exact solution with mutation during directional selection, long-dashed curves the exact solution of the truncated initial distribution with mutation (of the same rate as during preceeding stabilizing selection), short-dashed curves the exact solution without mutation during directional selection, dotted curves the Γ-approximation (only in **a** and **b**), and dash-double-dotted curves the Gaussian approximation. For $\mu = 0.0002$, the initial variance of the equilibrium distribution is $c_2(0) = 0.00365$ (which is close to the value 0.004 from the HC-approximation) and its kurtosis is 13.3; for its truncated sibling, $c_2(0) = 0.00345$ and the kurtosis is 10.1. For $\mu = 0.01$, the initial variance of the equilibrium distribution is $c_2(0) = 0.0629$ and the kurtosis is 0.65, for the truncated initial distribution the respective values are 0.0623 and 0.50.

and variance $\gamma^2 = 0.05$. (This mutation distribution is, of course, approximated by a matrix.) Numerical results show that for exponential selection the continuous-time model provides a very accurate approximation to the discrete-time model for a wide range of parameters including those used here (Bürger 1993, and unpublished).

Figure 7.1 displays the evolution of the standardized mean and the standardized variance for various initial distributions under a low (Figures 7.1a and 7.1b) and a high mutation-rate scenario (7.1c and 7.1d). The first may be applicable to a haploid locus (or factor) in a multilocus system in linkage equilibrium, the second to an asexually reproducing population if the mutation rate is interpreted as total mutation rate affecting the trait. In each case, an initial distribution was generated that was in equilibrium between mutation and stabilizing selection. From this a 'truncated' distribution was obtained by removing all alleles with frequencies of 5×10^{-5} and less, and subsequent renormalization. (Thus, extremely rare alleles of large effect are absent in this initial distribution.) For the low-mutation-rate scenario, these initial distributions are leptokurtic; for the high-mutation rate scenario, they are close to Gaussian (see the figure legend). Evolution of these two distributions was then simulated by iterating the recursion equation. To provide insight into the role of newly occuring mutations for the selection response, the evolution of the exact equilibrium distribution, but without mutation during directional selection, was also simulated. These numerically determined solutions are compared with the Gaussian and the Γ-approximation.

In the low mutation-rate scenario, and for the range of values st considered, new mutations occuring during the directional selection phase contribute very little to the selection response of both the mean and the variance, and the Γ-approximation is very accurate. In the high mutation-rate case, the Γ-approximation is not displayed because it gives a bimodal initial distribution. The Gaussian approximation always underestimates the response. Still, for $t \leq 5/s$, it predicts the mean with reasonable accuracy. The dynamics of the truncated initial distribution, which lacks alleles of large effects, shows indeed a much lower increase in variance than the full initial distribution. This is more pronounced for $\mu = 0.0002$, because the initial distribution is much more leptokurtic. Such rare alleles would be absent in populations of size $\leq 2 \times 10^4$. Truncation of alleles with less extreme effects further reduces the increase in variance and the resulting response of the mean (results not shown). As in most selection experiments population sizes are very low, this provides an explanation why typically no increase in variance is observed. After more than $5/s$ generations have elapsed, a complex restructuring of the distribution occurs in the low-mutation-rate scenario that is qualitatively similar to that displayed in Figure 7.3 for the moving-optimum model. However, under exponential selection and the current mutation model, the variance of the distribution continues to increase after the restructuring. In the absence of mutation, an asymptotic, apparently Gaussian, distribution with a high variance, σ_∞^2, evolves, whose mean increases at the constant rate $s\sigma_\infty^2$ (see Section 7.4 for more results). Only distributions that are nearly Gaussian do not change their shape radically but evolve in a 'smooth' way.

We note that the results of Figure 7.1 extend to the response of the mean and variance of the distribution of genotypic values, because, with multiple loci in linkage equilibrium, they are the sum of the respective single-locus quantities.

Eshel (1971, 1972) performed a rigorous analysis of the long-term response of an infinite asexual population to exponential selection. He used a similar setting as that in Chapter IV.4. He assumed, as we have done above, the random-walk mutation model and proved that if the mutation distribution has an upper bound u_b (thus, mutations cannot have arbitrarily large effects) and if the initial distribution also has an upper bound, then the mean will eventually evolve at the constant rate u_b, i.e., $\lim_{t\to\infty} \overline{G}(t)/t = u_b$. Moreover, the centered distribution has a limiting distribution with a finite variance, i.e., there is an asymptotically stable travelling-wave solution.

This asymptotic distribution is dependent only on the mutation distribution, but not on the initial distribution. If the assumption of a finite upper bound of the initial distribution is omitted, and the initial distribution is Gaussian with variance σ^2, then the asymptotic rate is $u_b + s\sigma^2$. If $u_b = \infty$, as for the Gaussian mutation distribution used in the analyses above, then the mean evolves asymptotically as $\overline{G}(t) = at^2$, where a is a positive constant (Karlin 1988). It is disturbing that these asymptotic results depend on the maximum possible mutation effect, but not on the variance of the mutation distribution. Karlin states further that under algebraically increasing fitness (e.g., $W(x) = x^k$) and a not necessarily bounded mutation distribution, the mean and the variance of the asymptotic distribution increase linearly, and the standardized distribution converges to the standard normal distribution. Some of these results carry over to randomly mating diploid populations if alleles act 'nearly' additive and fitness depends on a single locus with a continuum of possible types (cf., Eshel 1971, Karlin 1988). These results show that the long-term response to directional selection may depend on several model assumptions. In particular, the response to exponential selection in an infinite population may depend crucially on assumptions about the mutation distribution, but also on certain features of the initial distribution.

7.3 THE MOVING-OPTIMUM MODEL

Truncation and exponential selection are likely to be unrealistic in nature. More realistic may be the so-called moving-optimum model, according to which a quantitative trait is under stabilizing selection but the optimum phenotype changes as a function of time. A simple fitness function describing this is

$$W(G,t) = \exp\left\{-\frac{[G - P_O(t)]^2}{2V_s}\right\} . \tag{7.15}$$

Here, we study the case of an optimum that moves at a constant rate, k, per generation,

$$P_O(t) = kt . \tag{7.16}$$

Then the trait experiences a mixture of stabilizing and directional selection. It has been introduced as a model for sustained environmental change, such as global warming (Lynch *et al.* 1991, Lynch and Lande 1993). This and related models, for instance with a periodic optimum or an optimum fluctuating randomly around its expected (moving) position, were investigated by several authors (Charlesworth 1993, Bürger and Lynch 1995, Lande and Shannon 1996, Kondrashov and Yampolsky 1996a,b, Bürger 1999, Waxman and Peck 1999). All these investigations are based either on simplifying assumptions, such as a Gaussian phenotypic distribution with a constant variance, or on simulations. No general analysis is yet available.

Here, we investigate the dynamics of an infinitely large asexual population. This is of independent interest and yields some insight into the multilocus, linkage equilibrium case. Finite sexual populations and the role of recombination are considered in Section 7.4.

We denote the cumulants of the allele-frequency distribution by c_n, $n \geq 1$. Since after the onset of directional selection, the mean phenotype will lag behind the optimum phenotype, it will be convenient to describe the position of the distribution by the lag, $\lambda = \lambda(t) = kt - c_1(t)$. We assume a symmetric mutation distribution with mean zero, variance γ^2, and fourth moment v_4. Under the assumption of weak selection, the Gaussian fitness function can be approximated by a quadratic function, and the evolution of the cumulants can be approximated by the weak-selection (or continuous-time) model of Chapters V.2.2 and V.3.2, i.e., $\Delta_s c_n \approx \tilde{\Delta}_s c_n = \dot{c}_n$ (see V (3.12) and the paragraph following). Moreover, the allele-frequency distribution in the new generation is the same as among selected parents, i.e., $\tilde{\Delta} = \tilde{\Delta}_s$. Employing V (1.14), we set $s_1 = kt/V_s$, $s_2 = -1/(2V_s)$, and obtain from V (2.16) the dynamics of the lag, the variance, and the third and fourth cumulants:

$$\tilde{\Delta}\lambda = k - V_s^{-1}\left(\lambda c_2 - \tfrac{1}{2}c_3\right) \tag{7.17a}$$

$$\tilde{\Delta}c_2 = V_s^{-1}\left(\lambda c_3 - \tfrac{1}{2}c_4 - c_2^2\right) + \mu\gamma^2 \tag{7.17b}$$

$$\tilde{\Delta}c_3 = V_s^{-1}\left(\lambda c_4 - \tfrac{1}{2}c_5 - 3c_2c_3\right) \tag{7.17c}$$

$$\tilde{\Delta}c_4 = V_s^{-1}\left(\lambda c_5 - \tfrac{1}{2}c_6 - 4c_2c_4 - 3c_3^2\right) + \mu v_4 \,. \tag{7.17d}$$

Unfortunately, (7.17) does not produce a closed system of equations and can, therefore, not be used to calculate the dynamics of the distribution under the moving optimum model. However, it does provide some important qualitative insights into the relation between the lag, the variance, and the higher cumulants. For a Gaussian distribution, (7.17a,b) yield

$$\tilde{\Delta}\lambda = k - V_s^{-1}\lambda c_2 \tag{7.18a}$$

$$\tilde{\Delta}c_2 = -V_s^{-1}c_2^2 + \mu\gamma^2 \,. \tag{7.18b}$$

As (7.17d) shows, mutation drives a Gaussian distribution away from Gaussian. Yet, we shall compare the prediction ensuing from the assumption that (7.18) is exact, henceforth called the two-moments prediction, with the actual dynamics. (Note that this two-moments prediction is conceptually different from the Gaussian prediction used for exponential selection, because it ignores the contribution of mutation through the higher-order cumulants.)

Since under the moving-optimum model a population experiences a mixture of directional and stabilizing selection, and as an increasing lag leads to much stronger directional selection, one may expect that the lag of the mean behind the optimum phenotype stabilizes and a limiting distribution emerges. This is, indeed, borne out by numerical calculations that were performed for infinitely large asexual populations by iteration of a discrete version of IV (2.8), as described in the previous section.

Figure 7.2 displays the evolution of the lag and the variance for various initial distributions under a low (Figures 7.2a and 7.2b) and a high mutation-rate scenario (Figures 7.2c and 7.2d). The scenarios are analogous to those of Figure 7.1, and the initial distributions are also identical to those used there. Moreover, the two-moments approximation is shown, and an approximation based on the assumption that all cumulants of order greater than four can be neglected, i.e., that the recursion relations (7.17a–d) with $c_5 = c_6 = 0$ describe the dynamics accurately. As the figure shows, the initial response is independent of the details of the distribution, and both approximations are very accurate. Then, however, an increase of variance occurs in all exact

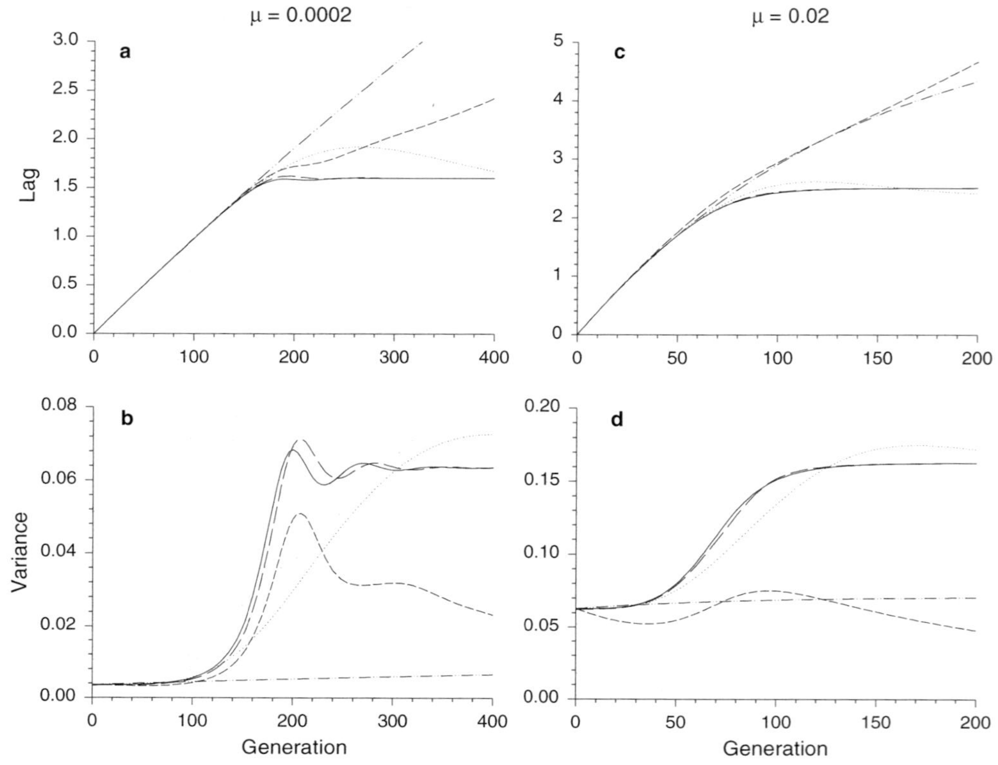

Figure 7.2 Evolution of the lag and the variance under the moving-optimum model. Figures **a** and **c** display the lag of the mean phenotype, **b** and **d** the variance. The left-hand side figures are for $\mu = 0.0002$ and $k = 0.01$, the right-hand side figures are for $\mu = 0.01$ and $k = 0.04$. In both cases the mutation distribution is Gaussian with variance $\gamma^2 = 0.05$. Initial distributions were in mutation-selection equilibrium for these mutation parameters and stabilizing selection of strength $V_s = 10$. Solid curves indicate the exact solution with mutation during directional selection (of the same rate as during preceeding stabilizing selection), long-dashed curves the exact solution for the truncated initial distribution with mutation, short-dashed curves the exact solution without mutation during directional selection, dotted curves the four-cumulants approximation, and dash-double-dotted curves the two-moments approximation. For $\mu = 0.0002$, the skewness and the kurtosis of the asymptotic distribution are 0.163 and 0.078, respectively, the mean fitness is 0.88. For $\mu = 0.01$, the skewness and the kurtosis of the asymptotic distribution are 0.135 and 0.069, respectively, the mean fitness is 0.72. The values of the initial distributions are given in the legend of Figure 7.1.

solutions that is very large in the low-mutation-rate case. Both approximations break down during the phase of rapid increase of variance. As expected, the increase without mutation is lower. However, in contrast to exponential selection, the difference in the dynamics between the full and the truncated initial distribution is small. Eventually, the two exact solutions with mutation settle down to a stationary distribution that has a constant lag and a constant shape, whereas the variance of the solution without mutation slowly declines to zero. Additional simulations (not shown) suggest, that for a given set of parameters, all initial distributions settle down to the same stationary distribution. For the parameters of Figure 7.2, the trajectories of a Gaussian initial distribution with the same variance as the mutation-selection-balance distribution, are almost indistinguishable from that one and, hence, are not displayed.

For the low-mutation-rate scenario, Figure 7.3 displays the evolution of the complete distribution of the mutation-selection-balance initial population with and without mutation during directional selection. It shows that a complex restructuring occurs while the variance increases to its asymptotic level. The complexity of this process makes it very unlikely that simple and accurate analytical approximations can be found. Such a restructuring also occurs for other initial distributions, including Gaussians, if their initial variance is much below the asymptotic value. The asymptotic distribution is close to Gaussian, but it is slightly skewed to the right (see Table 7.1). If, as in the case of a high mutation rate, the initial distribution is nearly Gaussian and has a variance that is not so much lower than the asymptotic variance, then no such restructuring occurs (results not shown). Also, a Gaussian distribution remains Gaussian under the moving optimum if there is no mutation. This can be proved analytically and in much greater generality (see the following subsection).

An asymptotic distribution of constant shape and constant lag must satisfy $\tilde{\Delta}\lambda = 0$ and $\tilde{\Delta}c_n = 0$ for all $n \geq 2$. It follows immediately from (7.17a), (7.17b), and (7.17c) that the equilibrium values (indicated by $\hat{\lambda}$ and $\hat{c}_n$) satisfy the relations

$$\hat{\lambda} = \frac{kV_s}{\hat{c}_2} + \frac{\hat{c}_3}{2\hat{c}_2}\,, \tag{7.19a}$$

$$\hat{c}_2 = \sqrt{\mu V_s \gamma^2 + \hat{\lambda}\hat{c}_3 - \tfrac{1}{2}\hat{c}_4}\,, \tag{7.19b}$$

$$3\hat{c}_2\hat{c}_3 = \hat{\lambda}\hat{c}_4 - \hat{c}_5\,. \tag{7.19c}$$

There are a number of important points to note. The first term in the square root of (7.19b) is the squared Gaussian allelic approximation, $\hat{\sigma}^4(\mathrm{G})$, for the equilibrium variance under mutation-selection balance; cf. IV(1.8). Numerical results suggest that for a wide range of parameters, but not for extremely small k (in the sense of a small fraction of a standard deviation of the initial distribution), the fifth cumulant in (7.19c) is much smaller than the other two terms, so that $\hat{c}_4 \approx 3\hat{c}_2\hat{c}_3/\hat{\lambda}$. Since $\hat{\lambda}\hat{c}_3 \gg \hat{c}_4$ holds, except for extremely small k, we obtain from (7.19b) the approximation

$$\hat{c}_2 \approx \sqrt{\hat{\sigma}^4(\mathrm{G}) + \hat{\lambda}\hat{c}_3}\,. \tag{7.20}$$

Since directional selection induces skewness, it follows that the asymptotic variance is typically larger than the Gaussian allelic approximation $\hat{\sigma}^2(\mathrm{G})$. Table 7.1 shows that it can be much larger if the mutation rate is small and k is not extremely small. Extracting $\hat{c}_3$ from (7.20) and substituting it into (7.19a), we find the following relation between the lag and the variance:

$$\hat{\lambda} = \frac{kV_s}{\hat{c}_2} + \frac{1}{2kV_s}\left[\hat{c}_2^2 - \hat{\sigma}^4(\mathrm{G})\right]\,. \tag{7.21}$$

Table 7.1 shows that the first term explains most of the lag and, by simple numerical calculations, that (7.21) is very accurate. Further, it is evident from Table 7.1 that

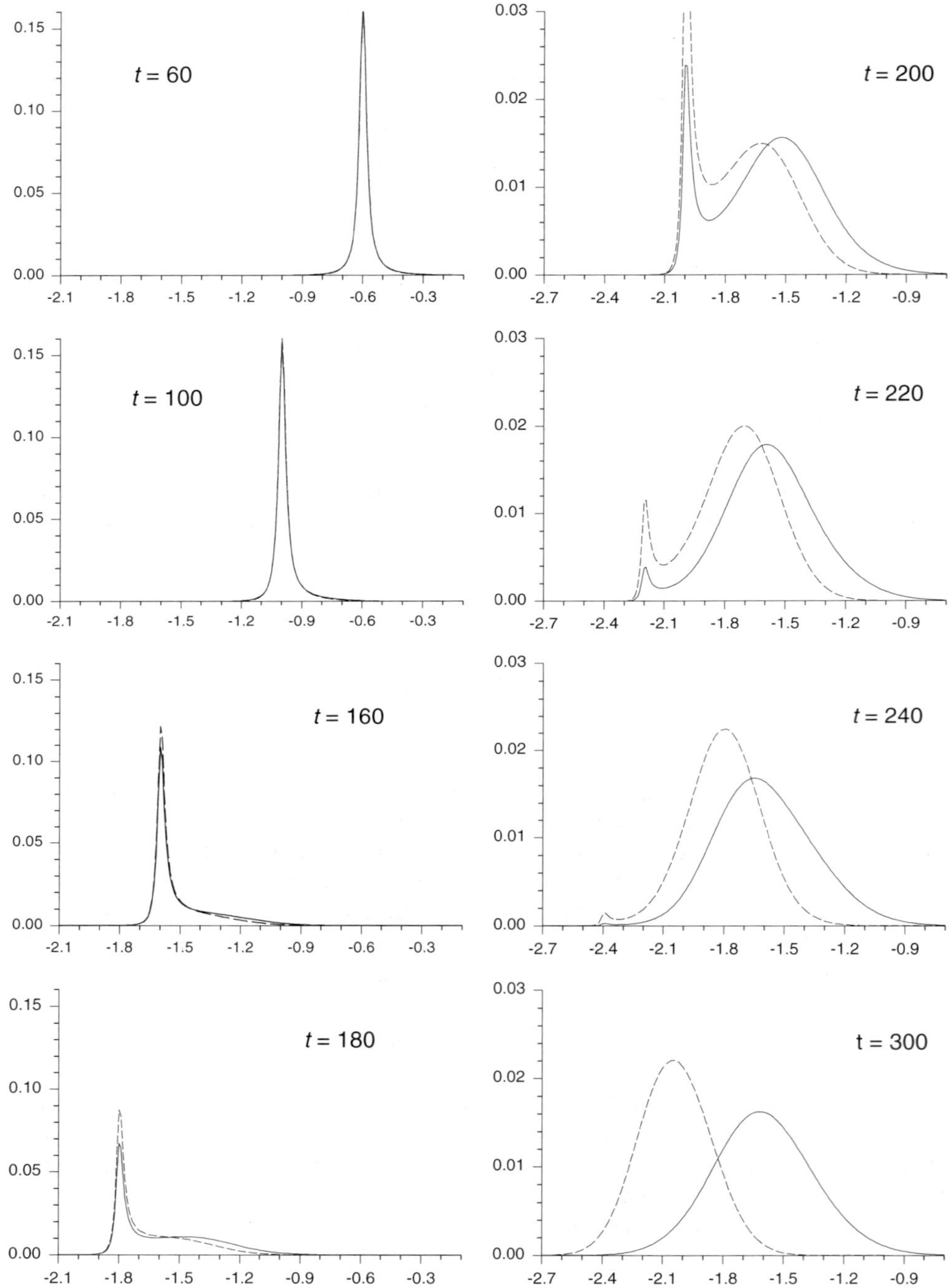

Figure 7.3 Evolution under the moving-optimum model of the initial equilibrium distribution for $\mu = 0.0002$ of Figure 7.2, with (solid lines) and without (dashed lines) mutation.

Table 7.1 Asymptotic distribution in the moving-optimum model. The variance and kurtosis of the initial distribution (which is in mutation-selection balance), the lag, variance, skewness, kurtosis, and mean fitness of the asymptotic distribution, and the leading term in the approximation (7.21) for the lag are all shown. The numbers in parentheses below $\hat{\sigma}^2$ are the HC-approximation and the Gaussian allelic approximation, respectively. The mutation distribution is normal with variance $\gamma^2 = 0.05$.

		INITIAL DISTR.			ASYMPTOTIC DISTRIBUTION					
μ	V_s	$\hat{\sigma}^2$	kurt	k	$\hat{\lambda}$	$\hat{c}_2$	skew	kurt	$\hat{\overline{W}}$	$kV_s/\hat{c}_2$
10^{-4}	200	0.0228	1.93	0.005	4.90	0.205	0.09	0.03	0.94	4.88
		(0.04)		0.01	7.19	0.279	0.07	0.02	0.88	7.17
		(0.0316)		0.02	10.52	0.381	0.06	0.01	0.76	10.50
	20	0.00365	13.4	0.0025	1.10	0.046	0.20	0.13	0.97	1.09
		(0.004)		0.005	1.60	0.063	0.16	0.08	0.94	1.59
		(0.01)		0.01	2.34	0.086	0.13	0.05	0.87	2.33
				0.02	3.41	0.118	0.10	0.03	0.75	3.39
				0.04	4.97	0.162	0.08	0.02	0.54	4.94
	2	0.0004	130	0.005	0.56	0.019	0.27	0.24	0.92	0.53
		(0.0004)		0.01	0.79	0.026	0.22	0.15	0.85	0.77
		(0.0032)		0.02	1.13	0.036	0.18	0.10	0.73	1.11
				0.04	1.62	0.050	0.14	0.06	0.52	1.60
2×10^{-4}	10	0.00365	13.3	0.005	1.10	0.047	0.20	0.13	0.94	1.06
		(0.004)		0.01	1.60	0.064	0.16	0.08	0.88	1.56
		(0.01)		0.02	2.33	0.087	0.13	0.05	0.76	2.30
				0.04	3.39	0.119	0.10	0.03	0.56	3.36
				0.08	4.93	0.164	0.08	0.02	0.30	4.88
10^{-3}	2	0.00366	13.4	0.0025	0.32	0.017	0.40	0.69	0.97	0.29
		(0.004)		0.005	0.46	0.023	0.33	0.41	0.94	0.44
		(0.01)		0.01	0.67	0.031	0.27	0.25	0.89	0.64
				0.02	0.98	0.042	0.21	0.15	0.78	0.95
				0.04	1.44	0.057	0.17	0.09	0.60	1.40
10^{-2}	20	0.0928	0.43	0.005	0.87	0.120	0.16	0.20	0.98	0.83
		(0.4)		0.01	1.42	0.144	0.15	0.13	0.95	1.39
		(0.1)		0.02	2.27	0.180	0.13	0.08	0.88	2.22
				0.04	3.53	0.230	0.11	0.05	0.73	3.48
				0.08	5.40	0.300	0.09	0.03	0.49	5.33
	2	0.0231	0.94	0.005	0.31	0.036	0.30	0.66	0.97	0.28
		(0.04)		0.01	0.50	0.044	0.23	0.42	0.93	0.45
		(0.0316)		0.02	0.77	0.055	0.24	0.26	0.86	0.73
				0.04	1.17	0.071	0.20	0.16	0.71	1.13
				0.08	1.78	0.094	0.16	0.10	0.46	1.70

the skew and kurtosis of the asymptotic distribution decrease with increasing rate of change of the optimum. For almost all data shown, the asymptotic distribution is very close to normal, and its variance is always larger than the Gaussian allelic approximation. Clearly, the asymptotic variance is an increasing function of k. For given k, the ratio of asymptotic variance to initial variance, $\hat{c}_2/\hat{\sigma}^2$, is smallest if the initial equilibrium distribution is close to normal, because nearly normal distributions do not change their shape much.

It is of interest to note from (7.17), and in full generality from V(2.16), that in the weak-selection approximation every set of parameters ($a\mu$, V_s/a, ak) gives rise to the same asymptotic distribution, independently of $a > 0$. Numerical results indicate that this is also true to a very good approximation for the discrete-time model (Table 7.1, and unpublished).

The above results are also valid for the distribution at a single haploid locus in an ℓ-locus model in global linkage equilibrium. Indeed, for exchangeable loci, the rate of change of the optimum realized by a single haploid locus is $k/(2\ell)$, and the assertion follows from V(2.16) because $C_1 = 2\ell c_1$.

*A multivariate Gaussian distribution of genotypes

Here we show that in the absence of mutation the multilocus dynamics under the moving-optimum model can be solved explicitly if the initial distribution of genotypes is multivariate Gaussian. Selection changes a multivariate distribution of diploid genotypes, $P(\boldsymbol{z})$, according to $P_s(\boldsymbol{z}) = P(\boldsymbol{z})W(\boldsymbol{z})/\overline{W}$, where $\boldsymbol{z} = (x_1, \dots, x_\ell, x_1^*, \dots, x_\ell^*)$ represents the genotype of a zygote (cf. Chapters V.1 and V.3.1). Under the additive model, and a Gaussian fitness function with optimum P_O and strength V_s on genotypic values, we obtain for the fitness function

$$W(\boldsymbol{z}) = \exp[-\tfrac{1}{2}(\boldsymbol{z} - \tfrac{1}{\ell}P_O\mathbf{1})\mathsf{V}_s^{-1}(\boldsymbol{z} - \tfrac{1}{\ell}P_O\mathbf{1})^\top] ,$$

where $\mathbf{1}$ denotes a vector of 1's of length 2ℓ, $\mathsf{V}_s^{-1} = V_s^{-1}\mathsf{U}$ and U is a $2\ell \times 2\ell$ matrix consisting only of 1's. Suppose that the distribution $P(\boldsymbol{z})$ is Gaussian with mean vector $\overline{\boldsymbol{Z}} = (\kappa_1, \dots, \kappa_\ell, \kappa_1, \dots, \kappa_\ell)$, where κ_j denotes the mean of the paternal and of the maternal allelic effects at locus j (Chapter V.3.1), and covariance matrix $\Sigma_{\boldsymbol{Z}}$. (Note that the mean genotypic value is $\overline{G} = 2\sum_{i=j}^{\ell}\kappa_j$ and the genetic variance is $\sigma_G^2 =$ sum of all entries of $\Sigma_{\boldsymbol{Z}}$.)

Then, it can be derived from the properties of a multivariate normal distribution that after selection the distribution is Gaussian with mean

$$\overline{\boldsymbol{Z}}_s = \overline{\boldsymbol{Z}} + \frac{P_O - \overline{G}}{\sigma_G^2 + V_s}\Sigma_{\boldsymbol{Z}}\mathbf{1} \tag{7.22a}$$

and covariance matrix

$$[\Sigma_{\boldsymbol{Z}}]_s = (\Sigma_{\boldsymbol{Z}}^{-1} + \mathsf{W}^{-1})^{-1} = \Sigma_{\boldsymbol{Z}} - \frac{1}{\sigma_G^2 + V_s}\Sigma_{\boldsymbol{Z}}\mathsf{U}\Sigma_{\boldsymbol{Z}} . \tag{7.22b}$$

This is analogous to Lande's (1980a) result (4.8) for selection on a multivariate phenotype. In terms of multivariate cumulants [V(3.4)], these relations can be written as

$$\Delta_s\kappa_j = \frac{P_O - \overline{G}}{\sigma_G^2 + V_s}\kappa_{j\cdot} , \tag{7.23a}$$

$$\Delta_s\kappa_{jk} = -\frac{1}{\sigma_G^2 + V_s}\kappa_{j\cdot}\kappa_{k\cdot} . \tag{7.23b}$$

Apart from the factor $V_s/(\sigma_G^2 + V_s)$, these equations are identical to those for the weak-selection response under quadratic selection [V(3.40)]. Recombination, however, induces departures from normality, although for a population in quasi-linkage equilibrium, these departures may be negligible if recombination is strong relative to selection.

For the distribution of breeding values, one obtains from (7.23) by summation, or by direct calculation,

$$\Delta_s \overline{G} = \frac{\sigma_G^2}{\sigma_G^2 + V_s} (P_O - \overline{G}) , \tag{7.24a}$$

$$\Delta_s \sigma_G^2 = -\frac{\sigma_G^4}{\sigma_G^2 + V_s} . \tag{7.24b}$$

For an asexually reproducing population, these equations already describe the dynamics between generations. In this case, they can be solved explicitly and, with $\overline{G}(0) = 0$, the solution is

$$\overline{G}(t) = \frac{kt(t+1)}{2V_s} \sigma_G^2(t) , \tag{7.25a}$$

$$\sigma_G^2(t) = \frac{\sigma_G^2(0)}{1 + \sigma_G^2(0)t/V_s} . \tag{7.25b}$$

Interestingly, the lag evolves, asymptotically for large t, as $kt(V_s - 1)/(2V_s) \approx \frac{1}{2}kt$, i.e., $\lim_{t\to\infty}[kt - \overline{G}(t)]/t \approx \frac{1}{2}k$. The recursion relations (7.24) can also be solved if mutation is included by assuming, as in the two-moments approximation, that the departures from normality introduced by mutation can be ignored (cf. Lynch *et al.* 1991, who introduced this moving-optimum model for the response of plankton populations to environmental change).

For sexual populations with recombination, the change of the variance between generations may differ substantially from (7.24b). Indeed, by a derivation analogous to the one leading to V(6.21), and by noting from V(3.16b) and V(3.17a) that for a multivariate Gaussian distribution $\Delta_s \kappa_{jk} = \Delta_s \kappa_{j,k}$ holds, we obtain in the case of free recombination

$$\Delta \sigma_G^2 = \tfrac{1}{2}(\sigma_{\mathrm{G,LE}}^2 - \sigma_G^2) + \tfrac{1}{2}\Delta_s \sigma_G^2 , \tag{7.26}$$

where $\sigma_{\mathrm{G,LE}}^2 = 2\sum_j \kappa_{jj}$ denotes the linkage-equilibrium variance and $\Delta_s \sigma_G^2$ is given by (7.24b). Thus, (7.26) coincides formally with Bulmer's (1971a) equation for the infinitesimal model [see also Turelli and Barton 1994, Eq. (45)].

7.4 THE LONG-TERM RESPONSE IN FINITE POPULATIONS

After a sufficiently long period of directional selection, all *initially* present genetic variation will be exhausted. If variation is not replenished by new mutations, a plateau, or selection limit, will be reached at which the response will finally cease. A well-known result about truncation selection of Robertson (1960) states that, if the population size is small so that random drift dominates selection, then with a large number of additive genes of small effects, the theoretical maximum (cumulative) response is

$$R_{\max} = 2N_e i h^2 \sigma_P . \tag{7.27}$$

This is $2N_e$ times the response in the first generation. (Unless otherwise mentioned, we assume a randomly mating diploid population with no sex differences that has constant effective size N_e.) Hill and Rasbash (1986) studied analytically and numerically how the selection limit depends on population size, selection intensity, and number, initial distribution, and effects of genes influencing the trait. In general, this dependence is rather complex, but Robertson's result is recovered as a valid approximation in the case of equal initial gene frequencies and if $N_e i\sqrt{h^2/\ell} < 0.2$, i.e., if random drift dominates selection.

Besides a lack of new mutations (and hence of genetic variation) in the direction of selection, there are several other reasons why a selection limit may be reached: for example, physiological constraints, or because natural selection acts on pleiotropically connected characters, thus causing a low viability or fertility of extreme phenotypes. A fairly comprehensive discussion of this topic may be found in Falconer and Mackay (1996, Chapter 12).

Zeng and Hill (1986) showed that truncation selection on a polygenic trait under stabilizing selection leads to a selection limit in an infinite population even if there is recurrent mutation. Further, assuming the HC-mutation model, they derived an approximation for the resulting equilibrium distribution in an infinite population. Clearly, the variance is smaller than the HC-approximation and declines with increasing strength of truncation selection. Further models in which the selection response may cease are briefly discussed in Sections 7.5 and 7.6.

However, in several experiments and on evolutionary time scales, mutations have contributed substantially to the selection response. Hill (1982a,b) derived general approximations for the response to directional selection that comes from fixation of new mutations, i.e., for the response that is obtained asymptotically for large t. He assumed that selection coefficients on alleles are related to the selection intensity i on the trait via (7.5), as is approximately true for truncation and exponential selection. In Hill's model, it is further assumed that the population is in linkage equilibrium and that the mutation rate per locus is sufficiently small that simultaneous segregation of more than two alleles can be ignored (this will be satisfied if $4N_e\mu < 1$, because the expected fixation time of a neutral allele is approximately $4N_e$ and nonneutral alleles are fixed more rapidly; see Appendix E.2). For additive genes, Hill (1982a) showed that the asymptotic response from new mutations at all loci in any generation is given approximately by

$$\Delta\bar{P}_\infty = 2N_e U \int_{-\infty}^{\infty} x\pi_{\text{fix}}(x)u(x)\,\mathrm{d}x\ , \tag{7.28}$$

where U denotes the total mutation rate, $u(x)$ the density function of mutational effects, and $\pi_{\text{fix}}(x)$ is the fixation probability of mutants of (heterozygous) effect x and, hence, selection coefficient $s = ix/\sigma_P$. If mutational effects are distributed symmetrically about zero with variance γ^2, then (7.28) can be approximated by

$$\Delta\bar{P}_\infty = 2N_e i\sigma_{\text{m}}^2/\sigma_P\ , \tag{7.29}$$

where σ_{m}^2 is the mutational variance. This is obtained from (7.28) by approximating the diffusion approximation (E.6) for π_{fix} (but here, $s > 0$ means an advantageous allele) as follows (Hill 1982a):

$$\pi_{\text{fix}} \approx \begin{cases} 2(N_e/N)s\,, & \text{if } 2N_e s > 1\,, \\ 1/(2N) + (N_e/N)s\,, & \text{if } |2N_e s| \leq 1\,, \\ 0\,, & \text{if } 2N_e s < 1\,, \end{cases}$$

where it is assumed that the initial frequency of the mutation is $1/(2N)$. From (7.29), together with (7.1b), we infer that the genetic variance approaches the asymptotic value

$$\sigma_G^2(\infty) = 2N_e\sigma_m^2 \tag{7.30}$$

if the phenotypic distribution is approximately Gaussian. Somewhat surprisingly, this asymptotic variance coincides with that of a neutral character under a balance between mutation and drift, (2.6a). In particular, the asymptotic variance is independent of the selection intensity.

If the distribution of mutations is not symmetric, and population size and mutational effects are sufficiently large, then the response can be approximated by multiplying (7.29) by the factor $2\gamma_+^2/\gamma^2$, where $\gamma_+^2 = \int_0^\infty x^2 u(x)\,dx$ denotes the mean squared effect of mutations having positive effect (Hill 1982a). The intuitive reason for this is that if random genetic drift is weak, then only mutations of positive effect on the trait contribute to the long-term response.

For completely dominant mutations and a symmetric mutation distribution, Hill (1982a) showed that $\Delta\bar{P}_\infty \approx N_e i\sigma_m^2/\sigma_P$, whereas for completely recessive mutations the asymptotic rate of response is proportional to $\sqrt{N_e i}$.

Extensive simulation studies of an additive genetic model by Hill (1982b) and Keightley and Hill (1983, 1987) showed that both under truncation and a form of artificial fertility selection, a travelling wave distribution (i.e., a stationary distribution advancing at a constant rate) is approached. For symmetric mutation distributions and given σ_m^2, the influence of linkage on the asymptotic rate of response depends only weakly on the variance and shape of the mutation distribution. The approximations (7.29) and (7.30) are quite accurate, although they tend to slightly overestimate the true asymptotic response and variance. Tight linkage leads to a significant reduction of the variance and the rate of response. Keightley and Hill (1987) investigated the influence of asymmetry and skewness of the mutation distribution and showed that, unless mutational effects or the population size is very small, it is indeed the proportion and distribution of mutants having positive effects that is crucial. In general, (7.28) provides a good approximation to their numerically observed responses, but in some cases deviations of more than 30% occur.

Hill (1982a) applied his result (7.29) to the data of Yoo (1980a), who had observed a nearly linear response between generations 50 to 80 of 0.3 bristle per generation. Substituting the data of Yoo's experiment into (7.29) yields a predicted response of 0.4 bristle per generation.

Exponential selection

We consider a finite sexual population in linkage equilibrium and assume that the distribution of genotypic values is approximately Gaussian. Then, by (7.13), the genetic variance remains unchanged under selection, and only mutation and random drift contribute to its change, i.e., $\Delta\sigma_G^2 = \sigma_{\rm m}^2 - \sigma_G^2/(2N_{\rm e})$. Thus, the variance remains constant if and only if it satisfies (7.30). It is an easy exercise to show that

$$\sigma_G^2(t) - 2N_{\rm e}\sigma_{\rm m}^2 = (\sigma_G^2(0) - 2N_{\rm e}\sigma_{\rm m}^2)e^{-t/(2N_{\rm e})} \,. \tag{7.31}$$

Hence, the asymptotic variance is approached at a geometric rate, which, however, is very low in large populations. The numerical results in Bürger (1993) show that with a Gaussian mutation distribution a travelling wave solution is indeed approached, and that its variance is approximated accurately by (7.30). This asymptotic distribution is close to Gaussian, and very little linkage disequilibrium is generated, unless recombination is very weak. Also, the approach to this travelling wave solution is well approximated by (7.31), except when the initial distribution is leptokurtic, in which case convergence is faster than predicted.

The moving optimum

Little satisfactory theory is available for this model, but extensive simulations using the computer model of Section 2.2 have provided some insight (Bürger and Lynch 1995, Bürger 1999). Before presenting some of these, as well as new numerical results, we outline some of the analytical results of Lynch and Lande (1993) and Bürger and Lynch (1995), which are based on the assumption of a Gaussian distribution of phenotypes with a constant genetic variance, and provide important qualitative insights. For a more general treatment and for more detailed proofs consult Bürger and Lynch (1995).

Let us assume discrete generations, viabilitiy selection according to (7.15) and (7.16), that on average each individual has B offspring, and that population regulation acts as described in Section 2.2. This has the effect that the population size remains constant at its carrying capacity if the multiplicative growth rate, $B\overline{W}$, is larger than one, and it decays to zero otherwise. The mean fitness, $\overline{W}$, depends on the lag of the mean behind the optimum. Assuming a normal distribution of phenotypes with constant variance, and applying Lande's (1976) analysis, it can be shown that, asymptotically for large t, the sampling distribution, ϕ, of the mean $\overline{G}$ settles down to a Gaussian travelling wave with expected value

$$\mathrm{E}[\overline{G}(t)] = kt - \frac{k(\sigma_G^2 + V_{\rm s})}{\sigma_G^2} \,, \tag{7.32}$$

and variance given by (2.9). The reader may note that for weak selection ($\sigma_G^2 \ll V_{\rm s}$) the lag agrees with the first term in (7.19a).

Evaluating the integral $\int \overline{W}(\overline{G})\phi(\overline{G})\,\mathrm{d}\overline{G}$, by using (2.13) and recalling that $\overline{P} = \overline{G}$ and $V_{\rm s} = \omega^2 + \sigma_E^2$, the expected asymptotic mean fitness (viability) can be written as

$$\mathrm{E}[\overline{W}_{\rm move}] = \frac{\omega}{\nu}\exp\left[-\frac{k^2(\sigma_{\rm move}^2 + V_{\rm s})^2}{2\nu^2\sigma_{\rm move}^4}\right] \,. \tag{7.33}$$

Here, $\sigma_{\rm move}^2$ denotes the expected genetic variance of the asymptotic distribution of genotypic values, and $\nu^2 = V_{\rm s}[1 + 1/(2N_{\rm e})] + \sigma_{\rm move}^2$, where the term $V_{\rm s}/(2N_{\rm e})$ arises from variation in $\overline{P}$ [cf. (2.9)]. Comparison with (2.14a) shows that ω/ν is the mean fitness for a resting optimum ($k = 0$) if the mean coincides with it ($\overline{G} = P_O$).

Lynch and Lande (1993) identified a critical rate of environmental change, k_c, beyond which extinction of the population is certain, because the lag increases so far that the expected multiplicative growth rate decreases below one, i.e., k_c is defined such that $B\,\mathrm{E}[\overline{W}_{\text{move}}] = 1$. For lower growth rates, the population will decline, whereas for higher growth rates, it will stay at the carrying capacity. If the population size is not too small and stabilizing selection neither extremely strong nor very weak, so that we can use the approximations $\omega^2 \approx V_s \approx V_s + \sigma^2_{\text{move}} \approx \nu^2$, then a simple calculation yields the approximation

$$k_c \approx \sigma^2_{\text{move}} \sqrt{2(\ln B)/V_s} \tag{7.34}$$

for the critical rate of environmental change. Without these approximations a somewhat more complicated expression is obtained. Equation (7.34) is deceptively simple because the genetic variance σ^2_{move} depends on several demographic and genetic parameters of the model, e.g., on k, V_s, μ, ℓ, and N_e, as well as on recombination rates and the reproductive mode. However, it clearly demonstrates that the genetic variance is the major limiting factor for the rate of environmental change that can be tolerated. If the numerically observed variance is plugged into (7.34), a reasonably accurate approximation is obtained (Bürger and Lynch 1995).

Numerical investigations based on the model of Section 2.2 have shown that populations that are initially in equilibrium between mutation, stabilizing selection, and random drift respond to a moving optimum not only by shifting their mean, but also by an increase of genetic variance, except when their size is small (say $N_e < 100$) or stabilizing selection is very weak. If the rate of change is such that the population survives for a long time and achieves a steady-state distribution that advances in parallel with the optimum, then the asymptotic variance is always below the neutral expectation (2.6), which is the asymptotic variance for truncation and exponential selection (see above). This is not surprising, because under the moving-optimum model, the stabilizing-selection component tends to erode the genetic variance. Moreover, for small populations only a small increase of variance can be expected, because their initial variance is close to the neutral prediction and approximately given by the SHC-approximation (2.8) or the stochastic version of the Gaussian, or Fleming's, approximation. In general, however, the increase is very difficult to predict.

For various levels of recombination, Figure 7.4 displays the functional relationship between the rate of environmental change k and the mean time to extinction, the asymptotic mean fitness, and the asymptotic genetic variance. This relationship is also depicted for two asexually reproducing diploid populations, one with a ten times higher growth rate B. The figure clearly demonstrates that an increase in the level of recombination leads to an increase of the asymptotic variance, of the mean fitness, and of the mean time to extinction of a population. Also, the critical rate of environmental change is increased. Drawing the recombination frequencies from an exponential distribution leads to a slightly reduced variance and mean fitness relative to the case of identical recombination frequencies (results unpublished). Asexually reproducing, but otherwise equivalent, populations have a lower mean fitness and a lower mean time to extinction than nonrecombining, but (segregating) sexual populations, even if their growth rate is much higher.

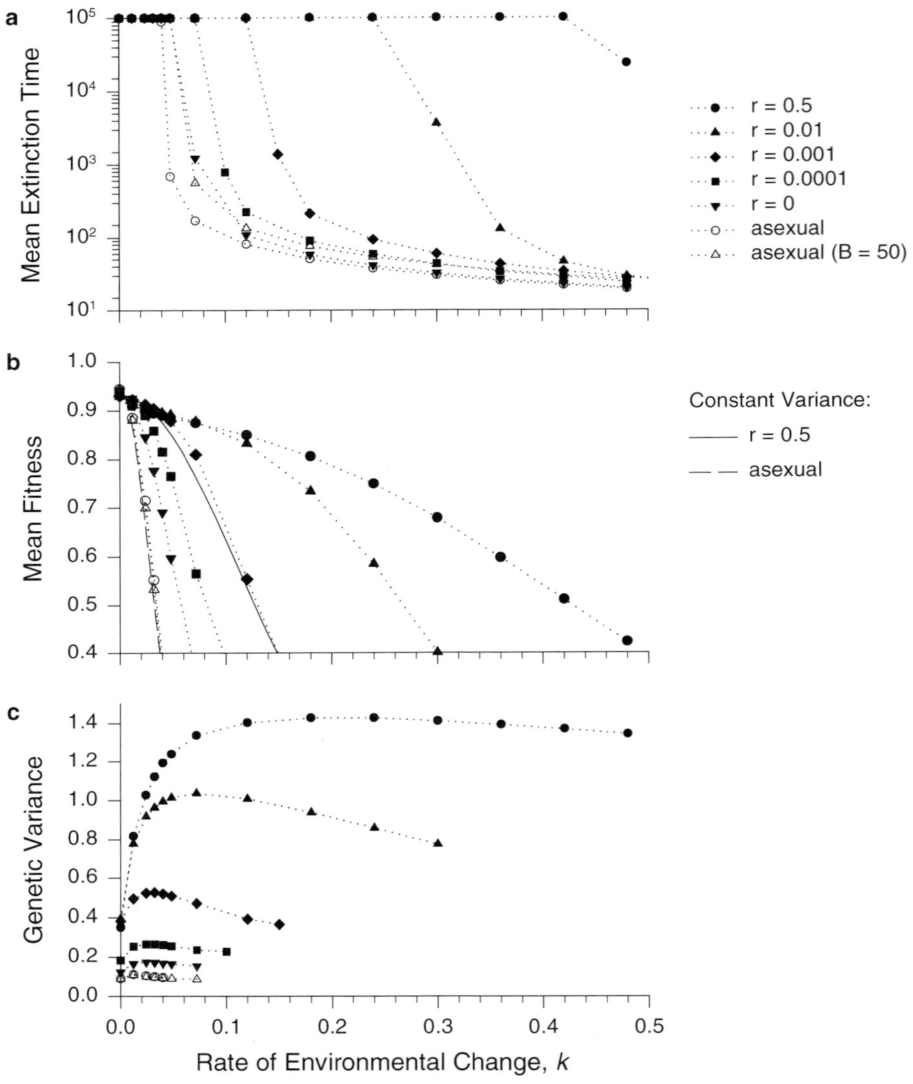

Figure 7.4 Evolution and extinction under the moving-optimum model. For all data, the genomic mutation rate is $U = 0.02$, the number of diploid loci is $\ell = 50$, the mutational variance is $\sigma_{\rm m}^2 = 0.001$, and the mutation distribution is normal with mean zero and variance $\gamma^2 = 0.05$. The strength of stabilizing selection is $V_{\rm s} = 10$, the effective population size is $N_{\rm e} = 2276$. Figure **a** displays the mean time to extinction as a function of the rate of environmental change k, Figure **b** displays the asymptotic mean fitness, and Figure **c** the average asymptotic genetic variance. Filled symbols are for sexually reproducing populations that differ in the level of recombination. Open symbols are for mutationally equivalent diploid asexual populations, one with a growth rate of $B = 5$, as the sexuals, the other with $B = 50$. The solid and dashed lines in the middle panel are obtained from (7.33) by assuming that the variance remains at its initial equilibrium value during evolution. The solid line is for the freely recombining population (thus, it should be compared with the filled circles), the dashed line is for the asexual populations (to be compared with the open symbols).

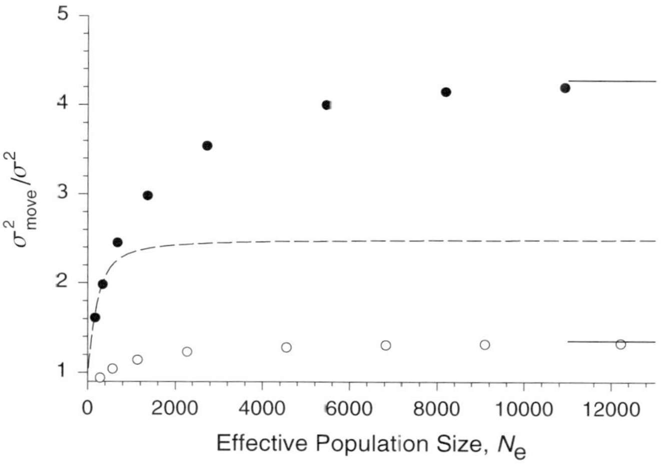

Figure 7.5 Genetic variance in the moving-optimum model as a function of population size. The mutational parameters are as in Figure 7.4, the filled symbols are for a sexual, freely recombining population ($V_s = 10$, $k = 0.04$, $B = 2$), the open symbols for an asexual population ($V_s = 10$, $k = 0.012$, $B = 5$). The lowest population size is $K = 2^7$ ($N_e \approx 170$). For populations less than $N_e \approx 150$, rapid extinction occurs for these parameter values. The lines on the right-hand side are for $N_e = \infty$ and calculated by assuming linkage equilibrium and extrapolating from the asexual values whose numerical generation is described Section 7.3. The dashed line represents the ratio of the stochastic Gaussian approximation (2.5) and the SHC-approximation, for the freely recombining population, as is suggested by (7.20) to be a lower bound for the expected increase.

In particular, the mean fitness and the mean time to extinction depend strongly on the asymptotic variance σ^2_{move}, as is also shown by the theoretical predictions (solid and dashed lines in Figure 7.4b), which are calculated by assuming that the variance does not increase, but remains at its initial equilibrium value. The asymptotic distribution of breeding values is very close to normal under almost any parameters that have been examined, and (7.33) produces a very accurate estimate of the average fitness if the average observed value of the variance is substituted into this formula. This holds despite the (small) deviations from normality and despite the numerically observed fact that the variance of the mean is much larger than the value (2.9), which is predicted by the Gaussian theory (results not shown, but see Bürger and Lande 1994 for some data). The shape of the mutation distribution (low or high kurtosis) appears to have little effect on the response to the moving optimum (Bürger and Lynch 1997).

Figure 7.5 shows how the increase of variance depends on population size. As expected from the theoretical and numerical results for asexual, infinite populations (Table 7.1), the stochastic version of the multilocus Gaussian approximation provides a lower bound for the asymptotic variance in freely recombining populations. For small populations, it provides a good approximation. Both statements are true for all other simulations that have been performed, but will not be valid for extremely small values of k, because for a resting optimum the equilibrium variance is lower than the stochastic Gaussian approximation.

All these simulations, as well as those of Keightley and Hill (1983, 1988), show high fluctuations between generations of the mean phenotype, the genetic variance, and other quantities, and substantial autocorrelation (see e.g., Bürger and Lynch 1995). Therefore, standard errors of average genetic variances and other quantities can be kept on the order of one percent of the mean (as in the figures) only by averaging over several or many (depending on population size) replicate runs, each lasting for 10^5 generations.

The above results clearly demonstrate that sexual reproduction and a high level of recombination may provide a substantial advantage to a population in comparison with, otherwise equivalent, sexual populations with restricted recombination, and with asexual populations (even if they have a much higher growth rate, far in excess of the so-called two-fold advantage of asexual reproduction). It is likely that this would lead to efficient selection of modifiers favoring higher recombination rates (cf. Charlesworth 1993, who investigated this issue for a Gaussian phenotypic model with a constant genetic variance).

For a periodically changing optimum similar phenomena occur, provided the period of the cycle is not too short and the amplitude not too small (Kondrashov and Yampolsky 1996a,b, Bürger 1999). For periodical changes with a short period (< 50) and a small amplitude ($< 2\omega$), even in large, freely recombining populations, little or no increase of variance occurs. Indeed, a Gaussian phenotypic theory shows that under such conditions an increase of variance does not lead to an increase of mean fitness (see Lande and Shannon 1996, Bürger 1999). Similarly, if in each generation the position of the optimum is drawn randomly from a normal distribution, an increase of variance is neither predicted by theory nor observed in simulations, except when the fluctuations are very large (Charlesworth 1993, Bürger 1999) or autocorrelated (Charlesworth 1993, Lande and Shannon 1996). In populations with age structure and overlapping generations, these conditions are less stringent (Ellner 1996). In such unpredictably varying environments, increasing the growth rate is a much more efficient strategy to enhance population persistence (Bürger 1999).

For analyses and discussions of various issues of the moving-optimum model with respect to population persistence and extinction risk, we refer to Lynch and Lande (1993), Huey and Kingsolver (1993), Bürger and Lynch (1995, 1997), and Bürger and Krall (2000).

Summarizing, it may be concluded that an increase of genetic variance occurs if there are prolonged periods of directional selection, so that adaptation is essential, if the population size is sufficiently large, and if there is enough recombination between the loci affecting the trait under selection. Therefore, directional selection may be an important mechanism in elevating levels of genetic variation beyond those predicted by mutation-selection-balance models, in particular, if it is taken into account that such selection could also increase genetic variability in pleiotropically connected traits, and that selection may act on different traits during differing time epochs.

7.5 MULTIVARIATE PHENOTYPES AND CORRELATED RESPONSES

Lande (1979, 1980b, 1982) developed a theory for the dynamics of multivariate Gaussian phenotypes that extends the classical univariate theory. This theory has had a huge impact on evolutionary biology, because it initiated the integration of quantitative-genetic models and methods into evolutionary genetics. Since a proper representation is beyond the scope of this book, we shall briefly outline the basic features of this approach and give some hints to the literature.

As in Section 4.1, we consider an effectively infinite, randomly mating diploid population with discrete generations and indistinguishable sexes. The phenotype, $\boldsymbol{P}$, of an individual is characterized by a vector of measurements of n quantitative traits, $\boldsymbol{P} = (P_1, \dots, P_n)^\top$, its genetic and environmental component by $\boldsymbol{G} = (G_1, \dots, G_n)^\top$ and $\boldsymbol{E} = (E_1, \dots, E_n)^\top$, respectively, and $\boldsymbol{P} = \boldsymbol{G} + \boldsymbol{E}$. The distributions of $\boldsymbol{G}$ and $\boldsymbol{E}$ are assumed to be mutually independent and multivariate normal (on an appropriate scale) with mean vectors $\overline{\boldsymbol{G}}$ and $\boldsymbol{0}$, respectively, and non-singular covariance matrices $\Sigma_{\boldsymbol{G}}$ and $\Sigma_{\boldsymbol{E}}$. Then the phenotype distribution, $f_P(\boldsymbol{P})$, is multivariate normal with mean $\overline{\boldsymbol{P}} = \overline{\boldsymbol{G}}$ and covariance matrix $\Sigma_{\boldsymbol{P}} = \Sigma_{\boldsymbol{G}} + \Sigma_{\boldsymbol{E}}$ [cf. (4.1) and (4.2)].

If the fitness of individuals with phenotype $\boldsymbol{P}$ is $W_P(\boldsymbol{P})$, then the mean fitness of the population is $\overline{W} = \int f_P(\boldsymbol{P}) W_P(\boldsymbol{P})\, \mathrm{d}\boldsymbol{P}$. After selection, the mean vector is

$$\overline{\boldsymbol{P}}_s = \overline{W}^{-1} \int \boldsymbol{P} f_P(\boldsymbol{P}) W_P(\boldsymbol{P})\, \mathrm{d}\boldsymbol{P} \,,$$

and the vector of selection differentials is $\boldsymbol{S} = \Delta_s \overline{\boldsymbol{P}} = \overline{\boldsymbol{P}}_s - \overline{\boldsymbol{P}}$. Lande (1979) showed that the gradient of the logarithm of the mean fitness can be written as

$$\nabla \ln \overline{W} = \Sigma_{\boldsymbol{P}}^{-1} \boldsymbol{S} = \Sigma_{\boldsymbol{G}}^{-1} \Delta \overline{\boldsymbol{P}} \,, \tag{7.35}$$

where the partial derivatives in the selection gradient, $\nabla \ln \overline{W}$, are taken with respect to $\overline{P}_1, \dots, \overline{P}_n$. Therefore, the evolution of the mean phenotype vector has the dynamics of a generalized gradient system, i.e.,

$$\Delta \overline{\boldsymbol{P}} = \Sigma_{\boldsymbol{G}} \nabla \ln \overline{W} \tag{7.36a}$$

$$= \Sigma_{\boldsymbol{G}} \Sigma_{\boldsymbol{P}}^{-1} \boldsymbol{S} \,. \tag{7.36b}$$

This is the multivariate generalization of the breeder's equation (7.1a).

Under weak selection, (7.36) can be approximated by the differential equation

$$\frac{\mathrm{d}}{\mathrm{d}t} \overline{\boldsymbol{P}} = \Sigma_{\boldsymbol{G}} \nabla \ln \overline{W} \,, \tag{7.37}$$

which also describes the dynamics in a population with overlapping generations and a stable age distribution if $\ln \overline{W}$ is interpreted as Malthusian fitness (Lande 1982). The reader may observe the structural similarity between (7.37) and I(10.14). As noted by Lande, (7.37) implies that mean fitness is nondecreasing,

$$\frac{\mathrm{d}}{\mathrm{d}t} \ln \overline{W} = (\nabla \ln \overline{W})^\top \, \Sigma_{\boldsymbol{G}} \, (\nabla \ln \overline{W}) \geq 0 \,. \tag{7.38}$$

Thus, Gaussian multivariate phenotypic evolution can be represented as an adaptive topography with dimensions $\overline{P}_1, \dots, \overline{P}_n$ and height $\overline{W}$. However, evolution of the vector of mean phenotypes does not proceed in the direction of steepest ascent in the Euclidean sense, which would be along the selection gradient $\nabla \ln \overline{W}$. Its direction and speed is modified by the genetic covariance matrix $\Sigma_{\boldsymbol{G}}$ (cf. Chapter I.10.3). For a

Gaussian fitness function, mean fitness is nondecreasing even under strong selection, i.e., under the discrete-time dynamics (7.36) (Lande 1979).

It follows that genetic correlations can alter the long-term result of selection by influencing the direction of evolution near saddle points in the adaptive landscape, and by reducing the rate of response to directional selection. In particular, because of genetically correlated responses to selection on other characters, a single character, k, may evolve maladaptively, i.e., in opposite direction as signified by the sign of $\partial \ln \overline{W}/\partial \overline{P}_k$. Specific models demonstrating the constraints on phenotypic evolution caused by genetic and phenotypic covariances were investigated by Wagner (1984, 1988) and Bürger (1986a,b) for a set of correlated traits, in which one is under directional selection whereas the others are under stabilizing selection.

The multivariate theory has been applied to develop methods for the measurement of natural selection, in particular, to determine which of a set of correlated characters are the targets of natural selection, and what the mode of selection is, i.e., whether it is stabilizing, directional, or disruptive (Lande and Arnold 1983; cf. Sections 1.4 and 4.2). Lande (1979, 1980b) also generalized his multivariate theory to include random genetic drift. He applied his theory to numerous problems in evolutionary biology, for instance to life-history evolution (Lande 1982), to problems of sexual selection (Lande 1981), and of speciation (Lande 1980c), to mention only some. Initiated by these methods and models, numerous empirical studies on quantitative inheritance in natural populations have been performed, especially genetic correlations between characters and the action of natural selection have been measured. For a review, see Lande (1988).

The application of this theory, in particular of the dynamical equations (7.36) and (7.37), to long-term evolution requires constancy of genetic and phenotypic variances and covariances, or at least that they change over a longer time scale than the mean vector. Little is known about that, but the genetic models investigated in this section show that variances may change quite rapidly in response to selection. In contrast to the Gaussian genetic models that have been investigated in the context of stabilizing selection, the assumption of a Gaussian phenotypic distribution is empirically testable and, in the univariate case, has received much support. The assumption of a multivariate Gaussian distribution is, of course, stronger. Lande's theory has not only initiated many new empirical and theoretical approaches to evolutionary theory, but has also been subject to intense discussion and criticism (e.g., Turelli 1988a,b; Barton and Turelli 1989). Certainly, it does have its limitations and many models in this book demonstrate this, but nevertheless it often provides the only available theoretical tool to approach problems in evolutionary biology, and it has led to many new theoretical developments, some of them covered in this book. Often it will provide a qualitatively correct guide, and predictions for the mean on short or medium time scales may often be even quantitatively correct.

7.6 SELECTION RESPONSE UNDER PLEIOTROPY

All models analyzed so far in this section have assumed that the selection on the genes affecting the trait(s) is caused only by the directional selection acting on the trait(s). If, however, these genes have pleiotropic effects on fitness, as in the models of Section 6, then the response to selection cannot be predicted by taking into account only direct selection on the trait. Unfortunately, very little is known in this case.

If the genetic variance of a trait is maintained by pleiotropic overdominance (Sections 3.1 and 6.3), then the initial response to directional selection will be similar to that in the absence of pleiotropic effects. Indeed, Gillespie (1984) showed that, in his model, initially $\Delta_s \overline{G} \approx s\sigma_G^2$ and $\Delta_s \sigma_G^2 \approx sC_3$ hold. However, a large shift in the optimum may lead to the fixation of alleles and, therefore, to a decrease in variance. In general, directional selection can overwhelm balancing selection only if its effect on the loci is stronger than that of balancing selection. As discussed in Section 6, the number of polymorphisms maintained by overdominance is unlikely to be very large. If there are not many overdominant loci contributing to the trait, then directional selection is likely to dominate, but the selection limit will be only on the order of a few genetic standard deviations, which is much less than typically observed (Barton 1990). This provides a further argument against the hypothesis that genetic variation of quantitative traits is frequently maintained as a side-effect of polymorphisms maintained by balancing selection.

If genetic variation is maintained through pleiotropic effects of unconditionally deleterious mutations (Section 6.1), then the following reasoning, due to Barton (1990), indicates that artificial selection can easily overcome the intrinsic deleterious effects, unless they are very large. According to (7.5), directional selection of intensity i on a trait induces a selection coefficient, s_*, on a mutation with average effect γ that is approximately $s_* \approx i\gamma/\sigma_P = i\sqrt{\sigma_{\mathrm{m}}^2/(2\ell\mu\sigma_P^2)}$. Since we know from (6.3) that for this model $\sigma_{\mathrm{m}}^2 = s\sigma_G^2$, where s is the deleterious mutational effect, it follows that for exchangeable loci $s_* \approx i\sqrt{h^2 s/(2\ell\mu)}$. If the heritability is not too low (i.e., $h^2 \geq 0.1$) and the total deleterious mutation rate for the trait is not too high (i.e., $2\ell\mu \leq 1$), then $s_* \geq 0.3i\sqrt{s}$. For appreciable artificial selection, this will be much higher than natural selection against deleterious alleles if the effect of the latter is on the order of a few to several percent. Therefore, the selection response should be similar to that expected if alleles have no deleterious pleiotropic effects. However, directional selection will lead to a fitness decline as deleterious mutations gain high frequencies or become fixed. Indeed, this has been frequently observed in artificial selection experiments. Unfortunately, no rigorous analysis of this interesting model has yet been performed.

Finally, we briefly discuss models with direct selection on two pleiotropically connected traits that are uncorrelated, i.e., there are hidden pleiotropic effects.

On the basis of their asexual, pleiotropic, stepwise-mutation model (see the end of Section 4.3), Slatkin and Frank (1990) explored numerically directional selection acting on a linear combination of two characters according to $W_P(P_1, P_2) = 1 + s(P_1 + P_2)$. For parameter sets such that under stabilizing selection the Gaussian allelic approximation applies (i.e., most variance is maintained by alleles of intermediate frequency and effect), they observed only a slight increase in genetic correlation after the onset of directional selection, whereas under conditions in favor of the rare-alleles approximation, a very strong increase was observed. Thus, not only the response of the variance depends on the shape of the initial distribution at individual loci, but also the response of the correlation between pleiotropically connected, but initially uncorrelated traits. In particular, this shows that in general it cannot be expected that genetic covariance matrices maintain their shape under selection.

Baatz and Wagner (1997) investigated a model of constrained pleiotropy (cf. Section 4.4), in which all genes affect two traits, one under directional selection, the other under stabilizing selection. The fitness function they employed,

$$W_P(P_1, P_2) = \exp\left(sP_1 - \frac{P_2^2}{2\omega^2}\right) ,$$

can be viewed as a ridge in direction of trait 1 with its flanks defined by stabilizing selection of width ω, and the top of the ridge at $P_2 = 0$. Baatz and Wagner explored evolution up to and along the ridge by deterministic two- and four-locus models, and by stochastic simulations of a 50-locus model, using a multivariate extension of the program described in Section 2.2. Their deterministic analysis shows that for sufficiently strong stabilizing selection relative to directional selection, and for sufficiently high recombination frequencies, equilibria on the ridge may be asymptotically stable that do not have maximum fitness. Populations, initially in the vicinity of such an equilibrium, are prevented from evolution uphill the ridge. The reason is that the hidden pleiotropic effects (on the ridge, the genetic and phenotypic correlations between the characters are zero) are inducing stabilizing selection on the first character and, if this is strong enough, it can overwhelm the effects of directional selection.

In finite populations, the asymptotic rate of response to exponential selection along the ridge is reduced by pleiotropic effects, as compared with the single-trait model of exponential selection, because the genetic variance of P_1 is lower than the single-trait asymptotic variance $2N_e\sigma_m^2$ (7.30). The reason is that the pleiotropic effects translate any increase of variance in P_1 into an increase of variance in P_2. Since P_2 is under stabilizing selection, such an increase of variance is selectively unfavorable and an intermediate asymptotic variance results. The simulations of Baatz and Wagner show that for some mutation parameters, and a range of selection parameters ($s = 0.1, 0.2, 0.4$, $\omega^2 = 9, 99$) and population sizes (20, 100, 200), the rate of response along the ridge is reduced by 50% or less below the single-trait rate of $2N_e\sigma_m^2 s$ (7.10). However, if stabilizing selection is very strong ($\omega^2 = 1$), then the rate of response is less than 50% of the single-trait case and drops down to 4% if directional selection is weak ($s = 0.1$) and population size large ($N = 200$). These results demonstrate that hidden pleiotropic effects may lead to 'adaptive inertia', even in the absence of genetic and phenotypic correlations (Baatz and Wagner 1997).

8. CONCLUSIONS

As the amount of genetic variation differs greatly between characters in each single population, as well as between homologous characters in different populations or species, a monocausal explanation for the maintenance of genetic variation cannot be expected. Such differences may occur because characters vary in the amount of generated mutational variability, or because the mutational variability is differentially transformed into existing variation. The amount of new genetic variability may be higher for traits with complex functional architecture, which are larger mutational targets, and for traits that are expressed later in life, because they inherit variation from earlier stages (Houle 1998). Also, canalization and decanalization, i.e., selection that reduces or increases the mutational effects might play a role (Stearns and Kawecki 1994, Wagner *et al.* 1997), although the data reviewed by Houle *et al.* (1996) and

Houle (1998) strongly support the mutational target hypothesis as an important factor responsible for variation in genetic variances of quantitative traits. In particular, as noted in Section 1, life-history traits have lower heritabilities than morphological traits, but much higher additive genetic coefficients of variation and lower ratios $\sigma_G^2/\sigma_{\rm m}^2$.

The models presented in this chapter may serve to evaluate how much genetic variability is maintained for a given mutational input. Thus, they may reveal the limits of certain evolutionary and genetic forces in determining heritable variation. The investigated models can be classified according to four different hypotheses on which they are based. (i) Quantitative variation is maintained by a balance between direct selection on one or several traits and recurrent mutation affecting the traits. (ii) Quantitative variation is a mere side effect of a balance between pleiotropic mutations affecting the trait and their unconditionally deleterious fitness effects. (iii) Most variation is maintained by a balance of selective forces acting on the trait either directly or through pleiotropic side effects; (iv) Populations are not at equilibrium, but are subject to periods of directional selection, during which frequent selective sweeps lead to elevated genetic variances. Clearly, these hypotheses are not mutually exclusive, but models incorporating more than one may be analytically intractable.

Both direct and pleiotropic mutation-selection-balance models predict that in infinitely large populations, $\sigma_G^2/\sigma_{\rm m}^2 = 1/s_{\rm T}$, where $s_{\rm T}$ is the total deleterious effect of the average mutant [cf. (6.8)]. As discussed in Section 2.4, a balance between direct stabilizing selection and mutation can account simultaneously for the low observed values of $\sigma_G^2/\sigma_{\rm m}^2$ and reasonably high values of heritability only if most variability is due to mutations of large effect. If several pleiotropically related traits are under direct stabilizing selection, then predicted heritabilities are much lower than in the univariate case. Models in which variation is maintained by recurrent deleterious mutations that have pleiotropic side effects on an otherwise neutral trait can more easily explain values of $\sigma_G^2/\sigma_{\rm m}^2$ of 100 or below, provided average deleterious effects are on the order of one or a few percent. They cannot, however, explain the frequently observed strong stabilizing selection. The observed low values of $\sigma_G^2/\sigma_{\rm m}^2$ (mostly between 25 and 500) and the high mutational variances ($10^{-3} \leq h_{\rm m}^2 \leq 10^{-2}$) suggest that most mutants are under strong selection, and some experimental findings are consistent with that (e.g., Nuzhdin *et al.* 1995). However, most experiments cannot detect mutants of very small effect (on either the trait or fitness), although they may constitute a large proportion of all mutations.

The numerical example in Section 6.2 suggests a scenario in which the heritability of an additive trait under direct stabilizing selection is on the order of $\frac{1}{3}$ and, simultaneously, the ratio $\sigma_G^2/\sigma_{\rm m}^2$ is less than 100. It involves two 'classes' of mutations: one with large effects on the trait and pleiotropic deleterious effects on fitness of appreciable magnitude, and a larger class of mutations with small effects both on the trait and fitness. These considerations indicate that a combination of direct and pleiotropic mutation-selection balance may be a viable mechanism for maintaining a substantial fraction of the commonly observed variation. This requires that only one or a few of a set of pleiotropically connected traits are under direct selection, and that a large proportion of mutations has only a small effect on the traits and, especially, on fitness. In terms of detectable mutational heritability, this proportion could be vanishingly small. Clearly, further analyses of such models would be highly desirable.

Several of the mechanisms involving balancing selection that were discussed in

Section 3 have the potential to maintain high levels of genetic variation. However, with a balance of selective forces, much variation can be maintained with a very low (or even no) input of mutation, so that $\sigma_G^2/\sigma_{\mathrm{m}}^2$ could be arbitrarily large. This, together with the fact that often special conditions have to be fullfilled for much variation to be maintained, suggests that balancing selection can be ruled out as a widespread explanation for quantitative variation.

There are additional reasons why it would be important to know how large the fraction of mutations contributing to quantitative variation is that have unconditionally deleterious effects. Such mutations cannot easily be used for adaptation. However, if there is a large class of mutations of very small effect that may remain undetected in the laboratory, these could contribute substantially to the existing genetic variation, as in the above-mentioned numerical example, and they could provide the raw material for adaptive evolution. Indeed, as shown in Section 7, a sustained input of new mutations can lead to a large long-term response to directional selection.

Most models designed for explaining the amount of quantitative variation assume that populations are close to equilibrium. However, is this assumption justified? Given the many changes which have continuously occured in our biosphere, in particular, the frequent and rapid climate changes, it is likely that populations often experience directional selection. During such a period, a substantial increase in genetic variation may occur if there is a sustained supply of favorable mutations (Section 7). After reaching a new optimum, the genetic variance decays only slowly, so that a new period of selection, possibly on a pleiotropically connected trait, might act again to increase the genetic variance. Clearly, this is speculative, because of the lack of theoretical studies, and the scarcity of data about the patterns of selection caused by environmental change and about the fraction of potentially adaptive mutations.

VIII

Appendix

A STABILITY IN DYNAMICAL SYSTEMS

The study of the evolution of gene frequencies requires the mathematical investigation of difference or differential equations that describe gene-frequency change across generations. Here, we summarize the basic concepts needed for exploring the dynamical and equilibrium properties of solutions of such equations. A highly recommended elementary introduction to the theory of differential equations and dynamical systems is the book by Hirsch and Smale (1974). A concise introduction, focused on stability of difference and differential equations, is LaSalle's (1976) text. Hofbauer and Sigmund (1998) develop the theory of dynamical systems hand in hand with topics from evolutionary biology.

1. DIFFERENCE EQUATIONS

Let $\mathbf{R}^k$ denote the k-dimensional Euclidean space and let X be a subset of $\mathbf{R}^k$. A *discrete dynamical system* consists of a map T of X into itself, $\boldsymbol{x} \to T\boldsymbol{x}$, where $\boldsymbol{x}$ is a vector in X. We shall call X the state space. The map T can be iterated, and the sequence $\boldsymbol{x}$, $T\boldsymbol{x}$, $T^2\boldsymbol{x} = T(T\boldsymbol{x}), \ldots, T^n\boldsymbol{x}, \ldots$, is called the *orbit*, or *trajectory*, of $\boldsymbol{x}$. The vector $\boldsymbol{x}$ may be interpreted as the initial state and n as the number of time intervals elapsed. We assume that T is differentiable. Associated with the map $\boldsymbol{x} \to T\boldsymbol{x}$ is the *difference equation*, or *recursion relation*,

$$\boldsymbol{x}' = T\boldsymbol{x} \,, \tag{A.1}$$

which stands for $\boldsymbol{x}(n+1) = T(\boldsymbol{x}(n))$.

Of primary interest is the behavior of $T^n\boldsymbol{x}$ for large values of n. A point $\boldsymbol{y}$ is a *limit point* (accumulation point) of $T^n\boldsymbol{x}$ if there is a sequence of integers n_i ($n_i \to \infty$) such that $T^{n_i}\boldsymbol{x} \to \boldsymbol{y}$. The *$\omega$-limit* $\omega(\boldsymbol{x})$ of (the orbit $T^n\boldsymbol{x}$ of) $\boldsymbol{x}$ is the set of all limit points of $T^n\boldsymbol{x}$. An orbit $T^n\boldsymbol{x}$ is called *periodic* (or *cyclic*) if for some $k > 0$, $T^k\boldsymbol{x} = \boldsymbol{x}$. The least such integer is the *period* of the cycle. A point $\boldsymbol{x}$ such that $T\boldsymbol{x} = \boldsymbol{x}$ is called an *equilibrium*, a *fixed point*, or a *stationary state*. Frequently, ω-limits consist of periodic orbits or equilibria. To describe the basic properties of ω-limits, we need two further notions. A set H is said to be *positively invariant* if $T(H) \subseteq H$, and *invariant* if $T(H) = H$. Finally, a closed invariant set H is said to be *invariantly connected* if it is not the union of two nonempty disjoint closed invariant sets. An invariant set with a finite number of elements is invariantly connected if and only if it is a periodic motion.

ω-limits have the following two important properties (LaSalle 1976).

Theorem A.1 1. *Every ω-limit is closed and positively invariant.*
2. *If $T^n\boldsymbol{x}$, $n \geq 1$, is bounded, then $\omega(\boldsymbol{x})$ is nonempty, compact, invariant, invariantly connected, and is the smallest closed set that $T^n\boldsymbol{x}$ approaches as $n \to \infty$.*

This implies, for example, that if $T^n\boldsymbol{x}$ approaches a set with finitely many elements, $\omega(\boldsymbol{x})$ is a periodic set. In particular, if $T^n\boldsymbol{x}$ converges (i.e., to a single point $\boldsymbol{y}$), then $\boldsymbol{y} = \omega(\boldsymbol{x})$ is an equilibrium.

The central tool for obtaining information about the location of ω-limits are *Lyapunov functions.* Let $V : X \to \mathbf{R}$ and define $\Delta V(\boldsymbol{x}) = V(\boldsymbol{x}') - V(\boldsymbol{x})$. This can be computed without knowing the solution of (A.1). The function V is called a Lyapunov function of (A.1) on a subset $Y \subseteq X$ if (i) V is continuous in $\boldsymbol{x}$ and (ii) $\Delta V(\boldsymbol{x}) \leq 0$ for all $\boldsymbol{x} \in Y$. (Condition (ii) could be replaced by $\Delta V(\boldsymbol{x}) \geq 0$.) We denote by $\overline{Y}$ the closure of Y, i.e., the smallest set containing Y and all of its accumulation points. Now we can formulate an extended version of Lyapunov's invariance principle (LaSalle 1976).

Theorem A.2 *If V is a Lyapunov function of (A.1) on Y and if $T^n\boldsymbol{x}$ is bounded and in Y for every $n \geq 0$, then $\omega(\boldsymbol{x})$ is contained in the maximal invariant subset of $\{\boldsymbol{y} \in \overline{Y} : \Delta V(\boldsymbol{y}) = 0\}$. In particular, there is a number c such that $V(\boldsymbol{y}) = c$ for all $\boldsymbol{y} \in \omega(\boldsymbol{x})$.*

Next we consider questions of stability. A set H is said to be *stable* if, given a neighborhood U of H (i.e., an open set containing the closure $\overline{H}$), there is a neighborhood W of H such that $T^n(W) \subseteq U$ for every $n \geq 0$. This means that any orbit through W remains in U. This is a fairly weak concept of stability. A stronger and more important notion of stability is that of asymptotic stability. First, a set H is an *attractor* if there is a neighborhood U of H such that $x \in U$ implies $\omega(\boldsymbol{x}) \subseteq \overline{H}$. The set H is said to be *asymptotically stable* if it is stable and an attractor. H is called *globally asymptotically stable* if $T^n\boldsymbol{x} \to \overline{H}$ as $n \to \infty$ for all $x \in X$. The *basin of attraction* of a set H is the set of all $\boldsymbol{x}$ such that $\omega(\boldsymbol{x}) \in H$. The following criterion shows how a Lyapunov function can be used to prove (global) asymptotic stability and to obtain estimates for the basin of attraction.

Theorem A.3 *Let Y be a bounded open positively invariant set in X, and denote by M the largest invariant set in $\{\boldsymbol{y} \in \overline{Y} : \Delta V(\boldsymbol{y}) = 0\}$. If* (i) *$V$ is a Lyapunov function of (A.1) on Y and* (ii) *$M \subset Y$, then M is an attractor and the basin of attraction contains $\overline{Y}$. If, in addition,* (iii) *V is constant on M, then M is asymptotically stable and globally stable relative to Y.*

We notice that condition (iii) is automatically satisfied if M is a single point or if M is an invariantly connected set with a finite number of elements. The main difficulty in applying the above results consists, of course, in finding an appropriate Lyapunov function. Asymptotic stability can also be inferred by linear approximation of (A.1). Suppose that $\hat{\boldsymbol{x}}$ is an equilibrium of (A.1) and denote by A the $k \times k$ matrix that is the linear approximation to T at $\hat{\boldsymbol{x}}$. Thus, $\mathsf{A} = D_{\boldsymbol{x}}T(\hat{\boldsymbol{x}})$ is the *Jacobian matrix*[1] of T,

[1] For a continuously differentiable map $f : \mathbf{R}^k \to \mathbf{R}^m$ the Jacobian (matrix) $D_{\boldsymbol{x}}f(\boldsymbol{y})$ is the $m \times k$ matrix of first-order partial derivatives $\partial f_i/\partial x_j$ of $f = (f_1, \ldots, f_m)$ evaluated at $\boldsymbol{y}$.

evaluated at $\hat{\boldsymbol{x}}$. Then (A.1) can be written as

$$\boldsymbol{x}' = \hat{\boldsymbol{x}} + \mathsf{A}(\boldsymbol{x} - \hat{\boldsymbol{x}}) + \boldsymbol{h}(\boldsymbol{x} - \hat{\boldsymbol{x}}) , \tag{A.2}$$

where $\boldsymbol{h}(\boldsymbol{x} - \hat{\boldsymbol{x}})$ is the remainder term. Denote by $r(\mathsf{A})$ the spectral radius of A (cf. Appendix B). Then the following holds:

Theorem A.4 *If $\boldsymbol{h}(\boldsymbol{x})$ is $o(\boldsymbol{x})$ as $\boldsymbol{x} \to 0$ and if $r(\mathsf{A}) < 1$ (i.e., the modulus of all eigenvalues is less than one), then $\hat{\boldsymbol{x}}$ is an asymptotically stable equilibrium of (A.2). If $r(\mathsf{A}) > 1$, then $\hat{\boldsymbol{x}}$ is unstable.*

If no eigenvalue of A has absolute value one, then the equilibrium is called *hyperbolic*. In this case, the orbits of (A.1) near an equilibrium $\hat{\boldsymbol{x}}$ look like those of the linearization $\boldsymbol{x}' = \hat{\boldsymbol{x}} + \mathsf{A}(\boldsymbol{x} - \hat{\boldsymbol{x}})$ near $\hat{\boldsymbol{x}}$. Hyperbolic equilibria are either sinks, sources, or saddles.

A powerful tool to prove that a given function is a Lyapunov function is the following result of Baum and Eagon (1967):

Theorem A.5 *Let $P(\boldsymbol{x})$ be a polynomial with nonnegative coefficients, homogeneous of degree d in its variables $x_1, \ldots, x_k$. Let $\boldsymbol{x} = (x_1, \ldots, x_k)$ be any point satisfying $x_i \geq 0$ for every i and $\sum_{i=1}^{k} x_i = 1$, and let $\boldsymbol{y}(\boldsymbol{x}) = (y_1(\boldsymbol{x}), \ldots, y_k(\boldsymbol{x}))$ be given by*

$$y_i(\boldsymbol{x}) = x_i \frac{\partial P}{\partial x_i} \Big/ \sum_{j=1}^{k} x_j \frac{\partial P}{\partial x_j} \tag{A.3}$$

Then $P(\boldsymbol{y}(\boldsymbol{x})) > P(\boldsymbol{x})$ holds unless $\boldsymbol{y}(\boldsymbol{x}) = \boldsymbol{x}$.

2. *DIFFERENTIAL EQUATIONS*

Let Y be an open set in $\mathbf{R}^k$, $f : Y \to \mathbf{R}^k$ a (sufficiently often) differentiable function, and let the state space X be contained in Y. We consider time-independent ordinary differential equations of the form

$$\frac{\mathrm{d}\boldsymbol{x}}{\mathrm{d}t} = \dot{\boldsymbol{x}} = f(\boldsymbol{x}) , \tag{A.4}$$

where $f(\boldsymbol{x}) = (f_1(\boldsymbol{x}), \ldots , f_k(\boldsymbol{x}))$. If for all $\boldsymbol{x} \in X$ and all $t \in \mathbf{R}$, the solution $\boldsymbol{x}(t)$ with $\boldsymbol{x}(0) = \boldsymbol{x}$ is defined and lies in X, then (A.4) determines a *continuous dynamical system* in X. Many of the concepts and results encountered above for difference equations possess straightforward analogues for such differential equations.

To each $\boldsymbol{x} \in X$ corresponds the *orbit* $\{\boldsymbol{x}(t) : t \in \mathbf{R}\}$. Sometimes, one considers semi-dynamical systems defined only for $t \geq 0$ and *positive semi-orbits* $\{\boldsymbol{x}(t) : t \geq 0\}$. Let $\boldsymbol{x}(t)$ be a solution of (A.4) defined for all $t \geq 0$, satisfying the initial condition $\boldsymbol{x}(0) = \boldsymbol{x}$. The ω-limit of $\boldsymbol{x}$ is the set of all accumulation points of $\boldsymbol{x}(t)$ for $t \to \infty$, i.e.,

$$\omega(\boldsymbol{x}) = \{\boldsymbol{y} \in \mathbf{R}^n : \boldsymbol{x}(t_k) \to \boldsymbol{y} \text{ for some sequence } t_k \to \infty\} . \tag{A.5}$$

The ω-limit may be empty. However, if the positive semi-orbit remains in some compact set, the ω-limit cannot be empty. Every point on the orbit of $\boldsymbol{x}$ has the same ω-limit. A set is called *(positively) invariant* if every solution which starts in it remains there for all $t \in \mathbf{R}$ (for all $t \geq 0$). α-limits and negatively invariant sets are defined in the same way, but for $t \to -\infty$. The following theorem is analogous to Theorem A.1.

Theorem A.6 1. *Every ω-limit is closed and invariant.*
2. *If $\boldsymbol{x}(t)$ remains in a compact set for all $t \geq 0$, then $\omega(\boldsymbol{x})$ is nonempty, compact, connected, invariant, and is the smallest closed set that $\boldsymbol{x}(t)$ approaches as $t \to \infty$.*

A point $\boldsymbol{x}$ is called an *equilibrium* if $\boldsymbol{x}(t) = \boldsymbol{x}$ for all $t \in \mathbf{R}$. These points are characterized by the condition $f(\boldsymbol{x}) = \mathbf{0}$. A point $\boldsymbol{x}$ is called a *periodic point* if $\boldsymbol{x}(\tau) = \boldsymbol{x}(0)$ for some $\tau > 0$, but $\boldsymbol{x}(t) \neq \boldsymbol{x}$ for all $t \in (0, \tau)$. The number τ is called the period. Such a motion describes a periodic oscillation. Equilibria and periodic orbits constitute their own ω-limit.

A continuously differentiable function $V : Y \to \mathbf{R}$ is called a *Lyapunov function* if V is nondecreasing (or nonincreasing) along orbits, i.e., if the time derivative $\dot{V}$ of the map $t \to V(\boldsymbol{x}(t))$ satisfies $\dot{V} \geq 0$ (or $\dot{V} \leq 0$). This derivative can be calculated without knowing the solutions, because

$$\dot{V}(\boldsymbol{x}) = \nabla V(\boldsymbol{x})^{\top} \dot{\boldsymbol{x}} = \sum_{i=1}^{k} \frac{\partial V(\boldsymbol{x})}{\partial x_i} f_i(\boldsymbol{x}) , \tag{A.6}$$

where $\nabla V(\boldsymbol{x})$ denotes the gradient vector; cf. I(10.10). Now we formulate a generalization of Lyapunov's invariance principle.

Theorem A.7 *Let S and Y_1 be positively invariant subsets of Y with respect to (A.4) such that $S \subset Y_1 \subseteq Y$.*
1. *If $\dot{V} < 0$ on $Y_1 \setminus S$ (thus V is Lyapunov function on $Y_1 \setminus S$), then for all $\boldsymbol{x} \in Y_1$ the ω-limit $\omega(\boldsymbol{x})$ is contained in S.*
2. *If $\dot{V} \leq 0$ on Y_1, then every such ω-limit is contained in the maximal invariant subset of $\{\boldsymbol{y} \in Y_1 : \dot{V}(\boldsymbol{y}) = 0\}$.*

Proofs and more details may be found in LaSalle (1976) and Hofbauer and Sigmund (1998) (cf. also Bürger 1983b).

For differential equations, the notions of attractor, asymptotic stability, etc. are defined in the same way as for difference equations. Thus, Theorem A.7 provides an important tool for proving global stability of equilibria relative to some subset of the state space.

In many cases the local behavior of solutions near an equilibrium $\hat{\boldsymbol{x}}$ of (A.4) can be determined by studying the approximating linear differential equation

$$\dot{\boldsymbol{x}} = \mathsf{A}\boldsymbol{x} , \tag{A.7}$$

where $\mathsf{A} = D_{\boldsymbol{x}} f(\hat{\boldsymbol{x}})$ is the Jacobian matrix of f. In an informal way, the Theorem of Hartman and Grobman states that for any equilibrium $\hat{\boldsymbol{x}}$ of (A.4) that has no eigenvalues on the imaginary axis (such equilibria are called *hyperbolic*), the orbits of (A.4) near $\hat{\boldsymbol{x}}$ look like those of (A.7) near $\mathbf{0}$. In particular, if all eigenvalues of A have (strictly) negative real part, then $\hat{\boldsymbol{x}}$ is (locally) asymptotically stable; the equilibrium $\hat{\boldsymbol{x}}$ is a *saddle* if some eigenvalues are in the left half and some in the right half of the complex plane, but none on the imaginary axis. The orbits whose ω-limit is $\{\mathbf{0}\}$ form a linear submanifold of $\mathbf{R}^k$, called the *stable manifold*; those whose α-limit is $\{\mathbf{0}\}$ form the *unstable manifold*. The local behavior of orbits of (A.4) near equilibria with eigenvalues on the imaginary axis depends on higher-order terms of the Taylor expansion of f.

In population genetics, as well as in several other fields of evolutionary biology, (systems of) differential equations occur, with the simplex S_k [cf. I(9.24)] as their state space, that are of the form

$$\dot{x}_i = x_i[f_i(\boldsymbol{x}) - \bar{f}(\boldsymbol{x})] , \quad i = 1, \ldots, k , \tag{A.8}$$

with $\bar{f}(\boldsymbol{x}) = \sum_{i=1}^{k} x_i f_i(\boldsymbol{x})$. These are called replicator equations (Hofbauer and Sigmund 1998). A typical example is the selection equation I(10.6). Equilibria of such equations are given by the conditions $x_i = 0$ or $f_i(\boldsymbol{x}) = 0$, $i = 1, \ldots, k$. An equilibrium point $\hat{\boldsymbol{x}}$ of (A.8) is called *saturated*, or *externally stable*, if $f_i(\hat{\boldsymbol{x}}) \leq \bar{f}(\hat{\boldsymbol{x}})$ whenever $\hat{x}_i = 0$. Every equilibrium in the interior of S_k is trivially saturated. For an equilibrium on the boundary, the condition means that if a missing type is introduced at low frequency it will be lost. Typical examples of nonsaturated, or externally unstable, equilibria are the boundary equilibria in the one-locus two-allele selection model with overdominance [see I(9.11) and I(9.13)]. If mutation is added to a model like (A.8), then a saturated equilibrium at the boundary will be pushed into the interior of the state space. For a boundary equilibrium $\hat{\boldsymbol{x}}$, $f_i(\hat{\boldsymbol{x}}) - \bar{f}(\hat{\boldsymbol{x}})$ is an eigenvalue of the Jacobian if $x_i = 0$. It is called a transversal eigenvalue and measures the rate of approach to the face $x_i = 0$ near $\hat{\boldsymbol{x}}$. A boundary equilibrium is saturated if and only if all its transversal eigenvalues are nonpositive (see Hofbauer and Sigmund (1998) for more results).

3. GRADIENT SYSTEMS

A differential equation of the form (A.4) is called a *gradient system* if there exists a function $V : Y \to \mathbf{R}$ with continuous partial second-order derivatives, such that

$$\dot{\boldsymbol{x}} = -\nabla V(\boldsymbol{x}) . \tag{A.9}$$

The function V is called the *potential*. Its derivative is

$$D_{\boldsymbol{x}} V(\boldsymbol{y}) = \nabla V(\boldsymbol{x})^{\top} \boldsymbol{y} . \tag{A.10}$$

By (A.6) and (A.9), this implies that the time derivative of V along orbits is given by

$$\dot{V}(\boldsymbol{x}) = \nabla V(\boldsymbol{x})^{\top} \dot{\boldsymbol{x}} = - |\nabla V(\boldsymbol{x})|^2 . \tag{A.11}$$

Hence, V is a Lyapunov function, and $\dot{V}(\boldsymbol{x}) = 0$ holds if and only if $\boldsymbol{x}$ is an equilibrium. Our next theorem summarizes the basic properties of gradient systems.

Theorem A.8 1. *At regular points ($\nabla V(\boldsymbol{x}) \neq 0$), the orbits cross the level surfaces ($V(\boldsymbol{x}) = const.$) orthogonally.*
2. *Nonregular points are equilibria.*
3. *The α- and ω-limits of any orbit are critical points (where $\nabla V(\boldsymbol{x}) = 0$) of V.*
4. *Isolated minima of V are asymptotically stable.*
5. *The differential equation $\dot{\boldsymbol{x}} = -f(\boldsymbol{x})$ is a gradient system if and only if the integrability conditions $\partial f_i/\partial x_j = \partial f_j/\partial x_i$ hold for every i, j. In particular, the Jacobian*

matrix $D_{\boldsymbol{x}} f = (\partial^2 V/(\partial x_i \partial x_j))$ *is symmetric and all eigenvalues at an equilibrium are real.*

In I(10.11)–I(10.14) it is shown that the selection equation I(10.6) can be written as the generalized gradient system

$$\dot{\boldsymbol{p}} = \tfrac{1}{2}\, \mathrm{G}_{\boldsymbol{p}} \nabla \overline{m}(\boldsymbol{p}) \,, \tag{A.12}$$

where the entries of the (genetic covariance) matrix $\mathrm{G}_{\boldsymbol{p}}$ are given by

$$g^{ij}(\boldsymbol{p}) = p_i(\delta_{ij} - p_j) \,. \tag{A.13}$$

If, more generally, $\boldsymbol{z} = (z_1, \ldots, z_k)$ is any function of $p_1, \ldots, p_k$ with Jacobian $\mathrm{A} = D_{\boldsymbol{p}}\boldsymbol{z}$ and $\tilde{\mathrm{G}}_{\boldsymbol{z}} = \mathrm{A}\mathrm{G}_{\boldsymbol{p}}\mathrm{A}^\top$, then (Barton and Turelli 1987)

$$\dot{\boldsymbol{z}} = \tfrac{1}{2}\, \tilde{\mathrm{G}}_{\boldsymbol{z}} \nabla_{\boldsymbol{z}} \overline{m}(\boldsymbol{p}(\boldsymbol{z})) \,. \tag{A.14}$$

Equation (A.12) shows that replicator equations, in particular the selection equation, are not gradient systems. However, the definition of a gradient depends on the underlying geometry, i.e., on distances and angles. From an intuitive point of view it is obvious that small changes in gene frequencies near the boundary of the simplex may have much more important consequences than small changes near the center. Actually, a small change in one of the components x_i may lead to a negative value if x_i was close to zero. Clearly, such changes are not admissible. Therefore, it is natural to replace the Euclidean geometry on the simplex by another one (Svirezhev 1972, Shahshahani 1979). For the biological relevance of the matrix $\mathrm{G}_{\boldsymbol{p}}$, we refer to Chapter II.6.6.

Let $\boldsymbol{x}$ be a point in the interior of S_k and let $\boldsymbol{x}(t)$ be a path through $\boldsymbol{x}$ with $\boldsymbol{\xi} = \dot{\boldsymbol{x}}(0)$ denoting the 'velocity'. Then, because $\sum_i x_i(t) = 1$, we have $\sum_i \xi_i = 0$. Conversely, to any vector $\boldsymbol{\xi}$ in

$$\mathbf{R}_0^k = \{\boldsymbol{\xi} = (\xi_1, \ldots, \xi_k) \in \mathbf{R}^k : \sum_{i=1}^k \xi_i = 0\} \,, \tag{A.15}$$

we can assign a path $\boldsymbol{x}(t)$ in S_k with $\dot{\boldsymbol{x}}(0) = \boldsymbol{\xi}$ by setting $\boldsymbol{x}(t) = \boldsymbol{x} + t\boldsymbol{\xi}$. Therefore, the tangent space of S_k at $\boldsymbol{x}$, which we shall denote by $T_{\boldsymbol{x}}S_k$, is the linear space $\mathbf{R}_0^k$. On this space, we define the so-called *Shahshahani inner product* by

$$\langle \boldsymbol{\xi}, \boldsymbol{\eta} \rangle_{\boldsymbol{x}} = \boldsymbol{\xi}^\top \mathrm{G}_{\boldsymbol{x}}^{-1} \boldsymbol{\eta} = \sum_{i=1}^k \frac{1}{x_i} \xi_i \eta_i \,, \quad \boldsymbol{\xi}, \boldsymbol{\eta} \in \mathbf{R}_0^k \,, \tag{A.16}$$

where the coefficients of the matrix $\mathrm{G}_{\boldsymbol{x}}^{-1}$ are defined by

$$c + \delta_{ij} x_i^{-1} \tag{A.17}$$

for some $c = c(\boldsymbol{x}) \geq 0$. The matrix defined by (A.17) is invertible for all $c \neq -1$ and, hence, positive definite. The entry (i, j) of the inverse is given by $p_i(\delta_{ij} + hp_j)$, where

$h = -c/(1+c)$. All matrices of the form (A.17) define the same inner product $\langle \boldsymbol{\xi}, \boldsymbol{\eta} \rangle_{\boldsymbol{x}}$ and are generalized inverses of $\mathrm{G}_{\boldsymbol{x}}$, i.e., $\mathrm{G}_{\boldsymbol{x}}^{-1}\mathrm{G}_{\boldsymbol{x}}\boldsymbol{\xi} = \boldsymbol{\xi}$ for every $\boldsymbol{\xi} \in \mathbf{R}_0^k$ satisfying $\sum_i \xi_i x_i = 0$.

In complete analogy to the Euclidean inner product $\boldsymbol{\xi}^\top \boldsymbol{\eta}$, the Shahshahani inner product defines the distance, $\| \ \|_{\boldsymbol{x}}$, and the angle, θ, between two vectors, $\boldsymbol{\xi}$ and $\boldsymbol{\eta}$, by $\|\boldsymbol{\xi} - \boldsymbol{\eta}\|_{\boldsymbol{x}} = \sqrt{\langle \boldsymbol{\xi} - \boldsymbol{\eta}, \boldsymbol{\xi} - \boldsymbol{\eta} \rangle_{\boldsymbol{x}}}$ and $\cos\theta = \langle \boldsymbol{\xi}, \boldsymbol{\eta} \rangle_{\boldsymbol{x}} / (\|\boldsymbol{\xi}\|_{\boldsymbol{x}} \|\boldsymbol{\eta}\|_{\boldsymbol{x}})$, respectively. It differs from the Euclidean one by attaching more weight to changes near to the boundary of S_k. Therefore, the length of a (velocity) vector is $\|\boldsymbol{\xi}\|_{\boldsymbol{x}} = (\sum_i \xi_i^2/x_i)^{1/2}$; cf. I(10.15).

By a general theorem of linear algebra, to any linear map $L : T_{\boldsymbol{x}}S_k \to \mathbf{R}$ there exists a unique vector $\boldsymbol{\ell} \in T_{\boldsymbol{x}}S_k$ such that $L(\boldsymbol{\xi}) = \boldsymbol{\ell}^\top \boldsymbol{\xi}$ for all $\boldsymbol{\xi} \in T_{\boldsymbol{x}}S_k$. By the same theorem, there also exists a unique vector $\boldsymbol{\ell}_{\boldsymbol{x}} \in T_{\boldsymbol{x}}S_k$ such that $L(\boldsymbol{\xi}) = \langle \boldsymbol{\ell}_{\boldsymbol{x}}, \boldsymbol{\xi} \rangle_{\boldsymbol{x}}$. Applying this theorem to the linear map $D_{\boldsymbol{x}}V$ and the Euclidean inner product, we obtain (A.10) with $\boldsymbol{y} \in T_{\boldsymbol{x}}S_k$. Similarly, denoting the vector $\boldsymbol{\ell}_{\boldsymbol{x}}$ corresponding to $D_{\boldsymbol{x}}V$ for the Shahshahani product by $\tilde{\nabla} V(\boldsymbol{x})$, we obtain

$$D_{\boldsymbol{x}}V(\boldsymbol{\xi}) = \langle \tilde{\nabla} V(\boldsymbol{x}), \boldsymbol{\xi} \rangle_{\boldsymbol{x}} \,. \tag{A.18}$$

An immediate consequence of (A.18), the uniqueness of the vector $\tilde{\nabla} V(\boldsymbol{x})$, and of (A.10) and (A.16), is the identity

$$\tilde{\nabla} V(\boldsymbol{x}) = \mathrm{G}_{\boldsymbol{x}} \nabla V(\boldsymbol{x}) \,. \tag{A.19}$$

These considerations yield the following result:

Theorem A.9 *If the differential equation* $\dot{x}_i = f_i(\boldsymbol{x}) = \partial V/\partial x_i$ *is a Euclidean gradient system on* $\mathbf{R}^k$, *then the replicator equation* (A.8) *is a so-called* Svirezhev–Shahshahani gradient *with potential* V, *i.e.,*

$$\dot{\boldsymbol{x}} = \tilde{\nabla} V(\boldsymbol{x}) \,. \tag{A.20}$$

In particular, for the diploid selection equation (10.6) *the potential is* $V = \frac{1}{2}\overline{m}$ *and, hence,* (10.6) *and* (10.14) *can be written as*

$$\dot{\boldsymbol{p}} = \tfrac{1}{2} \tilde{\nabla} \overline{m}(\boldsymbol{p}) \,. \tag{A.21}$$

A further example of a generalized gradient appears in Chapter III.2.3, where it is shown that the mutation-selection equation III(2.2) gives rise to such a gradient if the mutation rates satisfy the HC-condition III(2.8).

Generalized gradients with respect to other Riemannian geometries (i.e., an inner product defined as in (A.16), but with some other symmetric positive definite matrix $\mathrm{G}_{\boldsymbol{x}}$) share with Euclidean gradients the properties 1.–4. of Theorem A.8. The triangular integrability condition

$$\frac{\partial f_i}{\partial x_j} + \frac{\partial f_j}{\partial x_k} + \frac{\partial f_k}{\partial x_i} = \frac{\partial f_i}{\partial x_k} + \frac{\partial f_k}{\partial x_j} + \frac{\partial f_j}{\partial x_i} \quad \text{for every } i, j, k \tag{A.22}$$

ensures that (A.8) is a Svirezhev–Shahshahani gradient. By an argument analogous to (A.11), any potential is a Lyapunov function.

Finally, we note that the Svirezhev–Shahshahani metric is equivalent to the following metric in the interior of S_k:

$$d(\boldsymbol{p}, \boldsymbol{q}) = 2 \arccos\left(\sum_i \sqrt{p_i q_i}\right) . \tag{A.23}$$

This is shown by recognizing that the change of coordinates $p_i = y_i^2/4$ transforms S_k with the Svirezhev–Shahshahani metric into the part of the $(k-1)$-dimensional sphere of radius 2 lying in the positive orthant, equipped with the Euclidean metric. On this sphere, the distance between two points $\boldsymbol{y}, \boldsymbol{z}$ is the geodesic distance $\arccos(\sum_i y_i z_i/4)$, i.e., the length of the great circle through these points. Interestingly, (A.23) has been proposed as a measure of genetic distance by Cavalli-Sforza and Edwards (1967) (see also Jacquard 1974, Antonelli and Strobeck 1977, and Akin 1979).

Various criteria for and applications of generalized gradients, as well as proofs of the above results may be found in Hofbauer and Sigmund (1998).

B PERRON–FROBENIUS THEORY OF NONNEGATIVE MATRICES

In this appendix we summarize some important results from the spectral theory of nonnegative matrices. These were discovered by Perron and Frobenius around 1910 and are useful tools in proving existence, uniqueness, positivity, and stability of equilibrium solutions in mutation-selection models. For a more complete account of the spectral theory of nonnegative matrices, including proofs, the reader is referred to Gantmacher (1959), Schaefer (1974, Chapter I), or Seneta (1981). The latter reference contains, in particular, a detailed treatment of countably infinite matrices.

1. FINITE MATRICES

A $k \times k$ matrix $\mathsf{A} = (a_{ij})$ is called *nonnegative*, $\mathsf{A} \geq 0$, if $a_{ij} \geq 0$ for every i, j. It is called *positive*, $\mathsf{A} > 0$, if $a_{ij} > 0$ for every i, j. Similarly, a vector $\boldsymbol{x} = (x_1, \ldots, x_k)^\top$ is said to be nonnegative (positive) if $x_i \geq 0$ ($x_i > 0$) for every i.

The *spectral radius* $r = r(\mathsf{A})$ of an arbitrary matrix A is the radius of the smallest circle in the complex plain that contains all eigenvalues of A, i.e., $|\lambda| \leq r$ for all eigenvalues λ of A. It can be shown that $r = \lim_n ||\mathsf{A}^n||^{1/n}$, where $||\mathsf{A}||$ is an arbitrary norm of the matrix A, e.g., $||\mathsf{A}|| = \max_i \sum_{j=1}^k |a_{ij}|$. (Throughout this appendix, $\lim_n$ denotes the limit for $n \to \infty$.) Since the sequence $||\mathsf{A}^n||^{1/n}$ is monotone decreasing, $r \leq ||\mathsf{A}^n||^{1/n}$ holds for every $n \geq 1$. Nonnegative matrices have the following important property:

Theorem B.1 *Let $\mathsf{A} \geq 0$. Then the spectral radius r of A is an eigenvalue and there is at least one nonnegative eigenvector $\boldsymbol{x} \geq 0$ ($\boldsymbol{x} \neq \boldsymbol{0}$), i.e., $\mathsf{A}\boldsymbol{x} = r\boldsymbol{x}$. In addition, if A has an eigenvalue λ with an associated positive eigenvector, then $\lambda = r$.*

We use the notation $\mathsf{A}^n = (a_{ij}^{(n)})$ for nth powers. A nonnegative matrix A is called *irreducible* if for every pair of indices (i, j) an integer $n = n(i, j) \geq 1$ exists such that $a_{ij}^{(n)} > 0$. Now we state the *Theorem of Perron–Frobenius*.

Theorem B.2 *If* A *is irreducible, then the following hold:*
1. *The spectral radius* r *is positive and a simple root of the characteristic equation.*
2. *To* r *there corresponds a positive right eigenvector* $\boldsymbol{x} > 0$ *such that* $\mathsf{A}\boldsymbol{x} = r\boldsymbol{x}$*, and* $\boldsymbol{x}$ *is unique except for multiplication by a positive constant.*
3. *No other eigenvalue of* A *is associated with a nonnegative eigenvector.*

This theorem is sufficient to prove our existence, uniqueness, and stability results for the haploid mutation-selection model in continuous time. As noted below III(1.6), in discrete time a stronger condition than irreducibility is needed.

A nonnegative matrix A is called *primitive* if an integer $n \geq 1$ exists such that $\mathsf{A}^n > 0$. Obviously, every positive matrix is primitive, and every primitive matrix is irreducible.

Theorem B.3 *For an irreducible matrix* A *with spectral radius* r*, the following assertions are equivalent:*
1. A *is primitive.*
2. $|\lambda| < r$ *for all eigenvalues* $\lambda \neq r$ *of* A*.*
3. $\lim_n \left(r^{-1}\mathsf{A}\right)^n$ *exists.*

Concerning property 3, it is readily shown that for an arbitrary matrix A, $\lim_n \mathsf{A}^n = 0$ is equivalent to $r(\mathsf{A}) < 1$, and that $r(\mathsf{A}) > 1$ always implies that $\lim_n \mathsf{A}^n$ does not exist. If $r(\mathsf{A}) = 1$, then $\lim_n \mathsf{A}^n$ exists if and only if $r(\mathsf{A}) = 1$ is a simple root of the minimal polynomial and all other eigenvalues satisfy $|\lambda| < 1$.

A stronger and more precise statement of Theorem B.3.3. is the following:

Theorem B.4 *Let* A *be primitive with spectral radius* r *and corresponding eigenvector* $\boldsymbol{x} > 0$*. Then there exists a decomposition* $\mathsf{A} = r\mathsf{P} + \mathsf{B}$*, where* P *is a projection on the eigenspace spanned by* $\boldsymbol{x}$ *(i.e., for every* $\boldsymbol{y} \in \mathbf{R}^k$ *there is a constant* c *such that* $\mathsf{P}\boldsymbol{y} = c\boldsymbol{x}$*, and* $\mathsf{P}\boldsymbol{x} = \boldsymbol{x}$*),* $\mathsf{PB} = \mathsf{BP} = 0$*, and* $r(\mathsf{B}) < 1$*. Consequently,*

$$\lim_n \left(r^{-1}\mathsf{A}\right)^n \boldsymbol{y} = c\boldsymbol{x} + \lim_n \left(r^{-1}\mathsf{B}\right)^n \boldsymbol{y} = c\boldsymbol{x} \tag{B.1}$$

holds for all $\boldsymbol{y} \in \mathbf{R}^k$*.*

Finally, the exponential

$$e^{\mathsf{A}} = \sum_{n=0}^{\infty} \frac{1}{n!}\mathsf{A}^n \tag{B.2}$$

of an irreducible matrix A is always positive and, hence, primitive. It follows that

$$\lim_{t\to\infty} e^{-rt}e^{\mathsf{A}t}\boldsymbol{y} = c\boldsymbol{x} \tag{B.3}$$

for some constant c depending on $\boldsymbol{y}$.

2. COUNTABLE STOCHASTIC MATRICES

Most of the above results for finite matrices can be extended to infinite matrices with a countable index set, and even further (see Appendix C). However, some complications occur. For instance, the notion of primitivity has to be defined in a different way, and, in part, additional assumptions are necessary.

We say that the index i *leads* to j, $i \to j$, if there is an integer $n \geq 1$ such that $a_{ij}^{(n)} > 0$. Equivalently, there exists a chain of indices $(i, i_1, \ldots, i_l, j)$ such that $a_{ii_1} a_{i_1 i_2} \cdot \ldots \cdot a_{i_l j} > 0$. Two indices i, j are said to be *communicating*, $i \leftrightarrow j$, if $i \to j$ and $j \to i$. If $i \to i$, the *period* $d(i)$ of i is defined as the greatest common divisor of all n for which $a_{ii}^{(n)} > 0$. An index is called *aperiodic* if $d(i) = 1$. This is the case, of course, if $a_{ii} > 0$. It is easy to show that communicating indices form classes and that all indices in a communicating class have the same period.

Irreducibility of a matrix is defined as in the finite case and means that all indices are communicating. An irreducible matrix is said to be *aperiodic* (*acyclic*) if the period of any one and, hence, of each one of its indices is 1. For finite matrices it can be shown that this is equivalent to primitivity as defined above. A countable matrix is called *primitive* if it is irreducible and aperiodic.

However, even primitive infinite matrices do not necessarily possess a nonnegative eigenvector associated with the spectral radius (and then the corresponding mutation-selection equation does not have an equilibrium solution). We need to introduce the notion of *recurrence*. We restrict our attention to *stochastic* matrices, i.e., to matrices satisfying $\sum_i a_{ij} = 1$ for every j. The spectral radius of a stochastic matrix is always 1. Examples of stochastic matrices are the mutation matrices $\tilde{\mathsf{U}}$ defined in III(1.2) and III(4.3). For a stochastic matrix A, an index i is *recurrent* if and only if $\sum_{n=0}^{\infty} a_{ii}^{(n)} = \infty$. In the terminology of Markov chains, this is equivalent to saying that, starting from state i, the probability of revisiting the state i is one. Otherwise, an index is called *transient*. An irreducible stochastic matrix is called *recurrent* if one, and hence all of its indices are recurrent. Otherwise, it is called *transient*. Furthermore, a recurrent aperiodic index i is *positive recurrent* if $\lim_{n\to\infty} a_{ii}^{(n)} = \tau_i^{-1} > 0$. It is *null-recurrent* if this limit is zero. τ_i is called the mean recurrence time. Correspondingly, an irreducible matrix is called *positive recurrent* (*null-recurrent*) if any one of its indices is positive recurrent (null-recurrent). Then the following holds (e.g., Seneta 1981, Chapter 5):

Theorem B.5 1. *If* A *is recurrent and stochastic, then Theorem B.1 and statements 2. and 3. of Theorem B.2 remain valid. The spectral radius is* $r(\mathsf{A}) = 1$.

2. *If, in addition,* A *is aperiodic and positive recurrent, then* $\lim_{n\to\infty} \mathsf{A}^n$ *exists,*

$$\lim_{n\to\infty} a_{ij}^{(n)} = \tau_j^{-1} > 0 \quad \textit{for every } i, j\,, \tag{B.4}$$

and the normalized eigenvector to the spectral radius is $(\tau_1^{-1}, \tau_2^{-1}, \ldots)$.

3. *If* A *is transient or null-recurrent, then* $\lim_{n\to\infty} a_{ij}^{(n)} = 0$ *for each pair* i, j.

C SPECTRAL THEORY OF NONNEGATIVE INTEGRAL OPERATORS

The existence, uniqueness, and stability results for the mutation-selection models in Chapter IV rest on the spectral theory of nonnegative integral operators, in particular on theorems generalizing the Perron–Frobenius theory of matrices. Since the spectral theory for integral operators is much more complex than that for matrices (finite or infinite), we shall only present those results that are of immediate relevance for the proofs in Chapter IV. The best single source for the theory of nonnegative operators is probably the book of Schaefer (1974), who uses the terms positive and strictly positive instead of nonnegative and positive, respectively. However, many of the results needed in the present context are scattered throughout the literature. We shall not set out the theory in full generality, but rather restrict the presentation to nonnegative integral operators on the space L^1 of absolutely integrable functions. Compared with the case of finite (and infinite) matrices, one of the main complications is that the spectrum of an integral operator does in general not only consist of eigenvalues, but may be a closed subset in the complex plane. Therefore, compactness conditions have to be imposed on the operators occurring in the general mutation-selection model.

We begin by settling the notation and introducing the basic concepts. Let $\mathcal{X}$ denote a locally compact space endowed with a positive σ-finite measure ν. (A typical example is $\mathbf{R}^k$ with the ordinary (Lebesgue) measure.) For notational simplicity, we will always write $\int \ldots \mathrm{d}x$ instead of $\int_{\mathcal{X}} \ldots \nu(\mathrm{d}x)$. $L^1(\nu)$ is the Banach space of absolutely ν-integrable (equivalence classes of) complex-valued functions on $\mathcal{X}$ that satisfy $\|f\|_1 = \int |f(x)| \,\mathrm{d}x < \infty$. For an essentially bounded measurable function f, we define $\|f\|_\infty = \operatorname{ess\,sup}_{x\in\mathcal{X}} |f(x)|$.

Let $k(x,y) \geq 0$ be a measurable nonnegative function on $\mathcal{X} \times \mathcal{X}$, such that the function $\kappa(y) = \int k(x,y)\, dx$ is measurable and bounded on $\mathcal{X}$. Then

$$Kf(x) = \int k(x,y) f(y) \,\mathrm{d}y \tag{C.1}$$

defines a nonnegative bounded integral operator K on $L^1(\nu)$, i.e., if $f \geq 0$ then $Kf \geq 0$, and $\|Kf\|_1 \leq \|\kappa\|_\infty \|f\|_1$. The function $k(x,y)$ is called the kernel of K. If $\mathcal{X}$ is a finite or countable infinite discrete set and ν is the counting measure, then K can be interpreted as a matrix.

The *spectrum*, $\sigma(K)$, of an (arbitrary) bounded operator K on an (arbitrary) Banach space E is defined as the set $\{\lambda \in \mathbb{C} : \lambda - K \text{ is not a bijection on } E\}$ ($\lambda - K$ means $\lambda I_E - K$, where I_E is the identity operator on E). The spectrum is a compact and non-void set. The number $r(K) = \sup\{|\lambda| : \lambda \in \sigma(K)\}$ is called the *spectral radius* of K. The spectral radius satisfies the inequality

$$r(K) \leq \|K\| \,, \tag{C.2}$$

where

$$\|K\| = \sup_{\|f\|_1 = 1} \|Kf\|_1 \tag{C.3}$$

is the operator norm. If K is a nonnegative bounded operator, the spectral radius is an element of the spectrum (Schaefer 1974, V.4.1 Proposition) but, in contrast to the case of a matrix (cf. Theorem B.1), it is not necessarily an eigenvalue. The spectral radius is an eigenvalue if some power of K is compact. Additionally, to achieve uniqueness, irreducibility of K is required (as in the case of matrices). We define both concepts.

The integral operator K defined by (C.1) is *irreducible* if and only if for every measurable set $S \subseteq \mathcal{X}$, the inequalities $\nu(S) > 0$ and $\nu(\mathcal{X} \setminus S) > 0$ imply

$$\int_{\mathcal{X}\setminus S} \int_S k(x,y) \,\mathrm{d}x \,\mathrm{d}y > 0 \tag{C.4}$$

(cf. Schaefer 1974, V.6 Example 4). Condition (C.4) means that for some $f \in L^1(\nu)$, $f \geq 0$ and f living on S, the function Kf is positive on a set of positive measure disjoint from S. Obviously, any integral operator with positive kernel is irreducible. If $\mathcal{X}$ is a group and $k(x, y) = k_1(x - y)w(y)$ for some integrable nonnegative function k_1 on $\mathcal{X}$ and bounded nonnegative w (as in the random-walk mutation model), then K is irreducible if k_1 is positive in a neighborhood of zero.

A bounded operator K on a Banach space is *compact* if for every bounded sequence f_n the sequence Kf_n contains a convergent subsequence. The composition of a bounded and a compact operator is again compact. Here is a list of conditions leading to compactness of (some power of) a nonnegative integral operator on $L^1(\nu)$.

Theorem C.1 *Let K be a nonnegative bounded integral operator on $L^1(\nu)$ as defined by (C.1).*

1. *Let $k(x, y) = k_2(x) \geq 0$ for all $y \neq x$ and let $\int k_2(x)\,\mathrm{d}x < \infty$. Then K is compact.*
2. *Let $\int \operatorname{ess\,sup}_{y \in \mathcal{X}} k(x, y)\,\mathrm{d}x < \infty$. Then K^2 is compact. This condition is always satisfied if $\mathcal{X}$ is compact and $k(x, y)$ is bounded.*
3. *Let $\mathcal{X}$ be a locally compact commutative group and let $k(x, y) = k_1(x - y)w(y)$ for nonnegative k_1 and nonnegative bounded w vanishing at infinity. Then K is compact if and only if $\int k_1(x)\,\mathrm{d}x < \infty$.*
4. *Let $\mathcal{X}$ be a countably infinite discrete set and let $k(x, y) = \tilde{k}(x, y)w(y)$, where $\tilde{k}$ satisfies the hypothesis for (C.1) and w vanishes at infinity. Then K is compact.*

Proof. 1. is obvious because K is of rank one.

2. is Satz 11.9 in Jörgens (1970). Such operators are called Hille–Tamarkin operators.

3. is a special case of Corollary 3.6 of Feichtinger (1984).

4. is trivial, because the operator $Tf(x) = w(x)f(x)$ is compact on $L^1(\nu)$ if w vanishes at infinity, $\tilde{K}$ is bounded, and therefore $K = \tilde{K}T$ is compact. ◁

The following theorem is sometimes called the Theorem of Jentzsch (cf. Schaefer 1974, V.6.6 Theorem). It generalizes Theorem B.2 and forms the basis of the existence and uniqueness results in Chapters IV.2 and IV.3.

Theorem C.2 *Let K be a nonnegative integral operator on $L^1(\nu)$ as defined in (C.1). If K is irreducible and if some power of K is compact, then:*

1. *The spectral radius $r(K) > 0$ is an eigenvalue of K (actually, a pole of the resolvent) with a unique normalized eigenfunction f satisfying $f(x) > 0$ almost everywhere.*
2. *There is no other eigenvalue with a nonnegative eigenfunction.*
3. *If, additionally, for each $f \in L^1(\nu)$ there exists an integer n such that $K^n f$ is positive, then all eigenvalues $\lambda \neq r(K)$ satisfy $|\lambda| < r(K)$.*

For the proof of global stability in Theorem 3.2 of Chapter IV, a stronger result is needed, namely a generalization of Theorem B.4 (see Schaefer 1974, Corollary to V.5.4 Theorem).

Theorem C.3 *Let K satisfy the hypothesis of Theorem C.2. In addition, assume that all eigenvalues $\lambda \neq r(K)$ satisfy $|\lambda| < r(K)$. Then there is a decomposition*

$$L^1(\nu) = E_1 \oplus E_2$$

such that the spectra of the restrictions K_1 and K_2 to E_1 and E_2 of the operator K satisfy

$$\sigma(K_1) = r(K) \quad \textit{and} \quad \sigma(K_2) = \sigma(K) \setminus r(K)\,.$$

The subspace E_1 is spanned by the unique positive normalized eigenfunction of $r(K)$, and the spectral radius of K_2 satisfies $r(K_2) < r(K)$.

In addition the following results are needed in Chapter IV.3.

Lemma C.1 (Aliprantis and Burkinshaw 1980, Theorem 2.2). *Let K be a nonnegative compact operator and let the operator $\tilde{K}$ satisfy $0 \leq \tilde{K} \leq K$. Then $\tilde{K}^2$ is compact.*

Lemma C.2 (Newburgh 1951, Corollary to Theorem 3). *Denote by $\mathcal{S}$ the set of all compact subsets of $\mathbb{C}$ endowed with the Hausdorff distance. Then the mapping $K \to \sigma(K)$, from a Banach algebra of bounded operators to $\mathcal{S}$, is continuous at K if K is a compact operator.*

Lemma C.3 (Bürger 1988a, Lemma 3). *Consider the family of operators C_α defined in* IV(3.10). *Then the function $\alpha \to r(C_\alpha)$ is strictly monotone decreasing.*

D MULTIVARIATE MOMENTS AND CUMULANTS

Moments and cumulants are convenient sets of parameters for characterizing probability distributions. Here, we summarize the basic definitions, some of their properties, and the relations between them, as far as needed in the main text. For additional properties, we refer to Lederman (1980), Speed (1983, 1986), and Stuart and Ord (1994).

1. *GENERAL THEORY AND PROPERTIES*

We begin by setting up a convenient multivariate notation. Bold-face letters always denote vectors, in particular, $\boldsymbol{x} = (x_1, \ldots, x_\ell)$ denotes a vector in $\mathbf{R}^\ell$ (similarly $\boldsymbol{y}$, $\boldsymbol{\xi}$, $\boldsymbol{\eta}$), whereas $\boldsymbol{n} = (n_1, \ldots, n_\ell)$ (as well as $\boldsymbol{j}$, $\boldsymbol{k}$, etc.) are vectors of nonnegative integers, so-called multi-indices (of length ℓ), and $\mathbf{0} = (0, \ldots, 0)$. Furthermore, we set $\boldsymbol{x}^{\boldsymbol{n}} = x_1^{n_1} \cdot \ldots \cdot x_\ell^{n_\ell}$, $|\boldsymbol{n}| = n_1 + \cdots + n_\ell$, $\boldsymbol{n}! = n_1! \cdot \ldots \cdot n_\ell!$, and

$$\binom{\boldsymbol{n}}{\boldsymbol{k}} = \binom{n_1}{k_1} \cdot \ldots \cdot \binom{n_\ell}{k_\ell} = \frac{\boldsymbol{n}!}{\boldsymbol{k}!(\boldsymbol{n}-\boldsymbol{k})!} .$$

Finally, $\boldsymbol{j} \leq \boldsymbol{n}$ means $j_i \leq n_i$ for every i, and $\boldsymbol{j} < \boldsymbol{k}$ means $\boldsymbol{j} \leq \boldsymbol{k}$ but $\boldsymbol{j} \neq \boldsymbol{k}$. Then the polynomial theorem can be written as

$$(x_1 + \cdots + x_\ell)^k = \sum_{\boldsymbol{k}:|\boldsymbol{k}|=k} \frac{k!}{\boldsymbol{k}!} \boldsymbol{x}^{\boldsymbol{k}} , \tag{D.1}$$

where the summation is over all (nonnegative) multi-indices $\boldsymbol{k}$ of length ℓ satisfying $|\boldsymbol{k}| = k$. Partial derivatives of order $\boldsymbol{n}$ of a real-valued function $g(\boldsymbol{x})$ are denoted by $D^{\boldsymbol{n}} = D^{\boldsymbol{n}}_{\boldsymbol{x}}$, i.e.,

$$D^{\boldsymbol{n}}_{\boldsymbol{x}} g(\boldsymbol{x}) = D^{n_1}_{x_1} \ldots D^{n_\ell}_{x_\ell} g(\boldsymbol{x}) .$$

With this notation, the multivariate version of Leibniz's product rule reads

$$D^{\boldsymbol{n}}(gh) = \sum_{\boldsymbol{k}=\mathbf{0}}^{\boldsymbol{n}} \binom{\boldsymbol{n}}{\boldsymbol{k}} D^{\boldsymbol{k}} g \, D^{\boldsymbol{n}-\boldsymbol{k}} h , \tag{D.2}$$

and for functions $g_i = g_i(\boldsymbol{x})$, the generalized product rule,

$$D^{\boldsymbol{n}}(g_1 \cdot \ldots \cdot g_\nu) = \sum_{\boldsymbol{k}_1+\cdots+\boldsymbol{k}_\nu=\boldsymbol{n}} \frac{\boldsymbol{n}!}{\boldsymbol{k}_1! \cdot \ldots \cdot \boldsymbol{k}_\nu!} D^{\boldsymbol{k}_1} g_1 \cdot \ldots \cdot D^{\boldsymbol{k}_\nu} g_\nu , \tag{D.3}$$

is easily derived, where the sum is over all multi-indices $\boldsymbol{k}_i \geq 0$ that have the same length as $\boldsymbol{n}$ and satisfy the indicated constraint. (As usual, the summation over all partitions $\boldsymbol{k}_1 + \cdots + \boldsymbol{k}_\nu = \boldsymbol{n}$ is only over different partitions, i.e., partitions differing in the order of the $\boldsymbol{k}_i$ are identified and counted only once.)

Let $\boldsymbol{X} = (X_1, \ldots, X_\ell)$ be a random vector with probability density $p(\boldsymbol{x})$. Then the *moment generating function*, $\psi = \psi_{\boldsymbol{X}}$, is defined by

$$\psi(\boldsymbol{\xi}) = \mathrm{E}[\exp(\boldsymbol{\xi} \bullet \boldsymbol{X})] = \int \exp(\boldsymbol{\xi} \bullet \boldsymbol{x}) p(\boldsymbol{x}) \, \mathrm{d}\boldsymbol{x} , \tag{D.4a}$$

where $\boldsymbol{\xi} \bullet \boldsymbol{x} = \sum_i \xi_i x_i$ and integration is performed over all admissible types, e.g. over $\mathbf{R}^\ell$ if the X_i are real-valued continuous random variables. If the set of admissible types (values of X_i) is discrete, the integral has to be replaced by a sum. The (multivariate) *moment about zero of order* $\boldsymbol{n}$ is

$$m^0_{\boldsymbol{n}} = \mathrm{E}[\boldsymbol{X}^{\boldsymbol{n}}] = (D^{\boldsymbol{n}}_{\boldsymbol{\xi}} \psi)(\boldsymbol{0}) . \tag{D.4b}$$

Thus, provided all moments exist, the moment generating function has the Taylor-series expansion

$$\psi(\boldsymbol{\xi}) = \sum_{\boldsymbol{n} \geq \boldsymbol{0}} m^0_{\boldsymbol{n}} \frac{\boldsymbol{\xi}^{\boldsymbol{n}}}{\boldsymbol{n}!} . \tag{D.4c}$$

The (multivariate) *central moments* are defined and denoted by

$$m_{\boldsymbol{n}} = \mathrm{E}[(\boldsymbol{X} - \overline{\boldsymbol{X}})^{\boldsymbol{n}}] , \tag{D.5}$$

where $\overline{\boldsymbol{X}}$ is the mean vector of $\boldsymbol{X}$. We set $m_{\boldsymbol{0}} = m^0_{\boldsymbol{0}} = 1$, and observe that $m_{\boldsymbol{n}} = 0$ if $|\boldsymbol{n}| = 1$.

It is easy to convert central moments into moments about zero, and vice versa. Indeed, applying the (multivariate) binomial theorem,

$$\boldsymbol{x}^{\boldsymbol{n}} = [(\boldsymbol{x} - \boldsymbol{y}) + \boldsymbol{y}]^{\boldsymbol{n}} = \sum_{\boldsymbol{j} \leq \boldsymbol{n}} \binom{\boldsymbol{n}}{\boldsymbol{j}} (\boldsymbol{x} - \boldsymbol{y})^{\boldsymbol{j}} \, \boldsymbol{y}^{\boldsymbol{n}-\boldsymbol{j}} , \tag{D.6}$$

to $(\boldsymbol{X} - \overline{\boldsymbol{X}}) + \overline{\boldsymbol{X}}$ and taking the expectation, we obtain

$$m^0_{\boldsymbol{n}} = \sum_{\boldsymbol{j} \leq \boldsymbol{n}} \binom{\boldsymbol{n}}{\boldsymbol{j}} m_{\boldsymbol{j}} \, \overline{\boldsymbol{X}}^{\boldsymbol{n}-\boldsymbol{j}} . \tag{D.7}$$

The *cumulant generating function*, $\Psi = \Psi_{\boldsymbol{X}}$, is defined by

$$\Psi(\boldsymbol{\xi}) = \ln \psi(\boldsymbol{\xi}) , \tag{D.8a}$$

and the (multivariate) *cumulant of order* $\boldsymbol{n}$ is

$$c_{\boldsymbol{n}} = D_{\boldsymbol{\xi}}^{\boldsymbol{n}} \Psi(\mathbf{0}) \,. \tag{D.8b}$$

Naturally, we set $c_{\mathbf{0}} = 0$. The cumulants have the property that all but the first (the mean) are unchanged by a translation of coordinates. In view of this invariance property, they were originally called seminvariants.

The conversion of moments to cumulants, and vice versa, follows by differentiation from the identity

$$\exp\Bigl(\sum_{\boldsymbol{n}:|\boldsymbol{n}|\geq 1} c_{\boldsymbol{n}} \frac{\boldsymbol{\xi}^{\boldsymbol{n}}}{\boldsymbol{n}!} \Bigr) = \sum_{\boldsymbol{n}\geq \mathbf{0}} m_{\boldsymbol{n}}^0 \frac{\boldsymbol{\xi}^{\boldsymbol{n}}}{\boldsymbol{n}!} \,. \tag{D.9}$$

This is readily automated using a formula manipulation program, such as *Mathematica* (Wolfram 1996). Analytically, the conversion is best accomplished by first letting $\boldsymbol{n}$ be a multi-index of length ℓ whose entries are 0 or 1. Then, differentiation of (D.9) and evaluation at $\boldsymbol{\xi} = \mathbf{0}$ yields

$$m_{\boldsymbol{n}}^0 = \sum_{\nu=1}^{|\boldsymbol{n}|} \sum_{\boldsymbol{k}_1+\cdots+\boldsymbol{k}_\nu=\boldsymbol{n}} c_{\boldsymbol{k}_1} \cdot \ldots \cdot c_{\boldsymbol{k}_\nu} \,, \tag{D.10}$$

where the multi-indices $\boldsymbol{k}_i$ can be assumed to be nonzero, because $c_{\mathbf{0}} = 0$.

At several occasions, a different terminology will be convenient. Let $\boldsymbol{e}_i$ denote the vector consisting of zeros except for a one at position i. Then we use the notation $\kappa_i = c_{\boldsymbol{e}_i}$ for the mean of X_i, $\kappa_{ii} = c_{2\boldsymbol{e}_i}$ for the second cumulant (=variance) of X_i, and, more generally, κ_{ijkl} for the cumulant of order $\boldsymbol{n} = \boldsymbol{e}_i + \boldsymbol{e}_j + \boldsymbol{e}_k + \boldsymbol{e}_l$, etc. The univariate cumulant of order n will simply be denoted by c_n. Similarly, we write μ_{ij}^0 for $m_{\boldsymbol{e}_i+\boldsymbol{e}_j}^0$, etc., and m_n^0 for a univariate moment of order n. In contrast to moments, there is no simple direct procedure, such as (D.5), to obtain the cumulants.

With this terminology, and according to (D.10), the moments around zero can be expressed in terms of cumulants as

$$\begin{aligned}
\mu_i^0 &= \kappa_i \\
\mu_{ij}^0 &= \kappa_{ij} + \kappa_i\kappa_j \\
\mu_{ijk}^0 &= \kappa_{ijk} + \kappa_{ij}\kappa_k + \kappa_{ik}\kappa_j + \kappa_{jk}\kappa_i + \kappa_i\kappa_j\kappa_k \\
\mu_{ijkl}^0 &= \kappa_{ijkl} + \kappa_{ij}\kappa_{kl} + \kappa_{ik}\kappa_{jl} + \kappa_{il}\kappa_{jk} \\
&\quad + \kappa_i\kappa_{jkl} + \kappa_j\kappa_{ikl} + \kappa_k\kappa_{ijl} + \kappa_l\kappa_{ijk} \\
&\quad + \kappa_i\kappa_j\kappa_{kl} + \text{five permuted terms} \\
&\quad + \kappa_i\kappa_j\kappa_k\kappa_l \,,
\end{aligned} \tag{D.11}$$

where i, j, k, l indicate distinct loci.

The relations (D.10) and (D.11) extend to arbitrary multi-indices, simply by lumping together identical cumulants on the right-hand side. Thus, for instance, $\mu_{iii}^0 = \kappa_{iii} + 3\kappa_{ii}\kappa_i + \kappa_i^3$. For arbitrary multi-indices $\boldsymbol{n}$ in (D.10), however, the $\boldsymbol{k}_i$ in a given decomposition of $\boldsymbol{n}$ are not necessarily distinct.

From (D.10) or (D.11), the central moments are obtained formally by setting $\kappa_i = \kappa_j = \kappa_k = \kappa_l = 0$. In particular, κ_i is the mean, κ_{ij} the covariance, and κ_{ijk} the general central moment of order three.

Equation (D.10) expresses moments in terms of cumulants. Conversely, cumulants can be expressed in terms of moments by taking logarithms on both sides of (D.9) and differentiating:

$$c_{\boldsymbol{n}} = \sum_{\nu=1}^{|\boldsymbol{n}|} (-1)^{\nu-1}(\nu-1)! \sum_{\boldsymbol{k}_1+\cdots+\boldsymbol{k}_\nu=\boldsymbol{n}} m^0_{\boldsymbol{k}_1} \cdot \ldots \cdot m^0_{\boldsymbol{k}_\nu} \,. \tag{D.12}$$

Sometimes, this is taken as a definition of multivariate cumulants (cf. Speed 1983, 1986).

Cumulants have the following notable properties:

(i) Suppose that $Z = \sum_i X_i$, where $(X_1, \ldots, X_\ell)$ has multivariate cumulants $c_{\boldsymbol{n}}$. Then C_n, the nth cumulant of Z, is

$$C_n = \sum_{\boldsymbol{n}:|\boldsymbol{n}|=n} \frac{n!}{\boldsymbol{n}!} c_{\boldsymbol{n}} \,. \tag{D.13}$$

This additivity property follows from the fact that the cumulant generating function Φ_Z of Z satisfies $\Phi_Z(\zeta) = \Psi_{\boldsymbol{X}}(\zeta, \ldots, \zeta)$, and from the chain rule of differentiation. (D.13) can be rewritten using the alternative notation for the low-order cumulants; for instance,

$$C_3 = \sum_{i,j,k=1}^{\ell} \kappa_{ijk} \,. \tag{D.14}$$

(ii) Suppose that a subset of the random variables $\{X_1, \ldots, X_\ell\}$, say $\{X_1, \ldots, X_j\}$, is independent from the complement, $\{X_{j+1}, \ldots, X_\ell\}$. Then all cross cumulants are zero, i.e.,

$$c_{\boldsymbol{j},\boldsymbol{k}} = 0 \quad \text{for every } \boldsymbol{j} > \boldsymbol{0} \text{ and } \boldsymbol{k} > \boldsymbol{0} \,, \tag{D.15}$$

where $\boldsymbol{j}$ and $\boldsymbol{k}$ are multi-indices corresponding to $\{X_1, \ldots, X_j\}$ and $\{X_{j+1}, \ldots, X_\ell\}$, respectively. This follows from the definition (D.8b), because the moment generating function can be decomposed into a product of the two marginal generating functions corresponding to the two independent sets of variables. As a consequence, under the present assumption of independence, (D.13) simplifies to

$$C_n = \sum_{\boldsymbol{j}:|\boldsymbol{j}|=n} \frac{n!}{\boldsymbol{j}!} c_{\boldsymbol{j},\boldsymbol{0}} + \sum_{\boldsymbol{k}:|\boldsymbol{k}|=n} \frac{n!}{\boldsymbol{k}!} c_{\boldsymbol{0},\boldsymbol{k}} \,, \tag{D.16}$$

where $\boldsymbol{j}$ and $\boldsymbol{k}$ denote multi-indices of length j and $\ell - j$, respectively. In particular, if the random variables $X_1, \ldots, X_\ell$ are pairwise independent, then we obtain

$$C_n = \sum_{i=1}^{\ell} c_{n\boldsymbol{e}_i} = \sum_{i=1}^{\ell} c_n \,. \tag{D.17}$$

(iii) Multivariate Gaussian random variables have cumulants satisfying $c_{\boldsymbol{n}} = 0$ for every $|\boldsymbol{n}| \geq 3$. The Gaussian is the unique distribution having only a finite number of non-zero cumulants.

These properties are an enormous advantage for studying additive polygenic models. Moments share the property (D.13) with the cumulants, i.e.,

$$M_n^0 = \sum_{\boldsymbol{n}:|\boldsymbol{n}|=n} \frac{n!}{\boldsymbol{n}!} m_{\boldsymbol{n}}^0 , \tag{D.18}$$

where M_n^0 denotes the moments about zero of Z. (D.18) follows immediately from the definition of the moments and (D.1). However, neither the additivity property (ii) nor (iii) are valid for the moments.

2. MOMENTS AND CUMULANTS OF SOME PROBABILITY DISTRIBUTIONS

We summarize some important properties of distributions used throughout the main text. An important dimensionless measure for the shape of a distribution is the *kurtosis*, defined as c_4/c_2^2. The kurtosis is zero for a Gaussian distribution and is often used to quantify the departure from a Gaussian. Distributions with positive kurtosis are called leptokurtic, those with negative kurtosis, platykurtic.

The normal (Gaussian) distribution

is defined by

$$f(x) = \frac{1}{\sqrt{2\pi\gamma^2}} \exp\left\{-\frac{(x-\bar{x})^2}{2\gamma^2}\right\} , \tag{D.19}$$

where $\bar{x}$ is the mean and γ^2 the variance. The central moments of odd order vanish, and the even moments are

$$m_{2n} = \gamma^{2n} \cdot 1 \cdot 3 \cdot 5 \cdot \ldots \cdot (2n-1) = \gamma^{2n} \frac{(2n)!}{2^n n!} . \tag{D.20}$$

The cumulants are

$$c_1 = \bar{x} , \quad c_2 = \gamma^2 , \quad c_n = 0 \quad \text{for every } n \geq 3 . \tag{D.21}$$

The normal distribution is the only probability distribution with a finite number of non-zero cumulants. Obviously, its kurtosis is zero.

The reflected Γ-distribution

or, more precisely, the Gamma distribution reflected about zero is defined by

$$f(x) = \frac{d^\theta}{2\Gamma(\theta)} |x|^{\theta-1} \exp(-d|x|) , \tag{D.22}$$

where $d, \theta > 0$ and $\Gamma(\theta)$ denotes the gamma function. If $\theta \leq 1$, the distribution is unimodal. The mean of the reflected Γ-distribution is zero, its variance is

$$\gamma^2 = \frac{\theta(\theta+1)}{d^2} , \tag{D.23}$$

and its moment generating function is given by

$$\psi_\Gamma(\xi) = \frac{d^\theta}{2(d-\xi)^\theta} + \frac{d^\theta}{2(d+\xi)^\theta} \,. \tag{D.24}$$

The odd central moments and cumulants vanish. The first even cumulants are

$$\begin{aligned} c_2 &= \gamma^2 \\ c_4 &= 2\gamma^2(3+\theta-\theta^2)/d^2 \\ c_6 &= 8\gamma^2(15+8\theta-8\theta^2-2\theta^3+2\theta^4)/d^4 \\ c_8 &= 16\gamma^2(315+213\theta-213\theta^2+95\theta^3-95\theta^4+17\theta^5-17\theta^6)/d^6 \,. \end{aligned} \tag{D.25}$$

Its kurtosis is $2(3+\theta-\theta^2)/[\theta(\theta+1)]$.

For $\theta = 1$, (D.22) represents an exponential distribution reflected about zero. Its odd cumulants vanish, and its even cumulants are

$$c_{2n} = \frac{(2n)!}{n\, d^{2n}} = \gamma^{2n}\frac{(2n)!}{2^n n} \,. \tag{D.26}$$

The kurtosis is 3.

For leptokurtic distributions, large amounts of variance are due to large values x of the underlying random variable. For example, for a normal distribution (of mean zero and variance γ^2), values $|x| \geq 3\gamma$ contribute only 2.9% to the variance, whereas for an exponential distribution (kurtosis 3) such values contribute 20.5%, and for a reflected Γ-distribution with variance γ^2 and kurtosis 26/3 ($\theta = \frac{1}{2}$), such values contribute 39.2% to the variance.

E MISCELLANEOUS RESULTS FOR FINITE POPULATIONS

First, we review some basic concepts and formulas that are of importance in the theory of finite populations, and used in Chapter VII. Then, we deal with stationary distributions under mutation, selection, and drift in the multinomial Wright–Fisher model, and their relations to diffusion processes and generalized gradient systems.

1. EFFECTIVE POPULATION SIZE

In a finite population, random genetic drift leads to a decrease of the genetic variability and to the eventual fixation or loss of alleles. In Chapter I.5, it was shown that in an ideal population of constant size N that evolves according to the simple Wright–Fisher model, in which there is only one sex, two alleles per locus, no selection or mutation, etc., the heterozygosity is reduced each generation by the factor $1-(2N)^{-1}$. Similarly, it follows from I(5.4) that the variance of allele frequencies caused by reproduction (the reproductive variance) is

$$\mathrm{Var}[p(t+1)|p(t)] = \frac{1}{2N}p(t)(1-p(t)) \,, \tag{E.1}$$

where $p(t)$ is the frequency of allele $\mathcal{A}_1$ in generation t. In this model, the probability π_2 that two genes taken at random in any generation are descendents of the same parent gene is $(2N)^{-1}$, implying that the inbreeding coefficient increases according to I(5.7).

In general, the breeding structure of a population does not conform to that of an idealized population, so that the reproductive variance differs from (E.1), and the probability π_2 is not $(2N)^{-1}$. The concept of the effective population size is an attempt to relate the effects of random drift in a population of size N with a general breeding structure to that of an idealized population.

If one is mainly interested in the effects of random drift on the genetic variance, then the appropriate concept is the (variance) effective population size, $N_\mathrm{e}(= N_\mathrm{e}^{(v)})$, defined by

$$N_\mathrm{e} = \frac{p(t+1)(1-p(t+1))}{2\,\mathrm{Var}[p(t+1)|p(t)]}\,, \tag{E.2}$$

where $\mathrm{Var}[p(t+1)|p(t)]$ is the actual reproductive variance in the given population. Thus, a population of (variance) effective size N_e has the same properties with respect to the change of variance as an ideal population of actual size $N = N_\mathrm{e}$. This, however, does not imply that in this population random drift has the same consequences with respect to inbreeding, as in an idealized population of size N_e. If one is interested in inbreeding, then the (inbreeding) effective size, defined by

$$N_\mathrm{e}^{(i)} = (2\pi_2)^{-1}\,, \tag{E.3}$$

is the appropriate quantity. In general, these effective sizes may be very different, but in a number of important special cases they are identical or similar. For a rigorous and general theory of the concept of the effective population size that allows for an arbitrary number of alleles at a locus, consult Nagylaki (1992, Chapter 9). For a treatment of the binomial sampling model, but including the eigenvalue effective size, refer to Ewens (1979, Chapter 3.9).

For numerous important situations, explicit expressions or approximations have been derived for the various effective population sizes. For instance, in a randomly mating population of constant size that has separate sexes, both having the same number and the same progeny distribution, the variance effective size is approximately given by

$$N_\mathrm{e} = \frac{4(N-1)}{V_k+2}\,, \tag{E.4}$$

where V_k is the variance of the number of gametes contributed per individual to the next generation (Crow and Kimura 1970, Table 7.6.4.1). This situation applies to the simulation model used in Chapter VII. We refer to Crow and Kimura (1970) and Caballero (1994) for many worked-out specific cases, and to Crow and Denniston (1988), Caballero (1995), Nagylaki (1995), Wang (1997a,b), and Whitlock and Barton (1997) for recent developments.

2. *FIXATION PROPERTIES*

In Chapter I.5, it was shown that in the binomial Wright–Fisher model the fixation probability of a neutral gene is equal to its initial frequency. This theory can be generalized to a diallelic locus under selection. We summarize some of its aspects and refer to Ewens (1979, Chapters 3–5) for a comprehensive treatment and proofs. As in the neutral case, one of the alleles is lost with probability one, because the number X of $\mathcal{A}_1$ genes is a Markovian random variable with two absorbing states, $X = 0$ and $X = 2N$. Employing the theory of diffusion approximations, a general approximation can be derived for the probability of fixation of, say allele $\mathcal{A}_1$, given its initial frequency, p_0, and the fitness values. Let the genotypes $\mathcal{A}_1\mathcal{A}_1$, $\mathcal{A}_1\mathcal{A}_2$, and $\mathcal{A}_2\mathcal{A}_2$ have the fitnesses $1 - 2s$, $1 - 2hs$, and 1, respectively, so that $\mathcal{A}_1$ is deleterious if $0 < s < \frac{1}{2}$ and $\mathcal{A}_1$ is advantageous if $s < 0$. The neutral case ($s = 0$) is excluded. Then, the diffusion approximation for the probability of fixation of $\mathcal{A}_1$ is

$$\pi_{\text{fix}} = \int_0^{p_0} f(y)\,\mathrm{d}y \Big/ \int_0^1 f(y)\,\mathrm{d}y\,, \tag{E.5}$$

where $f(y) = \exp\{-4Ns[(2h-1)y^2) - 2hy]\}$. In the absence of dominance ($h = \frac{1}{2}$), the integrals in (E.5) can be evaluated explicitly, and the following simple expression is obtained:

$$\pi_{\text{fix}} = \frac{e^{4Np_0s} - 1}{e^{4Ns} - 1}\,. \tag{E.6}$$

If $\mathcal{A}_1$ arises as a new mutation, then its initial frequency will be $p_0 = 1/(2N)$.

The validity of the approximations (E.5) and (E.6) requires that s is of order N^{-1}. We owe (E.5) and (E.6) to Malécot (1952) and Kimura (1957). It can be proved that the diffusion approximation (E.6) is always an upper bound for the true fixation probability and that for a single mutant ($p_0 = (2N)^{-1}$), π_{fix} converges to the true fixation probability as $N \to \infty$ if $s = \tilde{s}/(4N)$ and $\tilde{s}$ is fixed. Also, simple higher-order approximations can be derived (Bürger and Ewens 1995). Ethier (1979) proved that if p_0 is fixed, then the diffusion approximation converges to the true fixation probability as $N \to \infty$.

Approximations for the mean time to absorption (loss or fixation) of $\mathcal{A}_1$ have been derived (e.g., Ewens 1979), as well as for this mean time conditional on either loss or fixation. Interestingly, the conditional mean fixation properties of a favored allele are the same as those for the corresponding disadvantageous allele, provided the dominance relations are reversed. In particular, the conditional mean fixation times of advantageous or disadvantageous alleles are the same, and lower than that of a neutral allele. For a neutral allele of initial frequency p_0, the diffusion approximation for the conditional mean time to fixation in generations is

$$\bar{t}^* = -4Np_0^{-1}(1 - p_0)\ln(1 - p_0)\,. \tag{E.7}$$

This is always less than $4N$ and approximately equal to $4N$ if $p_0 = 1/(2N)$. Therefore, the mean time to fixation of a single mutant that is destined to becoming fixed is always less than $4N$ generations. If it is advantageous or deleterious, this mean time may be much smaller (depending on $|Ns|$).

For populations whose effective size N_e differs from N, the formulas (E.6) and (E.7) still apply if N is replaced by N_e (see Crow and Kimura 1970, Chapter 8). Clearly, if a mutation arises as a single mutant, then $p_0 = 1/(2N)$ has to be chosen.

The above results are based on approximations to the Wright–Fisher model with selection, which assumes binomial transition probabilities for the allele frequencies.

This is a mathematical assumption and not derived from biological principles. Ethier and Nagylaki (1980) and Nagylaki (1997) developed models for random genetic drift from explicit biological assumptions and investigated the conditions under which they yield good approximations to the multinomial Wright–Fisher model (see also Nagylaki (1992, Chapter 9) and Appendix E.3). The basic problem is that, under a variety of biological assumptions, multinomial sampling of genotypes obtains, but usually not of genes.

3. STATIONARY DISTRIBUTIONS AND GRADIENTS

Let us consider a finite monoecious diploid population of size N, in which selection and mutation occur at a given gene locus with k alleles. In a corresponding infinite population with discrete generations, the dynamics of allele frequencies would be given by the selection-mutation equation III(2.1). In a finite population however, sampling of alleles occurs. Let $\boldsymbol{p} = (p_1, \dots, p_k) \in S_k$ denote the frequency distribution in some given generation and let $\boldsymbol{p}^* = \mathrm{E}[p']$ denote the expected allele frequencies in the next generation, as given by the right-hand side of III(2.1). According to the so-called multinomial Wright–Fisher model, the vector $2N\boldsymbol{p}'$ of allele numbers in the next generation has a multinomial distribution with sample size $2N$ and mean vector $2N\boldsymbol{p}^*$, i.e.,

$$\Pr[2N\boldsymbol{p}' = \boldsymbol{j}|\boldsymbol{p}] = \frac{(2N)!}{j_1! \cdot \ldots \cdot j_k!} (p_1^*)^{j_1} \cdot \ldots \cdot (p_k^*)^{j_k} , \tag{E.8}$$

where the coefficients of $\boldsymbol{j} = (j_1, \dots, j_k)$ sum up to $2N$. This model generalizes the simple binomial models introduced in Chapter I.5 and Appendix E.2.

If the three forces selection, mutation, and random genetic drift are of comparable strength, then a diffusion approximation can be derived by assuming that the fitnesses and mutation rates have the form

$$W_{ij} = 1 + \omega_{ij}/(2N) , \quad \mu_{ij} = \nu_{ij}/(2N) \tag{E.9}$$

where ω_{ij} and ν_{ij} denote constant, scaled selection coefficients and mutation rates. Time t will be measured in units of $2N$ generations, i.e., $n = [2Nt]$. Let $\{\boldsymbol{p}^N(n), n = 0, 1, \dots\}$ designate the Markov chain of allele frequencies specified by (E.8). Then it can be proved that in the limit of large population size, $N \to \infty$, $\boldsymbol{p}^N([2Nt])$ converges weakly to the diffusion process $\boldsymbol{p}(t)$ that has the state space

$$\Big\{\boldsymbol{p} = (p_1, \dots , p_{k-1}) : p_i \geq 0 , \quad \sum_{i=1}^{k-1} p_i \leq 1\Big\} ,$$

and generator (or backward operator)

$$\tfrac{1}{2} \sum_{i,j=1}^{k-1} a_{ij}(\boldsymbol{p}) \frac{\partial^2}{\partial p_i \, \partial p_j} + \sum_{i=1}^{k-1} b_i(\boldsymbol{p}) \frac{\partial}{\partial p_i} \tag{E.10a}$$

(Ethier and Nagylaki 1980, Nagylaki 1990).

The diffusion and drift coefficients are

$$a_{ij}(\boldsymbol{p}) = p_i(\delta_{ij} - p_j) = g^{ij}(\boldsymbol{p}) \tag{E.10b}$$

[cf. (A.13)], and

$$b_i(\boldsymbol{p}) = \dot{p}_i \ , \tag{E.10c}$$

respectively, where $\dot{p}_i$ is the rate of allele-frequency change in the continuous-time mutation-selection equation III(2.2) (with ν_{ij} replacing μ_{ij}).

We designate by $\varphi(\boldsymbol{p}, \boldsymbol{x}; t)$ the probability density that the frequencies of the alleles become $\boldsymbol{p}$ at generation t, given that their frequencies are $\boldsymbol{x}$ at $t = 0$. Then, this transition density satisfies the *forward equation* of Kolmogorov

$$\frac{\partial \varphi}{\partial t}(\boldsymbol{p}, \boldsymbol{x}; t) = F\varphi(\boldsymbol{p}, \boldsymbol{x}; t) \ , \tag{E.11}$$

where F denotes the forward operator

$$F\varphi(\boldsymbol{p}, \boldsymbol{x}; t) = \tfrac{1}{2} \sum_{i,j=1}^{k-1} \frac{\partial^2 [a_{ij}(\boldsymbol{p})\, \varphi(\boldsymbol{p}, \boldsymbol{x}; t)]}{\partial p_i\, \partial p_j} - \sum_{i=1}^{k-1} \frac{\partial [b_i(\boldsymbol{p})\, \varphi(\boldsymbol{p}, \boldsymbol{x}; t)]}{\partial p_i} \ . \tag{E.12}$$

If a *stationary density* $\psi = \psi(\boldsymbol{p})$ exists, it necessarily satisfies

$$\psi(\boldsymbol{p}) = \int \psi(\boldsymbol{x}) \varphi(\boldsymbol{p}, \boldsymbol{x}; t)\, \mathrm{d}\boldsymbol{x}$$

for all $t > 0$. From this, one can deduce that ψ satisfies

$$F\psi = 0 \ . \tag{E.13}$$

Now we show that a stationary density exists if the mutation-selection equation III(2.2) is a Svirezhev–Shahshahani gradient. From Chapter III.2.3 we already know that the latter is the case if and only if the mutation rates satisfy the HC-condition III(2.8), i.e., $\nu_{ij} = \nu_j$ if $i \neq j$. Let us express p_k as

$$p_k = 1 - \sum_{i=1}^{k-1} p_i \ , \quad \text{and denote } h_i = f_i - f_k \text{ and } \bar{h} = \sum_{i=1}^{k-1} h_i p_i = \bar{f} - f_k \ ,$$

where $f_i = m_i + \nu_i / p_i$ and $\bar{f} = \bar{m} + \sum_{i=1}^{k} \nu_i$ [cf. III(2.9), III(2.14)]. The potential V is given by III(2.15) or, by expressing the m_{ij} and ν_j by the original parameters W_{ij} and μ_{ij} (E.9),

$$V(\boldsymbol{p}) = N(\overline{W} - 1) + 2N \sum_{i=1}^{k} \mu_i \ln p_i \ . \tag{E.14}$$

Then V satisfies $\partial V / \partial p_i = h_i$. Hence, by Theorem A.9, we can write III(2.2) as

$$\dot{p}_i = p_i(h_i - \bar{h}) = \sum_{j=1}^{k-1} g^{ij} \frac{\partial V}{\partial p_j} \ , \quad i = 1, \ldots, k-1 \ . \tag{E.15}$$

We assert that the stationary density of the diffusion process is given by

$$\psi = c_0 \gamma^{-1} \exp(2V) \,, \tag{E.16a}$$

where c_0 is the normalization factor,

$$\gamma = \gamma(\boldsymbol{p}) = \det \mathsf{G}_{\boldsymbol{p}} = p_1 \cdot \ldots \cdot p_{k-1} \cdot \Big(1 - \sum_{i=1}^{k-1} p_i\Big) \,, \tag{E.16b}$$

and $\mathsf{G}_{\boldsymbol{p}}$ is as in (A.13), but with the last row and column omitted. Then ψ can be written as

$$\psi(\boldsymbol{p}) = c_0 e^{2N(\overline{W}-1)} \prod_{i=1}^{k} p_i^{4N\mu_i - 1} \approx c_0 \overline{W}^{2N} \prod_{i=1}^{k} p_i^{4N\mu_i - 1} \,. \tag{E.17}$$

This is (a generalization of) Wright's (1935b, 1949, 1969) formula; cf. Ethier and Nagylaki (1988) and Barton (1989). Typically, the stationary density (E.17) is dominated by sharp peaks at the stable equilibria. The normalization constant c_0 can be evaluated in terms of parabolic cylinder functions (cf. Abramowitz and Stegun 1965, Chapter 19, and Barton 1989).

To prove our assertion, it is sufficient to show that

$$\sum_{j=1}^{k-1} \frac{\partial(g^{ij}\psi)}{\partial p_j} = 2\psi b_i \,. \tag{E.18}$$

Clearly, this condition is much stronger than (E.13), as it is equivalent to saying that the probability flux is zero at every point of the state space. First, by (E.10c) and (E.15), the drift coefficients b_i are given by

$$b_i = \sum_{j=1}^{k-1} g^{ij} \frac{\partial V}{\partial p_j} \,, \quad i = 1, \ldots, k-1 \,, \tag{E.19a}$$

i.e., the vector $\boldsymbol{b}$ is

$$\boldsymbol{b} = \mathsf{G}_{\boldsymbol{p}} \nabla V = \tilde{\nabla} V \,. \tag{E.19b}$$

Therefore, the right-hand side of (E.18) is

$$2\psi b_i = 2\psi \sum_{j=1}^{k-1} g^{ij} \frac{\partial V}{\partial p_j} \,. \tag{E.20}$$

Secondly, a straightforward calculation shows that

$$\sum_j \frac{\partial}{\partial p_j} \frac{g^{ij}}{\gamma} = 0 \quad \text{for every } i \,. \tag{E.21}$$

Using (E.16) and (E.21), the left-hand side of (E.18) can be transformed as follows:

$$\begin{aligned}\sum_{j=1}^{k-1} \frac{\partial(g^{ij}\psi)}{\partial p_j} &= c_0 \sum_j \frac{\partial}{\partial p_j}\left(\frac{g^{ij}}{\gamma}\exp(2V)\right) \\ &= c_0 \sum_j \frac{g^{ij}}{\gamma}\,\frac{\partial \exp(2V)}{\partial p_j} \\ &= 2c_0 \sum_j \frac{g^{ij}}{\gamma}\exp(2V)\,\frac{\partial V}{\partial p_j} \\ &= 2\psi \sum_j g^{ij}\frac{\partial V}{\partial p_j}\,, \end{aligned} \tag{E.22}$$

which agrees with (E.20).

The validity of the simple formula (E.16) for the stationary distribution hinges on the fact that the covariance matrix of the diffusion (E.10) is $\mathrm{G}_{\boldsymbol{p}}$, the same matrix that defines the Svirezhev–Shahshahani metric on the simplex S_k and with respect to which the drift term $\boldsymbol{b}$ is a gradient. This fact was already used by Antonelli and Strobeck (1977) in their investigation of the geometry of random genetic drift.

With scaling assumptions about the parameters that differ from (E.9), different diffusion approximations and a different stationary density may be derived (cf. Ethier and Nagylaki 1988, Nagylaki 1990).

References

Abramowitz, M., and Stegun, I.A., (eds.) 1965. *Handbook of Mathematical Functions.* New York: Dover.

Akin, E. 1979. *The Geometry of Population Genetics.* Lect. Notes in Biomath. **31**. Berlin, Heidelberg, New York: Springer.

Akin, E. 1982. Cycling in simple genetic systems. J. Math. Biol. **13**, 305–324.

Akin, E. 1983. Hopf bifurcation in the two-locus genetic model. Mem. Amer. Math. Soc. **44**, No. 284.

Akin, E., and Szucs. J.M. 1994. Approaches to the Hardy–Weinberg manifold. J. Math. Biol. **32**, 633–643.

Aliprantis, C.D., and Burkinshaw, O. 1980. Positive compact operators on Banach lattices. Math. Z. **174**, 289–298.

Antonelli, P.L., and Strobeck, C. 1977. The geometry of random drift I. Stochastic distance and diffusion. Adv. Appl. Prob. **9**, 238–249.

Apostol, T.M. 1974. *Mathematical Analysis,* 2nd ed. Reading, MA: Addison-Wesley.

Atkinson, F.V., Watterson, G.A., and Moran, P.A.P. 1960. A matrix inequality. Quart. J. Math. **11**, 137–140.

Avery, P.J., and Hill, W.G. 1978. The effect of linkage disequilibrium on the genetic variance of a quantitative trait. Adv. Appl. Prob. **10**, 4–6.

Baake, E. 2000. Mutation and recombination with tight linkage. Preprint.

Baake, E., and Gabriel, W. 1999. Biological evolution through mutation, selection, and drift: an introductory review. Annu. Rev. Comp. Phys. **7**, 203–267.

Baake, E., and Wiehe, T. 1997. Bifurcations in haploid and diploid sequence space models. J. Math. Biol. **35**, 321–343.

Baatz, M., and Wagner, G.P. 1997. Adaptive inertia caused by hidden pleiotropic effects. Theor. Pop. Biol. **51**, 49–66.

Barton, N.H. 1986. The maintenance of polygenic variation through a balance between mutation and stabilizing selection. Genet. Res. **47**, 209–216.

Barton, N.H. 1989. Divergence of a polygenic system subject to stabilizing selection, mutation and drift. Genet. Res. **54**, 59–77.

Barton, N.H. 1990. Pleiotropic models of quantitative variation. Genetics **124**, 773–782.

Barton, N.H. 1995. A general model for the evolution of recombination. Genet. Res. **65**, 123–144.

Barton, N.H. 1999. Clines in polygenic traits. Genet. Res. **74**, 223–236.

Barton, N.H., and Charlesworth, B. 1998. Why sex and recombination? Science **281**, 1986–1990.

Barton, N.H., and Turelli, M. 1987. Adaptive landscapes, genetic distance and the evolution of quantitative characters. Genet. Res. **49**, 157–173.

Barton, N.H., and Turelli, M. 1989. Evolutionary quantitative genetics: how little do we know? Annu. Rev. Genet. **23**, 337–370.

Barton, N.H., and Turelli, M. 1991. Natural and sexual selection on many loci. Genetics **127**, 229–255.

Baum, L.E., and Eagon, J.A. 1967. An inequality with applications to statistical estimation for probability functions of Markov processes and to a model for ecology. Bull. Amer. Math. Soc. **73**, 360–363.

Bennett, J.H. 1954. On the theory of random mating. Ann. Eugen. **18**, 311–317.

Bennett, J.H. 1983. (ed.) *Natural Selection, Heredity, and Eugenics*. Including selected correspondence of R.A. Fisher with Leonhard Darwin and others. Oxford: Clarendon Press.

Boag, P.T. 1983. The heritability of external morphology in Darwin's ground finches (Geospiza) on Isla Daphne Major, Galápagos. Evolution **37**, 877–894.

Bodmer, W.F., and Felsenstein, J. 1967. Linkage and selection: theoretical analysis of the deterministic two locus random mating model. Genetics **57**, 237–265.

Bohuon, E.J.R., Ramsay, L.D., Craft, J.A., Arthur, A.E., Marshall, D.F., Lydiate, D.J., and Kearsey, M.J. 1998. The association of flowering time quantitative trait loci with duplicated regions and candidate loci in *Brassica oleracea*. Genetics **150**, 393–401.

Brooke, M. 1977. The breeding biology of the Manx Shearwater. Ph. D. Diss. Oxford, UK: University Oxford.

Broom, M., Cannings, C., and Vickers, G.T. 1993. On the number of local maxima of a constrained quadratic form. Proc. R. Soc. Lond. A **443**, 573–584.

Bulmer, M.G. 1971a. The effect of selection on genetic variability. Am. Nat. **105**, 201–211.

Bulmer, M.G. 1971b. The stability of equilibria under selection. Heredity **27**, 157–162.

Bulmer, M.G. 1972. The genetic variability of polygenic characters under optimising selection, mutation and drift. Genet. Res. **19**, 17–25.

Bulmer, M.G. 1973. The maintenance of genetic variability of polygenic characters by heterzygous advantage. Genet. Res. **22**, 9–12.

Bulmer, M.G. 1974. Linkage disequilibrium and genetic variability. Genet. Res. **23**, 281–289.

Bulmer, M.G. 1980. *The Mathematical Theory of Quantitative Genetics*. Oxford: Clarendon Press.

Burch, C.L., and Chao, L. 1999. Evolution by small steps and rugged landscapes in the RNA virus $\phi 6$. Genetics **151**, 921–927.

Bürger, R. 1983a. On the evolution of dominance modifiers I. A nonlinear analysis. J. Theor. Biol. **101**, 585–598.

Bürger, R. 1983b. Nonlinear analysis of some models for the evolution of dominance. J. Math. Biol. **16**, 269–280.

Bürger, R. 1986a. Constraints for the evolution of functionally coupled characters: a nonlinear analysis of a phenotypic model. Evolution **40**, 182–193.

Bürger, R. 1986b. Evolutionary dynamics of functionally constrained phenotypic characters. IMA J. of Math. Applied in Medicine & Biology **3**, 265–287.

Bürger, R. 1986c. On the maintenance of genetic variation: Global analysis of Kimura's continuum-of-alleles model. J. Math. Biol. **24**, 341–351.

Bürger, R. 1988a. Perturbations of positive semigroups and applications to population genetics. Math. Z. **197**, 259–272.

Bürger, R. 1988b. Mutation-selection balance and continuum-of-alleles models. Math. Biosci. **91**, 67–83.

Bürger, R. 1989. Linkage and the maintenance of heritable variation by mutation-selection balance. Genetics **121**, 175–184.

Bürger, R. 1991a. Moments, cumulants, and polygenic dynamics. J. Math. Biol. **30**, 199–213.

Bürger, R. 1991b. An integro-differential equation from population genetics and perturbations of differentiable semigroups in Fréchet spaces. Proc. Royal Soc. Edinb., **118A**, 63–73.

Bürger, R. 1991c. Unbounded one-parameter semigroups, Fréchet spaces, and Applications. In: Clément, P., Mitidieri, E., and dePagter, B. (eds.) *Semigroup Theory and Evolution Equations*. New York: Marcel Dekker.

Bürger, R. 1993. Predictions for the dynamics of a polygenic character under directional selection. J. Theor. Biol. **162**, 487–513.

Bürger, R. 1999. Evolution of genetic variability and the advantage of sex and recombination in changing environments. Genetics **153**, 1055–1069.

Bürger, R., and Bomze, I.M. 1996. Stationary distributions under mutation-selection balance: structure and properties. Adv. Appl. Prob. **28**, 227–251.

Bürger, R., and Ewens, W.J. 1995. Fixation probabilities of additive alleles in diploid populations. J. Math. Biol. **33**, 557–575.

Bürger, R., and Gimelfarb, A. 1999. Genetic variation maintained in multilocus models of additive quantitative traits under stabilizing selection. Genetics **152**, 807–820.

Bürger, R., and Hofbauer, J. 1994. Mutation load and mutation-selection-balance in quantitative genetic traits. J. Math. Biol. **32**, 193–218.

Bürger, R., and Krall, C. 2000. Genetic variation and population persistence in changing environments. In: Ferriere, R., Dieckmann, U., and Couvet, D. (eds.) *Evolutionary Conservation Biology*. Cambridge: University Press, to appear.

Bürger, R., and Lande, R. 1994. On the distribution of the mean and variance of a quantitative trait under mutation-selection-drift balance. Genetics **138**, 901–912.

Bürger, R., and Lynch, M. 1995. Evolution and extinction in a changing environment: a quantitative-genetic analysis. Evolution **49**, 151–163.

Bürger, R., and Lynch, M. 1997. Adaptation and extinction in changing environments. In: Bijlsma, R., and Loeschke, V. (eds.) *Environmental Stress, Adaptation and Evolution*. Basel: Birkhäuser.

Bürger, R., Wagner, G.P., and Stettinger, F. 1989. How much heritable variation can be maintained in finite populations by mutation-selection balance? Evolution **43**, 1748–1766.

Butcher, D. 1995. Muller's ratchet, epistasis and mutation effects. Genetics **141**, 431–437.

Caballero, A. 1994. Developments in the prediction of effective population size. Heredity **73**, 657–679.

Caballero, A. 1995. On the effective size of populations with separate sexes, with particular reference to sex-linked genes. Genetics **139**, 1007–1011.

Caballero, A., and Keightley, P.D. 1994. A pleiotropic nonadditive model of variation in quantitative traits. Genetics **138**, 883–900.

Caballero, A., and Keightley, P.D. 1998. Inferences on genome-wide deleterious mutation rates in inbred populations of *Drosophila* and mice. Genetica **102/103**, 229–239.

Caballero, A., Toro, M.A., and López-Fanjul, C. 1991. The response to artificial selection from new mutations in *Drosophila melanogaster*. Genetics **127**, 89–102.

Castilloux, A.-M., Lessard, S. 1995. The fundamental theorem of natural selection in Ewens' sense (case of many loci). Theor. Pop. Biol. **48**, 306–315.

Castle, W.E. 1903. The laws of Galton and Mendel and some laws governing race improvement by selection. Proc. Amer. Acad. Arts Sci. **39**, 233–242.

Cavalli-Sforza, L.L., and Edwards, A.F.W. 1967. Phylogenetic analysis: models and estimation procedures. Amer. J. Hum. Genet. **19**, 233–257.

Charlesworth, B. 1990. Mutation-selection balance and the evolutionary advantage of sex and recombination. Genet. Res. **55**, 199-221.

Charlesworth, B. 1993. Directional selection and the evolution of sex and recombination. Genet. Res. **61**, 205-224.

Charlesworth, B. 1994. *Evolution in Age-Structured Populations*. 2nd ed. Cambridge: University Press.

Charlesworth, B., and Barton, N.H., 1996. Recombination load associated with selection for increased recombination. Genet. Res. **67**, 27–41.

Charlesworth, B., and Charlesworth, D. 1997. Rapid fixation of deleterious alleles can be caused by Muller's ratchet. Genet. Res. **70**, 63–73.

Charlesworth, B., and Charlesworth, D. 1998. Some evolutionary consequences of deleterious mutations. Genetica **102/103**, 3–19.

Charlesworth, B., and Charlesworth, D. 1999. The genetic basis of inbreeding depression. Genet. Res. **74**, 329–340.

Charlesworth, B., and Hughes, K.A. 1999. The maintenance of genetic variation in life-history traits. In: Singh, R.S., and Krimbas, C.B. (eds.) *Evolutionary Genetics from Molecules to Morphology*, pp. 369–392. Cambridge: University Press.

Charlesworth, B., Charlesworth, D., and Morgan, M.T. 1990. Genetics loads and estimates of mutation rates in very inbred plant populations. Nature **347**, 380–382.

Charlesworth, B., Lande, R., and Slatkin, M. 1982. A neo-Darwinian commentary on macro-evolution. Evolution **36**, 474–498.

Charlesworth, D., Morgan, M.T., and Charlesworth, B. 1993. Mutation accumulation in finite outbreeding and inbreeding populations. Genet. Res. **61**, 39–56.

Cheverud, J.M., and J.E. Buikstra. 1982. Quantitative genetics of skeletal nonmetric traits in the Rhesus Macaque of Cayo Santiago. III. Relative heritability of skeletal nonmetric and metric traits. Amer. J. Phys. Anthropol. **59**, 151–155.

Cheverud, J.M., Routman, E.J., Durante, F.A.M., van Swinderen, B., Cothran, K., and Perel, C. 1996. Quantitative trait loci for murine growth. Genetics **142**, 1305–1319.

Cheverud, J.M., Routman, E.J., and Irschick, D.J. 1997. Pleiotropic effects of individual gene loci on mandibular morphology. Evolution **51**, 2006–2016.

Christiansen, F.B. 1987. The deviation from linkage equilibrium with multiple loci varying in a stepping-stone cline. J. Genet. **66**, 45–67.

Christiansen, F.B. 1988. Epistasis in the multiple locus symmetric viability model. J. Math. Biol. **26**, 595–618.

Christiansen, F.B. 1990. The generalized multiplicative model for viability selection. J. Math. Biol. **29**, 99–129.

Christiansen, F.B. 1999. *The Population Genetics of Multiple Loci.* Chichester: Wiley.

Clark, A.G. 1991. Mutation-selection balance and metabolic control theory. Genetics **129**, 909–923.

Clark, A.G., and Wang, L. 1997. Epistasis in measured genotypes: *Drosophila P*-element insertions. Genetics **147**, 157–163.

Clark, A.G., Wang, L., and Hulleberg, T. 1995. *P*-element-induced variation in metabolic regulation in *Drosophila.* Genetics **139**, 337–348.

Clayton, G.A., and Robertson, A. 1955. Mutation and quantitative variation. Am. Nat. **89**, 151–158.

Cockerham, C.C. 1954. An extension of the concept of partitioning hereditary variance for analysis of covariances among relatives when epistasis is present. Genetics **39**, 859–882.

Coppersmith, S.N., Blank, R.D., and Kadanoff, L.P. 1999. Analysis of a population genetics model with mutation, selection, and pleiotropy. J. Statistical Physics **97**, 429–457.

Cotterman, C.W. 1940. A calculus for statistico-genetics. Ph.D. thesis, Ohio State Univ. Reprinted in *Genetics and Social Structures* (P. Ballonoff, ed.), pp. 157–272. Stroudsburg, Pa.: Dowden, Hutchinson and Ross (1974).

Crow, J.F. 1970. Genetic loads and the cost of natural selection. In: Kojima, K. (ed.), *Mathematical Topics in Population Genetics*, pp. 128–177. Berlin, Heidelberg, New York: Springer.

Crow, J.F. 1986. *Basic Concepts in Population, Quantitative and Evolutionary Genetics.* New York: W.H. Freeman.

Crow, J.F. 1992. Mutation, mean fitnes, and genetic load. Oxford Surveys Evol. Biol. **9**, 3–42.

Crow, J.F. 1999. The odds of losing at genetic roulette. Nature **397**, 293–294.

Crow, J.F., and Denniston, C. 1988. Inbreeding and variance effective population numbers. Evolution **42**, 482–495.

Crow, J.F., and Kimura, M. 1964. The theory of genetic loads. In: *Proc. XI Int. Congr. Genetics, vol. 2*, pp. 495–505. Oxford: Pergamon Press.

Crow, J.F., and Kimura, M. 1970. *An Introduction to Population Genetics Theory.* New York: Harper and Row.

Crow, J.F., and Simmons, M.J. 1983. The mutation load in *Drosophila.* In: Ashburner, M., Carson, H.L., and Thompson, J.N. (eds.) *The Genetics and Biology of Drosophila*, pp. 1–35. London, New York: Academic Press.

Darwin, C. 1859. *The Origin of Species.* London: John Murray.

Davies, E.K., Peters, A.D., and Keightley P.D. 1999. High frequency of cryptic deleterious mutations in *Caenorhabditis elegans.* Science **285**, 1748–1751.

Dawson, K.J. 1999. The dynamics of infinitesimally rare alleles, applied to the evolution of mutation rates and the expression of deleterious mutations. Theor. Pop. Biol. **55**, 1–22.

Dawson, K.J. 2000. The decay of linkage disequilibrium under random union of gametes: How to calculate Bennett's principal components. Theor. Pop. Biol., in press.

de Jong, G. 1999. Unpredictable selection in a structured population leads to local differentiation in evolved reaction norms. J. Evol. Biol. **12**, 839–851.

Dempster, E.R. 1955. Maintenance of genetic heterogeneity. *Cold Spring Harbor Symp. Quant. Biol.* **20**, 25–32.

Deng, H.-W., and Fu, Y.-X. 1998. On the three methods for estimating deleterious genomic mutation parameters. Genet. Res. **71**, 223–236.

de Visser, J.A.G.M., Hoekstra, R.F., and van den Ende, H. 1997a. An experimental tst for synergistic epistasis and its application in *Chlamydomonas*. Genetics **145**, 815–819.

de Visser, J.A.G.M., Hoekstra, R.F., and van den Ende, H. 1997b. Test of interaction between genetic markers that affect fitness in *Aspergillus niger*. Evolution **51**, 1499-1505.

Drake, J.W., Charlesworth, B., Charlesworth, D., and Crow, J.F. 1998. Rates of spontaneous mutation. Genetics **148**, 1667–1686.

Dudley, J.W. 1977. 76 generations of selection for oil and protein percentage in maize. In: Pollak, E., Kempthorne, O., and Bailey, T.B. (eds.) *Proceedings of the International Conference on Quantitative Genetics*, pp. 459–473. Ames, Iowa: Iowa State University Press.

Dudley, J.W., and Lambert, J.R. 1992. Ninety generations of selection for oil and protein content in maize. Maydica **37**, 1–7.

Edwards, A.W.F. 1994. The fundamental theorem of natural selection. Biol. Rev. **69**, 443–474.

Eigen, M. 1971. Selforganization of matter and the evolution of biological macromolecules. Naturwissenschaften **58**, 465–523.

Eigen, M., and Schuster, P. 1977. The hypercycle. A principle of natural self-organization. Part A: Emergence of the hypercycle. Naturwissenschaften **64**, 541–565.

Elena, A.F., and Lenski, R.E. 1997. Test of synergistic interactions among deleterious mutations in bacteria. Nature **390**, 395–398.

Ellner, S. 1996. Environmental fluctuations and the maintenance of genetic diversity in age or stage-structured populations. Bull. Math. Biol. **58**, 102–127.

Endler, J.A. 1986. *Natural Selection in the Wild.* Princeton, N.J.: University Press.

Eshel, I. 1971. On evolution in a population with an infinite number of types. Theor. Pop. Biol. **2**, 209–236.

Eshel, I. 1972. Evolution processes with continuity of types. Adv. Appl. Prob. **4**, 475–507.

Ethier, S.N. 1979. Limit theorems for absorption times in a genetic process. Ann. Prob. **7**, 622-638.

Ethier, S.N., and Nagylaki, T. 1980. Diffusion approximations of Markov chains with two time scales and applications to population genetics. Adv. Appl. Prob. **12**, 14–49.

Ethier, S.N., and Nagylaki, T. 1988. Diffusion approximations of Markov chains with two time scales and applications to population genetics, II. Adv. Appl. Prob. **20**, 525–545.

Ewens, W.J. 1968. A genetic model having complex linkage behavior. Theor. Appl. Genet. **38**, 140–143.

Ewens, W.J. 1969a. A generalized fundamental theorem of natural selection. Genetics **63**, 531–537.

Ewens, W.J. 1969b. Mean fitness increases when fitnesses are additive. Nature **221**, 1076.

Ewens, W.J. 1976. Remarks on the evolutionary effect of natural selection. Genetics **83**, 601–607.

Ewens, W.J. 1979. *Mathematical Population Genetics.* Berlin, Heidelberg, New York: Springer.

Ewens, W.J. 1989. An interpretation and proof of the fundamental theorem of natural selection. Theor. Pop. Biol. **36**, 167–180.

Ewens, W.J. 1990. Population genetics theory – the past and the future. In: Lessard, S. (ed.) *Mathematical and Statistical Developments of Evolutionary Theory*, pp. 177–227. NATO Advanced Study Institut Series. Kluwer: Dordrecht.

Ewens, W.J. 1992. An optimizing principle of natural selection in evolutionary population genetics. Theor. Pop. Biol. **42**, 167–180.

Ewens, W.J. 1995. Addendum to "The fundamental theorem of natural selection" by Castilloux and Lessard. Theor. Pop. Biol. **48**, 316–317.

Ewens, W.J., and Thomson, G. 1977. Properties of equilibria in multilocus genetic systems. Genetics **87**, 807–819.

Eyre-Walker, A., and Keightley, P.D. 1999. High genomic deleterious mutation rates in hominids. Nature **397**, 344–347.

Falconer, D.S. 1952. The problem of environment and selection. Am. Nat. **86**, 293–298.

Falconer, D.S., and Mackay, T.F.C 1996. *Introduction to Quantitative Genetics,* 4th ed. Harlow, Essex, UK: Addison Wesley Longman.

Fedorowicz, G.M., Fry, J.D., Anholt, R.H.A., and Mackay, T.F.C. 1998. Epistatic interactions between smell-impaired loci in *Drosophila melanogaster.* Genetics **148**, 1885–1891.

Feichtinger, H.G. 1984. Compactness in translation invariant Banach spaces of distributions and compact multipliers. J. Math. Anal. Appl. **102**, 289–327.

Feldman, M.W. 1972. Selection for linkage modification. I. Random mating populations. Theor. Pop. Biol. **3**, 324–346.

Feldman, M.W., and Liberman, U. 1979. On the number of stable equilibria and the simultaneous stability of fixation and polymorphism in two-locus models. Genetics **92**, 1355–1360.

Feldman, M.W., Otto, S.P., and Christiansen, F.B. 1997. Population genetic perspectives on the evolution of recombination. Ann. Rev. Genet. **30**, 261–295.

Felsenstein, J. 1974. The evolutionary advantage of recombination. Genetics **78**, 737–756.

Felsenstein, J. 1976. The theoretical population genetics of variable selection and migration. Ann. Rev. Genet. **10**, 253–280.

Felsenstein, J. 1977. Multivariate normal genetic models with a finite number of loci. In: Pollak, E., Kempthorne, O., and Bailey, T.B. (eds.) *Proceedings of the International Conference on Quantitative Genetics*, pp. 227–245. Ames, Iowa: Iowa State University Press.

Fernández, J., and López-Fanjul, C. 1997. Spontaneous mutational genotype-environment interaction for fitness-related traits in *Drosophila melanogaster.* Evolution **51**, 856–864.

Feller, W. 1971. *An Introduction to Probability Theory and Its Applications, Vol. II.* 2nd edn. New York: Wiley.

Fisher, R.A. 1918. The correlation between relatives on the supposition of Mendelian inheritance. Trans. Roy. Soc. Edinb. **52**, 399–433.

Fisher, R.A. 1922. On the dominance ratio. Proc. Roy. Soc. Edinb. **42**, 321–341.

Fisher, R.A. 1930. *The Genetical Theory of Natural Selection.* Oxford: Clarendon Press.

Fisher, R.A. 1941. Average excess and average effect of a gene substitution. Ann. Eugen. **11**, 53–63.

Fisher, R.A. 1958. *The Genetical Theory of Natural Selection.* 2nd edn. New York: Dover.

Fisher, R.A. 1958a. Polymorphism and natural selection. J. Ecology **46**, 289–293.

Fleming, W.H. 1979. Equilibrium distributions of continuous polygenic traits. SIAM J. Appl. Math. **36**, 148–168.

Foley, P. 1992. Small population genetic variability at loci under stabilizing selection. Evolution **46**, 763–774.

Frank, S.A., and Slatkin, M. 1990. The distribution of allelic effects under mutation and selection. Genet. Res. **55**, 111–117.

Frankham, R. 1995. Effective population size/adult population size ratios in wildlife: a review. Genet. Res. **66**, 95–107.

Frankham, R., Briscoe, D.A., and Nurthen, R.K. 1980. Unequal crossing over at the rRNA tandon as a source of quantitative genetic variation in *Drosophila.* Genetics **95**, 727–742.

Franklin, I.R., and Lewontin, R.C. 1970. Is the gene the unit of selection? Genetics **65**, 701–734.

Franklin, I.R., and Mayo, O. 1998. The place of QTL in the basis of quantitative genetics. II. Mapping QTL from a wide cross. In: Proceedings of the 6th World Congress on Genetics Applied to Livestock Production, 26, pp. 273–276.

Fraser, G.R., and Mayo, O. 1974. Genetical load in man. Humangenetik **23**, 83–110.

Fry, J.D. 2000. Mutational parameters for viability in *Drosophila melanogaster*: new estimates, reanalyses of the previous studies, and an attempt at a synthesis. Preprint.

Fry, J.D., deRonde, K.A., and Mackay, T.F.C. 1995. Polygenic mutation in *Drosophila melanogaster*: genetic analysis of selection lines. Genetics **139**, 1293–1307.

Fry, J.D., Heinsohn, S.L., and Mackay, T.F.C. 1996. The contribution of new mutations to genotype-environment interaction for fitness in *Drosophila melanogaster*. Evolution **50**, 2316–2327.

Fry, J.D., Keightley, P.D., Heinsohn, S.L., and Nuzhdin, S.V. 1999. New estimates of the rates and effects of mildly deleterious mutation in *Drosophila melanogaster*. Proc. Natl. Acad. Sci., USA **96**, 574–579.

Gabriel, W., and Bürger, R. 1994. Extinction risk by mutational meltdown: synergistic effects between population regulation and genetic drift. In: Loeschke, V., Tomiuk, J., and Jain, S.K. (eds.) *Conservation Genetics*, pp. 69–84. Basel: Birkhäuser.

Gabriel, W., and Bürger, R. 2000. Fixation of clonal lineages under Muller's ratchet. Evolution **54**, in press.

Gabriel, W., Lynch, M., and Bürger, R. 1993. Muller's ratchet and mutational meltdowns. Evolution **47**, 1744–1757.

Gale, J.S., and Kearsey, M.J. 1968. Stable equilibria under stabilising selection in the absence of dominance. Heredity **23**, 553–561.

Galton, F. 1889. *Natural Inheritance*. London: MacMillan.

Gantmacher, F.R. 1959. *The Theory of Matrices*. 2 vol. New York: Chelsea.

García-Dorado, A. 1997. The rate and effects distribution of viability mutation in *Drosophila*: minimum distance estimation. Evolution **51**, 1130–1139.

García-Dorado, A., López-Fanjul, C., and Caballero, A. 1999. Properties of spontaneous mutations affecting quantitative traits. Genet. Res. **74**, 341–350.

Gavrilets, S. 1993. Equilibria in an epistatic viability model under arbitrary strength of selection. J. Math. Biol. **31**, 397–410.

Gavrilets, S., and de Jong, G. 1993. Pleiotropic models of polygenic variation, stabilizing selection, and epistasis. Genetics **134**, 609–625.

Gavrilets, S., and Hastings, A. 1993. Maintenance of genetic variability under strong stabilizing selection: a two-locus model. Genetics **134**, 377–386.

Gavrilets, S., and Hastings, A. 1994a. Maintenance of multilocus variability under strong stabilizing selection. J. Math. Biol. **32**, 287–302.

Gavrilets, S., and Hastings, A. 1994b. Dynamics of genetic variability in two-locus models of stabilizing selection. Genetics **138**, 519–532.

Gavrilets, S., and Hastings, A. 1995. Dynamics of polygenic variability under stabilizing selection, recombination, and drift. Genet. Res. **65**, 63–74.

Gavrilets, S., and Scheiner, S.M. 1993. The genetics of phenotypic plasticity. V. Evolution of reaction norm shape. J. Evol. Biol. **6**, 31–48.

Geiringer, H. 1944. On the probability theory of linkage in Mendelian heredity. Ann. Math. Stat. **15**, 25–57.

Gessler, D.D.G. 1995. The constraints of finite size in asexual populations and the rate of the ratchet. Genet. Res. **66**, 241–253.

Gessler, D.D.G., and Xu, S. 1999. On the evolution of recombination and meiosis. Genet. Res. **73**, 119–131.

Gibbs, H.L., and Grant, P.R. 1987. Oscillating selection on Darwin's finches. Nature **327**, 511–513.

Gillespie, J.H. 1984. Pleiotropic overdominance and the maintenance of genetic variation in polygenic characters. Genetics **107**, 321–330.

Gillespie, J.H., and Turelli, M. 1989. Genotype-environment interaction and the maintenance of polygenic variation. Genetics **121**, 129–138.

Gillespie, J.H., and Turelli, M. 1990. Genotype-environment interactions: a reply. Genetics **124**, 447.

Gimelfarb, A. 1986. Additive variation maintained under stabilizing selection: a two-locus model of pleiotropy for two quantitative characters. Genetics **112**, 717–725.

Gimelfarb, A. 1989. Genotypic variation for a quantitative character maintained under stabilizing selection without mutations: epistasis. Genetics **123**, 217–227.

Gimelfarb, A. 1990. How much genetic variation can be maintained by genotype-environment interactions? Genetics **124**, 443–445.

Gimelfarb, A. 1992. Pleiotropy and multilocus polymorphisms. Genetics **130**, 223–227.

Gimelfarb, A. 1996a. Some additional results about polymorphisms in models of an additive quantitative trait under stabilizing selection. J. Math. Biol. **35**, 88–96.

Gimelfarb, A. 1996b. Pleiotropy as a factor maintaining genetic variation in quantitative characters under stabilizing selection. Genet. Res. **68**, 65–73.

Gimelfarb, A. 1998. Stable equilibria in multilocus genetic systems: a statistical investigation. Theor. Pop. Biol. **54**, 133–145.

Gimelfarb, A., and Willis, J.H. 1994. Linearity *versus* nonlinearity of offspring-parent regression: an experimental study of *Drosophila melanogaster*. Genetics **138**, 343–352.

Gordo, I., and Charlesworth, B. 2000. The degeneration of asexual haploid populations and the speed of Muller's ratchet. Genetics **154**, 1379–1387.

Grant, P.R. 1983. Inheritance of size and shape in a population of Darwin's finches, *Geospiza conirostris*. Proc. R. Soc. Lond. B **220**, 219–236.

Grant, P.R., and Grant, B.R. 1995. Predicting microevolutionary responses to directional selection on heritable variation. Evolution **49**, 241–251.

Grant, P. R., and Price, T.D. 1981. Population variation in continuously varying traits as an ecological genetics problem. Amer. Zool. **21**, 795–811.

Griffing, B. 1960. Theoretical consequences of truncation selection based on the individual phenotype. Aust. J. Biol. Sci. **13**, 307–343.

Gröbner, W., and Hofreiter, N. 1975. *Integraltafel*. Zweiter Teil. Berlin: Springer.

Gurganus, M.C., Nuzhdin, S.V., Leips, J.W., and Mackay, T.F.C. 1999. High-resolution mapping of quantitative trait loci for sternopleural bristle number in *Drosophila melanogaster*. Genetics **152**, 1585–1604.

Gurganus, M.C., Fry, J.D., Nuzhdin, S.V., Pasyukova, E.G., Lyman, R.F., and Mackay, T.F.C. 1998. Genotype-environment interaction at quantitative trait loci affecting sensory bristle number in *Drosophila melanogaster*. Genetics **149**, 1883–1898.

Hadeler, K.P. 1981. Stable polymorphisms in a selection model with mutation. SIAM J. Appl. Math. **41**, 1–7.

Haigh, J. 1978. The accumulation of deleterious genes in a population – Muller's ratchet. Theor. Pop. Biol. **14**, 251–267.

Haldane, J.B.S. 1932. *The Causes of Evolution*. London: Longmans, Green. (Reprinted with a new introduction and afterword by E.G. Leigh, Jr., by Princeton University Press, 1992.)

Haldane, J.B.S. 1937. The effect of variation on fitness. Am. Nat. **71**, 337–349.

Haldane, J.B.S. 1954. The measurement of natural selection. Caryologia **6** (suppl.), 480–487.

Haldane, J.B.S. 1957. The cost of natural selection. J. Genet. **55**, 511–524.

Hardy, G.H. 1908. Mendelian proportions in a mixed population. Science **28**, 49–50.

Hartl, D.L. 1994. *Genetics*, 3rd ed. Boston: Jones and Bartlett.

Hartl, D.L., and Clark, A.G. 1997. *Principles of Population Genetics*, 3rd ed. Sunderland, Mass.: Sinauer.

Hastings, A. 1981a. Simultaneous stability of $D = 0$ and $D \neq 0$ for multiplicative viabilities at two loci: an analytical study. J. Theor. Biol. **89**, 69–81.

Hastings, A. 1981b. Marginal underdominance at a stable equilibrium. Proc. Natl. Acad. Sci. USA **78**, 6558–6559.

Hastings, A. 1981c. Stable cycling in discrete-time genetic models. Proc. Natl. Acad. Sci. USA **78**, 7224–7225.

Hastings, A. 1982. Unexpected behavior in two locus genetic systems: an analysis of marginal underdominance at a stable equilibrium. Genetics **102**, 129–138.

Hastings, A. 1985. Four simultaneously stable polymorphic equilibria in two-locus two-allele models. Genetics **109**, 255–261.

Hastings, A. 1987. Monotonic change of the mean phenotype in two-locus models. Genetics **117**, 583–585.

Hastings, A. 1988. Dependence of expected heterozygosity on locus number with stabilizing selection and drift. J. Theor. Biol. **134**, 103–112.

Hastings, A. 1989. Linkage disequilibrium and genetic variance under mutation-selection balance. Genetics **121**, 857–860.

Hastings, A. 1990a. Maintenance of polygenic variation through mutation-selection balance: bifurcation analysis of a biallelic model. J. Math. Biol. **28**, 329–340.

Hastings, A. 1990b. Second-order approximations for selection coefficients at polygenic loci. J. Math. Biol. **28**, 475–483.

Hastings, A., and Hom, C.L. 1989. Pleiotropic stabilizing selection limits the number of polymorphic loci to at most the number of characters. Genetics **122**, 459–463.

Hastings, A., and Hom, C.L. 1990. Multiple equilibria and maintenance of additive genetic variance in a model of pleiotropy. Evolution **44**, 1153–1163.

Hewitt, E., and Stromberg, K. 1965. *Real and Abstract Analysis.* Berlin, Heidelberg, New York: Springer.

Higgs, P.G. 1994. Error thresholds and stationary mutant distributions in multi-locus diploid genetics models. Genet. Res. **63**, 63–78.

Higgs, P.G., and Woodcock, G. 1995. The accumulation of mutations in asexual populations and the structure of genealogical trees in the presence of selection. J. Math. Biol. **33**, 677–702.

Hill, W.G. 1982a. Rates of change in quantitative traits from fixation of new mutations. Proc. Natl. Acad. Sci. USA **79**, 142–145.

Hill, W.G. 1982b. Predictions of response to artificial selection from new mutations. Genet. Res. **40**, 255–278.

Hill, W.G., and Keightley, P. 1988. Interrelations of mutation, population size, artificial and natural selection. In: Weir, B.S., Eisen, E.J., Goodman, M.M., and Namkoong, G. (eds.) *Proceedings of the Second International Conference on Quantitative Genetics*, pp. 57–70. Sunderland, MA: Sinauer.

Hill, W.G., and Rasbash. J. 1986. Models of long-term artificial selection in finite population. Genet. Res. **48**, 41–50.

Hirsch, M.W., Pugh, C., and Shub, M. 1977. Invariant Manifolds. Lect. Notes Math. **583**. Berlin, Heidelberg, New York: Springer.

Hirsch, M.W., and Smale, S. 1974. *Differential Equations, Dynamical Systems and Linear Algebra.* New York: Academic Press.

Hoekstra, R.F., Bijlsma, R., and Dolman, A.J. 1985. Polymorphism from environmental heterogeneity: models are only robust if the heterozygote is close in fitness to the favored homozygote in each environment. Genet. Res. **45**, 299–314.

Hofbauer, J. 1985. The selection mutation equation. J. Math. Biol. **23**, 41–53.

Hofbauer, J., and Sigmund, K. 1988. *The Theory of Evolution and Dynamical Systems.* Cambridge: University Press.

Hofbauer, J., and Sigmund, K. 1998. *Evolutionary Games and Population Dynamics.* Cambridge: University Press.

Houle, D. 1989. The maintenance of polygenic variation in finite populations. Evolution **43**, 1767–1780.

Houle, D. 1992. Comparing evolvability and variability of quantitative traits. Genetics **130**, 195–204.

Houle, D. 1998. How should we explain variation in the genetic variance of traits? Genetica **102/103**, 241–253.

Houle, D., Morikawa, B., and Lynch, M. 1996. Comparing mutational heritabilities. Genetics **143**, 1467–1483.

Huey, R.B., and Kingsolver, J.G. 1993. Evolution of resistance to high temperature in ectotherms. Amer. Natur. **142**, S21–S46.

Jacquard, A. 1974. *The Genetic Structure of Populations.* Berlin, Heidelberg, New York: Springer.

Johnston, M.O., and Schoen, D.J. 1995. Mutation rates and dominance levels of genes affecting total fitness in two angiosperm species. Science **267**, 226–229.

Jörgens, K. 1970. *Lineare Integraloperatoren.* Stuttgart: BG Teubner.

Karlin, S. 1975. General two-locus selection models: some objectives, results, and interpretations. Theor. Pop. Biol. **7**, 364–398.

Karlin, S. 1978. Theoretical aspects of multilocus selection balance I. In: Levin, S.A. (ed.), *Studies in Mathematical Biology. II. Populations and Communities*, pp. 503–587. Washington: Math. Assoc. of America.

Karlin, S. 1979. Principles of polymorphisms and epistasis for multilocus systems. Proc. Natl. Acad. Sci. USA **75**, 6332–6336.

Karlin, S. 1984. Mathematical models, problems, and controversies of evolutionary theory. Bull. Amer. Math. Soc. **10**, 221–273.

Karlin, S. 1988. Non-Gaussian phenotypic models of quantitative traits. In: Weir, B.S., Eisen, E.J., Goodman, M.M., and Namkoong, G. (eds.) *Proceedings of the Second International Conference on Quantitative Genetics*, pp. 123–144. Sunderland, MA: Sinauer.

Karlin, S., and Carmelli, D. 1975. Numerical studies on two-loci selection models with general viabilities. Theor. Pop. Biol. **7**, 399–421.

Karlin, S., and Feldman, M.W. 1970. Linkage and selection: two locus symmetric viability model. Theor. Pop. Biol. **1**, 39–71.

Karlin, S., and Feldman, M.W. 1978. Simultaneous stability of $D = 0$ and $D \neq 0$ for multiplicative viabilities at two loci. Genetics **90**, 813–825.

Karlin, S., and Lessard, S. 1984. On the optimal sex-ratio: A stability analysis based on a characterization for one-locus multiallele viability models. J. Math. Biol. **20**, 15–38.

Karlin, S., and Liberman, U. 1978. The two-locus multi-allele additive viability model. J. Math. Biol. **5**, 201–211.

Karlin, S., and Liberman, U. 1979. Representation of nonepistatic selection models and analysis of multilocus Hardy–Weinberg equilibrium configurations. J. Math. Biol. **7**, 353–374.

Karlin, S., and Liberman, U. 1982. The reduction property for central polymorphisms in nonepistatic systems. Theor. Pop. Biol. **22**, 69–95.

Karlin, S., and Liberman, U. 1990. Global convergence properties in multilocus viability selection models: the additive model and the Hardy–Weinberg law. J. Math. Biol. **29**, 161–176.

Karlin, S., and McGregor, J.L. 1972. Application of method of small parameters to multi-niche population genetic models. Theor. Pop. Biol. **3**, 186–208.

Karlin, S., and Taylor, H.M. 1975. *A First Course in Stochastic Processes.* Boston: Academic Press.

Kearsey, M.J., and Farquhar, A.G. 1998. QTL analysis in plants; where are we now? Heredity **80**, 137–142.

Kearsey, M.J., and Gale, J.S. 1968. Stabilising selection in the absence of dominance: an additional note. Heredity **23**, 617–620.

Keightley, P.D. 1996. Nature of deleterious mutation load in *Drosophila.* Genetics **144**, 1993-1999.

Keightley, P.D. 1998a. Genetic basis of response to 50 generations of selection on body weight in inbred mice. Genetics **148**, 1931–1939.

Keightley, P.D. 1998b. Inference of genome-wide mutation rates and distributions of mutation effects for fitness traits: a simulation study. Genetics **150**, 1283–1293.

Keightley, P.D., and Caballero, A. 1997. Genomic mutation rates for lifetime reproductive output and lifespan in *Caenorhabditis elegans.* Proc. Natl. Acad. Sci., USA **94**, 3823–3827.

Keightley, P.D., and Eyre-Walker, A. 1999. Terumi Mukai and the riddle of deleterious mutation rates. Genetics **153**, 515–523.

Keightley, P.D., and Hill, W.G. 1983. Effects of linkage on response to directional selection from new mutations. Genet. Res. **42**, 193–206.

Keightley, P.D., and Hill, W.G. 1987. Directional selection and variation in finite populations. Genetics **117**, 573–582.

Keightley, P.D., and Hill, W.G. 1988. Quantitative genetic variability maintained by mutation-stabilizing selection balance in finite populations. Genet. Res. **52**, 33–43.

Keightley, P.D., and Hill, W.G. 1990. Variation maintained in quantitative traits with mutation-selection balance: pleiotropic side effects on fitness traits. Proc. Roy. Soc. London **242**, 95–100.

Kempthorne, O. 1955. The theoretical values of the correlations between relatives in random mating populations. Genetics **40**, 153–167.

Kempthorne, O. 1957. *An Introduction to Genetic Statistics.* New York: Wiley.

Kibota, T., and Lynch, M. 1996. Estimate of the genomic mutation rate deleterious to overall fitness in *Escherichia coli.* Nature **381**, 694–696.

Kimmel, M., and Chakraborty, R. 1996. Measures of variation at DNA repeat loci under a general stepwise mutation model. Theor. Pop. Biol. **50**, 345–367.

Kimura, M. 1956. A model of a genetic system which leads to closer linkage by natural selection. Evolution **10**, 278–287.

Kimura, M. 1957. Some problems of stochastic processes in genetics. Ann. Math. Statistics **28**, 882–901.

Kimura, M. 1958. On the change of population fitness by natural selection. Heredity **12**, 145–167.

Kimura, M. 1965a. A stochastic model concerning the maintenance of genetic variability in quantitative characters. Proc. Natl. Acad. Sci. U.S.A. **54**, 731–736.

Kimura, M. 1965b. Attainment of quasi-linkage equilibrium when gene frequencies are changing by natural selection. Genetics **52**, 875–890.

Kimura, M. 1981. Possibility of extensive neutral evolution under stabilizing selection with special reference to nonrandom usage of synonymous codons. Proc Natl. Acad. Sci., USA **78**, 5773–5777.

Kimura, M. 1983. *The Neutral Theory of Molecular Evolution.* Cambridge: University Press.

Kimura, M., and Crow, J.F. 1978. Effect of overall phenotypic selection on genetic change at individual loci. Proc. Natl. Acad. Sci., USA **75**, 6168–6178.

Kimura, M., and Maruyama, T. 1966. The mutational load with epistatic gene interaction in fitness. Genetics **54**, 1337–1351.

Kimura, M., and Ohta, T. 1969. The average number of generations until fixation of a mutant gene in a finite population. Genetics **61**, 763–771.

King, J.L. 1966. The gene interaction component of the genetic load. Genetics **53**, 403–413.

Kingman, J.F.C. 1961a. An inequality in partial averages. Quart. J. Math. **12**, 78–80.

Kingman, J.F.C. 1961b. A mathematical problem in population genetics. Math. Proc. Camb. Phil. Soc. **57**, 574–582.

Kingman, J.F.C. 1977. On the properties of bilinear models for the balance between genetic mutation and selection. Math. Proc. Camb. Phil. Soc. **81**, 443–453.

Kingman, J.F.C. 1978. A simple model for the balance between selection and mutation. J. Appl. Prob. **15**, 1–12.

Kingman, J.F.C. 1980. *Mathematics of Genetic Diversity.* Philadelphia: SIAM.

Kingsolver, J.G., and Smith, S.G. 1995. Estimating selection on quantitative traits using capture-recapture data. Evolution **49**, 384–388.

Kojima, K.-I. 1959. Stable equilibria for the optimum model. Proc. Natl. Acad. Sci. USA **78**, 5773–5777.

Kondrashov, A.S. 1982. Selection against harmful mutations in large sexual and asexual populations. Genet. Res. **40**, 325–332.

Kondrashov, A.S. 1984. Deleterious mutations as an evolutionary factor. I. The advantage of recombination. Genet. Res. **44**, 199–217.

Kondrashov, A.S. 1988. Deleterious mutations and the evolution of sexual reproduction. Nature **336**, 435–440.

Kondrashov, A.S. 1993. Classification of hypotheses on the advantage of amphimixis. J. Hered. **84**, 372–387.

Kondrashov, A.S. 1994. Muller's ratchet under epistatic selection. Genetics **136**, 1469–1473.

Kondrashov, A.S. 1995. Contamination of genomes by very slightly deleterious mutations: why have we not died 100 times over? J. Theor. Biol. **175**, 583–594.

Kondrashov, A.S. 1998. Measuring spontaneous deleterious mutation process. Genetica **102/103**, 182–197.

Kondrashov, A.S., and Crow, J.F. 1988. King's formula for the mutation load with epistasis. Genetics **120**, 853–856.

Kondrashov, A.S., and Turelli, M. 1992. Deleterious mutations, apparent stabilizing selection and the maintenance of quantitative variation. Genetics **132**, 603–618.

Kondrashov, A.S., and Yampolsky, L.Yu. 1996a. High genetic variability under under the balance between symmetric mutation and fluctuating stabilizing selection. Genet. Res. **68**, 157–164.

Kondrashov, A.S., and Yampolsky, L.Yu. 1996b. Evolution of amphimixis and recombination under fluctuating selection in one and many traits. Genet. Res. **68**, 165–173.

Lande, R. 1975. The maintenance of genetic variability by mutation in a polygenic character with linked loci. Genet. Res. **26**, 221–235

Lande, R. 1976. Natural selection and random genetic drift in phenotypic evolution. Evolution **30**, 314–334.

Lande, R. 1977. The influence of the mating system on the maintenance of genetic variability in polygenic characters. Genetics **86**, 485–498.

Lande, R. 1979. Quantitative genetic analysis of multivariate evolution applied to brain : body size allometry. Evolution **34**, 402–416.

Lande, R. 1980a. The genetic covariance between characters maintained by pleiotropic mutations. Genetics **94**, 203–215.

Lande, R. 1980b. Genetic variation and phenotypic evolution during allopatric speciation. Am. Nat. **116**, 463–479.

Lande, R. 1980c. Sexual dimorphism, sexual selection and adaptation in polygenic characters. Evolution **34**, 292–305.

Lande, R. 1981. Models of speciation by sexual selection on polygenic traits. Proc. Natl. Acad. Sci., USA **78**, 3721–3725.

Lande, R. 1982. A quantitative genetic theory of life-history evolution. Ecology **63**, 292–305.

Lande, R. 1984. The genetic correlation between characters maintained by selection, linkage and inbreeding. Genet. Res. **44**, 309–320.

Lande, R. 1988. Quantitative genetics and evolutionary theory. In: Weir, B.S., Eisen, E.J., Goodman, M.M., and Namkoong, G. (eds.) *Proceedings of the Second International Conference on Quantitative Genetics*, pp. 71–84. Sunderland, MA: Sinauer.

Lande, R. 1991. Isolation by distance in a quantitative trait. Genetics **128**, 443–452.

Lande, R. 1992. Neutral theory of quantitative genetic variation in an island model with local extinction and colonization. Evolution **46**, 381–389.

Lande, R. 1994. Risk of population extinction from fixation of new deleterious mutations. Evolution **48**, 1460–1469.

Lande, R. 1995. Mutation and conservation. Conservation Biology **9**, 782–791.

Lande, R. 1998. Risk of population extinction from fixation of deleterious and reverse mutations. Genetica **102/103**, 21–27.

Lande, R., and Arnold, S.J. 1983. The measurement of selection on correlated characters. Evolution **37**, 1210–1226.

Lande, R., and Shannon, S. 1996. The role of genetic variation in adaptation and population persistence in a changing environment. Evolution **50**, 434–437.

Lander, E.S., and Botstein, D. 1989. Mapping Mendelian factors underlying quantitative traits using RFLP linkage maps. Genetics **121**, 185–199.

LaSalle, J.P. 1976. *The Stability of Dynamical Systems.* Regional Conf. Ser. Appl. Math. **25**. Philadelphia: SIAM.

Latter, B.D.H. 1960. Natural selection for an intermediate optimum. Aust. J. Biol. Sci. **13**, 30–35.

Latter, B.D.H. 1965. The response to artificial selection due to autosomal genes of large effect. Aust. J. Biol. Sci. **18**, 585–598.

Latter, B.D.H. 1970. Selection in finite populations with multiple alleles. II. Centripetal selection, mutation, and isoallelic variation. Genetics **66**, 165–186.

Latter, B.D.H., and Novitski, C.E. 1970. Selection in finite populations with multiple alleles. I. Limits to directional selection. Genetics **62**, 859–876.

Leamy, L.J., Routman, E.J., and Cheverud, J.M. 1999. Quantitative trait loci for early- and late-developing skull characters in mice: A test of the genetic independence model of morphological integration. Am. Nat. **153**, 201–214.

Lederman, W. (ed.) 1980. *Handbook of Applicable Mathematics. Vol. II: Probability.* Chichester: Wiley.

Lerner, I.M. 1954. *Genetic Homeostasis.* London: Oliver and Boyd.

Lessard, S. 1997. Fisher's fundamental theorem of natural selection revisited. Theor. Pop. Biol. **52**, 119–136.

Lessard, S., and Castilloux, A.-M. 1995. The fundamental theorem of natural selection in Ewens' sense (case of fertility selection). Genetics **141**, 733–742.

Lessells, C.M. 1982. Some causes and consequences of family size in the Canada Goose Branta canadensis. Ph. D. Diss. Oxford, UK: University Oxford.

Levene, H. 1953. Genetic equilibrium when more than one ecological niche is available. Am. Nat. **87**, 311–313.

Lewontin, R.C. 1964. The interaction of selection and linkage. II. Optimum models. Genetics **50**, 757–782.

Lewontin, R.C., and Kojima, K.-I. 1960. The evolutionary dynamics of complex polymorphisms. Evolution **14**, 458–472.

Lewontin, R.C. 1974. *The Genetic Basis of Evolutionary Change.* New York: Columbia University Press.

Li, C.C. 1967. Fundamental theorem of natural selection. Nature **214**, 505-506.

Li, J.-L., Li, J., and Deng, H.-W. 1999. The effects of overdominance on characterizing deleterious mutations in large natural populations. Genetics **151**, 895–913.

Li, Z., Pinson, S.R., Park, W.D., Paterson, A.H., and Stansel, J.W. 1997. Epistasis for three grain yield components in rice (*Oryza sativa L.*). Genetics **145**, 453–465.

Liberman, U. 1988. External stability and ESS: criteria for initial increase of new mutant allele. J. Math. Biol. **26**, 477–485.

Long, A.D., and Langley, C.H. 1999. The power of association studies to detect the contribution of candidate genetic loci to variation in complex traits. Genome Res. **9**, 720–731.

Long, A.D., Mullaney, S.L., Mackay, T.F.C., and Langley, C.H. 1996. Genetic interactions between naturally occurring alleles at quantitative trait loci and mutant alleles at candidated loci affecting bristle number in *Drosophila melanogaster.* Genetics **144**, 1497–1510.

Long, A.D., Lyman, R.F., Langley, C.H., and Mackay, T.F.C. 1998. Two sites in the *Delta* gene region contribute to naturally occurring variation in bristle number in *Drosophila melanogaster.* Genetics **149**, 999–1017.

López-Fanjul, C., and Hill, W.G. 1973. Genetic differences between populations of *D. melanogaster* for a quantitative trait. Genet. Res. **22**, 51-68.

López, M.A., and López-Fanjul, C. 1993a,b. Spontaneous mutation for a quantitative trait in *Drosophila melanogaster.* I. Response to artificial selection. Genet. Res. **61**, 107–116. II. Distribution of mutant effects on the trait and fitness. Genet. Res. **61**, 117–126.

Losert, V., and Akin, E. 1983. Dynamics of games and genes: discrete versus continuous time. J. Math. Biol. **17**, 241–251.

Lush, J.L. 1937. *Animal Breeding Plans.* Ames: Iowa State University Press.

Lyman, R.F., Lawrence, F., Nuzhdin, S.V., and Mackay, T.F.C. 1996. Effects of single *P* element insertions on bristle number and viability in *Drosophila melanogaster.* Genetics **143**, 277–292.

Lyman, R.F., and Mackay, T.F.C. 1998. Candidate quantitative trait loci and naturally occurring phenotypic variation for bristle number in *Drosophila melanogaster*: The *Delta-hairless* gene region. Genetics **149**, 983–998.

Lynch, M. 1984. The limits to life history evolution in *Daphnia.* Evolution **35**, 465–482.

Lynch, M., and Blanchard, J.L. 1998. Deleterious mutation accumulation in organelle genomes. Genetica **102/103**, 29–39.

Lynch, M., and Gabriel, W. 1990. Mutation load and the survival of small populations. Evolution **44**, 1725–1737.

Lynch, M., and Hill, W.G. 1986. Phenotypic evolution by neutral mutation. Evolution **40**, 915–935.

Lynch, M., and Lande, R. 1993. Evolution and extinction in response to environmental change. In: Kareiva, P.M., Kingsolver, J.G., and Huey, R.B. (eds.) *Biotic Interactions and Global Change*, pp. 234–250. Sunderland, MA: Sinauer.

Lynch, M., and Walsh, B. 1998. *Genetics and Analysis of Quantitative Traits*. Sunderland, Mass.: Sinauer.

Lynch, M., Connery, J., and Bürger, R. 1995a. Mutational meltdowns in sexual populations. Evolution **49**, 1067–1088.

Lynch, M., Connery, J., and Bürger, R. 1995b. Mutation accumulation and the extinction of small populations. Am. Nat. **146**, 489–518.

Lynch, M., Gabriel, W., and Wood, A.M. 1991. Adaptive and demographic responses of plankton populations to environmental change. Limnol. Oceanogr. **36**, 1301–1312.

Lynch, M., Bürger, R., Butcher, D., and Gabriel, W. 1993. The mutational meltdown in asexual populations. J. of Heredity **84**, 339-344.

Lynch, M., Latta, L., Hicks, J., and Giorgianni, M. 1998. Mutation, selection, and the maintenance of life-history variation in a natural population. Evolution **52**, 727–733.

Lynch, M., Blanchard, J., Houle, D., Kibota, T., Schultz, S., Vassilieva, L., and Willis, J. 1999. Perspective: spontaneous deleterious mutation. Evolution **53**, 645–663.

Lyubich, Yu.I. 1971. Basic concepts and theorems of evolutionary genetics of free populations. Russ. Math. Surv. **26**, 51–123.

Lyubich, Yu.I. 1992. *Mathematical Structures in Population Genetics*. Berlin, Heidelberg, New York: Springer.

Lyubich, Yu.I., Maĭstrovskiĭ, G.D., and Ol'khovskiĭ, Yu.G. 1976. Convergence to equilibrium under the action of selection in a single locus population. Soviet Math. Dokl. **17**, 55–58.

Lyubich, Yu.I., Maĭstrovskiĭ, G.D., and Ol'khovskiĭ, Yu.G. 1980. Selection-induced convergence to equilibrium in a single-locus autosomal population. Probl. Inform. Transm. **16**, 66–75.

Mackay, T.F.C. 1981. Genetic variation in varying environments. Genet. Res. **37**, 79–93.

Mackay, T.F.C., and Fry, J.D. 1996. Polygenic mutation in *Drosophila melanogaster*: Genetic interactions between selection lines and candidate quantitative trait loci. Genetics **144**, 671–688.

Mackay, T.F.C., and Lyman, R.F. 1998. Polygenic mutation in *Drosophila melanogaster*: genotype×environment interaction. Genetics for spontaneous mutations affecting bristle number. Genetica **102/103**, 199–215.

Mackay, T.F.C., Lyman, R.F., and Jackson, M.S. 1992. Effects of P elements on quantitative traits in *Drosophila melanogaster*. Genetics **130**, 315–332.

Malécot, G. 1948. *Les Mathématiques de l'Hérédité*. Paris: Masson. Extended translation: *The Mathematics of Heredity*. San Francisco: W.H. Freeman (1969).

Malécot, G. 1952. Les processus stochastiques et la méthode des fonctions génératrices ou caracteréstiques. Publ. Inst. Stat. University Paris **1**, Fasc. 3, 1–16.

Mandel, S.P.H. 1959. The stability of a multiple allelic system. Heredity **13**, 289–302.

Mather, K., and Jinks, J.L. 1982. *Biometrical Genetics*, 3rd ed. New York: Chapman & Hall.

May, R.M. 1976. Simple mathematical models with very complicated dynamics. Nature **261**, 459–467.

May, R.M., and Oster, G.F. 1976. Bifurcations and dynamic complexity in simple ecological models. Am. Nat. **110**, 573–599.

Maynard Smith, J. 1978. *The Evolution of Sex*. Cambridge: University Press.

Maynard Smith, J. 1983. The genetics of stasis and punctuation. Annu. Rev. Genet. **17**, 11–25.

Maynard Smith, J. 1998. *Evolutionary Genetics*, 2nd ed. Oxford: University Press.

Mayo, O. 1987. *The Theory of Plant Breeding*, 2nd ed. Oxford: Clarendon Press.

Mayo, O., and Franklin, I.R. 1998. The place of QTL in the basis of quantitative genetics. II. General considerations. In: Proceedings of the 6th World Congress on Genetics Applied to Livestock Production, 26, pp. 77–80.

Mendel, G.J. 1866. Versuche über Pflanzen-Hybriden. Verhandlungen des naturforschenden Vereines, Abhandlungen, Brünn **4**, 3–47.

Metz, J.A.J., and Diekmann, O. (eds.) 1986. The Dynamics of Physiologically Structured Populations. Lect. Notes in Biomath. **68**. Berlin, Heidelberg, New York: Springer.
Michod, R., and Levin, B. 1988. (eds.) *The Evolution of Sex: An Examination of Current Ideas*. Sunderland, MA: Sinauer.
Moran, P.A.P. 1964. On the nonexistence of adaptive topographies. Ann. Hum. Genet. **27**, 383–393.
Moran, P.A.P. 1967. Unsolved problems in evolutionary theory. Proc. Fifth Berkeley Symp. Math. Stat. Probab. **4**, 457–480.
Moran, P.A.P. 1968. On the theory of selection dependent on two loci. Ann. Hum. Genet. **32**, 183–190.
Moran, P.A.P. 1976. Global stability of genetic systems governed by mutation and selection. Math. Proc. Camb. Phil. Soc. **80**, 331–336.
Moran, P.A.P. 1977. Global stability of genetic systems governed by mutation and selection. II. Math. Proc. Camb. Phil. Soc. **81**, 435–441.
Mousseau, T.A., and Roff, D.A. 1987. Natural selection and the heritability of fitness components. Heredity **58**, 181–197.
Mukai, T. 1964. The genetic structure of natural populations of *Drosophila melanogaster*. I. Spontaneous mutation rate of polygenes controlling viability. Genetics **50**, 1–19.
Mukai, T., Chigusa, S.I., Mettler, L.E., and Crow, J.F. 1972. Mutation rate and dominance of genes affecting viability in *Drosophila melanogaster*. Genetics **72**, 335–355.
Mulholland, H.P., and Smith, C.A.B. 1959. An inequality arising in genetical theory. Amer. Math. Monthly **66**, 673–683.
Muller, H.J. 1950. Our load of mutations. Amer. J. Hum. Genet. **2**, 111–176.
Muller, H.J. 1964. The relation of recombination to mutational advance. Mutat. Res. **1**, 2–9.

Nagylaki, T. 1974. Continuous selective models with mutation and migration. Theor. Pop. Biol. **5**, 284–295.
Nagylaki, T. 1976. The evolution of one- and two-locus systems. Genetics **83**, 583–600.
Nagylaki, T. 1977. Selection in One- and Two-Locus Systems. Lect. Notes Biomath. **15**. Berlin, Heidelberg, New York: Springer.
Nagylaki, T. 1977a. The evolution of one- and two-locus systems. II. Genetics **85**, 347–354.
Nagylaki, T. 1984. Selection on a quantitative character. In: Chakravarti, A. (ed.) *Human Population Genetics: The Pittsburgh Symposium*, pp. 275–306. New York: Van Nostrand.
Nagylaki, T. 1989a. The maintenance of genetic variability in two-locus models of stabilizing selection. Genetics **122**, 235–248.
Nagylaki, T. 1989b. Rate of evolution of a character without epistasis. Proc. Natl. Acad. Sci. **86**, 1910–1913.
Nagylaki, T. 1990. Models and approximations for random genetic drift. Theor. Pop. Biol. **37**, 192–212.
Nagylaki, T. 1991. Error bounds for the fundamental and secondary theorems of natural selection. Proc. Natl. Acad. Sci. USA **88**, 2402–2406.
Nagylaki, T. 1992. *Introduction to Theoretical Population Genetics*. Berlin, Heidelberg, New York: Springer.
Nagylaki, T. 1992a. Rate of evolution of a quantitative character. Proc. Natl. Acad. Sci. USA **89**, 8121–8124.
Nagylaki, T. 1993. The evolution of multilocus systems under weak selection. Genetics **134**, 627–647.
Nagylaki, T. 1994. Geographical variation in a quantitative character. Genetics **136**, 361–381.
Nagylaki, T. 1995. The inbreeding effective population number in dioecious populations. Genetics **139**, 473–485.
Nagylaki, T. 1997. Multinomial-sampling models for random genetic drift. Genetics **145**, 485–491.
Nagylaki, T., and Crow, J. 1974. Continuous selective models. Theor. Pop. Biol. **5**, 257–283.
Nagylaki, T., Hofbauer, J., and Brunovský, P. 1999. Convergence of multilocus systems under weak epistasis or weak selection. J. Math. Biol. **38**, 103–133.

Nevo, E. 1978. Genetic variation in natural populations: patterns and theory. Theor. Pop. Biol. **13**, 121–177.

Newburgh, J.D. 1951. The variation of spectra. Duke Math. J. **18**, 165–176.

Norman, M.F. 1974. A central limit theorem for Markov processes that move by small steps. Ann. Prob. **2**, 1065–1074.

Nuzhdin, S.V., Dilda, C.L., and Mackay, T.F.C. 1999. The genetic architecture of selection response: inferences from fine-scale mapping of bristle number quantitative trait loci in *Drosophila melanogaster*. Genetics **153**, 1317–1331.

Nuzhdin, S.V., Fry, J.D., and Mackay, T.F.C. 1995. Polygenic mutation in *Drosophila melanogaster*: the causal relationship of bristle number to fitness. Genetics **139**, 861–872.

Ohta, T., and Kimura, M. 1973. A model of mutation appropriate to estimate the number of electrophoretically detectable alleles in a finite population. Genet. Res. **22**, 201–204.

Ohta, T., and Kimura, M. 1975. Theoretical analysis of electrophoretically detectable alleles: models of very slightly deleterious mutations. Am. Nat. **109**, 137–145.

Ohnishi, O. 1977. Spontaneous and ethyl methanesulfonate-induced mutations controlling viability in *Drosophila melanogaster*. II. Homozygous effect of polygenic mutations. Genetics **87**, 529–545.

Otto, S.P. 1997. Unravelling gene interactions. Nature **390**, 343.

Otto, S.P., and Feldman, M.W. 1997. Deleterious mutations, variable epistatic interactions, and the evolution of recombination. Theor. Pop. Biol. **51**, 134–147.

Pamilo, P., Nei, M., and Li, W.-H. 1987. Accumulations of mutations in sexual and asexual populations. Genet. Res. **49**, 135–146.

Passekov, V.P. 1978. On the relation between steady-state patterns of the gene frequencies' distribution and extension of the Fisher fundamental theorem for a single multi-allelic locus. Doklady MOIP, Obschaya Biologiya, 1978, pp. 135–138 (in Russian).

Passekov, V.P. 1984. Asymptotic analysis of selection in a multilocus, multiallelic population. Dokl. Akad. Nauk SSSR **277**, 1338–1341.

Pearson, K. 1903. Mathematical contributions to the theory of evolution. XI. On the influence of natural selection on the variability and correlation of organs. Phil. Trans. Roy. Soc. A **200**, 1–66.

Pearson, K. 1904. Mathematical contributions to the theory of evolution. XII. On a generalized theory of alternative inheritance, with special reference to Mendel's laws. Phil. Trans. Roy. Soc. A **203**, 53–86.

Peck, J.R. 1994. A ruby in the rubbish: beneficial mutations, deleterious mutations and the evolution of sex. Genetics **137**, 597–606.

Peck, J.R., Barreau, G., and Heath, S.C. 1997. Imperfect genes, Fisherian mutation and the evolution of sex. Genetics **145**, 1171–1199.

Piper, L.R., and Bindon, B.M. 1988. The genetics and endrocrinology of the Booroola sheep F gene. In: Weir, B.S., Eisen, E.J., Goodman, M.M., and Namkoong, G. (eds.) *Proceedings of the Second International Conference on Quantitative Genetics*, pp. 270–280. Sunderland, MA: Sinauer.

Price, G.R. 1970. Selection and covariance. Nature **227**, 520–521.

Price, G.R. 1972. Fisher's 'fundamental theorem' made clear. Ann. Hum. Genet. **36**, 129–140.

Provine, W.B. 1971. *The Origins of Theoretical Population Genetics.* Chicago: Univ. Press.

Rausher, M.D. 1992. The measurement of selection on quantitative traits: biases due to environmental covariances between traits and fitness. Evolution **46**, 616–626.

Riedl, R. 1978. *Order in Living Organisms.* New York: Wiley.

Robertson, A. 1956. The effect of selection against extreme deviants based on deviation or on homozygosis. J. Genet. **54**, 236–248.

Robertson, A. 1960. A theory of limits in artificial selection. Proc. Roy. Soc. London, B., **153**, 234–249.

Robertson, A. 1966. A mathematical model of the culling process in dairy cattle. Anim. Prod. **8**, 95–108.

Robertson, A. 1967. The nature of quantitative genetic variation. In: Brink, R.A. (ed.) *Heritage from Mendel*, pp. 265–280. Madison: University of Wisconsin Press.
Robertson, A. 1968. The spectrum of genetic variation. In: Lewontin, R.C. (ed.) *Population Biology and Evolution*, pp. 5–16. Syracuse: University Press.
Robbins, R.B. 1918. Some applications of mathematics to breeding problems. III. Genetics **3**, 375–389.
Roff, D.A. 1997. *Evolutionary Quantitative Genetics*. New York: Chapman & Hall.
Roff, D.A., and Mousseau, T.A. 1987. Quantitative genetics and fitness: Lessons from *Drosophila*. Heredity **58**, 103–118.
Rose, M.R. 1982. Antagonistic pleiotropy, dominance and genetic variation. Heredity **48**, 63–78.
Roughgarden, J. 1979. *Theory of Population Genetics and Evolutionary Ecology: An Introduction*. New York: Macmillan.
Routman, E.J., and Cheverud, J.M. 1997. Gene effects on a quantitative trait: two-locus epistatic effects measured at microsatellite markers and at estimated QTL. Evolution **51**, 1654–1662.
Roux, C.Z. 1974. Hardy–Weinberg equilibria in random mating populations. Theor. Pop. Biol. **5**, 393–416.

Santiago, E. 1998. Linkage and the maintenance of variation for quantitative traits by mutation-selection balance: an infinitesimal model. Genet. Res. **71**, 161–170.
Santiago, E., Albornoz, J., Dominguez, A., Toro, A.M., and López-Fanjul, C. 1992. The distribution of spontaneous mutations on quantitative traits and fitness in *Drosophila melanogaster*. Genetics **132**, 771–781.
Schaefer, H.H. 1974. *Banach Lattices and Positive Operators*. Berlin, Heidelberg, New York: Springer.
Scheiner, S.M. 1998. The genetics of phenotypic plasticity. VII. Evolution in a spatially-structured environment. J. Evol. Biol. **11**, 303-320.
Scheuer, P.A.G., and Mandel, S.P.H. 1959. An inequality in population genetics. Heredity **13**, 519–524.
Schluter, D. 1988. Estimating the form of natural selection on a quantitative trait. Evolution **42**, 849–861.
Schultz, S.T., and Lynch, M. 1997. Mutation and extinction: the role of variable mutational effects, synergistic epistasis, beneficial mutations, and degree of outcrossing. Evolution **51**, 1363–1371.
Schultz, S.T., Lynch, M., and Willis, J.H. 1999. Spontaneous deleterious mutation in *Arabidopsis thaliana*. Proc. Natl. Acad. Sci. USA **96**, 11393–11398.
Seneta, E. 1981. *Non-Negative Matrices*, 2nd edn. London: Allen and Unwin.
Shahshahani, S. 1979. A new mathematical framework for the study of linkage and selection. Memoirs Amer. Math. Soc. **211**. Providence, R.I.: Amer. Math. Soc.
Shook, D.R., and Johnson, T.E. 1999. Quantitative trait loci affecting survival and fertility-related traits in *Caenorhabditis elegans* show genotype-environment interactions, pleiotropy and epistasis. Genetics **153**, 1233–1243.
Shrimpton, A.E., and Robertson, A. 1988a. The isolation of polygenic factors controlling bristle score in *Drosophila melanogaster*. I. Allocation of third chromosome bristle effects to chromosome sections. Genetics **118**, 437–443.
Shrimpton, A.E., and Robertson, A. 1988b. The isolation of polygenic factors controlling bristle score in *Drosophila melanogaster*. II. Distribution of third chromosome bristle effects within chromosome sections. Genetics **118**, 445–459.
Simmons, M.J., and Crow, J.F. 1977. Mutations affecting fitness in *Drosophila* populations. Ann. Rev. Genet. **11**, 49–78.
Singh, M., and Lewontin, R.C. 1966. Stable equilibria under optimizing selection. Proc. Natl. Acad. Sci. USA **56**, 1345–1348.
Slatkin, M. 1972. On treating the chromosome as the unit of selection. Genetics **72**, 519–524.
Slatkin, M. 1978. Spatial patterns in the distribution of polygenic characters. J. Theor. Biol. **70**, 213–228.

Slatkin, M. 1979. Frequency- and density-dependent selection on a quantitative character. Genetics **93**, 755–771.

Slatkin, M. 1987. Heritable variation and heterozygosity under a balance between mutations and stabilzing selection. Genet. Res. **50**, 53–62.

Slatkin, M., and Lande, R. 1976. Niche width in a fluctuating environment-density independent model. Am. Nat. **110**, 31–55.

Slatkin, M., and Frank, S.A. 1990. The quantitative genetic consequences of pleiotropy under stabilizing and directional selection. Genetics **125**, 207–213.

Smith, H.F. 1936. A discriminant function for plant selection. Ann. Eugen. **7**, 240-250.

Speed, T.P. 1983. Cumulants and partition lattices. Austral. J. Statist. **25**, 378–388.

Speed, T.P. 1986. Applications of cumulants and their generalizations. In: Francis, I.S., Manly, B.F.J., and Lam, F.C. (ed.) *Pacific Statistical Congress*, pp. 12–20. New York: Elsevier (North Holland).

Stearns, S.C. 1992. *The Evolution of Life Histories.* Oxford: University Press.

Stearns. S.C., and Kawecki, T.J. 1994. Fitness sensitivity and the canalization of life-history traits. Evolution **48**, 1438–1450.

Stephan, W., Chao, L., and Smale, J.G. 1993. The advance of Muller's ratchet in a haploid asexual population: approximate solutions based on diffusion theory. Genet. Res. **61**, 225–232.

Stockard, C.R. 1941. *The Genetic and Endocrine Basis for Differences in Form and Behavior.* Philadelphia: Wistar Institute.

Stuart, A., and Ord, J.K. 1994. *Kendall's Advanced Theory of Statistics, Vol. 1. Distribution Theory.* 6th ed. London: Edward Arnold.

Svirezhev, Yu.M. 1972. Optimality principles in population genetics. In: *Studies in Theoretical Genetics*, pp. 86–102. Novisibirsk: Inst. of Cytology and Genetics. (In Russian).

Svirezhev, Yu.M., and Passekov V.P. 1990. *Fundamentals of Mathematical Evolutionary Genetics.* Dordrecht: Kluwer.

Swetina, J., and Schuster, P. 1982. Self-replication with errors. A model for poylnucleotide replication. Biophys. Chem. **16**, 329–345.

Tanaka, Y. The genetic variance maintained by pleiotropic mutations. Theor. Pop. Biol. **49**, 211–231.

Timofeeff-Ressovsky, N.V., and Svireshev, Yu.M. 1970. Population genetics and optimal processes. Genetika **6**, 10, 155–166. (In Russian).

Thompson, C.J., and McBride, J.L. 1974. On Eigen's theory of the self-organization of matter and the evolution of biological macromolecules. Math. Biosci. **21**, 127–142.

Thompson, J.N., Jr., Jeung, M., and Thoday, J.M. 1998. Environment-influenced expression of polygene mutations isolated from a natural population of *Drosophila melanogaster.* Genetica **102/103**, 217–228.

Turelli, M. 1984. Heritable genetic variation via mutation-selection balance: Lerch's zeta meets the abdominal bristle. Theor. Pop. Biol. **25**, 138–193.

Turelli, M. 1985. Effects of pleiotropy on predictions concerning mutation-selection balance for polygenic traits. Genetics **111**, 165–195.

Turelli, M. 1986. Gaussian versus non-Gaussian genetic analyses of polygenic mutation-selection balance. In: Karlin, S., and Nevo, E. (eds.) *Evolutionary Processes and Theory,* pp. 607–628. New York: Academic Press.

Turelli, M. 1988a. Phenotypic evolution, constant covariances and the maintenance of additive variance. Evolution **43**, 1342–1347.

Turelli, M. 1988b. Population genetic models for polygenic variation and evolution. In: Weir, B.S., Eisen, E.J., Goodman, M.M., and Namkoong, G. (eds.) *Proceedings of the Second International Conference on Quantitative Genetics*, pp. 601–618. Sunderland, MA: Sinauer.

Turelli, M., and Barton, N.H. 1990. Dynamics of polygenic characters under selection. Theor. Pop. Biol. **38**, 1–57.

Turelli, M., and Barton, N.H. 1994. Genetic and statistical analyses of strong selection on polygenic traits: What, me normal? Genetics **138**, 913–941.

Vassilieva, L.L., and Lynch, M. 1999. The rate of spontaneous mutation for life-history traits in *Caenorhabditis elegans*. Genetics **151**, 119–129.

Vaughn, T.T., Pletscher, L.S., Peripato, A., King-Ellison K., Adams, E., Erikson, C., and Cheverud, J.M. 1999. Mapping quantitative trait loci for murine growth: a closer look at genetic architecture. Genet. Res. **74**, 313–322.

Via, S., and Lande, R. 1987. Evolution of genetic variability in a spatially heterogeneous environment: effects of genotype-environment interaction. Genet. Res. **49**, 147–156.

Vickers, G.T., and Cannings, C. 1988. On the number of stable equilibria in a one-locus multi-allele system. J. Theor. Biol. **131**, 273–277.

Wagner, G.P. 1984. Coevolution of functionally constrained characters: Prerequisites for adaptive versatility: BioSystems **17**, 51–55.

Wagner, G.P. 1988. The influence of variation and of developmental constraints on the rate of multivariate phenotypic evolution. J. Evol. Biol. **1**, 45–66.

Wagner, G.P. 1989. Multivariate mutation-selection balance with constrained pleiotropic effects. Genetics **122**, 223–234.

Wagner, G.P. 1996a. Homology, natural kinds, and the evolution of modularity. Am. Zool. **36**, 36–43.

Wagner, G.P. 1996b. Apparent stabilizing selection and the maintenance of neutral genetic variation. Genetics **143**, 617–619.

Wagner, G.P. 1998. Complexity matters. Science **279**, 1158–1159.

Wagner, G.P., and Gabriel, W. 1990. Quantitative variation in finite parthenogenetic populations: what stops Muller's ratchet in the absence of recombination? Evolution **44**, 715–731.

Wagner, G.P., and Krall, P. 1993. What is the difference between models of error thresholds and Muller's ratchet? J. Math. Biol. **32**, 33–44.

Wagner, G.P., Booth, G., and Bagheri-Chaichian, H. 1997. A population genetic theory of canalization. Evolution **51**, 329–347.

Walsh, J.B. 1990. Inconsistencies in standard approximations for selection coefficients at polygenic loci. J. Math. Biol. **28**, 21–31.

Wang, J. 1997a. Effective size and F-statistics of subdivided populations. I. Monoecious species with partial selfing. Genetics **146**, 1453–1463.

Wang, J. 1997b. Effective size and F-statistics of subdivided populations. II. Dioecious species. Genetics **146**, 1465–1474.

Waxman, D., and Peck, J.R. 1998. Pleiotropy and the preservation of perfection. Science **279**, 1210-1213.

Waxman, D., and Peck, J.R. 1999. Sex and adaptation in a changing environment. Genetics **153**, 1041–1053.

Wayne, M.L., and Mackay, T.F.C. 1998. Quantitative genetics of ovariole number in *Drosophila melanogaster*. II. Mutational variation and genotype-environment interaction. Genetics **148**, 201–210.

Weinberg, W. 1908. Über den Nachweis der Vererbung beim Menschen. Jahreshefte des Vereins für vaterländische Naturkunde in Württemberg **64**, 368–382.

Weinberg, W. 1909. Über Vererbungsgesetze beim Menschen. Zeitschrift für induktive Abstammungs- und Vererbungslehre **1**, 377–392, 440–460; **2**, 276–330.

West, S.A., Peters, A.D., and Barton, N.H. 1998. Testing for epistasis between deleterious mutations. Genetics **149**, 435–444.

Whitlock, M.C., and Barton, N.H. 1997. The effective size of a subdivided population. Genetics **146**, 427–441.

Whitlock, M., Phillips, P., Moore, F.B.-G., and Tonsor, S.J. 1995. Multiple fitness peaks and epistasis. Annu. Rev. Ecol. Syst. **26**, 601–629.

Whitlock, M.C., and Bürger, R. 2000. The good, the bad, and the ugly: fixation of new mutations in small populations. In: Ferriere, R., Dieckmann, U., and Couvet, D. (eds.) *Evolutionary Conservation Biology*, Cambridge: University Press, to appear.

Wiehe, T. 1997. Model dependency of error thresholds: the role of fitness functions and contrasts between the finite and infinite sites models. Genet. Res. **69**, 127–136.

Williams, G.C. 1975. *Sex and Evolution*. Princeton, NJ: University Press.
Wolfram, S. 1996. *Mathematica*, 3rd ed. Cambridge: University Press.
Wright, F.A., and Kong, A. 1997. Linkage mapping in experimental crosses: the robustness of single-gene models. Genetics **146**, 417–425.
Wright, S. 1921. Systems of mating. Genetics **6**, 111–178.
Wright, S. 1922. Coefficients of inbreeding and relationship. Am. Nat. **56**, 330–338.
Wright, S. 1931. Evolution in Mendelian populations. Genetics **16**, 97–159.
Wright, S. 1932. The roles of mutation, inbreeding, crossbreeding and selection in evolution. In: *Proc. VI Int. Congr. Genetics* **1**, 356–366.
Wright, S. 1935a. The analysis of variance and the correlations between relatives with respect to deviations from an optimum. J. Genetics **30**, 243–256
Wright, S. 1935b. Evolution in populations in approximate equilibrium. J. Genetics **30**, 257–266.
Wright, S. 1949. Adaptation and Selection. In: Jepson, G.L., Simpson, G.G., and Mayr, E. (eds.) *Genetics, Paleontology, and Evolution*, pp. 365–89. Princeton: University Press.
Wright, S. 1952. The genetics of quantitative variability. In: Reeve, E.C.R., and Waddington, C.H. (eds.) *Quantitative Inheritance*, pp. 5–47. London: Her Majesty's Stationary Office.
Wright, S. 1968. *Evolution and the Genetics of Populations*. Vol. 1. *Genetic and Biometric Foundations*. Chicago: University Press.
Wright, S. 1969. *Evolution and the Genetics of Populations*. Vol. 2. *The Theory of Gene Frequencies*. Chicago: University Press.
Woodruff, R.C., and Thompson, Jr., J.N. (eds.) 1998. *Mutation and Evolution*. Dordrecht: Kluwer (reprinted from *Genetica*, Volumes 102/103, 1998).

Yoo, B.H. 1980a. Long-term selection for a quantitative character in large replicate populations of *Drosophila melanogaster*. I. Responses to selection. Genet. Res. **35**, 1–17.
Yoo, B.H. 1980b. Long-term selection for a quantitative character in large replicate populations of *Drosophila melanogaster*. II. Lethals and visible mutants with large effects. Genet. Res. **35**, 19–31.
Yule, G.U. 1902. Mendel's laws and their probable relations to intra-racial heredity. New Phytol. **1**, 192–207, 222–238.
Yule, G.U. 1906. On the theory of inheritance of quantitative compound characters on the basis of Mendel's laws – a preliminary note. In: *Report of the Third International Conference on Genetics*, pp. 140–142. London: Spottiswoode.

Zeng, Z.-B. 1994. Precision mapping of quantitative trait loci. Genetics **136**, 1457–1468.
Zeng, Z.-B., and Cockerham, C.C. 1991. Variance of neutral genetic variances within and between populations for a quantitative character. Genetics **129**, 335–553.
Zeng, Z.-B., and Cockerham, C.C. 1993. Mutation models and quantitative genetic variation. Genetics **133**, 729–736.
Zeng, Z.-B., and Hill, W.G. 1986. The selection limit due to the conflict between truncation and stabilizing selection with mutation. Genetics **114**, 1313–1328.
Zeng, Z.-B., Kao, C.-H., and Basten, C.J. 1999. Estimating the genetic architecture of quantitative traits. Genet. Res. **74**, 279–289.
Zhivotovsky, L.A., and Feldman, M.W. 1992. On models of quantitative genetic variability: a stabilizing selection-balance model. Genetics **130**, 947–955.
Zhivotovsky, L.A., and Gavrilets, S. 1992. Quantitative variability and multilocus polymorphism under epistatic selection. Theor. Pop. Biol. **42**, 254–283.

Notation Index

We have only included symbols that maintain their meaning through more than one section, except some symbols that are used in longer sections. A chapter or section number indicates that the meaning of the symbol is restricted to these chapters or sections. A symbol may have a different meaning, though only locally, before it is refered to in this index.

Author Index

Subject Index